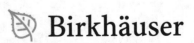

Frontiers in Mathematics

This series is designed to be a repository for up-to-date research results which have been prepared for a wider audience. Graduates and postgraduates as well as scientists will benefit from the latest developments at the research frontiers in mathematics and at the "frontiers" between mathematics and other fields like computer science, physics, biology, economics, finance, etc. All volumes are online available at SpringerLink.

More information about this series at https://link.springer.com/bookseries/5388

Leonid Positselski

Relative Nonhomogeneous
Koszul Duality

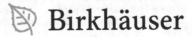

Leonid Positselski
Institute of Mathematics
Czech Academy of Sciences
Praha 1, Czech Republic

ISSN 1660-8046   ISSN 1660-8054   (electronic)
Frontiers in Mathematics
ISBN 978-3-030-89539-6   ISBN 978-3-030-89540-2   (eBook)
https://doi.org/10.1007/978-3-030-89540-2

This book is published under the imprint Birkhäuser, www.birkhauser-science.com, by the registered company Springer Nature Switzerland AG.
The registered company address is: Gewerbestrasse 11, 6330 Cham, Switzerland

Preamble

This book starts with a Prologue and an Introduction. The Prologue introduces the subject of derived Koszul duality, particularly derived nonhomogeneous Koszul duality. It is intended for the benefit of the reader not yet familiar with the major preceding results in the area, as presented, in particular, in the author's memoir [59].

The Introduction introduces the *relative* nonhomogeneous Koszul duality theory developed in this book.

For homogeneous Koszul duality, we refer the reader to the paper [7] and the memoir [59, Appendix A]. For underived Koszul duality, a suitable reference is the book [53].

Prologue

Koszul duality is a fundamental phenomenon in the "algebraic half of" mathematics, including such fields as algebraic topology, algebraic and differential geometry, and representation theory. The phenomenon is so general that it does not seem to admit a "maximal natural" generality. Manifestations of Koszul duality can be found everywhere. Noticing such manifestations is easy; interpreting them properly and distinguishing from other phenomena is harder.

Koszul duality is best explained as the relation connecting the homotopy groups of a topological space with its cohomology groups. Such an abstract topological setting is obviously very complicated; to consider something more approachable, one can specialize to rational homotopy theory or to stable homotopy theory.

In the context of rational homotopy theory, a version of Koszul duality connecting Lie DG-algebras computing the rational homotopy groups with commutative DG-algebras computing the cohomology spaces is due to Quillen [72]. The stable homotopy category, on the other hand, is additive, so its algebraic analogues are the derived categories of module-like objects. In the stable homotopy theory setting, the Adams spectral sequence can be understood as expressing the Koszul duality between the homotopy and cohomology groups.

The functor of stable homotopy groups is corepresented by the sphere spectrum, while the functor of cohomology groups is represented by the Eilenberg–MacLane spectrum. Morphisms from the sphere spectrum to the Eilenberg–MacLane spectrum are very easy to describe; in a sense, one can say that these are one-sided dual basis vectors with respect to the nonsymmetric graded Hom pairing. Graded endomorphisms of the Eilenberg–MacLane spectrum (known as the Steenrod algebra) are more complicated, and graded endomorphisms of the sphere spectrum (the stable homotopy groups) are much more complicated. The Adams spectral sequence [1] expresses the connection between these.

In representation theory, nonhomogeneous Koszul duality manifests itself as the connection between the universal enveloping algebra $U(\mathfrak{g})$ of a Lie algebra $\mathfrak{g}$ over a field k and the (co)homological Chevalley–Eilenberg complex $(\Lambda(\mathfrak{g}), d)$ or $(\Lambda(\mathfrak{g}^*), d)$ of $\mathfrak{g}$ with the coefficients in k [59, Example in Section 6.6]. In differential and algebraic geometry, relative nonhomogeneous Koszul duality appears as the duality between the ring of differential operators on a smooth variety and the de Rham complex of differential forms [36], [6, Section 7.2], [59, Appendix B]. Homogeneous Koszul duality plays a role in the theory of algebraic vector bundles over projective spaces [9], [51, Appendix A] and some other varieties.

The duality between the categories of Verma modules over the Virasoro algebra with the complementary central charges c and $26 - c$ (or the similar duality for the Kac–Moody algebras, etc.) [25, 74] is *not* an instance of Koszul duality. Rather, these are advanced manifestations of the comodule-contramodule correspondence or, more precisely, the *semimodule-semicontramodule correspondence* [58]. The MGM duality [19, 30, 55, 62] and the (covariant) Serre–Grothendieck duality [33, 47, 49, 63] are also instances of the comodule-contramodule correspondence (see the discussion in [62]) and *not* of Koszul duality.

A simple rule of thumb: a contravariant duality assigning projective modules to projective modules, or a covariant duality functor assigning projective modules to injective modules, or flat modules to injective modules, etc., is a comodule-contramodule correspondence. A duality assigning irreducible modules to projective modules, or irreducible modules to injective modules, etc., is a Koszul duality.

* * *

In the algebraic context, Koszul duality is most simply formulated as the connection between an augmented algebra A over a field k and the DG-algebra $\mathbb{R}\operatorname{Hom}_A(k, k)$ representing the graded algebra $\operatorname{Ext}_A^*(k, k)$. In particular, to a complex of A-modules $M^\bullet$ one can assign either one of the three DG-modules over $\mathbb{R}\operatorname{Hom}_A(k, k)$ representing $\operatorname{Ext}_A^*(M^\bullet, k)$, or $\operatorname{Tor}_A^*(k, M^\bullet)$, or $\operatorname{Ext}_A^*(k, M^\bullet)$. Naïvely, one would hope for such a functor to be an equivalence between the derived category of A-modules and the derived category of DG-modules over $\mathbb{R}\operatorname{Hom}_A(k, k)$.

More generally, one would start with an augmented DG-algebra $A = (A, d)$; then, the role of $\mathbb{R}\operatorname{Hom}_A(k, k)$ is played by the cobar construction of A. Once again, the naïve hope would be to have a triangulated equivalence between the derived categories of DG-modules over A and its cobar construction. An attempt to realize this hope was made in [37, Section 10], but it was possible to obtain a triangulated equivalence only under severe restrictions on the augmented DG-algebra (or DG-category). Under somewhat more relaxed but still quite restrictive assumptions, a fully faithful functor in one or other direction was constructed.

The following simple examples illustrate the situation and the lines of thought leading to the present-day approach. Let $S = k[x]$ be polynomial algebra in one variable x over a field k. Then, the DG-algebra $\operatorname{Hom}_S(k, k)$ is quasi-isomorphic to the ring of dual numbers, or in other words, to the exterior algebra $\Lambda = k[\epsilon]/(\epsilon^2)$, where ϵ is an element of cohomological degree 1. More generally, one can consider $S = k[x]$ as a DG-algebra with the generator x placed in the cohomological degree $n \in \mathbb{Z}$ and zero differential, and then the Koszul dual exterior algebra Λ has its generator ϵ situated in the cohomological degree $1 - n$ (and the differential on Λ is also zero).

The classical *homogeneous Koszul duality* context presumes that all the algebras and modules are endowed with an additional essentially positive (or negative) "internal" grading. Then one can say that the bounded derived category of finitely generated graded S-modules is equivalent to the bounded derived category of finite-dimensional graded Λ-modules. This result generalizes straightforwardly to the symmetric and exterior algebras in several variables [9], [51, Appendix A], and further to Noetherian Koszul graded algebras of finite homological dimension (in the role of S) and their finite-dimensional Koszul dual graded algebras (in the role of Λ) [7]. A formulation replacing all the finiteness assumptions with explicit assumptions on the sign of the additional grading is also possible [59, Appendix A].

In the nonhomogeneous setting (without an internal grading on the modules, but only with the cohomological one), it turns out that, for any $n \in \mathbb{Z}$ as above, the derived category of DG-modules over Λ is equivalent to a *full subcategory* in the derived category of DG-modules over S. If ones wishes, one can also view $D(\Lambda\text{–mod})$ as a *triangulated quotient category* of $D(S\text{–mod})$, but the two derived categories are decidedly *not equivalent*. There are two ways out of this predicament, each of which is further subdivided into two versions. Simply put, one can either shrink $D(S\text{–mod})$ (replacing it with a smaller category) or inflate $D(\Lambda\text{–mod})$ (replacing it with a larger category) to make the resulting versions of the derived categories of DG-modules over S and Λ equivalent.

The basic principle of derived nonhomogeneous Koszul duality, as developed in the paper [31], the dissertation [41], the note [38], and the memoir [59], is that it *connects algebras with coalgebras*. So the main decision one has to make is: Where is the algebra side, and where is the coalgebra side of the story? Replacing S with its dual coalgebra means shrinking the category $D(S\text{–mod})$. Replacing Λ with its dual coalgebra means inflating the category $D(\Lambda\text{–mod})$.

What does it mean to "replace S with its dual coalgebra"? When $n \neq 0$, the cohomological grading on S is nontrivial, the algebra S is locally finite-dimensional in this grading, and one can simply consider the graded dual coalgebra C to S. For $n = 0$, one starts with replacing the polynomial ring $S = k[x]$ with the ring of formal power series $k[[x]]$. Then C is the coalgebra over k whose dual algebra C^* is isomorphic to $k[[x]]$ (as a topological algebra). In both cases, C is viewed as a DG-coalgebra over k with the grading components or direct summands situated in the cohomological degrees $0, -n, -2n, \ldots$ and zero differential.

How does the passage from S to C affect the derived category of modules? The next guiding principle is that *there are two kinds of abelian module categories over a coalgebra*, the *comodules* and the *contramodules* [21, Section III.5], [61]. In the case of the polynomial algebra S and its dual coalgebra C, the C-comodules and the C-contramodules are two different full subcategories in S-modules. The abelian category $C\text{–comod}$ can be simply described as the category of *locally nilpotent* S-modules (i.e., the action of x must be locally nilpotent), while $C\text{–contra}$ is the category of $k[x]$-modules with *x-power infinite summation operations*. Both the derived categories $D(C\text{–comod})$ and $D(C\text{–contra})$ of

DG-comodules and DG-contramodules over C are full subcategories in $\mathsf{D}(S\text{–mod})$. Both $\mathsf{D}(C\text{–comod})$ and $\mathsf{D}(C\text{–contra})$ are equivalent to the derived category $\mathsf{D}(\Lambda\text{–mod})$.

What does it mean to "replace Λ with its dual coalgebra"? The graded algebra Λ is finite-dimensional, so its (graded) dual vector space Λ^* is a graded coalgebra over k. It is viewed as a DG-coalgebra with the grading components or direct summands situated in the cohomological degrees 0 and $n - 1$, and zero differential. Since Λ is finite-dimensional, both the Λ^*-comodules and Λ^*-contramodules are the same thing as Λ-modules, and both the DG-comodules and the DG-contramodules over Λ^* are the same thing as DG-modules over Λ. So, what's the difference, and what have we achieved by passing from Λ to Λ^*?

Here is the answer. Yet another guiding principle in Koszul duality is that *one should consider the derived categories of modules, the coderived categories of comodules, and the contraderived categories of contramodules*. The coderived and the contraderived categories, called collectively the *derived categories of the second kind*, are the categories of complexes or differential graded structures viewed up to a finer equivalence relation than the conventional quasi-isomorphism. So some acyclic complexes represent nonzero objects in the coderived category, while some other acyclic complexes represent nonzero objects in the contraderived category (or in both).

For example, let us take $n = 1$ in the notation above; so the k-algebra $\Lambda = k[\epsilon]/(\epsilon^2)$ is situated in the cohomological degree 0. Then the unbounded acyclic complex of free Λ-modules with one generator (which are both projective and injective as Λ-modules)

$$\cdots \longrightarrow \Lambda \xrightarrow{\ \epsilon* \ } \Lambda \xrightarrow{\ \epsilon* \ } \Lambda \longrightarrow \cdots \tag{*}$$

is neither coacyclic, nor contraacyclic. Cutting this complex in half by the canonical truncation, one obtains the bounded above complex

$$\cdots \xrightarrow{\ \epsilon* \ } \Lambda \xrightarrow{\ \epsilon* \ } \Lambda \longrightarrow k \longrightarrow 0 \tag{**}$$

which is contraacyclic but not coacyclic, and the bounded below complex

$$0 \longrightarrow k \longrightarrow \Lambda \xrightarrow{\ \epsilon* \ } \Lambda \xrightarrow{\ \epsilon* \ } \cdots \tag{***}$$

which is coacyclic but not contraacyclic [59, Examples in Section 3.3] (a detailed discussion of these examples can be found in [70, Section 5]).

For any $n \in \mathbb{Z}$, the coderived and the contraderived categories of DG-modules over Λ, denoted by $\mathsf{D}^{\mathsf{co}}(\Lambda\text{–mod}) = \mathsf{D}^{\mathsf{co}}(\Lambda^*\text{–comod})$ and $\mathsf{D}^{\mathsf{ctr}}(\Lambda\text{–mod}) = \mathsf{D}^{\mathsf{ctr}}(\Lambda^*\text{–contra})$, are naturally equivalent to each other. They are also equivalent to the derived category $\mathsf{D}(S\text{–mod})$ of DG-modules over S.

And what about the coalgebra C dual to the algebra S, why did we consider the derived categories of DG-comodules and DG-contramodules over C in the preceding discussion, seemingly contrary to the (subsequently introduced) guiding principle of taking the coderived categories of comodules and the contraderived categories of contramodules? Because, in this case, there is no difference: one has $D(C\text{–comod}) = D^{co}(C\text{–comod})$ and $D(C\text{–contra}) = D^{ctr}(C\text{–contra})$, since C is a graded coalgebra of finite homological dimension endowed with zero differential [39, Proposition 5.9].

Another way to explain the situation is to notice that the Koszul duality functor assigns the acyclic complex $(*)$ to the S-module $S[x^{-1}] = k[x, x^{-1}]$ of Laurent polynomials in the variable x over k (viewed as a DG-module with zero differential). So, in order to make this functor a triangulated equivalence, one has to either declare $(*)$ to be a nonzero object in an exotic derived category of Λ-modules, or otherwise prohibit the S-module $k[x, x^{-1}]$ (i.e., impose a restriction on the S-modules under consideration putting this module outside of the category of modules appearing in the duality). It turns out that there are two ways to do the former (basically, one can declare either the complex $(**)$ or the complex $(***)$ to be nonzero), and there are two ways to do the latter (one can either prohibit the free S-module $S = k[x]$ on the grounds of it being not a C-comodule, or prohibit the injective S-module $k[x, x^{-1}]/k[x]$ on the grounds of it being not a C-contramodule).

$$* * *$$

So, a more insightful formulation of Koszul duality for augmented DG-algebras over a field k presumes that it connects DG-algebras with DG-coalgebras. To an augmented DG-algebra $A = (A, d)$, a DG-coalgebra C computing $\text{Tor}^A_*(k, k)$ is assigned. To a coaugmented DG-coalgebra $C = (C, d)$, a DG-algebra A computing $\text{Ext}^*_C(k, k)$ is assigned. More specifically, given a DG-algebra A, the corresponding DG-coalgebra C can be produced as the bar construction of A, or, given a DG-coalgebra C, the corresponding DG-algebra A can be produced as the cobar construction of C. Both the constructions are covariant functors between the categories of augmented DG-algebras and augmented DG-coalgebras.

The next principle is that one should restrict oneself to *conilpotent* coalgebras. Here one observes that the nilpotency works better with coalgebras than with algebras inasmuch as the direct limits are more convenient to work with than the inverse limits. Given an augmented algebra A with the augmentation ideal A^+, one says that A is nilpotent if there exists an integer $n \geq 1$ such that $(A^+)^n = 0$. If one wants to let n approach infinity, as we do, then one has to speak of pronilpotency. The dual condition for coalgebras is conilpotency (which might as well be called ind-conilpotency, for it does not presume existence of a fixed finite nilpotency index n for the whole coalgebra C).

Furthermore, similarly to DG-comodules, one has to consider conilpotent DG-coalgebras up to a more delicate equivalence relation than the conventional quasi-isomorphism. This equivalence relation, called the *filtered quasi-isomorphism*, was discovered by Hinich in [31].

With these considerations in mind, the Koszul duality can be formulated quite generally as a category equivalence between the category of augmented DG-algebras up to quasi-isomorphism and the category of conilpotent DG-coalgebras up to filtered quasi-isomorphism [59, Theorem 6.10(b)]. Furthermore, whenever a DG-algebra $A = (A, d)$

and a DG-coalgebra $C = (C, d)$ correspond to each other under this equivalence, one has a triangulated equivalence between the derived category of DG-modules over A, the coderived category of DG-comodules over C, and the contraderived category of DG-contramodules over C ("Koszul triality")

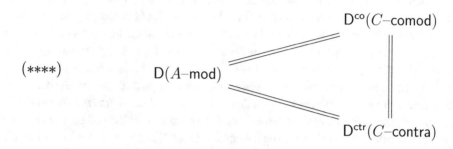

$(****)$

Here, the equivalence $\mathsf{D}^{\mathrm{co}}(C\text{–comod}) \simeq \mathsf{D}^{\mathrm{ctr}}(C\text{–contra})$, which holds quite generally for any (not necessarily coaugmented or conilpotent) DG-coalgebra C over a field k, is called the *derived comodule-contramodule correspondence*. The comodule side of the triangle $(****)$ can be found formulated in [38] based on the results of [41], who was following the approach of [31] (except that the definition of the coderived category in [38] is less intrinsic than the modern one in [59]).

* * *

A conilpotent coalgebra C is coaugmented by definition. But what if an algebra A is not augmented? Yet another guiding principle tells that *absence of a chosen augmentation on one side of the Koszul duality corresponds to presence of a curvature on the other side.* This observation goes back to the author's early paper [56].

A curved DG-ring is a very natural concept: in particular, the construction of the DG-category of DG-modules over a DG-ring extends naturally to CDG-rings. Curved DG-modules over a curved DG-ring form a DG-category. Moreover, the passage from DG-rings to CDG-rings involves not only adding new objects, but also new morphisms; the inclusion functor of $\mathsf{Rings}_{\mathrm{dg}}$ into $\mathsf{Rings}_{\mathrm{cdg}}$ is faithful, but not fully faithful. A CDG-isomorphism of (C)DG-rings (e.g., a change-of-connection isomorphism, otherwise known as a Maurer–Cartan twist) induces an equivalence of the DG-categories of (C)DG-modules.

Besides Koszul duality, curved DG-rings and curved A_∞-algebras play a fundamental role in the Fukaya theory [13, 27, 28] and deformation theory [16, 39], they appear in the theory of Legendrian knots [50], etc. A systematic treatment of *weakly curved* DG- and A_∞-algebras over topological local rings (i.e., curved algebras with the curvature element divisible by the maximal ideal of the local ring of coefficients) can be found in the memoir [60].

However important the coderived and contraderived categories of DG-modules (DG-comodules, or DG-contramodules) are, for curved DG-modules (curved DG-comodules, or curved DG-contramodules) they are more important still. The cohomology groups or modules of CDG-modules are *undefined*, as the differential does not square to zero (instead, it squares to the operator of multiplication with the curvature). So one cannot speak about acyclic CDG-modules or quasi-isomorphisms of CDG-modules in the usual sense of the word. But the full triangulated subcategories of *coacyclic* and *contraacyclic* CDG-modules in the homotopy category of CDG-modules are perfectly well-defined. Hence there is no alternative to derived categories of the second kind for CDG-modules, as the conventional derived category ("of the first kind") does not make sense for them. The weakly curved DG-modules (mentioned above) are the only known exception [60].

Matrix factorizations [12, 22] are an important and popular particular case of CDG-modules. Derived categories of the second kind, including specifically the coderived and "absolute derived" categories, play a crucial role in their theory [3, 20, 52, 54]. These are "strongly" (i.e., not weakly) curved.

Inverting the arrows in the definition of a CDG-algebra over a field, one obtains the definition of a *CDG-coalgebra*. Similarly to the CDG-modules, the differential does not square to zero either in CDG-rings or in CDG-coalgebras; so the conventional notion of quasi-isomorphism is undefined for them. But the definition of a filtered quasi-isomorphism makes perfect sense for conilpotent CDG-coalgebras.

The nonaugmented Koszul duality is an equivalence between the category of DG-algebras up to quasi-isomorphism and the category of conilpotent CDG-coalgebras up to filtered quasi-isomorphism (over a fixed field k) [59, Theorem 6.10(a)]. Whenever a DG-algebra $A = (A, d)$ and a CDG-coalgebra $C = (C, d, h)$ (where h is the curvature element) correspond to each other under this equivalence, the "Koszul triality" picture (∗∗∗∗) holds [59, Sections 6.3–6.6].

So the derived category of DG-modules over A, the coderived category of CDG-comodules over C, and the contraderived category of CDG-contramodules over C are equivalent to each other. Once again, the equivalence between the latter two categories is an instance of the derived comodule-contramodule correspondence (and it holds quite generally for any CDG-coalgebra C over a field k), while the other two equivalences in the triangle diagram are the comodule and the contramodule sides of the Koszul duality.

∗ ∗ ∗

The relative Koszul duality theory developed in this book is *both more and less general* than the derived nonhomogeneous Koszul duality of [31,38,41,59]. The theory in this book is more general in that it is relative, i.e., worked out over an arbitrary (noncommutative, nonsemisimple) base ring. It is a nontrivial generalization, both because the underived homogeneous Koszulity and Koszul duality theory over an arbitrary base ring is more complicated than over a field (or over a semisimple base ring, as in [7]), and because the differential operators (or the enveloping algebras of Lie algebroids) are nontrivially more complicated than the enveloping algebras of Lie algebras over fields. The de Rham DG-algebra of differential forms (or the Chevalley–Eilenberg complex of a Lie algebroid) is likewise more complicated than the Chevalley–Eilenberg complex of a Lie algebra over a field. Notice that the de Rham differential is not linear over the ring of functions.

Still the theory in this book is less general than in [59] in that the "algebra side of the story" is presumed to be just a ring (situated in the cohomological degree 0) rather than a DG-ring. In the notation above, this means that we restrict ourselves to the case $n = 0$ if S is chosen to be on the algebra side, or otherwise we take $n = 1$ if the algebra side is Λ. This restriction of generality is chosen in order not to make the exposition too complicated while including the most important examples (viz., various versions of the ring of differential operators, which are indeed just rings and not DG-rings). On the "coalgebra side of the story" we still obtain a (curved or uncurved) DG-ring, such as the de Rham DG-algebra.

Introduction

0.0 Let A be an associative ring and $R \subset A$ be a subring. *Derived Koszul duality* is the functor $\mathrm{Ext}_A^*(-, R)$, or $\mathrm{Tor}_*^A(R, -)$, or $\mathrm{Ext}_A^*(R, -)$, enhanced to an equivalence of derived categories of modules.

The above definition raises many questions. To begin with, R is not an A-module. So what does this Ext and Tor notation even *mean*?

Secondly, let us consider the simplest example where $R = k$ is a field and $A = k[x]$ is the algebra of polynomials in one variable. Then k indeed can be viewed as an A-module. There are many such module structures, indexed by elements a of the field k: given $a \in k$, one can let the generator $x \in A$ act in k by the multiplication with a. Denote the resulting A-module by k_a.

To be specific, let us choose $k = k_0$ as our preferred A-module structure on k. Then the functors $\mathrm{Ext}_A^*(-, k_0)$, $\mathrm{Tor}_*^A(k_0, -)$, and $\mathrm{Ext}_A^*(k_0, -)$ are indeed well-defined on the category of A-modules. But these functors are far from being faithful or conservative: all of them annihilate the A-modules k_a with $a \neq 0$. How, then, can one possibly hope to enhance such cohomological functors to derived equivalences?

0.1 Koszul duality has to be distinguished from the *comodule-contramodule correspondence*, which is a different, though related, phenomenon.

In the simplest possible form, the comodule-contramodule correspondence is the functor $\mathrm{Ext}_A^*(-, A)$ enhanced to a derived equivalence (while Koszul duality is $\mathrm{Ext}_A^*(-, k)$, where k is the ground field). In a more realistic covariant and relative situation, comparable to the discussion of Koszul duality in Sect. 0.0, the comodule-contramodule correspondence would be a derived equivalence enhancement of a functor like $\mathrm{Ext}_A^*(\mathrm{Hom}_R(A, R), -)$ or $\mathrm{Tor}_*^A(-, \mathrm{Hom}_R(A, R))$.

0.2 In the present author's research, the desire to understand Koszulity and Koszul duality was the starting point. Then the separate existence and importance of comodule-contramodule correspondence was realized, particularly in the context of semi-infinite homological algebra [58]. The derived nonhomogeneous Koszul duality over a field was formulated as a "Koszul triality" picture, which is a triangle diagram of derived equivalences with the comodule-contramodule correspondence present as one side of the triangle and two versions of Koszul duality as two other sides [59].

The comodule-contramodule correspondence, its various versions, generalizations, and philosophy, are now discussed in several books and papers of the present author, including [58,59,61,62,66,71] and others. On the other hand, the derived nonhomogeneous Koszul duality over a field attracted interest of a number of authors, starting from early works [31,37,38,41] and to very recent, such as [14,45]; there is even an operadic version of it in [32].

Still, there is a void in the literature concerning *relative* nonhomogeneous Koszul duality. Presently, the only source of information on this topic known to this author is his previous book [58], which contains an introductory discussion without proofs or details in [58, Section 0.4] and a heavily technical treatment in a very general and complicated setting in [58, Chapter 11]. (The memoir [60] represents a very different point of view.) This book is intended to fill the void by providing a reasonably accessible, detailed exposition on a moderate generality level.

Let us emphasize that relative nonhomogeneous Koszul duality is important. In addition to the presence of very natural examples such as the duality between the ring of differential operators and the de Rham DG-algebra (see Sect. 0.7 below), relative nonhomogeneous Koszul duality plays a crucial role in the semi-infinite (co)homology theory, as it was first pointed out in [2]. This idea was subsequently developed and utilized in [58, Section 11.9 and Appendix D].

The special case of triangulated equivalences between complexes of modules over the rings/sheaves of differential operators and DG-modules over the de Rham DG-algebra has been considered in [36] and [6, Section 7.2]. Our own treatment of it is presented in [59, Appendix B].

0.3 Let us start to explain the meaning of the terms involved. In the notation of Sect. 0.0, *relative* means that R is an arbitrary ring rather than simply the ground field. *Homogeneous* Koszul duality means that $A = \bigoplus_{n=0}^{\infty} A_n$ is a nonnegatively graded ring and $R = A_0$ is the degree-zero grading component. In this case, R is indeed naturally both a left and a right R-module, so the meaning of the Ext and Tor notation in Sect. 0.0 is clear. *Nonhomogeneous* Koszul duality is the situation when there is no such grading on the ring A.

The main specific aspect of the homogeneous case is that one can consider graded A-modules with a bounding condition on the grading, that is, only positively graded or only negatively graded modules. If M is a positively graded left A-module, then $R \otimes_A M = 0$ implies $M = 0$, while if P is a negatively graded left A-module, then $\mathrm{Hom}_A(R, P) = 0$ implies $P = 0$. Hence the second problem described in Sect. 0.0 does not occur, either.

In the nonhomogeneous situation, the solution to the second problem from Sect. 0.0 is to consider *derived categories of the second kind*. This means that certain complexes or DG-modules are declared to be nonzero objects in the derived category even though their cohomology modules vanish.

As to the first problem, it may well happen that R has a (left or right) A-module structure even though A is not graded. When such a module structure (extending the natural R-module structure on R) has been chosen, one says that the ring A is *augmented*. In this case, the related Ext or Tor functor is well-defined. One wants to enhance it to a functor with values in DG-modules over a suitable DG-ring in such a way that it would induce a triangulated equivalence.

Generally speaking, the solution to the first problem is to consider *curved DG-modules* (*CDG-modules*), whose cohomology modules are *undefined*. So the Ext or Tor itself has no meaning, but the related curved DG-module has. In the augmented case, this DG-module becomes uncurved, and indeed computes the related Ext or Tor.

0.4 Let us now begin to state what our assumptions and results are. We assume that a ring $\widetilde{A}$ is endowed with an increasing filtration $R = F_0\widetilde{A} \subset F_1\widetilde{A} \subset F_2\widetilde{A} \subset \cdots$ which is exhaustive ($\widetilde{A} = \bigcup_n F_n\widetilde{A}$) and compatible with the multiplication in $\widetilde{A}$. Furthermore, the successive quotients $\mathrm{gr}_n^F \widetilde{A} = F_n\widetilde{A}/F_{n-1}\widetilde{A}$ are assumed to be finitely generated projective left R-modules. Finally, the associated graded ring $A = \mathrm{gr}^F \widetilde{A} = \bigoplus_n \mathrm{gr}_n^F \widetilde{A}$ has to be *Koszul* over its degree-zero component $A_0 = R$; this means, in particular, that the ring A is generated by its degree-one component A_1 over A_0 and defined by relations of degree 2.

In these assumptions, we assign to $(\widetilde{A}, F)$ a *curved DG-ring* (*CDG-ring*) (B, d, h), which is graded by nonnegative integers, $B = \bigoplus_{n=0}^\infty B^n$, $B^0 = R$, has a differential (odd derivation) $d\colon B^n \longrightarrow B^{n+1}$ of degree 1, and a *curvature element* $h \in B^2$. The CDG-ring (B, d, h) is defined uniquely up to a unique isomorphism of CDG-rings, which includes the possibility of *change-of-connection* transformations. The grading components B^n are finitely generated projective right R-modules. In particular, one has $B^1 = \mathrm{Hom}_R(A_1, R)$ and $A_1 = \mathrm{Hom}_{R^{\mathrm{op}}}(B^1, R)$.

Furthermore, to any left $\widetilde{A}$-module P we assign a CDG-module structure on the graded left B-module $B \otimes_R P$, and to any right $\widetilde{A}$-module M we assign a CDG-module structure on the graded right B-module $\mathrm{Hom}_{R^{\mathrm{op}}}(B, M)$. These constructions are then extended to complexes of left and right $\widetilde{A}$-modules $P^\bullet$ and $M^\bullet$, assigning to them left and right CDG-modules $B \otimes_R P^\bullet$ and $\mathrm{Hom}_{R^{\mathrm{op}}}(B, M^\bullet)$ over (B, d, h). A certain (somewhat counterintuitive) way to totalize bigraded modules is presumed here. The resulting functors induce the derived equivalences promised in Sect. 0.0. The functor $P^\bullet \longmapsto B \otimes_R P^\bullet$ is a CDG-enhancement of the (possibly nonexistent) $\mathrm{Ext}_{\widetilde{A}}^*(R, P)$, and the functor $M^\bullet \longmapsto \mathrm{Hom}_{R^{\mathrm{op}}}(B, M^\bullet)$ is a CDG-enhancement of the (possibly nonexistent) $\mathrm{Tor}^{\widetilde{A}}_*(M, R)$. However, there are further caveats.

0.5 One important feature of the nonhomogeneous Koszul duality over a field, as developed in the memoir [59], is that it connects modules with comodules or contramodules. In fact, the "Koszul triality" of [59] connects modules with comodules *and* contramodules. In the context of relative nonhomogeneous Koszul duality theory in the full generality of this book, the Koszul triality picture splits into two separate dualities. A certain exotic derived category of right $\widetilde{A}$-modules is equivalent to an exotic derived category of *right*

B-comodules, while another exotic derived category of left $\widetilde{A}$-modules is equivalent to an exotic derived category of *left B-contramodules*. The triality picture is then restored under some additional assumptions (namely, two-sided locally finitely generated projectivity of the filtration F and finiteness of homological dimension of the base ring R).

What are the "comodules" and "contramodules" in our context? First of all, we have complexes of $\widetilde{A}$-modules on the one side and CDG-modules over B on the other side; so both the comodules and the contramodules are graded B-modules. In fact, the (graded) right B-comodules are a certain *full subcategory* in the graded right B-modules, and similarly the (graded) left B-contramodules are a certain *full subcategory* in the graded left B-modules.

Which full subcategory? A graded right B-module N is called a *graded right B-comodule* if for every element $x \in N$ there exists an integer $m \geq 1$ such that $x B^n = 0$ for all $n \geq m$. The definition of B-contramodules is more complicated and, as usually, involves certain infinite summation operations. A graded left B-module Q is said to be a *graded left B-contramodule* if, for every integer j, every sequence of elements $q_n \in Q^{j-n}$, $n \geq 0$, and every sequence of elements $b_n \in B^n$, an element denoted formally by $\sum_{n=0}^{\infty} b_n q_n \in Q^j$ is defined. One imposes natural algebraic axioms on such infinite summation operations, and then proves that an infinite summation structure on a given graded left B-module Q is unique if it exists.

In particular, this discussion implies that (somewhat counterintuitively), in the notation of Sect. 0.4, the bigraded module $\mathrm{Hom}_{R^{op}}(B, M^\bullet)$ has to be totalized by taking infinite *direct sums* along the diagonals (to obtain a graded right B-comodule), while the bigraded module $B \otimes_R P^\bullet$ needs to be totalized by taking infinite *products* along the diagonals (to obtain a graded left B-contramodule).

0.6 The explanation for the counterintuitive totalization procedures mentioned in Sect. 0.5, from our perspective, is that B is a "fake" graded ring. It really "wants" to be a coring, but this point of view is hard to fully develop. It plays a key role, however, in (at least) one of our two proofs of the Poincaré–Birkhoff–Witt theorem for nonhomogeneous Koszul rings.

The graded coring in question is $C = \mathrm{Hom}_{R^{op}}(B, R)$, that is, the result of applying the dualization functor $\mathrm{Hom}_{R^{op}}(-, R)$ to the graded ring B. The point is that we have already done one such dualization when we passed from the R-R-bimodule A_1 to the R-R-bimodule $B^1 = \mathrm{Hom}_R(A_1, R)$, as mentioned in Sect. 0.4. The two dualization procedures are essentially inverse to each other, so the passage to the coring C over R returns us to the undualized world, depending covariantly functorially on the ring A.

Experience teaches that the passage to the dual vector space is better avoided in derived Koszul duality. This is the philosophy utilized in the memoir [59] and the book [58]. This philosophy strongly suggests that the graded coring C is preferable to the graded ring B as a Koszul dual object to a Koszul graded ring A.

The problem arises when we pass to the nonhomogeneous setting. In the context of the discussion in Sect. 0.4, the odd derivation d, which is a part of the structure of a CDG-ring (B, d, h), is *not* R-linear. In fact, the restriction of d to the subring $R = B^0 \subset B$ may well be nonzero, and in the most interesting cases it is. This is a distinctive feature of the relative nonhomogeneous Koszul duality. So how does one apply the functor $\mathrm{Hom}_{R^{\mathrm{op}}}(-, R)$ to a non-R-linear map?

0.7 The duality between the ring of differential operators and the de Rham DG-algebra of differential forms is the thematic example of relative nonhomogeneous Koszul duality. Let X be a smooth affine algebraic variety over a field of characteristic 0 (or a smooth real manifold). Let $O(X)$ denote the ring of functions and $\mathrm{Diff}(X)$ denote the ring of differential operators on X. Endow the ring $\mathrm{Diff}(X)$ with an increasing filtration F by the order of the differential operators. So the associated graded ring $\mathrm{Sym}_{O(X)}(T(X)) = \mathrm{gr}^F \mathrm{Diff}(X)$ is the symmetric algebra of the $O(X)$-module $T(X)$ of vector fields on X.

In this example, $R = O(X)$ is our base ring, $\widetilde{A} = \mathrm{Diff}(X)$ is our nonhomogeneous Koszul ring over R, and $A = \mathrm{Sym}_{O(X)}(T(X))$ is the related homogeneous Koszul ring. The graded ring Koszul to A over R is the graded ring of differential forms $B = \Omega(X)$. There is no curvature in the CDG-ring (B, d, h) (one has $h = 0$; a nonzero curvature appears when one passes to the context of *twisted* differential operators, e.g., differential operators acting in the sections of a vector bundle E over X; see [58, Section 0.4.7] or [59, Appendix B]). The differential $d \colon B \longrightarrow B$ is the de Rham differential, $d = d_{dR}$; so (B, d) is a DG-algebra over k.

But the de Rham DG-algebra is not a DG-algebra over $O(X)$ (and neither the ring $\mathrm{Diff}(X)$ is an algebra over $O(X)$). In fact, the restriction of the de Rham differential to the subring $O(X) \subset \Omega(X)$ is quite nontrivial.

0.8 So the example of differential operators and differential forms is a case in point for the discussion in Sect. 0.6. In this example, $C = \mathrm{Hom}_{O(X)}(\Omega(X), O(X))$ is the graded coring of polyvector fields over the ring of functions on X. Certainly there is no de Rham differential on polyvector fields. What structure on polyvector fields corresponds to the de Rham differential on the forms?

Here is what we do. We adjoin an additional generator δ to the de Rham DG-ring $(\Omega(X), d_{dR})$, or more generally to the underlying graded ring B of a CDG-ring (B, d, h). The new generator δ is subject to the relations $[\delta, b] = d(b)$ for all $b \in B$ (where the bracket denotes the graded commutator) and $\delta^2 = h$. Then there is a new differential on the graded ring $\widehat{B} = B[\delta]$, which we denote by $\partial = \partial/\partial\delta$.

The differential ∂ is R-linear (and more generally, B-linear with signs), so we can dualize it, obtaining a coring $\widehat{C} = \mathrm{Hom}_{R^{\mathrm{op}}}(\widehat{B}, R)$ with the dual differential $\mathrm{Hom}_{R^{\mathrm{op}}}(\partial, R)$. This is the structure that was called a *quasi-differential coring* in [58]. It plays a key role in the exposition in [58, Chapter 11].

Of course, the odd derivation $\partial = \partial/\partial\delta$ is acyclic, and so is the dual odd coderivation on the coring $\widehat{C}$. This may look strange, but, in fact, this is how it should be. Recall that we started with a curved DG-ring (B, d, h). Its differential d does not square to zero, and its cohomology is undefined. So there is no cohomology ring in the game, and it is not supposed to suddenly appear from the construction.

0.9 Now, how does one assign a derived category of modules to the acyclic DG-ring $(\widehat{B}, \partial)$? The related constructions are discussed in [58, Section 11.7]. A *quasi-differential module* over $(\widehat{B}, \partial)$ is simply a graded $\widehat{B}$-module, without any differential. Such modules form a DG-category. In fact, a DG-module over $(\widehat{B}, \partial)$ is the same thing as a *contractible* object of the DG-category of quasi-differential modules. This point of view, adopted in [58, Chapter 11] in the context of quasi-differential comodules and contramodules over quasi-differential corings, is so counterintuitive that one is having a hard time with what otherwise are very simple constructions. We have none of that in this book, using the equivalent, but much more tractable concept of a CDG-module over the CDG-ring (B, d, h).

Some words about the *coderived* and *contraderived categories* are now in order. These are the most important representatives of the class of constructions known as the "derived categories of the second kind."

In the spirit of the discussion in Sect. 0.5, we consider right CDG-comodules and left CDG-contramodules over (B, d, h). These are certain full subcategories in the DG-categories of, respectively, right and left CDG-modules over the CDG-ring (B, d, h). Following the general definitions in [58, 59], the *coderived category of right CDG-comodules* over (B, d, h) is constructed as the Verdier quotient category of the homotopy category of CDG-comodules by its minimal triangulated subcategory containing the total CDG-comodules of all the short exact sequences of CDG-comodules and closed under infinite direct sums. Similarly, the *contraderived category of left CDG-contramodules* over (B, d, h) is the Verdier triangulated quotient category of the homotopy category of CDG-contramodules by its minimal triangulated subcategory containing the total CDG-contramodules of all the short exact sequences of CDG-contramodules over (B, d, h) and closed under infinite products.

0.10 When the base ring R has finite right homological dimension, our derived Koszul duality result simply states that the derived category of right $\widetilde{A}$-modules is equivalent to the coderived category of right CDG-comodules over (B, d, h). When the ring R has finite left homological dimension, one similarly has a natural equivalence between the derived category of left $\widetilde{A}$-modules and the contraderived category of left CDG-contramodules over (B, d, h).

The situation gets more complicated when the homological dimension of R is infinite. In this case, following the book [58] and the paper [63], one can consider the *semiderived categories* of right and left $\widetilde{A}$-modules, or more precisely the *semicoderived category* of right $\widetilde{A}$-modules relative to R and the *semicontraderived category* of left $\widetilde{A}$-modules relative to R. These are defined as the Verdier quotient categories of the homotopy categories of complexes of right and left $\widetilde{A}$-modules by the triangulated subcategories of complexes that are, respectively, coacyclic or contraacylic *as complexes of R-modules*.

Then the derived Koszul duality theorem tells that the semicoderived category of right $\widetilde{A}$-modules is equivalent to the coderived category of right CDG-comodules over (B, d, h), and the semicontraderived category of left $\widetilde{A}$-modules is equivalent to the contraderived category of left CDG-contramodules over (B, d, h).

One can also describe the derived category of right $\widetilde{A}$-modules as the quotient category of the coderived category of right CDG-comodules over (B, d, h) by its minimal triangulated subcategory closed under direct sums and containing all the CDG-comodules (N, d_N) such that $NB^i = 0$ for $i > 0$ and N is acyclic with respect to the differential d_N (where $d_N^2 = 0$ since $Nh = 0$). Similarly, the derived category of left $\widetilde{A}$-modules is equivalent to the quotient category of the contraderived category of left CDG-contramodules over (B, d, h) by its minimal triangulated subcategory closed under products and containing all the CDG-contramodules (Q, d_Q) such that $B^i Q = 0$ for $i > 0$ and Q is acyclic with respect to the differential d_Q.

0.11 A basic fact of the classical theory of modules over the rings of differential operators $\mathrm{Diff}(X)$ is that the abelian categories of left and right $\mathrm{Diff}(X)$-modules are naturally equivalent to each other. This is a rather nontrivial equivalence, in that the ring $\mathrm{Diff}(X)$ is *not* isomorphic to its opposite ring.

The classical *conversion functor* $\mathrm{Diff}(X)\text{–mod} \longrightarrow \mathrm{mod–Diff}(X)$ assigns to a left $\mathrm{Diff}(X)$-module M a natural right $\mathrm{Diff}(X)$-module structure on the tensor product $\Omega^m(X) \otimes_{O(X)} M$, where $m = \dim X$ and $\Omega^m(X)$ is the $O(X)$-module of global sections of the line bundle of differential forms of the top degree on X. The inverse conversion $\mathrm{mod–Diff}(X) \longrightarrow \mathrm{Diff}(X)\text{–mod}$ is performed by taking the tensor product over $O(X)$ with the (module of global sections of) the line bundle of top polyvector fields $\mathrm{Hom}_{O(X)}(\Omega^m(X), O(X)) = \Lambda^m_{O(X)}(T(X))$.

In this book we offer an interpretation of the conversion functor in the context of relative nonhomogeneous Koszul duality. Let (B, d, h) be a nonnegatively graded CDG-ring with the (possibly noncommutative) degree-zero component $B^0 = R$. Assume that the grading components of B are finitely generated projective left and right R-modules, there is an integer $m \geq 0$ such that $B^n = 0$ for $n > m$, the R-R-bimodule B^m is invertible, and the multiplication maps $B^n \otimes_R B^{m-n} \longrightarrow B^m$ are perfect pairings. Assume further that B is a Koszul graded ring over R. Then we say that $B = (B, d, h)$ is a *relatively Frobenius Koszul CDG-ring*.

As the grading components of B are finitely generated and projective over R on both sides, there are *two* nonhomogeneous Koszul dual filtered rings to (B, d, h), one on the left side and one on the right side; we denote them by $\widetilde{A}$ and $\widetilde{A}^{\#}$. Then the claim is that, whenever B is relatively Frobenius over R, the two rings $\widetilde{A}$ and $\widetilde{A}^{\#}$ are Morita equivalent. The tensor product with the invertible R-R-bimodule $T = B^m$ transforms any left $\widetilde{A}$-module into a left $\widetilde{A}^{\#}$-module, and any right $\widetilde{A}^{\#}$-module into a right $\widetilde{A}$-module.

The functors $\mathrm{Hom}_R(T, -)$ and $\mathrm{Hom}_{R^{\mathrm{op}}}(T, -)$ provide the inverse transformations. (When the graded ring B is graded commutative and $h = 0$, the ring $\widetilde{A}^{\#}$ is simply the opposite ring to the ring $\widetilde{A}$.)

In this context, assuming additionally that the ring R has finite left homological dimension, we even obtain a "Koszul quadrality" picture. This means a commutative diagram of triangulated equivalences between four (conventional or exotic) derived categories: the derived category of left $\widetilde{A}^{\#}$-modules, the derived category of left $\widetilde{A}$-modules, the coderived category of left CDG-modules over (B, d, h), and the contraderived category of left CDG-modules over (B, d, h).

0.12 Several examples of relative nonhomogeneous Koszul duality are considered in this book. These are various species of differential operators, to which correspond the related species of differential forms.

The words "differential operators" have many meanings. First of all, one has to choose the world one wants to live in: algebraic varieties, complex analytic manifolds, or smooth real manifolds? If one chooses algebraic varieties, does one care about finite characteristic, or only characteristic zero? Speaking of complex manifolds, there is also a choice: Is one interested in complex analytic functions/forms/operators only, or does one want to do Dolbeault theory?

On the other hand, does one want one's differential operators to act in functions, or does one also care about the differential operators acting in the sections of a vector bundle? Does one want to enter the universe of *twisted differential operators*?

There are further options: What about the relative or fiberwise differential operators and forms for a smooth morphism of algebraic varieties (or for a submersion of smooth real manifolds)? Does one want to depart the differential operators altogether, replacing them with the more general concept of Lie algebroids (Lie–Rinehart algebras)? Is there a notion of differential operators in noncommutative geometry?

We explore not all, but many of these possibilities. Notice first of all that the main setting of this book presumes rings rather than sheaves of rings; so we only consider *affine* algebraic varieties. For the same reason, discussing complex analytic differential operators in the context of this book would presume working with Stein manifolds; the author did not feel qualified to delve into this theory, so this setting is skipped. However, we do consider the Dolbeault DG-algebra and $\bar{\partial}$-differential operators on compact complex manifolds. Our smooth real manifolds are likewise assumed to be compact, for technical reasons.

Furthermore, over a field k of finite characteristic p one has $(\frac{d}{dx})^p(f) = 0$ for any polynomial $f \in k[x]$, while $\frac{1}{p}(\frac{d}{dx})^p$ is a well-defined operator $k[x] \longrightarrow k[x]$. For this reason, there are *two* notions of differential operators on a smooth (affine) algebraic variety in finite characteristic: the *Grothendieck differential operators*, which are acting faithfully in the functions, and the *crystalline differential operators*, which form a nonhomogeneous quadratic ring over the ring of functions. We discuss both, in order to provide context, but it is only the latter concept of differential operators that is relevant for our theory.

In each of the cases mentioned above, the ring of differential operators, filtered by the order of the differential operators, is a nonhomogeneous Koszul ring. The ring of differential operators acting in the functions (irrespectively of whether these differential operators are algebraic in characteristic zero, crystalline, real smooth, or Dolbeault) is *left augmented* over the ring of functions: the left ideal of all differential operators annihilating the constant functions is the augmentation ideal. Accordingly, the nonhomogeneous Koszul dual structure is a DG-ring; it is the de Rham DG-algebra of differential forms in the respective world (or the Dolbeault DG-algebra in the case of $\bar{\partial}$-differential operators).

The ring of differential operators acting in a vector bundle is usually *not* augmented: in fact, it admits a left augmentation if and only if the bundle has an integrable connection. Otherwise, one has to choose a (possibly nonintegrable) connection in the vector bundle, and the nonhomogeneous Koszul duality assigns to such ring of differential operators the curved DG-algebra of differential forms with the coefficients in the endomorphisms of the vector bundle, endowed with the de Rham differential depending on the connection. The curvature form of the connection becomes the curvature element in the resulting CDG-ring. We also consider the rings of differential operators twisted by an arbitrary closed 2-form, and then the nonhomogeneous Koszul dual structure is the usual de Rham DG-algebra viewed as a CDG-ring with the chosen 2-form playing the role of the curvature element.

Turning to noncommutative geometry, we discuss the DG-ring of noncommutative differential forms for a morphism of noncommutative rings. In this context, the related ring of "noncommutative differential operators" is simply the ring of all endomorphisms of the bigger ring as a module over the subring (endowed with the obvious two-step filtration). For all the classes of examples mentioned above, we formulate the related triangulated equivalences of relative derived nonhomogeneous Koszul duality (and, when relevant, draw the "Koszul quadrality" diagrams).

0.13 We discuss the homogeneous quadratic duality over a base ring in Chap. 1, flat and finitely projective Koszul graded rings over a base ring in Chap. 2, relative nonhomogeneous quadratic duality in Chap. 3, and the Poincaré–Birkhoff–Witt theorem for nonhomogeneous Koszul rings over a base ring in Chap. 4. The discussion of comodules and contramodules over graded rings in Chap. 5 prepares ground for the derived Koszul duality for module categories, which is worked out on the comodule side in Chap. 6 and on the contramodule side in Chap. 7. The comodule-contramodule correspondence, connecting the comodule and contramodule sides of the Koszul duality, is developed in Chap. 8. The interpretation of the conversion functor in terms of Koszul duality is discussed in Chap. 9.

Examples of relative nonhomogeneous Koszul duality are offered in Chap. 10. We consider algebraic differential operators over smooth affine varieties in characteristic 0, crystalline differential operators over smooth affine varieties in arbitrary characteristic, differential operators acting in the sections of a vector bundle, and differential operators twisted with a chosen closed 2-form. Passing from the algebraic to the analytic setting, we discuss smooth differential operators on a smooth compact real manifold and $\bar{\partial}$-differential operators in the Dolbeault theory on a compact complex manifold. Returning to the algebraic context, we consider relative differential forms and differential operators for a morphism of commutative rings, Lie algebroids with their enveloping algebras and cohomological Chevalley–Eilenberg complexes, and finally noncommutative differential forms for a morphism of noncommutative rings. For the benefit of the reader, we have tried to make our exposition of these examples from various areas of algebra and geometry reasonably self-contained with many background details included.

Acknowledgment

Parts of the material presented in this book go back more than a quarter century. This applies to the content of Chaps. 1–2 and the computations in Chap. 3 (with the notable exception of the 2-category story), which I worked out sometime around 1992. The particular case of duality over a field, which is much less complicated, was presented in the paper [56], and the possibility of extension to the context of a base ring was mentioned in [56, beginning of Section 4]. The main results in Chaps. 6–7 go back to Spring 2002. Subsequently, I planned and promised several times over the years to write up a detailed exposition. This book partially fulfills that promise. The book also contains some much more recent results; this applies, first of all, to the material of Chap. 5, which is largely based on [68, Section 6] or [67, Theorem 3.1]. I would like to thank all the people, too numerous to be mentioned here by name, whose help and encouragement contributed to my survival over the decades. Speaking of more recent events, I am grateful to Andrey Lazarev, Julian Holstein, and Bernhard Keller for stimulating discussions and interest to this work. The author was supported by research plan RVO: 67985840 and the GAČR project 20-13778S when writing this book.

Contents

All the *associative rings* in this book are unital. We will always presume unitality without mentioning it, so all the left and ring *modules* over associative rings are unital, all the *ring homomorphisms* take the unit to the unit, all the *subrings* contain the unit, and all the *gradings* and *filtrations* are such that the unit element belongs to the degree-zero grading/filtration component.

Given an associative ring R, we denote by $R\text{–mod}$ the abelian category of left R-modules and by $\text{mod–}R$ the abelian category of right R-modules.

Let R, S, and T be three associative rings. For any left R-modules L and M, we denote by $\text{Hom}_R(L, M)$ the abelian group of all left R-module morphisms $L \longrightarrow M$. If L is an R-S-bimodule and M is an R-T-bimodule, then the group $\text{Hom}_R(L, M)$ acquires a natural structure of S-T-bimodule. Similarly, for any right R-modules Q and N, the abelian group of all right R-module morphisms $Q \longrightarrow N$ is denoted by $\text{Hom}_{R^{\text{op}}}(Q, N)$ (where R^{op} stands for the ring opposite to R). If Q is an S-R-bimodule and N is a T-R-bimodule, then $\text{Hom}_{R^{\text{op}}}(Q, N)$ is a T-S-bimodule.

In particular, for any R-S-bimodule U, the abelian group $\text{Hom}_R(U, R)$ is naturally an S-R-bimodule. If U is a finitely generated projective left R-module, then $\text{Hom}_R(U, R)$ is a finitely generated projective right R-module. Similarly, for any S-R-bimodule M, the abelian group $\text{Hom}_{R^{\text{op}}}(M, R)$ is naturally an R-S-bimodule. If M is a finitely generated projective right R-module, then $\text{Hom}_{R^{\text{op}}}(M, R)$ is a finitely generated projective left R-module.

For any R-S-bimodule U, there is a natural morphism of R-S-bimodules $U \longrightarrow \text{Hom}_{R^{\text{op}}}(\text{Hom}_R(U, R), R)$, which is an isomorphism whenever the left R-module U is finitely generated and projective. For any S-R-bimodule M, there is a natural morphism of S-R-bimodules $M \longrightarrow \text{Hom}_R(\text{Hom}_{R^{\text{op}}}(M, R), R)$, which is an isomorphism whenever the right R-module M is finitely generated and projective.

© The Author(s), under exclusive license to Springer Nature Switzerland AG 2021

L. Positselski, *Relative Nonhomogeneous Koszul Duality*, Frontiers in Mathematics, https://doi.org/10.1007/978-3-030-89540-2_1

Let U be an R-S-bimodule and V be an S-T-bimodule. Then the left R-module $U \otimes_S V$ is projective whenever the left R-module U and the left S-module V are projective. The left R-module $U \otimes_S V$ is finitely generated whenever the left R-module U and the left S-module V are finitely generated. The similar assertions apply to the projectivity and finite generatedness on the right side.

Lemma 1.1

(a) *Let U be an R-S-bimodule and V be an S-T-bimodule. Then there is a natural morphism of T-R-bimodules*

$$\mathrm{Hom}_S(V, S) \otimes_S \mathrm{Hom}_R(U, R) \longrightarrow \mathrm{Hom}_R(U \otimes_S V, R),$$

which is an isomorphism whenever the left S-module V is finitely generated and projective.

(b) *Let M be an S-R-bimodule and N be a T-S-bimodule. Then there is a natural morphism of R-T-bimodules*

$$\mathrm{Hom}_{R^{\mathrm{op}}}(M, R) \otimes_S \mathrm{Hom}_{S^{\mathrm{op}}}(N, S) \longrightarrow \mathrm{Hom}_{R^{\mathrm{op}}}(N \otimes_S M, R),$$

which is an isomorphism whenever the right S-module N is finitely generated and projective.

Proof Part (a): the desired map takes an element $g \otimes f \in \mathrm{Hom}_S(V, S) \otimes_S \mathrm{Hom}_R(U, R)$ to the map $U \otimes_S V \longrightarrow R$ taking an element $u \otimes v$ to the element $f(ug(v)) \in R$, for any $g \in \mathrm{Hom}_S(V, S)$, $f \in \mathrm{Hom}_R(U, R)$, $u \in U$, and $v \in V$. The second assertion does not depend on the T-module structure on V, so one can assume $T = \mathbb{Z}$ and, passing to the finite direct sums and direct summands in the argument $V \in S\text{–mod}$, reduce to the obvious case $V = S$. Part (b): the desired map takes an element $h \otimes k \in \mathrm{Hom}_{R^{\mathrm{op}}}(M, R) \otimes_S \mathrm{Hom}_{S^{\mathrm{op}}}(N, S)$ to the map $N \otimes_S M \longrightarrow R$ taking an element $n \otimes m$ to the element $h(k(n)m) \in R$, for any $h \in \mathrm{Hom}_{R^{\mathrm{op}}}(M, R)$, $k \in \mathrm{Hom}_{S^{\mathrm{op}}}(N, S)$, $n \in N$, and $m \in M$. The second assertion does not depend on the T-module structure on N, so it reduces to the obvious case $N = S$. $\square$

Let R be an associative ring and V be an R-R-bimodule. The *tensor ring* of V over R (otherwise called the ring *freely generated* by an R-R-bimodule V) is the graded ring $T_R(V) = \bigoplus_{n=0}^{\infty} T_{R,n}(V)$ with the components $T_{R,0}(V) = R$, $T_{R,1}(V) = V$, $T_{R,2}(V) = V \otimes_R V$, and $T_{R,n}(V) = V \otimes_R \cdots \otimes_R V$ (n factors) for $n \geq 2$. The multiplication in $T_R(V)$ is defined by the obvious rules $r(v_1 \otimes \cdots \otimes v_n) = (rv_1) \otimes v_2 \otimes \cdots \otimes v_n$, $(v_1 \otimes \cdots \otimes v_n)s = v_1 \otimes \cdots \otimes v_{n-1} \otimes (v_n s)$, and $(v_1 \otimes \cdots \otimes v_n)(v_{n+1} \otimes \cdots \otimes v_{n+m}) = v_1 \otimes \cdots \otimes v_{n+m}$ for all r, $s \in R$ and $v_i \in V$.

Let $A = \bigoplus_{n=0}^{\infty} A_n$ be a nonnegatively graded ring with the degree-zero component $A_0 = R$. Denote by V the R-R-bimodule $V = A_1$. Then there exists a unique homomorphism of graded rings $\pi_A : T_R(V) \longrightarrow A$ acting by the identity maps on the components of degrees 0 and 1. Furthermore, denote by $J_A = \ker(\pi_A)$ the kernel ideal of the ring homomorphism π_A. Then J_A is a graded ideal in $T_R(V)$, so we have $J_A = \bigoplus_{n=2}^{\infty} J_{A,n}$, where $J_{A,n} \subset T_{R,n}(V)$. Set $I_A = J_{A,2} \subset V \otimes_R V$.

Definition 1.2 The ring A is said to be *generated by* A_1 (over A_0) if the map π_A is surjective. A graded ring A generated by A_1 over A_0 is said to be *quadratic* (over $R = A_0$) if the two-sided ideal $J_A \subset T_R(V)$ is generated by I_A, that is, $J_A = (I_A)$, or explicitly

$$J_{A,n} = \sum_{i=1}^{n-1} T_{R,i-1}(V) \cdot I_A \cdot T_{R,n-i-1}(V) \quad \text{for all } n \geq 3. \tag{1.1}$$

Examples 1.3

(1) The tensor ring $A = T_R(V)$ is quadratic over R for any associative ring R and any R-R-bimodule V. In this case, one has $J_A = 0$ and $I_A = 0$.
(2) The graded ring $A = R \oplus V$, with $A_0 = R$ and $A_1 = V$, where the multiplication in A is given by the ring structure on R and the R-R-bimodule structure on V, is quadratic over R for any ring R and any R-R-bimodule V. In this case, one has $J_A = \bigoplus_{n=2}^{\infty} T_{R,n}(V)$ and $I_A = V \otimes_R V$.
(3) The graded ring $A = R \oplus V \oplus \cdots \oplus T_{R,n}(V)$, defined as the quotient ring $A = T_R(V)/\bigoplus_{i=n+1}^{\infty} T_{R,i}(V)$, where $n \geq 2$, is generated by A_1 but usually *not* quadratic. In this case, one has $J_A = \bigoplus_{i=n+1}^{\infty} T_{R,i}(V)$ and $I_A = 0$. In fact, the graded ring A is quadratic if and only if $T_{R,n+1}(V) = 0$, or equivalently, $T_{R,i}(V) = 0$ for all $i > n$.
(4) Let R be a commutative ring and V be an R-module, viewed as an R-R-bimodule in the usual way (so that the left and the right actions of R in V coincide). The graded ring $\mathrm{Sym}_R(V)$ is defined as the quotient ring of the tensor ring $T_R(V)$ by the ideal generated by the elements $v \otimes w - w \otimes v \in V \otimes_R V$, where $v, w \in V$. The graded ring $\mathrm{Sym}_R(V)$, known as the *symmetric algebra* of the R-module V, is quadratic over R for any R-module V.
(5) Let R be a commutative ring and V be an R-module, viewed as an R-R-bimodule as in (4). The graded ring $\Lambda_R(V)$ is defined as the quotient ring of the tensor ring $T_R(V)$ by the ideal generated by the elements $v \otimes w + w \otimes v$ and $v \otimes v \in V \otimes_R V$, where v, $w \in V$. In fact, the same ideal is generated by the elements $v \otimes v$ only. When $1/2 \in R$, the same ideal is also generated by the elements $v \otimes w + w \otimes v$ only. The graded ring $\Lambda_R(V)$, known as the *exterior algebra* of the R-module V, is quadratic over R for any R-module V.

Definition 1.4 Let $A = \bigoplus_{n=0}^{\infty} A_n$ be a quadratic graded ring with the degree-zero component $R = A_0$. We will say that A is *2-left finitely projective* if the left R-modules

A_1 and A_2 are projective and finitely generated. Furthermore, A is said to be 3-*left finitely projective* if the same applies to the left R-modules A_1, A_2, and A_3. Similarly, a quadratic graded ring $B = \bigoplus_{n=0}^{\infty} B_n$ with the degree-zero component $R = B_0$ is 2-*right finitely projective* if the right R-modules B_1 and B_2 are finitely generated projective, and B is 3-*right finitely projective* if the same applies to the right R-modules B_1, B_2, and B_3.

Notice that, in a graded ring A generated by A_1 over $R = A_0$, the left R-modules A_n are finitely generated for all $n \geq 0$ whenever the left R-module A_1 is finitely generated. The nontrivial aspect of Definition 1.4 is the projectivity.

Examples 1.5

(1) The tensor ring $A = T_R(V)$ from Example 1.3(1) is 3-left finitely projective if and only if it is 2-left finitely projective and if and only if the left R-module V is finitely generated and projective. The same applies to the graded ring $A = R \oplus V$ from Example 1.3(2).

(2) The symmetric and exterior algebras $\mathrm{Sym}_R(V)$ and $\Lambda_R(V)$ (over a commutative ring R) from Examples 1.3(4–5) are algebras over R, so there is obviously no difference between the left and right projectivity conditions for them. The symmetric algebra $\mathrm{Sym}_R(V)$ is 3-finitely projective if and only if it is 2-finitely projective and if and only if the R-module V is projective. The same applies to the exterior algebra $\Lambda_R(V)$.

(3) To give an example of a quadratic graded ring A such that A_1 is a finitely generated projective R-module, but A_2 is not, take $R = k[\lambda]$ to be the ring of polynomials in one variable λ over a field k. Let $A_1 = V = Rx$ be the free R-module with one generator x, viewed as an R-R-bimodule so that the left and right actions of R in V coincide. Put $I_A = \lambda Rx^2 \subset Rx^2 = V \otimes_R V$ and $J_A = (I_A) = \lambda x^2 R[x] \subset R[x] = T_R(V)$. Then the quotient ring $A = T_R(V)/J_A$ is quadratic, but *not* 2-left (or 2-right) finitely projective quadratic over R. Indeed, the (finitely generated) R-module $A_n = Rx^n/\lambda Rx^n \simeq k$ is not projective for any $n \geq 2$.

(4) The "naïve exterior algebra," defined without due regard to the characteristic 2 issues, provides another example similar to (3). Let $R = \mathbb{Z}$ be the ring of integers, and let $A_1 = V$ be a finitely generated free abelian group viewed as an R-R-bimodule. Denote by I_A the subgroup in $V \otimes_R V$ spanned by the elements $v \otimes w + w \otimes v$, where $v, w \in V$. Put $J_A = (I_A) \subset T_R(V)$ and $A = T_R(V)/J_A$. Then the graded ring A is quadratic, but *not* 2-left (or 2-right) finitely projective quadratic over R (assuming that $V \neq 0$). Indeed, for any element $v \in V \setminus 2V$, one has $2v^2 = 0$ but $v^2 \neq 0$ in A_2, so A_2 is not a projective R-module.

(5) Let us give an example of a 2-left finitely projective, but not 3-left finitely projective quadratic graded ring. Take $R = k[\lambda]$ as in (3), and let $A_1 = V = Rx \oplus Ry$ be free R-module with two generators x and y, viewed as an R-R-bimodule so that the left and right actions of R in V coincide. Put $I_A = Rxy \oplus R(x^2 - \lambda y^2) \subset Rx^2 \oplus Rxy \oplus Ryx \oplus$

$Ry^2 = V \otimes_R V$ and $J_A = (I_A) \subset T_R(V)$. Then the quotient ring $A = T_R(V)/J_A$ is 2-left (and 2-right) finitely projective quadratic, as A_2 is a free R-module with two generators; e.g., the cosets of the elements yx and $y^2 \in T_{R,2}(V)$ in $A_2 = T_{R,2}(V)/I_A$ are free generators of the R-module A_2.

However, the graded ring A is *not* 3-left finitely projective quadratic over R, as the R-module A_n is not projective for any $n \geq 3$. Indeed, A is an associative algebra over the commutative ring R, which is contained in the centrum of A. For any element $c \in k$, denote by k_c the R-module k in which the element $\lambda \in R$ acts by the scalar c. Then the graded k-algebra $A(c) = A/(\lambda - c)A$ is the quadratic k-algebra with the generators x, y and the relations $xy = 0 = x^2 - cy^2$. Following [53, Section 5 in Chapter 6, Example 2], for every $n \geq 3$, one has $A(c)_n = 0$ for $c \neq 0$ and $\dim_k A(0)_n = 2$. In fact, one can easily check that $\lambda A_n = 0$, as $x^3 = \lambda xy^2 = 0$, $\lambda y^3 = x^2 y = 0$, $yx^2 = \lambda y^3 = 0$, and $\lambda y^2 x = x^3 = 0$ in A_3. So A_n is a torsion R-module isomorphic to k^2 as a k-vector space.

Now we fix an associative ring R and consider the *category of graded rings over R* R–$\mathsf{rings}_{\mathsf{gr}}$, defined as follows. The objects of R–$\mathsf{rings}_{\mathsf{gr}}$ are nonnegatively graded associative rings $A = \bigoplus_{n=0}^{\infty} A_n$ endowed with a fixed ring isomorphism $R \simeq A_0$. Morphisms $'A \longrightarrow ''A$ in R–$\mathsf{rings}_{\mathsf{gr}}$ are graded ring homomorphisms forming a commutative triangle diagram with the isomorphisms $R \simeq {'A}_0$ and $R \simeq {''A}_0$. Various specific classes of graded rings defined above in this chapter (and below in the next one) are viewed as full subcategories in R–$\mathsf{rings}_{\mathsf{gr}}$.

Proposition 1.6 *There is an anti-equivalence between the categories of 2-left finitely projective quadratic graded rings A over R and 2-right finitely projective quadratic graded rings B over R, called the* quadratic duality *and defined by the following rules. Given a ring A, the ring B is constructed as $B = T_R(B_1)/(I_B)$, where $B_1 = \mathrm{Hom}_R(A_1, R)$ and $I_B = \mathrm{Hom}_R(A_2, R) \subset B_1 \otimes_R B_1$. Then the R-R-bimodule B_2 is naturally isomorphic to $\mathrm{Hom}_R(I_A, R)$. Conversely, given a ring B, the ring A is constructed as $A = T_R(A_1)/(I_A)$, where $A_1 = \mathrm{Hom}_{R^{\mathrm{op}}}(B_1, R)$ and $I_A = \mathrm{Hom}_{R^{\mathrm{op}}}(B_2, R) \subset A_1 \otimes_R A_1$. Then the R-R-bimodule A_2 is naturally isomorphic to $\mathrm{Hom}_{R^{\mathrm{op}}}(I_B, R)$.*

Proof The category of quadratic graded rings A over R is equivalent to the category of R-R-bimodules $V = A_1$ endowed with a subbimodule $I = I_A \subset V \otimes_R V$. Here morphisms in the category of pairs $('V, 'I) \longrightarrow (''V, ''I)$ are defined as R-R-bimodule morphisms $f: {'V} \longrightarrow {''V}$ such that $(f \otimes f)('I) \subset ''I$.

A quadratic graded ring A is 2-left finitely projective if and only if in the related pair (V, I) the left R-modules V and $A_2 = (V \otimes_R V)/I$ are finitely generated and projective. Assuming the former condition, the left R-module $V \otimes_R V$ is then finitely generated and projective, too, so the latter condition is equivalent to the R-R-subbimodule $I \subset V \otimes_R V$ being *split as a left R-submodule*.

Now we have a short exact sequence of R-R-bimodules

$$0 \longrightarrow I_A \longrightarrow A_1 \otimes_R A_1 \longrightarrow A_2 \longrightarrow 0,$$

which is split as a short exact sequence of left R-modules. Applying the functor $\mathrm{Hom}_R(-, R)$ and taking into account Lemma 1.1(a), we obtain a short exact sequence of R-R-bimodules

$$0 \longrightarrow \mathrm{Hom}_R(A_2, R) \longrightarrow \mathrm{Hom}_R(A_1, R) \otimes_R \mathrm{Hom}_R(A_1, R) \longrightarrow \mathrm{Hom}_R(I_A, R) \longrightarrow 0,$$

which is split as a short exact sequence of right R-modules. It remains to set $B_1 = \mathrm{Hom}_R(A_1, R)$ and $I_B = \mathrm{Hom}_R(A_2, R)$, so that $B_2 = \mathrm{Hom}_R(I_A, R)$. According to the discussion in the beginning of this chapter, B_1 and B_2 are finitely generated projective right R-modules. It is clear that this construction is a contravariant functor between the categories of 2-left finitely projective and 2-right finitely projective quadratic graded rings over R and that the similar construction with the left and right sides switched provides the inverse functor in the opposite direction. $\square$

The 2-left finitely projective quadratic ring A and the 2-right finitely projective quadratic ring B as in Proposition 1.6 are said to be *quadratic dual* to each other.

Examples 1.7

(1) Let R be an associative ring and V be an R-R-bimodule. Assume that V is finitely generated and projective as a left R-module. Denote by $V^\vee$ the R-R-bimodule $\mathrm{Hom}_R(V, R)$, so the R-R-bimodule $V^\vee$ is finitely generated and projective as a right R-module. Then the 2-left finitely projective quadratic graded ring $T_R(V)$ from Examples 1.3(1) and 1.5(1) is quadratic dual to the 2-right finitely projective quadratic graded ring $R \oplus V^\vee$ from Examples 1.3(2) and 1.5(1). Similarly, the 2-left finitely projective quadratic ring $R \oplus V$ is quadratic dual to the 2-right finitely projective quadratic ring $T_R(V^\vee)$.

(2) Let R be a commutative ring and A be a nonnegatively graded R-algebra with $A_0 = R$. Obviously, in this case, A is 2-left finitely projective quadratic if and only if it is 2-right finitely projective quadratic. Moreover, the 2-right finitely projective quadratic graded ring quadratic dual to the 2-left finitely projective quadratic graded ring A coincides with the 2-left finitely projective quadratic graded ring quadratic dual to the 2-right finitely projective quadratic graded ring A.

(3) Let R be a commutative ring and V be a finitely generated projective R-module. Denote by $V^\vee$ the R-module $\mathrm{Hom}_R(V, R)$, so $V^\vee$ is also a finitely generated projective R-module. Then the 2-finitely projective quadratic ring $\mathrm{Sym}_R(V)$ from Examples 1.3(4) and 1.5(2) is quadratic dual to the 2-finitely projective quadratic ring $\Lambda_R(V^\vee)$ from Examples 1.3(5) and 1.5(2).

Proposition 1.8 *The anti-equivalence of categories from Proposition 1.6 takes* 3-*left finitely projective quadratic graded rings to* 3-*right finitely projective quadratic graded rings and vice versa.*

Proof For any quadratic graded ring A, the grading component A_3 is the cokernel of the map $(A_1 \otimes_R I_A) \oplus (I_A \otimes_R A_1) \longrightarrow A_1 \otimes_R A_1 \otimes_R A_1$ induced by the inclusion map $I_A \longrightarrow A_1 \otimes_R A_1$. When the components A_1 and A_2 are projective as (say, left) R-modules, the maps $A_1 \otimes_R I_A \longrightarrow A_1 \otimes_R A_1 \otimes_R A_1$ and $I_A \otimes_R A_1 \longrightarrow A_1 \otimes_R A_1 \otimes_R A_1$ are injective, so we have a four-term exact sequence of R-R-bimodules

$$0 \longrightarrow I_A^{(3)} \longrightarrow (A_1 \otimes_R I_A) \oplus (I_A \otimes_R A_1)$$
$$\longrightarrow A_1 \otimes_R A_1 \otimes_R A_1 \longrightarrow A_3 \longrightarrow 0, \qquad (1.2)$$

where $I_A^{(3)} = (A_1 \otimes_R I_A) \cap (I_A \otimes_R A_1) \subset A_1 \otimes_R A_1 \otimes_R A_1$. When the component A_3 is a projective left R-module, too, we observe that all the terms of this exact sequence, except perhaps the leftmost one, are projective left R-modules. It follows that the sequence (1.2) splits as an exact sequence of left R-modules, and the leftmost term $I_A^{(3)}$ is a projective left R-module, too.

Furthermore, when A_1, A_2, and A_3 are finitely generated projective left R-modules, all the terms of the sequence (1.2) are also finitely generated projective left R-modules. Applying the functor $\mathrm{Hom}_R(-, R)$ to (1.2), we obtain a four-term exact sequence of R-R-bimodules

$$0 \longrightarrow \mathrm{Hom}_R(A_3, R) \longrightarrow B_1 \otimes_R B_1 \otimes_R B_1$$
$$\longrightarrow (B_1 \otimes_R B_2) \oplus (B_2 \otimes_R B_1) \longrightarrow \mathrm{Hom}_R(I_A^{(3)}, R) \longrightarrow 0. \qquad (1.3)$$

Now for any quadratic graded ring B, the cokernel of the map $B_1 \otimes_R B_1 \otimes_R B_1 \longrightarrow (B_1 \otimes_R B_2) \oplus (B_2 \otimes_R B_1)$ induced by the (surjective) multiplication map $B_1 \otimes_R B_1 \longrightarrow B_2$ is the grading component B_3. Hence we have a natural isomorphism of R-R-bimodules $B_3 \simeq \mathrm{Hom}_R(I_A^{(3)}, R)$, and it follows that B_3 is a finitely generated projective right R-module. $\qquad\square$

Example 1.9 The statement similar to Proposition 1.8 does *not* hold in degrees higher than 3 for quadratic graded rings in general. It holds under the Koszulity assumption, though, as we will see in the next chapter.

To give a specific counterexample, take $R = k[\lambda]$ as in Examples 1.5(3) and (5). Let $A_1 = V = Rx \oplus Ry \oplus Rz$ be the free R-module with three generators x, y, and z, viewed as an R-R-bimodule in the usual way. Put $I_A = Rzx \oplus R(xz - xy) \oplus R(zy - yz - \lambda y^2) \subset V \otimes_R V$ and $J_A = (I_A) \subset T_R(V)$. Then the quotient ring $A = T_R(V)/J_A$ is 3-left (and 3-right) finitely projective quadratic, as A_2 is a free R-module with six generators and A_3 is a free R-module with ten generators.

The quadratic dual ring B to the quadratic ring A over R (cf. Example 1.7(2)) is also 3-left and 3-right finitely projective. Its grading components B_1 and B_2 are free R-modules with three generators, while B_3 is a free R-module with one generator and $B_n = 0$ for $n \geq 4$. So the graded R-module B is (finitely generated) projective.

But the graded R-module A is *not* projective. In fact, there is a nonzero element annihilated by $\lambda - 1$ in A_4, and furthermore, a nonzero element annihilated by $\lambda - 1/n$ in A_{n+3} for every $n \geq 1$ such that $1/n \in k$. Specifically, one has $0 = xy^n zx = xy^{n-1} zyx - \lambda xy^{n+1}x = xy^{n-2}zy^2x - 2\lambda xy^{n+1}x = \cdots = xzy^n x - n\lambda xy^{n+1}x = (1 - n\lambda)xy^{n+1}x$ in A_{n+3}. See the paper [26] and [53, Section 6 in Chapter 6] for a further discussion of this and similar examples.

Remark 1.10 The above discussion of the categories of nonnegatively graded and quadratic rings can be modified or expanded by including *non-identity isomorphisms* (in particular, automorphisms) in the degree-zero component. Denote by Rings_{gr} the category whose objects are nonnegatively graded associative rings $A = \bigoplus_{n=0}^{\infty} A_n$, and morphisms are defined as follows. A morphism $'A \longrightarrow {''}A$ in Rings_{gr} is a morphism of graded rings $f\colon {'}A \longrightarrow {''}A$ whose *degree-zero component $f_0\colon {'}A_0 \longrightarrow {''}A_0$ is an isomorphism*. Then the same classes of 2- and 3-left/right finitely projective quadratic rings as in Propositions 1.6 and 1.8 can be viewed as full subcategories in Rings_{gr}. The assertions of the two propositions remain valid with this modification.

The inclusion of the full subcategory of quadratic graded rings over R into the category of (nonnegatively) graded rings over R has a right adjoint functor, which we denote by $A \longmapsto qA$. For any nonnegatively graded ring A, the quadratic graded ring $A' = qA$ together with the graded ring homomorphism $A' \longrightarrow A$ is characterized by the properties that the maps $A'_0 \longrightarrow A_0$ and $A'_1 \longrightarrow A_1$ are isomorphisms and the map $A'_2 \longrightarrow A_2$ is injective. Explicitly, the ring qA is constructed as the ring with degree-one generators and quadratic relations $qA = T_{A_0}(A_1)/(I_A)$, where $I_A \subset A_1 \otimes_R A_1$ is the kernel of the multiplication map $A_1 \otimes_R A_1 \longrightarrow A_2$.

Flat and Finitely Projective Koszulity

2

2.1 Graded and Ungraded Ext and Tor

So far in this book, we denoted a graded ring by $A = \bigoplus_{n=0}^{\infty} A_n$ (and for the most part we will continue to do so in the sequel), but this is a colloquial abuse of notation. A graded abelian group U is properly thought of as a collection of abelian groups $U = (U_n)_{n \in \mathbb{Z}}$. Then there are several ways to produce an ungraded group from a graded one.

Two of them are important for us in this section. One can take the direct sum of the grading components, which we denote by $\Sigma U = \bigoplus_{n \in \mathbb{Z}} U_n$, or one can take the product of the grading components, which we denote by $\Pi U = \prod_{n \in \mathbb{Z}} U_n$.

In particular, let $A = (A_n)_{n \in \mathbb{Z}}$ be a graded ring and $M = (M_n)_{n \in \mathbb{Z}}$ be a graded left A-module. Then $\Sigma A = \bigoplus_{n \in \mathbb{Z}} A_n$ is the underlying ungraded ring of A, and there are *two* underlying ungraded ΣA-modules associated with M. Namely, both the abelian groups $\Sigma M = \bigoplus_{n \in \mathbb{Z}} M_n$ and $\Pi M = \prod_{n \in \mathbb{Z}} M_n$ have natural structures of left ΣA-modules. Denoting the category of graded left A-modules by $A\text{--mod}_{\mathsf{gr}}$, we have two forgetful functors Σ and $\Pi \colon A\text{--mod}_{\mathsf{gr}} \longrightarrow \Sigma A\text{--mod}$.

The tensor product of a graded right A-module N and a graded left A-module M is naturally a graded abelian group $N \otimes_A M$, and applying the functor Σ to $N \otimes_A M$ produces the tensor product of the ungraded ΣA-modules ΣN and ΣM,

$$\Sigma(N \otimes_A M) \simeq \Sigma N \otimes_{\Sigma A} \Sigma M. \tag{2.1}$$

Similarly, for any graded left A-modules L and M, one can consider the graded abelian group $\mathrm{Hom}_A(L, M)$ with the components $\mathrm{Hom}_{A,n}(L, M)$ consisting of all the homogeneous left A-module maps $L \longrightarrow M$ of degree n. The purpose of introducing the functor Π above was to formulate the comparison between the graded and ungraded Hom, which has the form

$$\Pi \operatorname{Hom}_A(L, M) \simeq \operatorname{Hom}_{\Sigma A}(\Sigma L, \Pi M). \tag{2.2}$$

Furthermore, the functor Σ takes projective graded A-modules to projective ΣA-modules, while the functor Π takes injective graded A-modules to injective ΣA-modules (as one can see from the description of projective and injective modules as the direct summands of the free and cofree modules, respectively). In addition, the functor Σ takes flat graded A-modules to flat ΣA-modules (as one can see from the Govorov–Lazard description of flat modules as the filtered direct limits of finitely generated free modules). We define the graded versions of Tor and Ext as the derived functors of the graded tensor product and Hom, computed in the abelian categories of graded (right and left) modules.

So, for any graded right A-module N and any graded left A-module M, there is a *bigraded* abelian group

$$\operatorname{Tor}^A(N, M) = (\operatorname{Tor}^A_{i,j}(N, M))_{i,j}, \qquad i \geq 0, \ j \in \mathbb{Z},$$

where i is the usual *homological* grading and j is the *internal* grading (induced by the grading of A, N, and M). In order to compute the bigraded group $\operatorname{Tor}^A(M, N)$, one chooses a *graded* projective (or flat) resolution of one of the A-modules M and N and takes its tensor product over A with the other module; then the grading i is induced by the homological grading of the resolution and the grading j comes from the grading of the tensor product of any two graded modules. In view of the above considerations concerning projective/flat graded modules, the formula (2.1) implies a similar formula for the Tor groups,

$$\Sigma \operatorname{Tor}^A_i(N, M) \simeq \operatorname{Tor}^{\Sigma A}_i(\Sigma N, \Sigma M) \quad \text{for every } i \geq 0,$$

or more explicitly,

$$\operatorname{Tor}^{\Sigma A}_i(\Sigma M, \Sigma N) \simeq \bigoplus_{j \in \mathbb{Z}} \operatorname{Tor}^A_{i,j}(M, N). \tag{2.3}$$

Similarly, for any graded left A-modules L and M there is a *bigraded* abelian group

$$\operatorname{Ext}_A(L, M) = (\operatorname{Ext}^i_{A,n}(L, M))_{i,n}, \qquad i \geq 0, \ n \in \mathbb{Z},$$

where i is the usual *cohomological* grading and n is the *internal* grading. In order to compute the bigraded group $\operatorname{Ext}_A(L, M)$, one chooses either a graded projective resolution of the A-module L or a graded injective resolution of the A-module M and takes the graded

Hom; then the grading i is induced by the (co)homological grading of the resolution and the grading n comes from the grading of the Hom groups.

In the context of the internal grading of the Ext, we will put $n = -j$ and use the notation $\mathrm{Ext}_A^{i,j}(L, M) = \mathrm{Ext}_{A,-j}^{i}(L, M)$. By abuse of terminology, the grading j will be also called the *internal grading* of the Ext.

In view of the above considerations concerning projective and injective graded modules, the formula (2.2) implies a similar formula for the Ext groups,

$$\Pi\, \mathrm{Ext}_A^i(L, M) \simeq \mathrm{Ext}_{\Sigma A}^i(\Sigma L, \Pi M) \quad \text{for every } i \geq 0,$$

or more explicitly,

$$\mathrm{Ext}_{\Sigma A}^i(\Sigma L, \Pi M) \simeq \prod_{j \in \mathbb{Z}} \mathrm{Ext}_A^{i,j}(L, M). \tag{2.4}$$

For any three graded left A-modules K, L, and M, there are natural associative, unital composition/multiplication maps

$$\mathrm{Ext}_A^{i',j'}(L, M) \times \mathrm{Ext}_A^{i',j'}(K, L) \longrightarrow \mathrm{Ext}_A^{i'+i'',j'+j''}(K, M), \quad i', i'' \geq 0, \ j', j'' \in \mathbb{Z} \tag{2.5}$$

on the bigraded Ext groups. Whenever the graded A-module L only has a finite number of nonzero grading components (so $\Sigma L = \Pi L$), the passage to the infinite products with respect to the internal gradings j' and j'' makes the multiplications (2.5) on the graded Ext groups agree with the similar multiplications on the ungraded Ext (between the ΣA-modules ΣK, $\Sigma L = \Pi L$, and ΠM).

2.2   Relative Bar Resolution

Given an R-R-bimodule V, we will use the notation $V^{\otimes_R n} = T_{R,n}(V)$ for the tensor product $V \otimes_R \cdots \otimes_R V$ (n factors).

Let $R \longrightarrow A$ be an injective homomorphism of associative rings. Denote by A_+ the R-R-bimodule A/R. Let L be a left A-module. The *reduced relative bar resolution* of L is the complex of left A-modules

$$\cdots \longrightarrow A \otimes_R A_+ \otimes_R A_+ \otimes_R L \longrightarrow A \otimes_R A_+ \otimes_R L \longrightarrow A \otimes_R L \longrightarrow L \longrightarrow 0 \tag{2.6}$$

with the differential given by the standard formula $\partial(a_0 \otimes \bar{a}_1 \otimes \cdots \otimes \bar{a}_n \otimes l) = a_0 a_1 \otimes a_2 \otimes \cdots \otimes a_n \otimes l - a_0 \otimes a_1 a_2 \otimes a_3 \otimes \cdots \otimes l + \cdots + (-1)^n a_0 \otimes a_1 \otimes \cdots \otimes a_{n-1} \otimes a_n l$. One can easily check that the image of the right-hand side in $A \otimes_R A_+^{\otimes_R n-1} \otimes_R L$ does not depend on the arbitrary choice of liftings $a_i \in A$ of the given elements $\bar{a}_i \in A_+$, $1 \leq i \leq n$, so the differential is well-defined.

The complex (2.6) is contractible as a complex of left R-modules; the contracting homotopy is given by the formulas $t(l) = 1 \otimes l$, $l \in L$, and $t(a_0 \otimes \bar{a}_1 \otimes \cdots \otimes \bar{a}_n \otimes l) = 1 \otimes \bar{a}_0 \otimes \bar{a}_1 \otimes \cdots \otimes \bar{a}_n \otimes l$, where $\bar{a}_i \in A_+$, $1 \leq i \leq n$, and $\bar{a}_0 \in A_+$ is the image of the element $a_0 \in A$ under the natural surjection $A \longrightarrow A_+$. Hence it follows that the complex (2.6) is acyclic.

If both L and A_+ are flat left R-modules, then all the left A-modules $A \otimes_R A_+^{\otimes_R n} \otimes_R L$ are flat, so (2.6) is a flat resolution of the left A-module L. Similarly, (2.6) is a projective resolution of the left A-module L whenever both the left R-modules L and A_+ are projective.

Assume that L and A_+ are flat left R-modules, and let N be an arbitrary right A-module. Then one can use the flat resolution (2.6) of the left A-module L in order to compute the groups $\mathrm{Tor}_i^A(N, L)$. Thus the groups $\mathrm{Tor}_i^A(N, L)$ are naturally isomorphic to the homology groups of the bar complex

$$\cdots \longrightarrow N \otimes_R A_+ \otimes_R A_+ \otimes_R L \longrightarrow N \otimes_R A_+ \otimes_R L \longrightarrow N \otimes_R L \longrightarrow 0. \qquad (2.7)$$

Switching the roles of the left and right modules and using the reduced relative bar resolution of N, we conclude that the same bar complex (2.7) computes the groups $\mathrm{Tor}_i^A(N, L)$ whenever a right A-module N is a flat right R-module, A_+ is a flat right R-module, and L is an arbitrary left A-module.

Let M be a left A-module. The *reduced relative cobar resolution* of M is the complex of left A-modules

$$0 \longrightarrow M \longrightarrow \mathrm{Hom}_R(A, M) \longrightarrow \mathrm{Hom}_R(A_+ \otimes_R A, M)$$

$$\longrightarrow \mathrm{Hom}_R(A_+ \otimes_R A_+ \otimes_R A, M) \longrightarrow \cdots \qquad (2.8)$$

with the differential given by the formula $(\partial f)(\bar{a}_n \otimes \cdots \otimes \bar{a}_1 \otimes a_0) = a_n f(a_{n-1} \otimes \cdots \otimes a_0) - f(a_n a_{n-1} \otimes a_{n-2} \otimes \cdots \otimes a_0) + \cdots + (-1)^n f(a_n \otimes \cdots \otimes a_2 \otimes a_1 a_0)$, where $f \in \mathrm{Hom}_R(A_+^{\otimes_R n-1} \otimes_R A, M)$ and $a_i \in A$ are arbitrary liftings of elements $\bar{a}_i \in A_+$, $1 \leq i \leq n$. One easily checks that the expression in the right-hand side vanishes on the kernel of the natural surjection $A^{\otimes_R n+1} \longrightarrow A_+^{\otimes_R n} \otimes_R A$, so the differential is well-defined. The left A-module structure on $\mathrm{Hom}_R(A_+^{\otimes_R n} \otimes_R A, M)$ is induced by the right A-module structure on A.

The complex (2.8) is contractible as a complex of left R-modules; the contracting homotopy is given by the formula $t(f)(\bar{a}_{n-1} \otimes \cdots \otimes \bar{a}_1 \otimes a_0) = (-1)^n f(\bar{a}_{n-1} \otimes \cdots \otimes \bar{a}_1 \otimes \bar{a}_0 \otimes 1)$. In particular, it follows that the complex (2.8) is acyclic. If M is an injective left R-module and A_+ is a flat right R-module, then all the left A-modules $\mathrm{Hom}_R(A_+^{\otimes_R n} \otimes_R A, M)$, $n \geq 0$, are injective; so (2.8) is an injective resolution of the left A-module M.

Let L and M be left A-modules. Assuming that L is a projective left R-module and A_+ is a projective left R-module, one can use the projective resolution (2.6) of the left A-module L in order to compute the groups $\mathrm{Ext}^i_A(L, M)$. Thus the groups $\mathrm{Ext}^i_A(L, M)$ are naturally isomorphic to the cohomology groups of the cobar complex

$$0 \longrightarrow \mathrm{Hom}_R(L, M) \longrightarrow \mathrm{Hom}_R(A_+ \otimes_R L, M)$$
$$\longrightarrow \mathrm{Hom}_R(A_+ \otimes_R A_+ \otimes_R L, M) \longrightarrow \cdots \qquad (2.9)$$

Alternatively, assuming that M is an injective left R-module and A_+ is a flat right R-module, one can use the injective resolution (2.8) of the left A-module M in order to compute the groups $\mathrm{Ext}^i_A(L, M)$. Under these assumptions, one comes to the same conclusion that the groups $\mathrm{Ext}^i_A(L, M)$ are naturally isomorphic to the cohomology groups of the cobar complex (2.9).

When R and A are graded rings, $R \longrightarrow A$ is a graded ring homomorphism, and L is a graded left A-module, one can interpret (2.6) as a graded resolution of the graded A-module L. When L and A_+ are flat graded left R-modules, (2.6) is a graded flat resolution of the graded A-module L, and when L and A_+ are projective graded left R-modules, (2.6) is a graded projective resolution.

For a graded left A-module M, one can also interpret the Hom notation in (2.8) as the graded Hom, obtaining a graded resolution of the graded A-module M. When M is an injective graded left R-module and A_+ is a flat graded right R-module, (2.8) is a graded injective resolution.

It follows that, under the graded versions of the above flatness, projectivity, and/or injectivity assumptions, the graded bar complex (2.7) computes the bigraded $\mathrm{Tor}^A(N, L)$, and the graded version of the cobar complex (2.9) computes the bigraded $\mathrm{Ext}_A(L, M)$.

The functor Σ transforms the graded versions of (2.6) and (2.7) (for graded rings R and A and graded modules L and N) into the ungraded ones (for the ungraded rings ΣR and ΣA and the ungraded modules ΣL and ΣN). The functor Π transforms the graded versions of (2.8) and (2.9) (for graded rings R and A and graded modules L and M) into the ungraded ones (for the ungraded rings ΣR and ΣA and the ungraded modules ΣL and ΠM).

The cobar complexes also compute the composition/multiplication on the Ext groups. Let K, L, and M be left A-modules; assume that the left R-modules K, L, and A_+ are projective. Then the natural composition/multiplication maps on the cobar complexes

$$\mathrm{Hom}_R(A_+^{\otimes_R i'} \otimes_R L, M) \times \mathrm{Hom}_R(A_+^{\otimes_R i''} \otimes_R K, L) \longrightarrow \mathrm{Hom}_R(A_+^{\otimes_R i'+i''} \otimes_R K, M)$$

agree with the cobar differentials and induce the Yoneda multiplication maps $\mathrm{Ext}^{i'}_A(L, M) \times \mathrm{Ext}^{i''}_A(K, L) \longrightarrow \mathrm{Ext}^{i'+i''}_A(K, M)$ on the Ext groups. In the case of graded rings R and A and graded A-modules K, L, and M, the same assertion applies to the graded Ext.

2.3   Diagonal Tor and Ext

Let $A = \bigoplus_{n=0}^{\infty} A_n$ be a nonnegatively graded ring with the degree-zero component $R = A_0$. Then the projection onto the degree-zero component is a graded ring homomorphism $A \longrightarrow R$. Using this homomorphism, one can consider R as a left and right graded module over A.

So we can consider the bigraded Tor groups $\mathrm{Tor}^A_{i,j}(R, R)$ and the bigraded Ext ring $\mathrm{Ext}^{i,j}_A(R, R)$, with the (co)homological grading i and the internal grading j. In fact, as R is an R-A-bimodule and an A-R-bimodule, the groups $\mathrm{Tor}^A_{i,j}(R, R)$ have natural structures of R-R-bimodules.

First of all, we notice the connection between the graded and ungraded Tor and Ext. As the graded (left or right) A-module R is concentrated in the internal grading 0, we have $\Sigma R = R = \Pi R$. Hence the formulas (2.3) and (2.4) reduce to

$$\mathrm{Tor}^{\Sigma A}_i(R, R) \simeq \bigoplus_{j \in \mathbb{Z}} \mathrm{Tor}^A_{i,j}(R, R)$$

$$\mathrm{Ext}^i_{\Sigma A}(R, R) \simeq \prod_{j \in \mathbb{Z}} \mathrm{Ext}^{i,j}_A(R, R).$$

Proposition 2.1 *Assume that the grading components A_n, $n \geq 1$, are flat left R-modules, and the multiplication map $A_1 \otimes_R A_1 \longrightarrow A_2$ is surjective with the kernel I_A. Then there are natural isomorphisms of R-R-bimodules*

(a) $\mathrm{Tor}^A_{i,j}(R, R) = 0$ *whenever $i < 0$, or $i = 0$ and $j > 0$, or $i > j$.*
(b) $\mathrm{Tor}^A_{0,0}(R, R) = R$, $\mathrm{Tor}^A_{1,1}(R, R) \simeq A_1$, $\mathrm{Tor}^A_{2,2}(R, R) \simeq I_A$, *and*

$$\mathrm{Tor}^A_{n,n}(R, R) \simeq \bigcap_{k=1}^{n-1} A_1^{\otimes_R k-1} \otimes_R I_A \otimes_R A_1^{\otimes_R n-k-1} \subset A_1^{\otimes_R n}, \qquad n \geq 2.$$

Proof The R-R-bimodules $\mathrm{Tor}^A_{i,j}(R, R)$ can be computed as the homology bimodules of the bar complex (2.7) for $N = R = L$,

$$\cdots \longrightarrow A_+ \otimes_R A_+ \otimes_R A_+ \longrightarrow A_+ \otimes_R A_+ \longrightarrow A_+ \xrightarrow{\;0\;} R \longrightarrow 0. \qquad (2.10)$$

More explicitly, this means that, for every fixed $n \geq 1$, the bimodules $\mathrm{Tor}^A_{i,n}(R, R)$ are the homology bimodules of the complex of R-R-bimodules

$$0 \longrightarrow A_1^{\otimes_R n} \longrightarrow \bigoplus_{k=1}^{n-1} A_1^{\otimes_R k-1} \otimes_R A_2 \otimes_R A_1^{\otimes_R n-k-1} \longrightarrow$$

$$\cdots \longrightarrow (A_+^{\otimes_R i})_n \longrightarrow \cdots$$

$$\longrightarrow \bigoplus_{k=1}^{n-1} A_k \otimes_R A_{n-k} \longrightarrow A_n \longrightarrow 0. \qquad (2.11)$$

The assertion (a) follows immediately, and to prove (b), it remains to compute, for every $1 \le k \le n - 1$, the kernel of the map $A_1^{\otimes_R n} \longrightarrow A_1^{\otimes_R k-1} \otimes_R A_2 \otimes_R A_1^{\otimes_R n-k-1}$ induced by the multiplication map $A_1 \otimes_R A_1 \longrightarrow A_2$.

Specifically, we have to check that the natural short sequence of R-R-bimodules

$$0 \longrightarrow A_1^{\otimes_R k-1} \otimes_R I_A \otimes_R A_1^{\otimes_R n-k-1} \longrightarrow A_1^{\otimes_R n} \longrightarrow A_1^{\otimes_R k-1} \otimes_R A_2 \otimes_R A_1^{\otimes_R n-k-1} \longrightarrow 0$$

is exact. Indeed, by assumption, we have a short exact sequence of R-R-bimodules

$$0 \longrightarrow I_A \longrightarrow A_1 \otimes_R A_1 \longrightarrow A_2 \longrightarrow 0, \tag{2.12}$$

whose terms are flat left R-modules. Furthermore, $A_1^{\otimes_R n-k-1}$ is a flat left R-module, too. It follows that tensoring (2.12) (with any right R-module, and in particular) with $A_1^{\otimes_R k-1}$ on the left does not affect exactness, and neither does tensoring (any exact sequence of right R-modules, and in particular) the resulting short sequence with $A_1^{\otimes_R n-k-1}$ on the right. $\qquad\square$

Concerning the Ext, our notation $\mathrm{Ext}_A(R, R)$ presumes, as above, that the Ext is taken in the category of *left* R-modules. So, in particular, the ring $\mathrm{Ext}_A^0(R, R) = \mathrm{Hom}_R(R, R) = R^{\mathrm{op}}$ is the opposite ring to R.

Proposition 2.2 *Assume that the grading components A_n, $n \ge 1$, are finitely generated projective left R-modules, and the multiplication map $A_1 \otimes_R A_1 \longrightarrow A_2$ is surjective with the kernel I_A. Then*

(a) $\mathrm{Ext}_A^{i,j}(R, R) = 0$ *whenever $i < 0$, or $i = 0$ and $j > 0$, or $i > j$.*
(b) *The diagonal Ext ring $\bigoplus_{n=0}^{\infty} \mathrm{Ext}_A^{n,n}(R, R)$ is naturally isomorphic, as a graded ring, to the opposite ring B^{op} to the 2-right finitely projective quadratic dual ring $B = T_R(B_1)/(I_B)$, $B_1 = \mathrm{Hom}_R(A_1, R)$, $I_B = \mathrm{Hom}_R(A_2, R)$ to the 2-left finitely projective quadratic ring $qA = T_R(A_1)/(I_A)$.*

Proof Part (a) does not depend on the finite generatedness assumptions on the R-modules A_n and requires only the projectivity assumptions. The bigraded ring $\mathrm{Ext}_A(R, R)$ can be computed as the cohomology ring of the DG-ring (2.9) for $L = R = M$. The latter can be obtained by applying the functor $\mathrm{Hom}_R(-, R)$ to the bar complex (2.10),

$$0 \longrightarrow R \xrightarrow{\ 0\ } \mathrm{Hom}_R(A_+, R) \longrightarrow \mathrm{Hom}_R(A_+ \otimes_R A_+, R) \longrightarrow \cdots \tag{2.13}$$

More specifically, for every fixed $n \ge 1$, the groups $\mathrm{Ext}_A^{i,n}(R, R)$ are the cohomology groups of the complex obtained by applying $\mathrm{Hom}_R(-, R)$ to the complex (2.11). This proves part (a).

When all the grading components A_j are finitely generated and projective left R-modules, the complex obtained by applying the functor $\mathrm{Hom}_R(-, R)$ to the complex (2.11) can be computed using Lemma 1.1(a) as

$$0 \longrightarrow A_n^\vee \longrightarrow \bigoplus_{k=1}^{n-1} A_{n-k}^\vee \otimes_R A_k^\vee \longrightarrow$$

$$\cdots \longrightarrow (A_+^{\vee \otimes_R i})^n \longrightarrow \cdots$$

$$\longrightarrow \bigoplus_{k=1}^{n-1} A_1^{\vee \otimes_R n-k-1} \otimes_R A_2^\vee \otimes_R A_1^{\vee \otimes_R k-1} \longrightarrow A_1^{\vee \otimes_R n} \longrightarrow 0, \qquad (2.14)$$

where the notation is $U^\vee = \mathrm{Hom}_R(U, R)$. Setting $B_1 = A_1^\vee$, $I_B = A_2^\vee$, and $B = T_R(B_1)/(I_B)$, we obtain the desired isomorphism $\mathrm{Ext}_A^{n,n}(R, R) \simeq B_n$. It follows immediately from the construction of the multiplication on the cobar complex that this is a graded ring isomorphism between $\bigoplus_n \mathrm{Ext}_A^{n,n}(R, R)$ and B^{op}. $\qquad \square$

Remark 2.3 The bar complex $\mathrm{Bar}_R(A)$ computing $\mathrm{Tor}^A(R, R)$ (2.10) has a natural structure of a (coassociative, counital) *DG-coring* over the ring R, with the obvious maps of counit $\mathrm{Bar}_R(A) \longrightarrow R$ and a comultiplication $\mathrm{Bar}_R(A) \longrightarrow \mathrm{Bar}_R(A) \otimes_R \mathrm{Bar}_R(A)$ compatible with the bar differential. However, as the functor of tensor product over R is not left exact, this DG-coring structure, generally speaking, does *not* descend to a coring structure on the homology modules. When A_+ is a flat left R-module *and* $\mathrm{Tor}^A(R, R)$ is a flat left R-module, the nonexactness problem does not interfere, and the (bi)graded R-R-bimodule $\mathrm{Tor}^A(R, R)$ is a (coassociative, counital) coring over R. In particular, the diagonal Tor bimodule $C = \bigoplus_{n=0}^\infty \mathrm{Tor}_{n,n}^A(R)$ becomes a graded coring over R in these assumptions, with the obvious counit $C \longrightarrow R$ and the induced comultiplication map $C \longrightarrow C \otimes_R C$.

2.4   Low-Dimensional Tor, Degree-One Generators, and Quadratic Relations

As in the previous section, we consider a nonnegatively graded ring $A = \bigoplus_{n=0}^\infty A_n$ with the degree-zero component $R = A_0$.

Proposition 2.4 *Assume that the grading components A_n, $n \geq 1$, are flat left R-modules. Then*

(a) *The graded ring A is generated by A_1 over R if and only if $\mathrm{Tor}_{1,j}^A(R, R) = 0$ for all $j > 1$.*

(b) *The graded ring A is quadratic if and only if $\mathrm{Tor}_{1,j}^A(R, R) = 0$ for all $j > 1$ and $\mathrm{Tor}_{2,j}^A(R, R) = 0$ for all $j > 2$.*

Proof Part (a): following the proof of Proposition 2.1, the group $\mathrm{Tor}^A_{1,n}(R, R)$ can be computed as the rightmost homology group of the complex (2.11), that is, the cokernel of the multiplication map $\bigoplus^n_{k=1} A_k \otimes_R A_{n-k} \longrightarrow A_n$. Now it is clear that a nonnegatively graded ring A is generated by A_1 over $R = A_0$ if and only if the latter map is surjective for all $n \geq 2$.

Part (b): by part (a), we can assume that A is generated by A_1; then we have to show that A is quadratic if and only if $\mathrm{Tor}^A_{2,j}(R, R) = 0$ for all $j > 2$. Once again, we compute the group $\mathrm{Tor}^A_{2,n}(R, R)$ using the complex (2.11). So we have to show that A is quadratic if and only if the short sequence

$$\bigoplus^{k,l,m \geq 1}_{k+l+m=n} A_k \otimes_R A_l \otimes_R A_m \longrightarrow \bigoplus^{k,l \geq 1}_{k+l=n} A_k \otimes_R A_l \longrightarrow A_n \longrightarrow 0 \qquad (2.15)$$

is right exact for all $n \geq 3$.

Given an R-R-bimodule V and an R-R-subbimodule $I \subset V \otimes_R V$, one can construct the quadratic ring $A' = T_R(V)/(I)$ by the following inductive procedure. Set $A'_0 = R$, $A'_1 = V$, and $A'_2 = (V \otimes_R V)/I$; then there are the obvious multiplication maps $A'_k \times A'_l \longrightarrow A'_{k+l}$ for $k, l \geq 0$, $k + l \leq 2$. For every $n \geq 3$, set A'_n to be the cokernel of the R-R-bimodule morphism

$$\bigoplus^{k,l,m \geq 1}_{k+l+m=n} A'_k \otimes_R A'_l \otimes_R A'_m \longrightarrow \bigoplus^{k,l \geq 1}_{k+l=n} A'_k \otimes_R A'_l.$$

Then we have biadditive multiplication maps $A'_k \times A'_l \longrightarrow A'_n$ defined for all $k + l = n$, $k, l \geq 0$, and satisfying the associativity equations $(ab)c = a(bc)$ for all $a \in A'_k$, $b = A'_l$, $c \in A'_m$, $k + l + m = n$, $k, l, m \geq 0$. So we obtain a graded ring A'. Obviously, A' is the graded associative ring freely generated by $A'_1 = V$ over $A'_0 = R$ with the relations $I \subset V \otimes_R V$, so, in other words, $A' \simeq T_R(V)/(I)$.

Returning to the original graded ring A, put $V = A_1$, and let I be the kernel of the multiplication map $A_1 \otimes_R A_1 \longrightarrow A_2$. Then there is a unique homomorphism of graded rings $A' \longrightarrow A$ acting by the identity map in degrees 0 and 1. The ring homomorphism $A' \longrightarrow A$ is surjective by assumption. Arguing by induction in n, one can easily see that this ring homomorphism is an isomorphism if and only if the short sequence (2.15) is right exact for every $n \geq 3$. $\qquad\Box$

2.5 First Koszul Complex

Let $A = \bigoplus^\infty_{n=0} A_n$ and $B = \bigoplus^\infty_{n=0} B_n$ be two nonnegatively graded rings with the same degree-zero component $A_0 = R = B_0$. Suppose that B_1 is a finitely generated projective right R-module and that we are given an R-R-bimodule morphism $\tau \colon \mathrm{Hom}_{R^{\mathrm{op}}}(B_1, R) \longrightarrow A_1$.

Equivalently, instead of a map τ, one can consider an element of the tensor product $B_1 \otimes_R A_1$ satisfying the equation spelled out in the next lemma (which is to be applied to the R-R-bimodules $V = \operatorname{Hom}_{R^{op}}(B_1, R)$ and $U = A_1$).

Lemma 2.5 *Let R and S be associative rings.*

(a) *Let U be a left R-module and V be a finitely generated projective left R-module. Then there is a natural isomorphism of abelian groups $\operatorname{Hom}_R(V, U) \simeq \operatorname{Hom}_R(V, R) \otimes_R U$.*

(b) *Let V and U be two R-S-bimodules such that V is a finitely generated projective left R-module. Then R-S-bimodule morphisms $f \colon V \longrightarrow U$ correspond to elements $e \in \operatorname{Hom}_R(V, R) \otimes_R U$ satisfying the equation $se = es$ for all $s \in S$ under the isomorphism of part (a). In other words, there is a natural bijective correspondence between R-S-bimodule morphisms $f \colon V \longrightarrow U$ and S-S-bimodule morphisms $e \colon S \longrightarrow \operatorname{Hom}_R(V, R) \otimes_R U$.*

Proof Both the assertions are easy. Let us only specify that the correspondence between the maps $f \colon V \longrightarrow U$ and the elements $e = e(1) \in \operatorname{Hom}_R(V, R) \otimes_R U$ is given by the rule $f(v) = \langle v, e \rangle$. Here the pairing notation (cf. Sect. 3.4 below) stands for the map

$$\langle \, , \, \rangle \colon V \times \operatorname{Hom}_R(V, R) \longrightarrow R, \qquad \langle v, b \rangle = b(v),$$

and, by extension, for the map

$$\langle \, , \, \rangle \colon V \times \operatorname{Hom}_R(V, R) \otimes_R U \longrightarrow U$$

given by the formulas $\langle v, b \otimes u \rangle = \langle v, b \rangle u = b(v) u$, where $v \in V$, $u \in U$, and $b \in \operatorname{Hom}_R(V, R)$. $\qquad \square$

Denote by $K^\tau(B, A)$ the following bigraded A-B-bimodule endowed with an A-B-bimodule endomorphism ∂^τ. The bigrading components of $K^\tau(B, A)$ are

$$K^\tau_{i,n}(B, A) = \operatorname{Hom}_{R^{op}}(B_i, A_{n-i}), \quad i \geq 0, \ n \geq i.$$

Here i is interpreted as the homological grading and n is the internal grading. The homogeneous A-B-bimodule endomorphism $\partial^\tau \colon K^\tau(B, A) \longrightarrow K^\tau(B, A)$ of bidegree $(i, n) = (-1, 0)$ is constructed as the composition

$$\operatorname{Hom}_{R^{op}}(B_i, A_j) \longrightarrow \operatorname{Hom}_{R^{op}}(B_{i-1} \otimes_R B_1, A_j)$$

$$\simeq \operatorname{Hom}_{R^{op}}(B_{i-1}, \operatorname{Hom}_{R^{op}}(B_1, A_j)) \simeq \operatorname{Hom}_{R^{op}}(B_{i-1}, A_j \otimes_R \operatorname{Hom}_{R^{op}}(B_1, R))$$

$$\xrightarrow{\ \tau\ } \operatorname{Hom}_{R^{op}}(B_{i-1}, A_j \otimes_R A_1) \longrightarrow \operatorname{Hom}_{R^{op}}(B_{i-1}, A_{j+1}).$$

Here the map $\mathrm{Hom}_{R^{\mathrm{op}}}(B_i, A_j) \longrightarrow \mathrm{Hom}_{R^{\mathrm{op}}}(B_{i-1} \otimes_R B_1, A_j)$ is induced by the multiplication map $B_{i-1} \otimes_R B_1 \longrightarrow B_i$, the map $\mathrm{Hom}_{R^{\mathrm{op}}}(B_{i-1}, A_j \otimes_R \mathrm{Hom}_{R^{\mathrm{op}}}(B_1, R)) \longrightarrow$ $\mathrm{Hom}_{R^{\mathrm{op}}}(B_{i-1}, A_j \otimes_R A_1)$ is induced by the given map $\tau \colon \mathrm{Hom}_{R^{\mathrm{op}}}(B_1, R) \longrightarrow A_1$, and the map $\mathrm{Hom}_{R^{\mathrm{op}}}(B_{i-1}, A_j \otimes_R A_1) \longrightarrow \mathrm{Hom}_{R^{\mathrm{op}}}(B_{i-1}, A_{j+1})$ is induced by the multiplication map $A_j \otimes_R A_1 \longrightarrow A_{j+1}$.

Lemma 2.6 *Assume that both B_1 and B_2 are finitely generated projective right R-modules. Then the following conditions are equivalent:*

(a) $(\partial^\tau)^2 = 0$ *on the whole bigraded A-B-bimodule $K^\tau(B, A)$.*

(b) *The composition $\mathrm{Hom}_{R^{\mathrm{op}}}(B_2, R) \xrightarrow{\partial^\tau} \mathrm{Hom}_{R^{\mathrm{op}}}(B_1, A_1) \xrightarrow{\partial^\tau} \mathrm{Hom}_{R^{\mathrm{op}}}(R, A_2) = A_2$ vanishes.*

(c) *The composition of maps $\mathrm{Hom}_{R^{\mathrm{op}}}(B_2, R) \longrightarrow \mathrm{Hom}_{R^{\mathrm{op}}}(B_1 \otimes_R B_1, R) \simeq \mathrm{Hom}_{R^{\mathrm{op}}}(B_1, R) \otimes_R \mathrm{Hom}_{R^{\mathrm{op}}}(B_1, R) \xrightarrow{\tau \otimes \tau} A_1 \otimes_R A_1 \longrightarrow A_2$ vanishes.*

Proof

(a) $\Longrightarrow$ (b) is obvious.

(b) $\Longleftrightarrow$ (c) holds because the two maps in question are the same.

(c) $\Longrightarrow$ (a) is straightforward, using the isomorphism $\mathrm{Hom}_{R^{\mathrm{op}}}(B_2, A_j) \simeq A_j \otimes_R \mathrm{Hom}_{R^{\mathrm{op}}}(B_2, R)$.

$\square$

Definition 2.7 When the equivalent conditions of Lemma 2.6 hold, the bigraded A-B-bimodule $K^\tau(B, A)$ with the differential ∂^τ can be viewed as a complex of graded left A-modules. We call it the *first Koszul complex* and denote by $K_\bullet^\tau(B, A)$. Assuming that B_i is a finitely generated projective right R-module for every $i \geq 0$, the first Koszul complex $K_\bullet^\tau(B, A)$ has the form

$$\cdots \longrightarrow A \otimes_R \mathrm{Hom}_{R^{\mathrm{op}}}(B_2, R) \longrightarrow A \otimes_R \mathrm{Hom}_{R^{\mathrm{op}}}(B_1, R) \longrightarrow A \longrightarrow 0. \qquad (2.16)$$

Specifically, the following particular case is important. Let A be a nonnegatively graded ring with the degree-zero component $R = A_0$. Assume that A_1 and A_2 are finitely generated projective left R-modules and the multiplication map $A_1 \otimes_R A_1 \longrightarrow A_2$ is surjective. Denote the kernel of the latter map by I_A, and consider the 2-left finitely projective quadratic graded ring $\mathrm{q}A = T_R(A_1)/(I_A)$. Let B be the 2-right finitely projective quadratic graded ring quadratic dual to $\mathrm{q}A$.

Then we have an isomorphism of R-R-bimodules $\mathrm{Hom}_{R^{\mathrm{op}}}(B_1, R) \simeq A_1$, which we use as our choice of the map τ. The assumptions of Lemma 2.6 are satisfied, and condition (c) holds by the construction of quadratic duality. Therefore, the first Koszul complex $K_\bullet^\tau(B, A)$ is well-defined.

Examples 2.8

(1) The terminology "Koszul complex" has a meaning in commutative algebra [23, 44] very different from the usage in the theory of quadratic algebras and Koszul duality presumed in this book. The two concepts intersect at the particular case of the symmetric algebra $A = \mathrm{Sym}_R(V)$ from Examples 1.3(4) and 1.5(2) (for a free, or more generally, projective R-module V) and the exterior algebra $B = \Lambda_R(V^{\vee})$ as per Example 1.7(3). In this case, the Koszul complex (2.16) is quasi-isomorphic to the graded A-module R situated in the homological degree 0 and the internal degree 0.

Essentially the same Koszul complex (2.16), isomorphic up to a change of the bigrading, corresponds to the quadratic dual situation with $A = \Lambda_R(V^{\vee})$ and $B = \mathrm{Sym}_R(V)$.

(2) It is a straightforward exercise for the reader to check that, in the case of the tensor ring $A = T_R(V)$ and its quadratic dual ring $B = R \oplus V^{\vee}$ from Example 1.7(1), for an R-R-bimodule V which is finitely generated and projective as a left R-module, the Koszul complex (2.16) is also quasi-isomorphic to the graded left A-module R situated in the cohomological degree 0 and the internal degree 0. The same description of the cohomology of the Koszul complex (2.16) holds in the quadratic dual case of $A = R \oplus V$ and $B = T_R(V^{\vee})$.

All these examples can be rephrased by saying that the quadratic graded rings involved are *finitely projective Koszul*, as we will see below.

2.6   Dual Koszul Complex

As in Sect. 2.5, we consider two nonnegatively graded rings $A = \bigoplus_{n=0}^{\infty} A_n$ and $B = \bigoplus_{n=0}^{\infty} B_n$ with the same degree-zero component $A_0 = R = B_0$. Let $e \in B_1 \otimes_R A_1$ be an element satisfying the equation $re = er$ for all $r \in R$, as in Lemma 2.5(b).

Denote by $K_e^{\vee}(B, A)$ the following bigraded B-A-bimodule endowed with a B-A-bimodule endomorphism d_e. The bigrading components of $K_e^{\vee}(B, A)$ are

$$K_e^{\vee\,i,n}(B, A) = B_i \otimes A_{n+i}, \quad i \geq 0, \ n \geq -i.$$

Here i is interpreted as the cohomological grading and n is the internal grading. The homogeneous B-A-bimodule endomorphism $d_e \colon K_e^{\vee}(B, A) \longrightarrow K_e^{\vee}(B, A)$ of bidegree $(i, n) = (1, 0)$ consists of the components $B_i \otimes_R A_j \longrightarrow B_{i+1} \otimes_R A_{j+1}$, which are constructed as the compositions

$$B_i \otimes_R A_j = B_i \otimes_R R \otimes_R A_j \xrightarrow{\ e\ } B_i \otimes_R B_1 \otimes_R A_1 \otimes_R A_j \longrightarrow B_{i+1} \otimes_R A_{j+1}.$$

Here the map $B_i \otimes_R R \otimes_R A_j \longrightarrow B_i \otimes_R B_1 \otimes_R A_1 \otimes_R A_j$ is induced by the R-R-bimodule map $e \colon R \longrightarrow B_1 \otimes_R A_1$, while the map $B_i \otimes_R B_1 \otimes_R A_1 \otimes_R A_j \longrightarrow B_{i+1} \otimes_R A_{j+1}$ is the tensor product of the multiplication maps $B_i \otimes_R B_1 \longrightarrow B_{i+1}$ and $A_1 \otimes_R A_j \longrightarrow A_{j+1}$.

Lemma 2.9 *Assume that both B_1 and B_2 are finitely generated projective right R-modules and that an R-R-bimodule map $\mathrm{Hom}_{R^{\mathrm{op}}}(B_1, R) \longrightarrow A_1$ and an element $e \in B_1 \otimes_R A_1$ correspond to each other under the construction of Lemma 2.5. Then the following conditions are equivalent:*

(a) $d_e^2 = 0$ *on the whole bigraded B-A-bimodule $K_e^{\vee}(B, A)$.*
(b) *The element $d_e^2(1 \otimes 1) \in B_2 \otimes_R A_2$ vanishes.*
(c) *The equivalent conditions of Lemma 2.6 hold.*

Furthermore, the bigraded A-B-bimodule $K^{\tau}(B, A)$ with the endomorphism ∂^{τ} can be obtained by applying the functor $\mathrm{Hom}_{A^{\mathrm{op}}}(-, A)$ to the bigraded B-A-bimodule $K_e^{\vee}(B, A)$ with the endomorphism d_e.

Proof The equivalence (a) $\Longleftrightarrow$ (b) does not depend on the assumptions of finite generatedness and projectivity of the right R-modules B_1 and B_2. The equivalence (b) $\Longleftrightarrow$ (c) appears to need these assumptions. The last assertion of the lemma is straightforward and only needs the assumption of B_1 being a finitely generated projective right R-module.

(a) $\Longrightarrow$ (b) is obvious.

(b) $\Longrightarrow$ (a) holds because d_e is a B-A-bimodule map.

(b) $\Longleftrightarrow$ (c) holds because the map $\mathrm{Hom}_{R^{\mathrm{op}}}(B_2, R) \longrightarrow A_2$ from Lemma 2.6(c) corresponds to the element $d_e^2(1 \otimes 1) \in B_2 \otimes_R A_2$ under the correspondence of Lemma 2.5.

(a) $\Longrightarrow$ (c): the condition (a) implies the condition of Lemma 2.6(a) in view of the last assertion of the present lemma. $\square$

Definition 2.10 When the equivalent conditions of the lemma hold, the bigraded B-A-bimodule $K_e^{\vee}(B, A)$ with the differential d_e can be viewed as a complex of graded right A-modules. We call it the *dual Koszul complex* and denote by $K_e^{\vee \bullet}(B, A)$. The dual Koszul complex has the form

$$0 \longrightarrow A \longrightarrow B_1 \otimes_R A \longrightarrow B_2 \otimes_R A \longrightarrow B_3 \otimes_R A \longrightarrow \cdots \tag{2.17}$$

According to the lemma, we have

$$K_{\bullet}^{\tau}(B, A) = \mathrm{Hom}_{A^{\mathrm{op}}}(K_e^{\vee \bullet}(B, A), A).$$

The graded Hom group $\mathrm{Hom}_{R^{op}}(B, R)$ with the components $\mathrm{Hom}_{R^{op}}(B_i, R)$, $i \geq 0$, has natural structures of a graded right B-module (induced by the graded left B-module structure on B) and a graded left R-module (induced by the left R-module structure on R). We denote the tensor product $\mathrm{Hom}_{R^{op}}(B, R) \otimes_B K_e^\vee(B, A)$ by

$$^\tau K(B, A) = \mathrm{Hom}_{R^{op}}(B, R) \otimes_B K_e^\vee(B, A) = \mathrm{Hom}_{R^{op}}(B, R) \otimes_R A.$$

$^\tau K(B, A)$ is a bigraded R-A-bimodule with the bigrading components

$$^\tau K_{i,n}(B, A) = \mathrm{Hom}_{R^{op}}(B_i, R) \otimes_R A_{n-i}, \quad i \geq 0, \ n \geq i.$$

Here i is the homological grading and n is the internal grading. Notice that there is *no* B-module structure on $^\tau K(B, A)$.

The B-A-bimodule endomorphism $d_e\colon K_e^\vee(B, A) \longrightarrow K_e^\vee(B, A)$ induces an R-A-bimodule endomorphism of bidegree $(i, n) = (-1, 0)$ on $^\tau K(B, A)$, which we denote by $^\tau \partial\colon {}^\tau K(B, A) \longrightarrow {}^\tau K(B, A)$. When the equivalent conditions of Lemma 2.9 hold, it follows that $(^\tau \partial)^2 = 0$.

Definition 2.11 In the assumptions above, the bigraded R-A-bimodule $^\tau K(B, A)$ with the differential $^\tau \partial$ can be viewed as a complex of graded right A-modules. We call it the *second Koszul complex* and denote by $^\tau K_\bullet(B, A)$. The second Koszul complex $^\tau K_\bullet(B, A)$ has the form

$$\cdots \longrightarrow \mathrm{Hom}_{R^{op}}(B_2, R) \otimes_R A \longrightarrow \mathrm{Hom}_{R^{op}}(B_1, R) \otimes_R A \longrightarrow A \longrightarrow 0. \qquad (2.18)$$

The particular case described at the end of Sect. 2.5 is important for us. In this case, qA is a 2-left finitely projective quadratic graded ring and B is the quadratic dual 2-right finitely projective quadratic graded ring. Choosing τ to be the natural isomorphism $\mathrm{Hom}_{R^{op}}(B_1, R) \simeq A_1$, we see that the equivalent conditions of Lemma 2.9 are satisfied (because the condition of Lemma 2.6(c) holds). Therefore, the dual Koszul complex $K_e^{\vee \bullet}(B, A)$ and the second Koszul complex $^\tau K_\bullet(B, A)$ are well-defined.

Examples 2.12 The description of the cohomology modules of the second Koszul complex (2.18) for the quadratic graded rings $\mathrm{Sym}_R(V)$, $\Lambda_R(V)$, $T_R(V)$, and $R \oplus V$ is similar to the description of the cohomology of the first Koszul complex (2.16) in Examples 2.8. This applies to finitely projective Koszul graded rings generally, as we will see below in Sects. 2.9–2.10. However, the cohomology of the dual Koszul complex (2.17) is very different.

(1) The example of the graded rings $A = \mathrm{Sym}_R(V)$ and $B = \Lambda_R(V^\vee)$ is somewhat misleading, as it partly conceals the difference between the Koszul complexes (2.16), (2.18) and the dual Koszul complex (2.17). Specifically, let R be a

commutative ring and V be a free R-module with m generators. Then the graded B-module $\mathrm{Hom}_{R^{op}}(B, R)$ is isomorphic to the graded B-module B, up to a grading shift by m (as "the exterior algebra is Frobenius"). Consequently, the dual Koszul complex (2.17) is naturally quasi-isomorphic to the free R-module B_m with one generator, placed in the cohomological grading m and the internal grading $-m$ and endowed with the action of A via the graded ring homomorphism $A \longrightarrow R$. (See Sect. 9.4 below for a general discussion of this situation.)

Similarly, for $A = \Lambda_R(V^\vee)$ and $B = \mathrm{Sym}_R(V)$, the dual Koszul complex (2.17) is naturally quasi-isomorphic to the free R-module A_m with one generator, placed in the cohomological grading 0 and the internal grading m, and endowed with the action of A via the graded ring homomorphism $A \longrightarrow R$.

(2) To see the difference, it is instructive to consider the example of the graded rings $T_R(V)$ and $R \oplus V^\vee$. Let for simplicity $R = k$ be a field, and let V be a k-vector space with $1 < m = \dim_k V < \infty$. Then the graded k-vector space of cohomology of the dual Koszul complex (2.17) for the graded algebras $A = T_k(V)$ and $B = k \oplus V^\vee$ is infinite-dimensional, with nonzero finite-dimensional components situated in all the internal degrees $n \geq -1$ and the cohomological degree 1.

Dually, for $A = k \oplus V^\vee$ and $B = T_k(V)$, the dual Koszul complex (2.17) has a nonzero cohomology vector space in every cohomological degree $i \geq 0$ and the internal degree $n = 1 - i$.

Remark 2.13 Assume that B_n is a finitely generated projective right R-module for every $n \geq 0$. Then the isomorphism of Lemma 1.1(b) provides the graded R-R-bimodule C with the components $C_n = \mathrm{Hom}_{R^{op}}(B_n, R)$ with the structure of a graded coring over R. The comultiplication maps $C_{i+j} \longrightarrow C_i \otimes_R C_j$ are obtained by applying the functor $\mathrm{Hom}_{R^{op}}(-, R)$ to the multiplication maps $B_j \otimes B_i \longrightarrow B_{i+j}$.

By construction, we have $B_n = \mathrm{Hom}_R(C_n, R)$. Put ${}^{\#}B_n = \mathrm{Hom}_{R^{op}}(C_n, R)$ (in the spirit of the notation in [2]). Then applying the functor $\mathrm{Hom}_{R^{op}}(-, R)$ to the comultiplication maps $C_{i+j} \longrightarrow C_i \otimes_R C_j$ and composing the resulting maps $\mathrm{Hom}_{R^{op}}(C_i \otimes_R C_j, R) \longrightarrow \mathrm{Hom}_{R^{op}}(C_{i+j}, R)$ with the natural R-R-bimodule morphisms $\mathrm{Hom}_{R^{op}}(C_j, R) \otimes_R \mathrm{Hom}_{R^{op}}(C_i, R) \longrightarrow \mathrm{Hom}_{R^{op}}(C_{i+j}, R)$ from Lemma 1.1(b) produce multiplication maps ${}^{\#}B_j \otimes_R {}^{\#}B_i \longrightarrow {}^{\#}B_{i+j}$. So ${}^{\#}B = \bigoplus_{n=0}^\infty {}^{\#}B_n$ is a nonnegatively graded ring with the degree-zero component ${}^{\#}B_0 = R$.

One easily observes that $C = \bigoplus_{n=-\infty}^0 C_{-n}$ is a graded ${}^{\#}B$-B-bimodule. In fact, the left action map ${}^{\#}B_i \otimes_R C_{i+j} \longrightarrow C_j$ can be constructed as the composition

$$ {}^{\#}B_i \otimes_R C_{i+j} \longrightarrow {}^{\#}B_i \otimes_R C_i \otimes_R C_j \longrightarrow {}^{\#}B_i \otimes_R C_j $$

of the map ${}^{\#}B_i \otimes_R C_{i+j} \longrightarrow {}^{\#}B_i \otimes_R C_i \otimes_R C_j$ induced by the comultiplication map $C_{i+j} \longrightarrow C_i \otimes_R C_j$ with the map ${}^{\#}B_i \otimes_R C_i \otimes_R C_j \longrightarrow {}^{\#}B_i \otimes_R C_j$ induced by the evaluation map ${}^{\#}B_i \otimes_R C_i = \mathrm{Hom}_{R^{op}}(C_i, R) \otimes_R C_i \longrightarrow R$.

Thus the tensor product $^\tau K(B, A) = C \otimes_B K_e^\vee(B, A) = C \otimes_R A$ is naturally a graded $^\#B$-A-bimodule, and $^\tau \partial : {}^\tau K(B, A) \longrightarrow {}^\tau K(B, A)$ is its graded $^\#B$-A-bimodule endomorphism. So, instead of an action of the graded ring B, there is an action of the graded ring $^\#B$ in $^\tau K(B, A)$.

We refer to Sect. 8.7 for a further discussion of the two Koszul complexes.

2.7 Distributive Collections of Subobjects

Let C be an abelian category in which all the subobjects of any given object form a set, and let $W \in C$ be an object. A set Ω of subobjects of W is said to be a *lattice of subobjects* if $0 \in \Omega$, $W \in \Omega$, and for any $X, Y \in \Omega$, one has $X \cap Y \in \Omega$ and $X + Y \in \Omega$. Any lattice of subobjects is *modular*, i.e., one has $(X + Y) \cap Z = X + (Y \cap Z)$ whenever X, $Y, Z \in \Omega$ and $X \subset Z$. A lattice of subobjects Ω is said to be *distributive* if the identity $(X + Y) \cap Z = X \cap Z + Y \cap Z$ holds for all $X, Y, Z \in \Omega$, or equivalently, the identity $(X + Y) \cap (X + Z) = X + (Y \cap Z)$ holds for all $X, Y, Z \in \Omega$.

Let $n \geq 2$ be an integer and $X_1, \ldots, X_{n-1} \subset W$ be a collection of $n-1$ subobjects in W. The lattice Ω of subobjects of W *generated by* $X_1, \ldots, X_{n-1}$ consists of all the subobjects of W that can be obtained from $X_1, \ldots, X_{n-1}$ by applying iteratively the operations of finite sum and finite intersection.

Definition 2.14 A collection of subobjects $(X_1, \ldots, X_{n-1})$ in W is said to be *distributive* if the lattice Ω of subobjects of W generated by $X_1, \ldots, X_{n-1}$ is distributive. Any pair of subobjects X_1, X_2 $(n = 3)$ forms a distributive collection, but a triple of subobjects X_1, X_2, X_3 $(n = 4)$ does not need to be distributive. A collection of subobjects $(X_1, \ldots, X_{n-1})$ in W is said to be *almost distributive* if all its proper subcollections $(X_1, \ldots, \widehat{X}_k, \ldots, X_{n-1})$, $1 \leq k \leq n - 1$, are distributive.

The following two lemmas hold in any modular lattice Ω, but we state them for lattices of subobjects only.

Lemma 2.15 *A triple of subobjects $X, Y, Z \subset W$ is distributive if and only if $(X + Y) \cap Z = X \cap Z + Y \cap Z$ and if and only if $(X + Y) \cap (X + Z) = X + (Y \cap Z)$. Any permutation of X, Y, Z replaces these equations by equivalent ones.*

Proof This is an easy exercise; see [34, Lemma 1] or [53, Lemma 6.1 in Chapter 1]. □

Example 2.16 The example of the lattice of vector subspaces in a vector space W (over a field k) is instructive. In this context, there is a well-known classification of triples of vector subspaces $X, Y, Z \subset W$ up to isomorphism. There are 9 indecomposable triples, which include 8 triples of subspaces in a one-dimensional vector space W and one indecomposable triple of three different lines in a plane ($\dim_k W = 2$). A given triple of subspaces X, Y, Z in an arbitrary vector space W is distributive if and only if its indecomposable decomposition consists of triples of subspaces in one-dimensional vector spaces only (i.e., the indecomposable configuration of three different lines in a plane does not occur).

Lemma 2.17 *An almost distributive collection of subobjects* $(X_1, \ldots, X_{n-1})$ *in* W *is distributive if and only if, for every* $2 \leq k \leq n - 2$, *the triple of subobjects*

$$X_1 + \cdots + X_{k-1}, \; X_k, \; X_{k+1} \cap \cdots \cap X_{n-1} \subset W$$

is distributive.

Proof This is the result of the paper [48], improving upon the previous work in [34]. See also [53, Theorem 6.3 in Chapter 1]. □

For any collection of subobjects $X_1, \ldots, X_{n-1} \subset W$, we consider the following three complexes in $\mathbf{C}$. The *bar complex* $B_\bullet = B_\bullet(W; X_1, \ldots, X_{n-1})$ has the form

$$W \longrightarrow \bigoplus_t W/X_t \longrightarrow \bigoplus_{t_1 < t_2} W/(X_{t_1} + X_{t_2}) \longrightarrow \cdots$$

$$\longrightarrow \bigoplus_{t_1 < \cdots < t_i} W/(X_{t_1} + \cdots + X_{t_i}) \longrightarrow \cdots \longrightarrow W/(X_1 + \cdots + X_{n-1}). \quad (2.19)$$

Here the indices t_s range from 1 to $n-1$. The leftmost term W is placed in the homological degree n, the term with the summation over $1 \leq t_1 < \cdots < t_i \leq n - 1$ is placed in the homological degree $n - i$, and the rightmost term $W/(X_1 + \cdots + X_{n-1})$ is placed in the homological degree 1. The component of the differential acting from the direct summand $W/(X_{t_1} + \cdots + \widehat{X}_{t_s} + \cdots + X_{t_i})$ in B_{n-i+1} to the direct summand $W/(X_{t_1} + \cdots + X_{t_i})$ in B_{n-i} is the natural epimorphism taken with the sign $(-1)^{s-1}$.

The *cobar complex* $B^\bullet = B^\bullet(W; X_1, \ldots, X_{n-1})$ has the form

$$X_1 \cap \cdots \cap X_{n-1} \longrightarrow \cdots \longrightarrow \bigoplus_{1 \leq t_1 < \cdots < t_i \leq n-1} X_{t_1} \cap \cdots \cap X_{t_i} \longrightarrow \cdots$$

$$\longrightarrow \bigoplus_{1 \leq t_1 < t_2 \leq n-1} X_{t_1} \cap X_{t_2} \longrightarrow \bigoplus_{1 \leq t \leq n-1} X_t \longrightarrow W. \quad (2.20)$$

Here the leftmost term $X_1 \cap \cdots \cap X_{n-1}$ is placed in the cohomological degree 1, the term with the summation over $1 \leq t_1 < \cdots < t_i \leq n - 1$ is placed in the cohomological degree $n - i$, and the rightmost term W is placed in the cohomological degree n. The component of the differential acting from the direct summand $X_{t_1} \cap \cdots \cap X_{t_i}$ in B^{n-i} to the direct summand $X_{t_1} \cap \cdots \cap \widehat{X}_{t_s} \cap \cdots \cap X_{t_i}$ in B^{n-i+1} is the natural monomorphism taken with the sign $(-1)^{s-1}$.

The *Koszul complex* $K_\bullet(W; X_1, \ldots, X_{n-1})$ is

$$X_1 \cap \cdots \cap X_{n-1} \longrightarrow X_2 \cap \cdots \cap X_{n-1} \longrightarrow X_3 \cap \cdots \cap X_{n-1}/X_1$$

$$\longrightarrow \cdots \longrightarrow (X_{i+1} \cap \cdots \cap X_{n-1})/(X_1 + \cdots + X_{i-1}) \longrightarrow \cdots \longrightarrow$$

$$X_{n-1}/(X_1 + \cdots + X_{n-3}) \longrightarrow W/(X_1 + \cdots + X_{n-2}) \longrightarrow W/(X_1 + \cdots + X_{n-1}).$$

$$(2.21)$$

Here the notation is $Y/Z = Y/(Y \cap Z) = (Y + Z)/Z$. The term $X_1 \cap \cdots \cap X_{n-1}$ is placed in the homological degree n, and the term $W/(X_1 + \cdots + X_{n-1})$ is placed in the homological degree 0.

Lemma 2.18 *Let $X_1, \ldots, X_{n-1}$ be an almost distributive collection of subobjects of an object $W \in \mathsf{C}$. Then the following conditions are equivalent:*

(a) *The collection $(X_1, \ldots, X_{n-1})$ is distributive.*
(b) *The Koszul complex $K_\bullet(W; X_1, \ldots, X_{n-1})$ is exact.*
(c) *The bar complex $B_\bullet(W; X_1, \ldots, X_{n-1})$ is exact everywhere except for its leftmost term W.*
(c*) *The cobar complex $B^\bullet(W; X_1, \ldots, X_{n-1})$ is exact everywhere except for its rightmost term W.*

Proof This is [53, Proposition 7.2 in Chapter 1]. The assertion in [53] is stated for collections of subspaces in vector spaces, but the same argument applies to subobjects of any object in an abelian category. □

Let $\mathsf{F} \subset \mathsf{C}$ be a class of objects closed under extensions and the passages to the kernels of epimorphisms. We will say that a lattice Ω of subobjects in an object W is an F-*lattice* if one has $Y/Z \in \mathsf{F}$ for any pair of subobjects $Y, Z \in \Omega$ such that $Z \subset Y$. In particular, the existence of an F-lattice of subobjects in W implies that $W = W/0 \in \mathsf{F}$. A collection of subobjects $X_1, \ldots, X_{n-1} \subset W$ is said to be F-*distributive* if the lattice Ω of subobjects in W generated by $X_1, \ldots, X_{n-1}$ is a distributive F-lattice.

Lemma 2.19 *A distributive collection of subobjects $X_1, \ldots, X_{n-1} \subset W$ is F-distributive if and only if one has $W/(X_{t_1} + \cdots + X_{t_i}) \in \mathsf{F}$ for all $1 \le t_1 < \cdots < t_i \le n-1$, $0 \le i \le n-1$.*

Proof This is a generalization of [58, Lemma 11.4.3.2], and the same proof applies. Alternatively, one can argue by induction in n in the following way. For any three subobjects $Z \subset Y$ and X in W, there is a short exact sequence $0 \longrightarrow (X \cap Y)/(X \cap Z) \longrightarrow$

$Y/Z \longrightarrow (X + Y)/(X + Z) \longrightarrow 0$. Taking $Y, Z \in \Omega$ and $X = X_{n-1}$, we observe that the object $(X_{n-1} + Y)/(X_{n-1} + Z)$ belongs to F by the induction assumption applied to the collection of subobjects $(X_t + X_{n-1})/X_{n-1} \subset W/X_{n-1}$, $1 \le t \le n - 2$. Since F is closed under extensions, it suffices to show that $(X_{n-1} \cap Y)/(X_{n-1} \cap Z) \in \mathsf{F}$. Applying the same argument to $X = X_{n-2}$, etc., we reduce the question to showing that the object $X_1 \cap X_2 \cap \cdots \cap X_{n-1}$ belongs to F. The argument finishes similarly to the proof in [58], by invoking Lemma 2.18 (a) $\Rightarrow$ (c) and the assumption that the class F is closed under the kernels of epimorphisms. $\square$

Example 2.20 It is easy to give a counterexample showing that, in the context of Lemma 2.19, the conditions that $W \in \mathsf{F}$ and $W/X_t \in \mathsf{F}$ for all $1 \le t \le n - 1$ are *not* sufficient for a distributive collection of subobjects to be F-distributive. Indeed, let $W \simeq \mathbb{Z}^2$ be the free abelian group $W = \mathbb{Z}(1, 0) + \mathbb{Z}(0, 1) + \mathbb{Z}(1/2, 1/2) \subset \frac{1}{2}(\mathbb{Z} \oplus \mathbb{Z})$, and let F be the class of all free (or torsionfree) abelian groups. Take $n = 3$, and let X_1 and $X_2 \subset W$ be the subgroups $X_1 = \mathbb{Z}(1, 0)$ and $X_2 = \mathbb{Z}(0, 1)$. Then both the quotient groups W/X_1 and W/X_2 are isomorphic to $\mathbb{Z}$ (hence torsionfree), but the quotient group $W/(X_1 + X_2)$ is isomorphic to $\mathbb{Z}/2\mathbb{Z}$ (not torsionfree). Still the collection of two subgroups is obviously distributive (indeed, any collection of two subobjects is distributive). This example can be found in [42, Example 2.5].

We will say that a lattice Ω of subobjects in an object W is *split* if for any pair of subobjects $Y, Z \in \Omega$ such that $Z \subset Y$, we have that Z is a split subobject of Y. Equivalently, Ω is split if every $Z \in \Omega$ is a split subobject of W. A collection of subobjects $X_1, \ldots, X_{n-1} \subset W$ is said to be *split distributive* if the lattice Ω of subobjects in W generated by $X_1, \ldots, X_{n-1}$ is split and distributive.

Lemma 2.21 *A collection of subobjects $X_1, \ldots, X_{n-1} \subset W$ is split distributive if and only if there exists a finite direct sum decomposition $W = \bigoplus_\eta W_\eta$ of the object W such that each of the subobjects X_i is the sum of a set of subobjects W_η.*

Proof The proof of [53, Proposition 7.1 (a) $\Leftrightarrow$ (b) in Chapter 1] is applicable. $\square$

2.8   Collections of Subbimodules

Let R and S be associative rings and W be an R-S-bimodule. Following the terminology of Sect. 2.7, we say that a collection of R-S-subbimodules $X_1, \ldots, X_{n-1} \subset W$ is *distributive* if the lattice of R-S-subbimodules Ω generated by $X_1, \ldots, X_{n-1}$ in W is distributive.

Definition 2.22 A collection of subbimodules $X_1, \ldots, X_{n-1} \subset W$ is said to be *left flat distributive* if it is distributive, and for every pair of subbimodules $Y, Z \in \Omega$ such that $Z \subset Y$, the quotient bimodule Y/Z is a flat left R-module.

Lemma 2.23 *Let R, S, and T be associative rings, W be an R-S-bimodule, and U be an S-T-bimodule. Let $X_1, \ldots, X_{n-1} \subset W$ be a left flat distributive collection of R-S-subbimodules, and let $Y_1, \ldots, Y_{m-1} \subset U$ be a left flat distributive collection of S-T-subbimodules. Then*

$$X_1 \otimes_S U, \ldots, X_{n-1} \otimes_S U, \ W \otimes_S Y_1, \ldots, W \otimes_S Y_{m-1} \subset W \otimes_S U$$

is a left flat distributive collection of R-T-subbimodules in the R-T-bimodule $W \otimes_S U$.

Proof First of all, the map $X_i \otimes_S U \longrightarrow W \otimes_S U$ induced by the inclusion $X_i \longrightarrow W$ is injective for every $1 \leq i \leq n-1$, since U is a flat left S-module. The map $W \otimes_S Y_j \longrightarrow W \otimes_S U$ induced by the inclusion $Y_j \longrightarrow U$ is also injective for every $1 \leq j \leq m-1$, since U/Y_j is a flat left S-module. So $X_i \otimes_S U$ and $W \otimes_S Y_j$ are indeed subbimodules in $W \otimes_S U$.

Furthermore, arguing by induction in $n + m$, we can assume that our collection of $n + m - 2$ subbimodules in $W \otimes_S U$ is almost distributive. Up to a homological shift by $[-1]$, the bar complex $B_\bullet = B_\bullet(W \otimes_S U; X_1 \otimes_S U, \ldots, X_{n-1} \otimes_S U,$ $W \otimes_S Y_1, \ldots, W \otimes_S Y_{m-1})$ (2.19) is isomorphic to the tensor product of two bar complexes $B_\bullet(W; X_1, \ldots, X_{n-1}) \otimes_S B_\bullet(U; Y_1, \ldots, Y_{m-1})$. The only nonzero homology bimodule of the bar complex $B_\bullet(U; Y_1, \ldots, Y_{m-1})$ is $Y_1 \cap \ldots \cap Y_{m-1} \subset U$, and it is a flat left S-module by assumption, and so are all the terms of the bar complex $B_\bullet(U; Y_1, \ldots, Y_{m-1})$. Thus the bar complex $B_\bullet$ is exact everywhere except for its leftmost term $W \otimes_S U$. Applying Lemma 2.18 (c) $\Rightarrow$ (a), we can conclude that out collection of $n + m - 2$ subbimodules in $W \otimes_S U$ is distributive.

Finally, the quotient bimodule of $W \otimes_S U$ by the sum of any subset of $X_1 \otimes_S U, \ldots,$ $X_{n-1} \otimes_S U$, $W \otimes_S Y_1, \ldots, W \otimes_S Y_{m-1}$ is isomorphic to the tensor product of the quotient bimodules of W and U by the sums of the respective subsets of $X_1, \ldots, X_{n-1}$ and $Y_1, \ldots, Y_{n-1}$. As the tensor product of an R-flat R-S-bimodule and an S-flat S-T-bimodule is an R-flat R-T-bimodule, any such quotient bimodule of $W \otimes_S U$ is a flat left R-module. It remains to apply Lemma 2.19 (for the class F of R-flat R-S-bimodules in the abelian category C of R-S-bimodules) in order to finish the proof of the lemma. $\square$

Lemma 2.24 *In the context of Lemma 2.23, for any two subbimodules X', $X'' \subset W$ belonging to the lattice of subbimodules generated by $X_1, \ldots, X_{n-1}$ in W and any two subbimodules Y', $Y'' \subset U$ belonging to the lattice of subbimodules generated by $Y_1, \ldots, Y_{m-1}$ in U, the following equations for subbimodules in $W \otimes_S U$ hold:*

(a) $(X' + X'') \otimes_S U = (X' \otimes_S U) + (X'' \otimes_S U)$.
(b) $(X' \cap X'') \otimes_S U = (X' \otimes_S U) \cap (X'' \otimes_S U)$.
(c) $W \otimes_S (Y' + Y'') = (W \otimes_S Y') + (W \otimes_S Y'')$.
(d) $W \otimes_S (Y' \cap Y'') = (W \otimes_S Y') \cap (W \otimes_S Y'')$.

(e) $X' \otimes_S Y' = (W \otimes_S Y') \cap (X' \otimes_S U)$.

(f) $(X' \cap X'') \otimes_S (Y' \cap Y'') = (X' \otimes_S Y') \cap (X'' \otimes_S Y'')$.

Proof The maps $X' \otimes_S U \longrightarrow W \otimes_S U$ and $W \otimes_S Y' \longrightarrow W \otimes_S U$ induced by the inclusions $X' \longrightarrow W$ and $Y' \longrightarrow U$ are injective, as explained in the proof of Lemma 2.23. Furthermore, the map $X' \otimes_S Y' \longrightarrow W \otimes_S U$ is injective as the composition of injective maps $X' \otimes_S Y' \longrightarrow W \otimes_S Y' \longrightarrow W \otimes_S U$ or $X' \otimes_S Y' \longrightarrow X' \otimes_S U \longrightarrow W \otimes_S U$. So these are indeed subbimodules in $W \otimes_S U$.

Now the equations (a) and (c) are obvious. The equation (b) holds since U is a flat left S-module. To prove (d), it suffices to consider the four-term exact sequence of flat left S-modules

$$0 \longrightarrow Y' \cap Y'' \longrightarrow Y' \oplus Y'' \longrightarrow U \longrightarrow U/(Y' + Y'') \longrightarrow 0$$

and tensor it with W over S on the left.

To prove (e), one has to check that the short sequence $0 \longrightarrow X' \otimes_S Y' \longrightarrow W \otimes_S U \longrightarrow (W/X' \otimes_S U) \oplus (W \otimes_S U/Y')$ is left exact. This follows from exactness of the sequences $0 \longrightarrow X' \otimes_S Y' \longrightarrow W \otimes_S Y' \longrightarrow (W/X') \otimes_S Y'$ and $0 \longrightarrow W \otimes_S Y' \longrightarrow W \otimes_S U \longrightarrow W \otimes_S (U/Y')$ and injectivity of the map $(W/X') \otimes_S Y' \longrightarrow (W/X') \otimes_S U$. The equation (f) follows from (b), (d), and (e).   □

Examples 2.25

(1) Without the flatness assumptions, the assertions of Lemmas 2.23–2.24 are *not* true, as the following counterexample illustrates. Let $R = k[\lambda]$ be the ring of polynomials in one variable λ over a field k, and let $W = Ra \oplus Rb \oplus Rc$ be the free R-module with three generators a, b, and c. Consider the submodules $X = Ra$, $Y = Rb$, and $Z = R(a + b + \lambda c) \subset W$. Then the lattice of submodules in W generated by X, Y, Z is distributive, and the quotient modules W/X, W/Y, W/Z, $W/(X+Y)$, $W/(X+Z)$, $W/(Y+Z)$ are free R-modules, but $W/(X+Y+Z) \simeq R/\lambda R$ is not a flat R-module. Let U be the R-module $R/\lambda R$; then the collection of submodules (essentially, vector subspaces) $X \otimes_R U$, $Y \otimes_R U$, $Z \otimes_R U$ in the R-module (essentially, k-vector space) $W \otimes_R U \simeq k^3$ is *not* distributive.

(2) To illustrate what is in some sense an opposite phenomenon, let $R = k[\lambda]_{(\lambda)}$ be the localization of the polynomial ring $k[\lambda]$ at the maximal ideal $(\lambda) \subset k[\lambda]$. Let $W = Ra \oplus Rb$ be the free R-module with two generators a and b. Consider the submodules $X = Ra$, $Y = Rb$, and $Z = R(a + \lambda b) \subset W$. Then the quotient modules W/X, W/Y, and W/Z are free R-modules; the collection of submodules X, Y, Z in W is *not* distributive, but the collection of submodules/vector subspaces $X \otimes_R U$, $Y \otimes_R U$, $Z \otimes_R U$ in $W \otimes_R U \simeq k^2$, where $U = R/\lambda R$, *is* distributive.

Definition 2.26 We will say that a collection of subbimodules $X_1, \ldots, X_{n-1} \subset W$ in an R-S-bimodule W is *left projective distributive* if it is distributive, and for every pair of subbimodules $Y, Z \in \Omega$ such that $Z \subset Y$, the quotient bimodule Y/Z is a projective left R-module. A collection of subbimodules $X_1, \ldots, X_{n-1} \subset W$ is said to be *left split distributive* if it is split distributive as a collection of submodules in the left R-module W (in the sense of the definition in Sect. 2.7). *Right projective distributive* and *right split distributive* collections of subbimodules are defined similarly.

Clearly, a collection of subbimodules in an R-S-bimodule W is left projective distributive if and only if it is left split distributive *and* the left R-module W is projective.

Lemma 2.27 *Let R, S, and T be associative rings, W be an R-S-bimodule, and U be an S-T-bimodule. Let $X_1, \ldots, X_{n-1} \subset W$ be a left projective distributive collection of R-S-subbimodules, and let $Y_1, \ldots, Y_{m-1} \subset U$ be a left projective distributive collection of S-T-subbimodules. Then*

$$X_1 \otimes_S U, \ldots, X_{n-1} \otimes_S U, \ W \otimes_S Y_1, \ldots, W \otimes_S Y_{m-1} \subset W \otimes_S U$$

is a left projective distributive collection of subbimodules in the R-T-bimodule $W \otimes_S U$.

Proof Similar to the proof of Lemma 2.23. □

Lemma 2.28 *Let $X_1, \ldots, X_{n-1} \subset W$ be a left split distributive collection of subbimodules in an R-S-bimodule W. Then $\mathrm{Hom}_R(W/X_1, R), \ldots, \mathrm{Hom}_R(W/X_{n-1}, R) \subset \mathrm{Hom}_R(W, R)$ is a right split distributive collection of subbimodules in the S-R-bimodule $\mathrm{Hom}_R(W, R)$. If the left R-module W is projective and finitely generated, then the above collection of subbimodules in $\mathrm{Hom}_R(W, R)$ is right projective distributive. In this case, the map $W \supset Y \longmapsto \mathrm{Hom}_R(W/Y, R) \subset \mathrm{Hom}_R(W, R)$ is an anti-isomorphism between the lattice of subbimodules in W generated by $X_1, \ldots, X_{n-1}$ and the lattice of subbimodules in $\mathrm{Hom}_R(W, R)$ generated by $\mathrm{Hom}_R(W/X_1, R), \ldots, \mathrm{Hom}_R(W/X_{n-1}, R)$ (i.e., a bijection of sets transforming the sums into the intersections and vice versa).*

Proof It follows easily from Lemma 2.21 (applied in the category of left R-modules $\mathsf{C} = R\text{–mod}$). □

2.9 Left Flat Koszul Rings

Let $A = \bigoplus_{n=0}^{\infty} A_n$ be a nonnegatively graded ring with the degree-zero component $R = A_0$.

Definition 2.29 The graded ring A is said to be *left flat Koszul* if one of the equivalent conditions of the next theorem is satisfied. (*Right flat Koszul* graded rings are defined similarly.)

Theorem 2.30 *The following three conditions are equivalent:*

(a) A_n *is a flat left R-module for every $n \geq 1$ and* $\mathrm{Tor}^A_{i,j}(R, R) = 0$ *for all $i \neq j$.*

(b) A_n *is a flat left R-module for every $n \geq 1$, the graded ring A is quadratic, and, setting $V = A_1$ and denoting by $I \subset V \otimes_R V$ the kernel of the multiplication map $A_1 \otimes_R A_1 \longrightarrow A_2$, for every $n \geq 4$, the collection of $n - 1$ subbimodules*

$$I \otimes_R V \otimes_R \cdots \otimes_R V, \ V \otimes_R I \otimes_R V \otimes_R \cdots \otimes_R V, \ \ldots, \ V \otimes_R \cdots \otimes_R V \otimes_R I \qquad (2.22)$$

in the R-R-bimodule $W = V^{\otimes_R n}$ is distributive.

(c) A_1 *and A_2 are flat left R-modules, the graded ring A is quadratic, and for every $n \geq 1$, the collection of R-R-subbimodules $V^{\otimes_R k-1} \otimes_R I \otimes_R V^{\otimes_R n-k-1}$, $1 \leq k \leq n - 1$ (2.22) in the R-R-bimodule $W = V^{\otimes_R n}$ is left flat distributive.*

Proof First of all, by Proposition 2.4(b), condition (a) implies that A is quadratic. So we can assume that $A = T_R(V)/(I)$. Furthermore, any collection of less than three subobjects is distributive, so the distributivity condition in (b) is trivial for $n \leq 3$. It is clear from the formula (1.1) from Chap. 1 that the left flat distributivity condition in (c) implies flatness of the left R-modules A_n for all $n \geq 1$. This suffices to prove the implication (c) $\Longrightarrow$ (b).

To deduce (b) $\Longrightarrow$ (c), we observe that the quotient bimodule of $V^{\otimes_R n}$ by any subset of the subbimodules (2.22) is isomorphic to the tensor product $A_{j_1} \otimes_R \cdots \otimes_R A_{j_s}$ for some $j_1, \ldots, j_s \geq 1$, $j_1 + \cdots + j_s = n$. Flatness of the left R-modules A_j implies flatness of such tensor products (as left R-modules), and it remains to invoke Lemma 2.19 for the abelian category C of R-R-bimodules and the class $\mathsf{F} \subset \mathsf{C}$ of all R-R-bimodules that are flat as left R-modules.

It remains to prove the equivalence (a) $\Longleftrightarrow$ (b). Arguing by induction in $n \geq 1$, we can assume that the collection of $j - 1$ subbimodules (2.22) in the R-R-bimodule $V^{\otimes_R j}$ is left flat distributive for all $1 \leq j \leq n - 1$ and $\mathrm{Tor}^A_{i,j}(R, R) = 0$ for all $i \neq j \leq n - 1$. Under this induction assumption, we will prove the equivalence of conditions (a) and (b) for the fixed value of $j = n$.

By Lemma 2.23 (applied to the lattice of $k - 1$ subbimodules in the R-R-bimodule $W = V^{\otimes_R k}$ and the lattice of $n - k - 1$ subbimodules in the R-R-bimodule $U = V^{\otimes_R n-k}$, $1 \leq k \leq n - 1$), the induction assumption implies that the collection of $n - 1$ subbimodules (2.22) in the R-R-bimodule $V^{\otimes_R n}$ is almost distributive.

Finally, the bar complex (2.11) computing the R-R-bimodules $\mathrm{Tor}^A_{i,n}(R, R)$ is isomorphic to the lattice bar complex $B_\bullet(W; X_1, \ldots, X_{n-1})$ (2.19) for the collection of $n - 1$ subbimodules $X_k = V^{\otimes_R k-1} \otimes_R I \otimes_R V^{\otimes_R n-k-1}$ in the R-R-bimodule $W = V^{\otimes_R n}$. Hence Lemma 2.18(a) $\Leftrightarrow$ (c) implies the desired equivalence (a) $\Longleftrightarrow$ (b). $\qquad\square$

Examples 2.31

(1) Let R be an associative ring and V be an R-R-bimodule that is flat as a left R-module. Then the tensor ring $A = T_R(V)$ and the graded ring $A = R \oplus V$ from Examples 1.3(1–2) are left flat Koszul. Indeed, conditions (b) and/or (c) of Theorem 2.30 are trivial to check for these graded rings.
(2) Let R be a commutative ring and V be a flat R-module. Then the symmetric algebra $\mathrm{Sym}_R(V)$ and the exterior algebra $\Lambda_R(V)$ are left and right flat Koszul graded rings. We refer to Sect. 10.1 below for a more detailed discussion.

The construction of the first Koszul complex in Sect. 2.5 was using double dualization: first we passed from A_1 to $B_1 = \mathrm{Hom}_R(A_1, R)$ and then set $K_1^\tau(B, A) = \mathrm{Hom}_{R^{\mathrm{op}}}(B_1, A)$. Therefore, the assumption that A_1 is a finitely generated projective R-module was needed. A similar double dualization was used in the construction of the second Koszul complex in Sect. 2.6. The following alternative approach allows to produce the two Koszul complexes for any left flat Koszul ring.

Let A be a left flat Koszul graded ring. Denote the kernel of the (surjective) multiplication map $A_1 \otimes_R A_1 \longrightarrow A_2$ by $I_A \subset A_1 \otimes_R A_1$. Set $I_A^{(0)} = R$, $I_A^{(1)} = A_1$, $I_A^{(2)} = I_A$, and

$$I_A^{(n)} = \bigcap_{k=1}^{n-1} A_1^{\otimes_R k-1} \otimes_R I_A \otimes_R A_1^{\otimes_R n-k-1} \subset A_1^{\otimes_R n}. \tag{2.23}$$

So $I_A^{(n)}$ is an R-R-subbimodule in $A_1^{\otimes_R n}$.

The intersection of all the subbimodules indexed by $k = 2, \ldots, n-1$ (i.e., of all but the first one) in (2.23) is the subbimodule

$$\bigcap_{k=2}^{n-1} A_1^{\otimes_R k-1} \otimes_R I_A \otimes_R A_1^{\otimes_R n-k-1} = A_1 \otimes_R I_A^{(n-1)} \subset A_1^{\otimes_R n}$$

by Lemma 2.24(d). Hence we obtain an injective R-R-bimodule morphism $I_A^{(n)} \longrightarrow A_1 \otimes_R I_A^{(n-1)}$, which is defined for all $n \geq 1$. For every $n \geq 2$, the image of the composition $I_A^{(n)} \longrightarrow A_1 \otimes_R I_A^{(n-1)} \longrightarrow A_1 \otimes_R A_1 \otimes_R I_A^{(n-2)}$ is contained in the subbimodule $I_A \otimes_R I_A^{(n-2)} \subset A_1 \otimes_R A_1 \otimes_R I_A^{(n-2)}$.

Put $K_{i,n}^\tau(A) = A_{n-i} \otimes_R I_A^{(i)}$ for every $i \geq 0$, $n \geq i$ (where τ is just a placeholder or a notation for the identity map $A_1 \longrightarrow A_1$). Define the differential $\partial \colon K_i^\tau(A) \longrightarrow K_{i-1}^\tau(A)$ as the composition

$$A_j \otimes_R I_A^{(i)} \longrightarrow A_j \otimes_R A_1 \otimes_R I_A^{(i-1)} \longrightarrow A_{j+1} \otimes_R I_A^{(i-1)}$$

of the map $A_j \otimes_R I_A^{(i)} \longrightarrow A_j \otimes_R A_1 \otimes_R I_A^{(i-1)}$ induced by the map $I_A^{(i)} \longrightarrow A_1 \otimes_R I_A^{(i-1)}$ and the map $A_j \otimes_R A_1 \otimes_R I_A^{(i-1)} \longrightarrow A_{j+1} \otimes_R I_A^{(i-1)}$ induced by the multiplication map

$A_j \otimes_R A_1 \longrightarrow A_{j+1}$. Since the composition $I_A \longrightarrow A_1 \otimes_R A_1 \longrightarrow A_2$ vanishes, we have $\partial^2 = 0$. So we obtain a complex of graded left A-modules $K_\bullet^\tau(A)$ with the homological grading i and the internal grading $n = i + j$.

Proposition 2.32 *For any left flat Koszul graded ring A, the first Koszul complex $K_\bullet^\tau(A)$ is a graded flat resolution of the left A-module R.*

Proof The graded left A-modules $A \otimes_R I_A^{(i)}$ are flat, because the left R-modules $I_A^{(i)}$ are. Furthermore, for every $n \geq 1$, the internal degree n component of the Koszul complex $K_\bullet^\tau(A)$ is isomorphic to the lattice Koszul complex $K_\bullet(W; X_1, \ldots, X_{n-1})$ (2.21) for the collection of $n-1$ submodules $X_k = A_1^{\otimes_R k-1} \otimes_R I_A \otimes_R A_1^{\otimes_R n-k-1}$ in the R-R-bimodule $W = A_1^{\otimes_R n}$. Thus it remains to apply Lemma 2.18 (a) $\Rightarrow$ (b). $\square$

Similarly, the intersection of all the submodules indexed by $k = 1, \ldots, n-2$ (i.e., of all but the last one) in (2.23) is the subsmodule

$$\bigcap_{k=1}^{n-2} A_1^{\otimes_R k-1} \otimes_R I_A \otimes_R A_1^{\otimes_R n-k-1} = I_A^{(n-1)} \otimes_R A_1 \subset A_1^{\otimes_R n}$$

by Lemma 2.24(b). Hence we obtain an injective R-R-bimodule morphism $I_A^{(n)} \longrightarrow I_A^{(n-1)} \otimes_R A_1$, which is defined for all $n \geq 1$.

Put $^\tau K_{i,n}(A) = I_A^{(i)} \otimes_R A_{n-i}$ for every $i \geq 0$, $n \geq i$. Define the differential $\partial : {}^\tau K_i(A) \longrightarrow {}^\tau K_{i-1}(A)$ as the composition

$$I_A^{(i)} \otimes_R A_j \longrightarrow I_A^{(i-1)} \otimes_R A_1 \otimes_R A_j \longrightarrow I_A^{(i-1)} \otimes_R A_{j+1}.$$

Similarly to the construction above, we obtain a complex of graded right A-modules $^\tau K_\bullet(A)$ with the homological grading i and the internal grading $n = i + j$.

Recall the definition of a *weakly A/R-flat right A-module* for a morphism of associative rings $R \longrightarrow A$ such that A is a flat left R-module [63, Section 5]. A right A-module F is said to be weakly A/R-flat (or *weakly flat relative to R*) if the functor $F \otimes_A -$ takes short exact sequences of R-flat left A-modules to short exact sequences of abelian groups. For any right R-module N, the right A-module $N \otimes_R A$ is weakly flat relative to R. The main properties of the class of weakly A/R-flat right A-modules are listed in [63, Lemma 5.3(b)].

Proposition 2.33 *For any left flat Koszul graded ring A, the second Koszul complex $^\tau K_\bullet(A)$ is a graded weakly A/R-flat resolution of the right A-module R.*

Proof In view of the discussion above, the (graded) right A-modules $I_A^{(i)} \otimes_R A$ are (graded) weakly A/R-flat. This assertion does not require any assumptions about the right R-modules $I_A^{(i)}$. Furthermore, for every $n \geq 1$, the internal degree n

component of the Koszul complex ${}^\tau K_\bullet(A)$ is isomorphic to the lattice Koszul complex $K_\bullet(W; X_1, \ldots, X_{n-1})$ (2.21) for the collection of $n - 1$ subbimodules $X_k = A_1^{\otimes_R n-k-1} \otimes_R I_A \otimes_R A_1^{\otimes_R k-1}$ in the R-R-bimodule $W = A_1^{\otimes_R n}$. This is the same collection of subbimodules as in the proof of Proposition 2.32, but numbered in the opposite order. So it remains to apply Lemma 2.18 (a) $\Rightarrow$ (b) in order to conclude that the internal degree n component of the complex ${}^\tau K_\bullet(A)$ is exact for $n > 0$. Thus $H_i({}^\tau K_\bullet(A)) = 0$ for $i > 0$ and $H_0({}^\tau K_\bullet(A)) = R$. $\square$

2.10 Finitely Projective Koszul Rings

Let $A = \bigoplus_{n=0}^\infty A_n$ be a nonnegatively graded ring with the degree-zero component $R = A_0$. Assume that A_1 and A_2 are finitely generated projective left R-modules. When the multiplication map $A_1 \otimes_R A_1 \longrightarrow A_2$ is surjective, we denote its kernel by $I \subset V \otimes_R V$, where $V = A_1$, and consider the 2-left finitely projective quadratic graded ring $qA = T_R(V)/(I)$ together with its quadratic dual 2-right finitely projective quadratic graded ring B.

As above in this book, we denote by $\mathrm{Ext}_A(R, R)$ the bigraded Ext ring of the graded *left* A-module R and by $\mathrm{Ext}_{B^{op}}(R, R)$ the bigraded Ext ring of the graded *right* B-module R.

Definition 2.34 The graded ring A is said to be *left finitely projective Koszul* if one of the equivalent conditions of the next theorem is satisfied.

Theorem 2.35 *For any nonnegatively graded ring A, the following four conditions are equivalent:*

(a) *A_n is a finitely generated projective left R-module for every $n \geq 1$ and the graded ring A is left flat Koszul.*

(b) *A_n is a finitely generated projective left R-module for every $n \geq 1$, the multiplication map $A_1 \otimes_R A_1 \longrightarrow A_2$ is surjective, B_n is a finitely generated projective right R-module for every $n \geq 1$, and $\mathrm{Ext}_A^{i,j}(R, R) = 0$ for all $i \neq j$.*

(c) *The graded ring A is quadratic, A_1 and A_2 are finitely generated projective left R-modules, and for every $n \geq 1$, the collection of R-R-subbimodules $V^{\otimes_R k-1} \otimes_R I \otimes_R V^{\otimes_R n-k-1}$, $1 \leq k \leq n - 1$, in the R-R-bimodule $W = V^{\otimes_R n}$ is left projective distributive.*

(d) *A_1 and A_2 are finitely generated projective left R-modules, the multiplication map $A_1 \otimes_R A_1 \longrightarrow A_2$ is surjective, B_i is a finitely generated projective right R-module for every $i \geq 1$, and the first Koszul complex $K_\bullet^\tau(B, A)$ (2.16) from Section 2.5 is exact in all the internal degrees $n \geq 1$.*

(e) *A_n is a finitely generated projective left R-module for every $n \geq 1$, the multiplication map $A_1 \otimes_R A_1 \longrightarrow A_2$ is surjective, and the second Koszul complex ${}^\tau K_\bullet(B, A)$ (2.18) from Sect. 2.6 is exact in all the internal degrees $n \geq 1$.*

Proof

(a) $\Longleftrightarrow$ (c) Take the condition of Theorem 2.30(b) as the definition of left flat Koszulity and argue similarly to the proof of Theorem 2.30 (b) $\leftrightarrow$ (c), using Lemma 2.19 for the abelian category C of R-R-bimodules and the class $\mathsf{F} \subset \mathsf{C}$ of all R-R-bimodules that are projective as left R-modules.

(a) $\Longleftrightarrow$ (b) Take the condition of Theorem 2.30(a) as the definition of left flat Koszulity. The R-R-bimodules $\mathrm{Tor}^A_{i,n}(R, R)$ can be computed as the homology of the complex (2.11), while the R-R-bimodules $\mathrm{Ext}_A^{i,n}(R, R)$ can be computed as the cohomology of the complex (2.14). Under (a), the graded ring A is quadratic, so any one of the conditions (a) or (b) implies surjectivity of the multiplication map $A_1 \otimes_R A_1 \longrightarrow A_2$. By Proposition 2.2, we have $\mathrm{Ext}_A^{n,n}(R, R) \simeq B_n$.

Now the following lemma shows that, given a nonnegatively graded algebra A with the degree-zero component $R = A_0$ such that A_j is a finitely generated projective left R-module for every $j \geq 1$, and given a fixed $n \geq 1$, one has $\mathrm{Tor}^A_{i,n}(R, R) = 0$ for all $1 \leq i \leq n-1$ if and only if $\mathrm{Ext}_A^{i,n}(R, R) = 0$ for all $1 \leq i \leq n - 1$ *and* the right R-module $\mathrm{Ext}_A^{n,n}(R, R)$ is projective.

Lemma 2.36 *Let* $0 \longrightarrow C_n \longrightarrow C_{n-1} \longrightarrow \cdots \longrightarrow C_1 \longrightarrow 0$ *be a complex of finitely generated projective left* R-*modules, and let* $0 \longrightarrow \mathrm{Hom}_R(C_1, R) \longrightarrow \cdots \longrightarrow \mathrm{Hom}_R(C_n, R) \longrightarrow 0$ *be the dual complex of finitely generated projective right* R-*modules. Then the following four conditions are equivalent:*

(a) $H_i(C_\bullet) = 0$ *for all* $1 \leq i \leq n - 1$.

(a′) *The complex of left* R-*modules* $C_\bullet$ *is homotopy equivalent to the one-term complex* $H_n(C_\bullet)[n]$.

(b) $H^i(\mathrm{Hom}_R(C_\bullet, R)) = 0$ *for all* $1 \leq i \leq n - 1$ *and* $H^n(\mathrm{Hom}_R(C_\bullet, R))$ *is a (finitely generated) projective right* R-*module;*

(b′) *the complex of right* R-*modules* $\mathrm{Hom}_R(C_\bullet, R)$ *is homotopy equivalent to the one-term complex* $H^n(\mathrm{Hom}_R(C^\bullet, R))[-n]$.

If any one of these equivalent conditions holds, then $H_n(C_\bullet)$ *is a finitely generated projective left* R-*module and* $H^n(\mathrm{Hom}_R(C_\bullet, R)) \simeq \mathrm{Hom}_R(H_n(C_\bullet), R)$. $\qquad\square$

(c) $\Longrightarrow$ (d), (e) Under (c), the results of Lemmas 2.24 and 2.28 are available for the collections of subbimodules $V^{\otimes_R k-1} \otimes_R I \otimes_R V^{\otimes_R j-k-1}$ in the R-R-bimodules $V^{\otimes_R j}$, $1 \leq k \leq j - 1$, $j \geq 1$. This allows to compute the internal degree n component of the Koszul complex $K_\bullet^\tau(B, A)$ as the lattice Koszul complex $K_\bullet(W; X_1, \ldots, X_{n-1})$ (2.21) for the collection of subbimodules $X_k = V^{\otimes_R k-1} \otimes_R I \otimes_R V^{\otimes_R n-k-1} \subset V^{\otimes_R n} = W$. Similarly, the internal degree n component of the Koszul complex ${}^\tau K_\bullet(B, A)$ is computed as the lattice Koszul complex for the collection of subbimodules $X_k = V^{\otimes_R n-k-1} \otimes_R I \otimes_R V^{\otimes_R k-1} \subset V^{\otimes_R n}$ (the same subbimodules numbered in the opposite order). Then, in both cases, it remains to apply Lemma 2.18 (a) $\Rightarrow$ (b).

(d) $\Longrightarrow$ (a), (b) Under (d), the Koszul complex $K_\bullet^\tau(B, A)$ is a graded projective resolution of the graded left A-module R. Computing $\mathrm{Tor}^A(R, R)$ and $\mathrm{Ext}_A(R, R)$ in terms of this resolution immediately yields (a) and (b) (where (a) is interpreted as the condition of Theorem 2.30(a)).

(e) $\Longrightarrow$ (a) Under (e), the Koszul complex $^\tau K_\bullet(B, A)$ is a graded weakly A/R-flat resolution of the graded right A-module R, as explained in Sect. 2.9 (see the proof of Proposition 2.33). Since the graded left A-module R is flat over R, one can compute $\mathrm{Tor}^A(R, R)$ in terms of this resolution (e.g., in view of [63, Lemma 5.3(b)]). This immediately implies (a) (interpreted as the condition of Theorem 2.30(a)).   □

The definition of *right finitely projective Koszul* graded ring is obtained from the above definition of a left finitely projective Koszul ring by switching the roles of the left and right sides.

Proposition 2.37 *Let A be a 2-left finitely projective quadratic graded ring with the degree-zero component $R = A_0$, and let B be the 2-right finitely projective quadratic graded ring quadratic dual to A. Then the ring A is left finitely projective Koszul if and only if the ring B is right finitely projective Koszul. If this is the case, one has*

$$B \simeq \mathrm{Ext}_A(R, R)^{\mathrm{op}} \quad and \quad A \simeq \mathrm{Ext}_{B^{\mathrm{op}}}(R, R).$$

Proof The first assertion is provable using the condition of Theorem 2.35(c) as the definition of projective Koszulity and the result of Lemma 2.28. Then, by Theorem 2.35(b), we have $\mathrm{Ext}_A^{i,j}(R, R) = 0$ for $i \neq j$, and similarly, $\mathrm{Ext}_{B^{\mathrm{op}}}^{i,j}(R, R) = 0$ for $i \neq j$. Finally, the graded ring isomorphism $B^{\mathrm{op}} \simeq \bigoplus_{n \geq 0} \mathrm{Ext}_A^{n,n}(R, R)$ is provided by Proposition 2.2. The graded ring isomorphism $A \simeq \bigoplus_{n \geq 0} \mathrm{Ext}_{B^{\mathrm{op}}}^{n,n}(R, R)$ can be obtained by applying the same proposition to the ring B^{op} and observing that the 2-left finitely projective quadratic graded ring B^{op} is quadratic dual to the 2-right finitely projective quadratic graded ring A^{op}.   □

Corollary 2.38 *The anti-equivalences of categories from Propositions 1.6–1.8 restrict to an anti-equivalence between the category of left finitely projective Koszul graded rings A and the category of right finitely projective Koszul graded rings B over any fixed base ring R.*   □

Examples 2.39

(1) Let R be an associative ring and V be an R-R-bimodule that is finitely generated and projective as a left R-module. Then the tensor ring $A = T_R(V)$ and the graded ring $A = R \oplus V$ from Example 1.5(1) are left finitely projective Koszul.

   The R-R-bimodule $V^\vee = \mathrm{Hom}_R(V, R)$ is finitely generated and projective as a right R-module, so the graded rings $B = T_R(V^\vee)$ and $B = R \oplus V^\vee$ from Example 1.7(1) are right finitely projective Koszul. Under the equivalence of categories from Corollary 2.38, the graded ring $T_R(V)$ corresponds to the graded ring $R \oplus V^\vee$, and the graded ring $R \oplus V$ corresponds to the graded ring $T_R(V^\vee)$.

(2) Let R be a commutative ring and A be an associative R-algebra with $A_0 = R$. Similarly to Example 1.7(2), in this case the graded ring A is left finitely projective Koszul if and only if it is right finitely projective Koszul. Besides, A is a left flat Koszul if and only if it is right flat Koszul. This is clear, e.g., from Theorem 2.30(b).

(3) Let R be a commutative ring and V be a finitely generated projective R-module. Then the symmetric algebra $\mathrm{Sym}_R(V)$ and the exterior algebra $\Lambda_R(V)$ are left and right finitely projective Koszul graded rings (as one can see from the discussion of Koszul complexes in Example 2.8(1); we refer to Sect. 10.1 for further details). The equivalence of categories from Corollary 2.38 assigns the graded ring $\Lambda_R(V^\vee)$ to the graded ring $\mathrm{Sym}_R(V)$ and vice versa.

Remark 2.40 Let us warn the reader once again that, even when the graded rings A and B are Koszul (say, A is left finitely projective Koszul and B is right finitely projective Koszul, as above), the dual Koszul complex $K_e^{\vee\bullet}(B, A)$ (2.17) from Section 2.6 has *no* good exactness properties, generally speaking. In fact, in this case, one has $K_e^{\vee\bullet}(B, A) = \mathrm{Hom}_A(K_\bullet^\tau(B, A), A)$, and hence it follows from Theorem 2.35(d) that the complex $K_e^{\vee\bullet}(B, A)$ computes the graded $\mathrm{Ext}_A^*(R, A)$. This Ext can be quite complicated even for a Koszul algebra A over a field $R = k$ (cf. Examples 2.12).

Remark 2.41 Deformations of Koszul algebras over fields can be viewed as Koszul graded rings central over their commutative base ring (cf. Examples 1.5(3,5) and 1.9 above). In particular, the deformation property of Koszul algebras [18] elaborated upon in [53, Section 2 of Chapter 6] can be expressed as a characterization of such Koszul graded rings.

   Specifically, let R be a Noetherian commutative ring, and let A be a nonnegatively graded associative R-algebra with $A_0 = R$. Assume that the graded ring A is quadratic over R and that the R-module A_1 is finitely generated. Then the following conditions are equivalent:

(a) The graded ring A is (left or right) flat Koszul.
(b) The graded ring A is (left or right) finitely projective Koszul.

(c) The R-module A_n is projective (or flat) for every $n \geq 1$ and the graded algebra $\varkappa(\mathfrak{p}) \otimes_R A$ over the field $\varkappa(\mathfrak{p})$ is Koszul for every prime ideal $\mathfrak{p} \subset R$ in the ring R with the residue field $\varkappa(\mathfrak{p})$.

(d) The R-module A_n is projective (or flat) for $1 \leq n \leq 3$ and the graded algebra $A/\mathfrak{m}A$ over the field $R/\mathfrak{m}$ is Koszul for every maximal ideal $\mathfrak{m} \subset R$ in the ring R.

   The equivalence of these properties is an easy corollary of [53, Theorem 7.3 in Chapter 6]. We leave the details to the reader.

Relative Nonhomogeneous Quadratic Duality

3

Nonhomogeneous quadratic rings $\widetilde{A}$ can be informally described as rings defined by nonhomogeneous quadratic relations over a fixed base ring R. Not every system of non-homogeneous quadratic relations is good enough to define a nonhomogeneous quadratic ring (see the general discussion in the introductions to [53, 56], [53, Chapter 5], and the counterexamples in [64, Section 5] and [56, Section 3.4] or [53, Section 2 in Chapter 5]). For a system of nonhomogeneous quadratic relations to "make sense," its coefficients must, in turn, satisfy a certain system of equations, called the *self-consistency equations*.

Dualizing the degree-one and degree-zero parts of the nonhomogeneous quadratic relations and imposing the equations dual to the self-consistency equations on the relations' coefficients produce a curved DG-ring structure (B, d, h) on the dual quadratic graded ring B to the quadratic graded ring A defined by the homogeneous quadratic parts of the nonhomogeneous quadratic relations.

3.1 Nonhomogeneous Quadratic Rings

Let $\widetilde{A}$ be an associative ring with a subring $R \subset \widetilde{A}$. Consider the R-R-bimodule $\widetilde{A}/R$ and suppose that we have chosen a subbimodule $V \subset \widetilde{A}/R$. Denote by $\widetilde{V} \subset \widetilde{A}$ the full preimage of V under the surjective R-R-bimodule morphism $\widetilde{A} \longrightarrow \widetilde{A}/R$.

Consider the tensor ring $T_R(\widetilde{V}) = \bigoplus_{n=0}^{\infty} T_{R,n}(\widetilde{V})$, $T_{R,n}(\widetilde{V}) = \widetilde{V}^{\otimes_R n}$, as in Chap. 1. Then the R-R-bimodule morphism of identity inclusion $\widetilde{V} \longrightarrow \widetilde{A}$ extends uniquely to a ring homomorphism $\pi_{\widetilde{A}} \colon T_R(\widetilde{V}) \longrightarrow \widetilde{A}$ forming a commutative triangle diagram with the subring inclusions $R \longrightarrow T_R(\widetilde{V})$ and $R \longrightarrow \widetilde{A}$.

Assume that the ring $\widetilde{A}$ is generated by its subgroup $\widetilde{V}$. In other words, the ring homomorphism $\pi_{\widetilde{A}} \colon T_R(\widetilde{V}) \longrightarrow \widetilde{A}$ is surjective. Let $\widetilde{J}_{\widetilde{A}} \subset T_R(\widetilde{V})$ be the kernel of $\pi_{\widetilde{A}}$.

© The Author(s), under exclusive license to Springer Nature Switzerland AG 2021
L. Positselski, *Relative Nonhomogeneous Koszul Duality*, Frontiers in Mathematics,
https://doi.org/10.1007/978-3-030-89540-2_3

Define an increasing filtration F on the ring $T_R(\widetilde{V})$ by the rule $F_n T_R(\widetilde{V}) = \bigoplus_{i=0}^{n} T_{R,i}(\widetilde{V}) \subset T_R(\widetilde{V})$. Furthermore, put $F_n \widetilde{A} = \pi_{\widetilde{A}}(F_n T_R(\widetilde{V})) \subset \widetilde{A}$. So we have $F_0 T_R(\widetilde{V}) = R$ and $F_1 T_R(\widetilde{V}) = R \oplus \widetilde{V}$, and hence $F_0 \widetilde{A} = R$ and $F_1 \widetilde{A} = \widetilde{V} \subset \widetilde{A}$.

Clearly, F is an exhaustive multiplicative filtration on the ring $\widetilde{A}$, that is, $\widetilde{A} = \bigcup_{n \geq 0} F_n \widetilde{A}$ and $F_n \widetilde{A} \cdot F_m \widetilde{A} \subset F_{n+m} \widetilde{A}$. In fact, the filtration F on $\widetilde{A}$ is *generated by* F_1, which means that the equation $F_n \widetilde{A} \cdot F_m \widetilde{A} = F_{n+m} \widetilde{A}$ holds for all $n, m \geq 0$. (Sometimes, we will say that the filtration F on $\widetilde{A}$ is *generated by* $F_1 \widetilde{A}$ *over* R, which means that F is generated by F_1 and $F_0 \widetilde{A} = R$.) The filtration F on the graded ring $T_R(\widetilde{V})$ has similar properties.

Definition 3.1 Put $\widetilde{I}_{\widetilde{A}} = F_2 T_R(\widetilde{V}) \cap \widetilde{J}_{\widetilde{A}}$, so $\widetilde{I}_{\widetilde{A}}$ is an R-R-subbimodule in $F_2 T_R(\widetilde{V}) = R \oplus \widetilde{V} \oplus \widetilde{V}^{\otimes_R 2}$. We will say that the ring $\widetilde{A}$ with a fixed subring $R \subset \widetilde{A}$ and a fixed R-R-subbimodule of generators $R \subset \widetilde{V} \subset \widetilde{A}$ is *weak nonhomogeneous quadratic* over R if the ideal $\widetilde{J}_{\widetilde{A}} \subset T_R(\widetilde{V})$ is generated by its subgroup $\widetilde{I}_{\widetilde{A}}$.

Let $'\widetilde{A}$ and $''\widetilde{A}$ be two weak nonhomogeneous quadratic rings over the same base ring R, and let $R \subset '\widetilde{V} \subset '\widetilde{A}$ and $R \subset ''\widetilde{V} \subset ''\widetilde{A}$ be their fixed subbimodules of generators. A *morphism* of weak nonhomogeneous quadratic rings $f \colon '\widetilde{A} \longrightarrow ''\widetilde{A}$ is a ring homomorphism forming a commutative triangle diagram with the subring inclusions $R \longrightarrow '\widetilde{A}$ and $R \longrightarrow ''\widetilde{A}$ and satisfying the condition that $f('\widetilde{V}) \subset ''\widetilde{V}$.

Conversely, suppose that we are given an associative ring $\widetilde{A}$ with an exhaustive multiplicative increasing filtration $0 = F_{-1}\widetilde{A} \subset F_0 \widetilde{A} \subset F_1 \widetilde{A} \subset F_2 \widetilde{A} \subset \cdots$. Then $R = F_0 \widetilde{A}$ is a subring in $\widetilde{A}$. Suppose that the filtration F on $\widetilde{A}$ is generated by F_1, and put $\widetilde{V} = F_1 \widetilde{A}$. Then we have $R \subset \widetilde{V} \subset \widetilde{A}$ and the ring $\widetilde{A}$ is generated by its subgroup $\widetilde{V}$.

Definition 3.2 Consider the associated graded ring $\mathrm{gr}^F \widetilde{A} = \bigoplus_{n=0}^{\infty} \mathrm{gr}_n^F \widetilde{A}$, where $\mathrm{gr}_n^F \widetilde{A} = F_n \widetilde{A}/F_{n-1}\widetilde{A}$. We will say that the filtered ring $\widetilde{A}$ is *nonhomogeneous quadratic* if the ring $A = \mathrm{gr}^F \widetilde{A}$ is quadratic (in the sense of Chap. 1).

Let $('\widetilde{A}, F)$ and $(''\widetilde{A}, F)$ be two nonhomogeneous quadratic rings with the same degree-zero filtration component $F_0 ''\widetilde{A} = R = F_0 '\widetilde{A}$. A *morphism* of nonhomogeneous quadratic rings $f \colon '\widetilde{A} \longrightarrow ''\widetilde{A}$ is a morphism of filtered rings (that is, a ring homomorphism such that $f(F_n '\widetilde{A}) \subset F_n ''\widetilde{A}$ for all $n \geq 0$) forming a commutative triangle diagram with the subring inclusions $R \longrightarrow '\widetilde{A}$ and $R \longrightarrow ''\widetilde{A}$.

Examples 3.3

(1) Let $\widetilde{A}$ be an associative ring with a subring $R \subset \widetilde{A}$. Put $F_0 \widetilde{A} = R$ and $F_n \widetilde{A} = \widetilde{A}$ for $n \geq 1$, so $\widetilde{V} = F_1 \widetilde{A} = \widetilde{A}$. Then the filtered ring $(\widetilde{A}, F)$ is nonhomogeneous quadratic. The corresponding quadratic graded ring $A = \mathrm{gr}^F \widetilde{A}$ has the form $A = R \oplus V$, where V is the R-R-bimodule A/R; more precisely, the grading components of A are $A_0 = R$, $A_1 = V$, and $A_n = 0$ for all $n \geq 2$, as in Example 1.3(2).

(2) Let k be a field and $\widetilde{A}$ be an associative k-algebra endowed with an exhaustive multiplicative increasing filtration F such that the associated graded algebra $\mathrm{gr}^F \widetilde{A}$ is isomorphic to the tensor algebra $T_k(V)$ of some vector space V. Then the algebra $\widetilde{A}$ itself is also isomorphic to $T_k(V)$, albeit not canonically: to construct such an isomorphism, one has to make an arbitrary choice of a vector subspace complementary to $k = F_0\widetilde{A}$ in $\widetilde{V} = F_1\widetilde{A}$.

The similar assertion is *not* true in the more general setting of a filtered ring $\widetilde{A}$ such that the graded ring $A = \mathrm{gr}^F \widetilde{A}$ is isomorphic to the tensor ring $T_R(V)$ of a bimodule V over some ring R (as in Example 1.3(1)). In this setting, the filtered ring $\widetilde{A}$ need not be isomorphic to the graded ring A at all. More precisely, one can see that the filtered ring $\widetilde{A}$ is uniquely determined by the short exact sequence of R-R-bimodules $0 \longrightarrow R \longrightarrow F_1\widetilde{A} \longrightarrow F_1\widetilde{A}/F_0\widetilde{A} = V \longrightarrow 0$. We refer to Examples 3.12(2) and 3.22 below for a further discussion.

Lemma 3.4 *The above construction assigning the subbimodule of generators $\widetilde{V} = F_1\widetilde{A}$ to a filtration F on a ring $\widetilde{A}$ defines a fully faithful functor from the category of nonhomogeneous quadratic rings to the category of weak nonhomogeneous quadratic rings (over any fixed base ring R). In particular, any nonhomogeneous quadratic ring is weak nonhomogeneous quadratic (so our terminology is consistent).*

Proof We will only prove the second assertion. Let (A, F) be a nonhomogeneous quadratic ring. We have to show that the ideal $\widetilde{J}_{\widetilde{A}} \subset T_R(\widetilde{V})$ is generated by $\widetilde{I}_{\widetilde{A}} = F_2 T_R(\widetilde{V}) \cap \widetilde{J}_{\widetilde{A}}$. Indeed, denote the ideal generated by $\widetilde{I}_{\widetilde{A}}$ by $\widetilde{J}' \subset T_R(\widetilde{V})$, and consider the ring $\widetilde{A}' = T_R(\widetilde{V})/\widetilde{J}'$. Then $\widetilde{J}' \subset \widetilde{J}_{\widetilde{A}}$, so there is a unique surjective ring homomorphism $\widetilde{A}' \longrightarrow \widetilde{A}$ forming a commutative triangle diagram with the surjective ring homomorphisms $T_R(\widetilde{V}) \longrightarrow \widetilde{A}'$ and $T_R(\widetilde{V}) \longrightarrow \widetilde{A}$.

For every $n \geq 0$, denote by $F_n\widetilde{A}' \subset \widetilde{A}'$ the image of the subgroup $F_n T_R(\widetilde{V}) \subset T_R(\widetilde{V})$ under the ring homomorphism $T_R(\widetilde{V}) \longrightarrow \widetilde{A}'$. Then F is an exhaustive multiplicative filtration on the ring $\widetilde{A}'$. Furthermore, the image of $F_n\widetilde{A}'$ under the ring homomorphism $\widetilde{A}' \longrightarrow \widetilde{A}$ coincides with $F_n\widetilde{A}$. It is clear that the maps $F_n\widetilde{A}' \longrightarrow F_n\widetilde{A}$ are isomorphisms for $n = 0$ and 1. Moreover, we have $\widetilde{I}_{\widetilde{A}} \subset F_2 T_R(\widetilde{V}) \cap \widetilde{J}'$ by construction, and hence $F_2 T_R(\widetilde{V}) \cap \widetilde{J}' = F_2 T_R(\widetilde{V}) \cap \widetilde{J}_{\widetilde{A}}$. Therefore, the map $F_2\widetilde{A}' \longrightarrow F_2\widetilde{A}$ is an isomorphism, too.

We need to show that $\widetilde{J}' = \widetilde{J}_{\widetilde{A}}$; equivalently, this means that the ring homomorphism $\widetilde{A}' \longrightarrow \widetilde{A}$ is an isomorphism. It suffices to check that the induced homomorphism of graded rings $\mathrm{gr}^F \widetilde{A}' \longrightarrow \mathrm{gr}^F \widetilde{A}$ is an isomorphism. Set $A' = \mathrm{gr}^F \widetilde{A}'$ and $A = \mathrm{gr}^F \widetilde{A}$. The graded ring A' is generated by A'_1 over $A'_0 = R$, since the filtration F on the ring $\widetilde{A}'$ is generated by F_1 by construction. The graded ring A is quadratic by assumption. The graded ring homomorphism $A' \longrightarrow A$ is an isomorphism in the degrees 0, 1, and 2, as we have shown. It remains to apply the next lemma. $\qquad\square$

Lemma 3.5 *Let $f: A' \longrightarrow A$ be a homomorphism of nonnegatively graded rings such that the maps $f_n: A'_n \longrightarrow A_n$ are isomorphisms for $n = 0$, 1, and 2. Assume that the graded ring A' is generated by A'_1 over A'_0, while the graded ring A is quadratic. Then the map f is an isomorphism of graded rings.*

Proof Set $A'_0 = R = A_0$ and $A'_1 = V = A_1$. Then there is a unique graded ring homomorphism $\pi_{A'}: T_R(V) \longrightarrow A'$ acting by the chosen isomorphisms $R \simeq A_0$ and $V \simeq A'_1$ on the components of degrees 0 and 1 and a similar unique graded ring homomorphism $\pi_A: T_R(V) \longrightarrow A$. The triangle diagram of ring homomorphisms $T_R(V) \longrightarrow A' \longrightarrow A$ is commutative, i.e., $\pi_A = f\pi_{A'}$. Furthermore, both the graded rings A and A' are generated by their degree-one components (over their degree-zero components) by assumption, and hence both the maps $\pi_{A'}$ and π_A are surjective.

Denote by $J_{A'}$ and $J_A \subset T_R(V)$ the kernels of the graded ring homomorphisms $\pi_{A'}$ and π_A. Then we have $J_{A'} \subset J_A$ and $J_{A',2} = J_{A,2}$, since the map $f_2: A'_2 \longrightarrow A_2$ is an isomorphism by assumption. The algebra A is quadratic, so the ideal J_A is generated by $I_A = J_{A,2}$. It follows that $J_{A'} = J_A$, and hence f_n is an isomorphism for all n. $\qquad\square$

Remark 3.6 We will see below in Sect. 4.6 that under the left finitely projective Koszulity assumption the classes of weak nonhomogeneous quadratic rings and nonhomogeneous quadratic rings coincide. Specifically, if $\widetilde{A}$ is a weak nonhomogeneous quadratic ring such that the quadratic graded ring $q\,\mathrm{gr}^F \widetilde{A}$ is left finitely projective Koszul (where the filtration F on $\widetilde{A}$ is generated by $F_1\widetilde{A} = \widetilde{V}$ over $F_0\widetilde{A} = R$, as above), then the graded ring $\mathrm{gr}^F \widetilde{A}$ is quadratic (so the filtered ring $\widetilde{A}$ is nonhomogeneous quadratic).

3.2   Curved DG-Rings

The following series of definitions is of key importance.

Definition 3.7 A *CDG-ring* (*curved differential graded ring*) $B = (B, d, h)$ is a graded associative ring $B = \bigoplus_{n\in\mathbb{Z}} B^n$ endowed with a sequence of additive maps $d_n: B^n \longrightarrow B^{n+1}$, $n \in \mathbb{Z}$, and an element $h \in B^2$ satisfying the following conditions:

(i) d is an odd derivation of B, that is, $d(bc) = d(b)c + (-1)^{|b|}bd(c)$ for all $b \in B^{|b|}$ and $c \in B^{|c|}$, $|b|, |c| \in \mathbb{Z}$.
(ii) $d^2(b) = [h, b]$ for all $b \in B$ (where $[h, b] = hb - bh$ is the commutator).
(iii) $d(h) = 0$.

In the context of this book, all CDG-rings will be nonnegatively graded, that is, $B = \bigoplus_{n=0}^{\infty} B^n$. We denote the grading of B by upper indices, because the differential d has degree 1.

Definition 3.8 Let $'B = ('B, d', h')$ and $''B = (''B, d'', h'')$ be two CDG-rings. A *morphism* of CDG-rings $''B \longrightarrow 'B$ is a pair (f, a) consisting of a morphism of graded rings $f \colon ''B \longrightarrow 'B$ and an element $a \in 'B^1$ such that

(iv) $f(d''(b)) = d'(f(b)) + [a, f(b)]$ for all $b \in ''B^{|b|}$ (where $[x, y] = xy - (-1)^{|x||y|}yx$, $x \in 'B^{|x|}$, $y \in 'B^{|y|}$, is the graded commutator).

(v) $f(h'') = h' + d'(a) + a^2$.

The composition of morphisms is defined by the formula $(f, a)(g, b) = (fg, a + f(b))$. The identity morphism is the morphism $(\mathrm{id}, 0)$. These rules define the *category of CDG-rings*.

The element $h \in B^2$ is called the *curvature element*. The element $a \in 'B^1$ is called the *change-of-connection element*.

For any CDG-ring $B = (B, d, h)$ and any element $a \in B^1$, the triple $'B = (B, d', h')$ with $d' = d + [a, -]$ and $h' = h + d(a) + a^2$ is also a CDG-ring. The CDG-rings B and $'B$ are connected by the isomorphism $(\mathrm{id}, a) \colon 'B \longrightarrow B$. Such isomorphisms will be called *change-of-connection isomorphisms*, while CDG-ring morphisms of the form $(f, 0)$ will be called *strict morphisms*. Any morphism of CDG-rings $(f, a) \colon ''B = (''B, d'', h'') \longrightarrow 'B = ('B, d', h')$ decomposes uniquely into a strict morphism followed by a change-of-connection isomorphism, $(f, a) = (\mathrm{id}, a)(f, 0)$.

We will denote the category of nonnegatively graded CDG-rings (B, d, h) with the fixed degree-zero component $B^0 = R$ by $R\text{–rings}_{\mathrm{cdg}}$. Morphisms $''B \longrightarrow 'B$ in $R\text{–rings}_{\mathrm{cdg}}$ are CDG-ring morphisms $(f, a) \colon ''B \longrightarrow 'B$ such that the graded ring homomorphism $f \colon ''B \longrightarrow 'B$ forms a commutative triangle diagram with the fixed isomorphisms $R \simeq ''B^0$ and $R \simeq 'B^0$.

Definition 3.9 One can define the *2-category of CDG-rings* as follows. Let (f, a) and $(g, b) \colon ''B = (''B, d'', h'') \longrightarrow 'B = ('B, d', h')$ be two CDG-ring morphisms with the same domain and codomain. A *2-morphism* $(f, a) \xrightarrow{z} (g, b)$ is an invertible element $z \in 'B^0$ satisfying the equations

(vi) $g(c) = zf(c)z^{-1}$ for all $c \in ''B$.

(vii) $b = zaz^{-1} - d'(z)z^{-1}$.

The element $z \in 'B^0$ is called the *gauge transformation element*.

The vertical composition of two 2-morphisms $(f', a') \xrightarrow{w} (f'', a'') \xrightarrow{z} (f''', a''')$ is the 2-morphism $(f', a') \xrightarrow{zw} (f''', a''')$. The identity 2-morphism is the 2-morphism $(f, a) \xrightarrow{1} (f, a)$. The horizontal composition of two 2-morphisms $(g', b') \xrightarrow{w} (g'', b'') \colon (C, d_C, h_C) \longrightarrow (B, d_B, h_B)$ and $(f', a') \xrightarrow{z} (f'', a'') \colon (B, d_B, h_B) \longrightarrow (A, d_A, h_A)$ is the 2-morphism $(f'g', a' + f'(b')) \xrightarrow{z \circ w} (f''g'', a'' + f''(b'')) \colon (C, d_C, h_C) \longrightarrow (A, d_A, h_A)$ with the element $z \circ w = zf'(w) = f''(w)z \in A^0$.

All the 2-morphisms of CDG-rings are invertible. If $(f, a) \colon {}''B \longrightarrow {}'B$ is a morphism of CDG-rings and $z \in {}'B^0$ is an invertible element, then the pair (g, b) defined by the formulas (vi–vii) is also a morphism of CDG-rings $(g, b) \colon {}''B \longrightarrow {}'B$. The morphisms (f, a) and (g, b) are connected by the 2-isomorphism $(f, a) \xrightarrow{z} (g, b)$.

The 2-category Rings_{cdg2} of nonnegatively graded CDG-rings is defined as the following subcategory of the 2-category of CDG-rings (cf. Remark 1.10). The objects of Rings_{cdg2} are nonnegatively graded CDG-rings (B, d, h), $B = \bigoplus_{n=0}^{\infty} B^n$. Morphisms $({}''B, d'', h'') \longrightarrow ({}'B, d', h')$ in Rings_{cdg2} are morphisms of CDG-rings $(f, a) \colon ({}''B, d'', h'') \longrightarrow ({}'B, d', h')$ such that the *degree-zero component* $f_0 \colon {}''B^0 \longrightarrow {}'B^0$ of the graded ring homomorphism f is an isomorphism ${}''B^0 \simeq {}'B^0$. 2-morphisms $(f, a) \xrightarrow{z} (g, b)$ in Rings_{cdg2} between morphisms (f, a) and (g, b) belonging to Rings_{cdg2} are arbitrary 2-morphisms from (f, a) to (g, b) in the 2-category of CDG-rings.

Examples 3.10

(0) The classical homological algebra notion of a differential graded algebra or a differential graded ring (DG-ring) provides the simplest class of examples of CDG-rings. We recall this classical concept immediately below.

(1) Some examples of CDG-algebras appearing in the nonhomogeneous quadratic duality theory over a field can be extracted from [56, Section 2.7] and [53, Section 5 in Chapter 5]. Examples of CDG-rings appearing in relative nonhomogenous Koszul duality can be found in [56, Section 4], [59, Appendix B], and Chap. 10 below.

(2) Let R be an associative ring and $w \in R$ be a central element, called the *potential*. Consider the graded ring B with $B^{2n} = R$ and $B^{2n+1} = 0$ for all $n \in \mathbb{Z}$. Endow B with the zero differential $d = 0$ and the curvature element $h = w \in B^2$. Then (B, d, h) is a CDG-ring. This CDG-ring is important in the context of the theory of *matrix factorizations* [12, 22]; in fact, a "matrix factorization of w" is the same thing as a CDG-module over (B, d, h) (see Sect. 6.1 below for the definition). This point of view is explicitly taken in the papers [20, 54].

(3) For a discussion of CDG-rings appearing in the context of deformation theory, see, e.g., the paper [39].

A *DG-ring* (B, d) is a graded associative ring $B = \bigoplus_{n \in \mathbb{Z}} B^n$ endowed with an odd derivation $d \colon B \longrightarrow B$ of degree 1 such that $d^2 = 0$. In other words, one can say that a DG-ring is a CDG-ring (B, d, h) with $h = 0$. Unless otherwise mentioned, we will usually consider nonnegatively graded DG-rings, that is, $B = \bigoplus_{n=0}^{\infty} B^n$.

A *morphism of DG-rings* $f \colon ({}''B, d'') \longrightarrow ({}'B, d')$ is a morphism of graded rings $f \colon {}''B \longrightarrow {}'B$ such that $f d'' = d' f$. In other words, one can say that a morphism of DG-rings is a morphism of CDG-rings $(f, a) \colon ({}''B, d'', 0) \longrightarrow ({}'B, d', 0)$ with $a = 0$. Notice that there exist CDG-ring morphisms (f, a) with $a \neq 0$, both the domain and codomain of which are DG-rings. In other words, DG-rings form a subcategory in CDG-rings, but it is *not* a full subcategory.

We will denote the category of nonnegatively graded DG-rings (B, d) with the fixed degree-zero component $B^0 = R$ by $R\text{–rings}_{\text{dg}}$. Morphisms $f: (''B, d'') \longrightarrow ('B, d')$ in $R\text{–rings}_{\text{dg}}$ are DG-ring morphisms such that the graded ring homomorphism $f: ''B \longrightarrow {}'B$ forms a commutative triangle diagram with the fixed isomorphisms $R \simeq {}''B^0$ and $R \simeq {}'B^0$.

One can define the 2-*category of DG-rings* as follows. Let $f, g: (''B, d'') \longrightarrow ('B, d')$ be a pair of parallel morphisms of DG-rings. A 2-morphism $f \xrightarrow{z} g$ is an invertible element $z \in {}'B^0$ such that $d'(z) = 0$ and $g(c) = zf(c)z^{-1}$ for all $c \in {}''B$.

The vertical composition of two 2-morphisms $f' \xrightarrow{w} f'' \xrightarrow{z} f'''$ is the 2-morphism $f' \xrightarrow{zw} f'''$. The identity 2-morphism is the 2-morphism $f \xrightarrow{1} f$. The horizontal composition of two 2-morphisms $g' \xrightarrow{w} g'': (C, d_C) \longrightarrow (B, d_B)$ and $f' \xrightarrow{z} f'': (B, d_B) \longrightarrow (A, d_A)$ is the 2-morphism $f'g' \xrightarrow{z \circ w} f''g'': (C, d_C) \longrightarrow (A, d_A)$ with the element $z \circ w = zf'(w) = f''(w)z \in A^0$.

All the 2-morphisms of DG-rings are invertible. If $f: (''B, d'') \longrightarrow ('B, d')$ is a morphism of DG-rings and $z \in {}'B^0$ is an invertible element such that $d'(z) = 0$, then $g: c \longmapsto zf(c)z^{-1}$ is also a morphism of DG-rings $g: (''B, d'') \longrightarrow ('B, d')$. The morphisms f and g are connected by the 2-isomorphism $f \xrightarrow{z} g$.

It is clear from these definitions that the 2-category of DG-rings is a 2-subcategory of the 2-category of CDG-rings. Notice the difference, however, in that the 2-morphisms of CDG-rings correspond to arbitrary invertible elements $z \in {}'B^0$. The 2-morphisms of DG-rings correspond to invertible *cocycles* $z \in {}'B^0$, $d'(z) = 0$.

The 2-category $\text{Rings}_{\text{dg2}}$ of nonnegatively graded DG-rings is defined as the following subcategory of the 2-category of DG-rings. The objects of $\text{Rings}_{\text{dg2}}$ are nonnegatively graded DG-rings (B, d), $B = \bigoplus_{n=0}^{\infty} B^n$. Morphisms $f: (''B, d'') \longrightarrow ('B, d')$ in $\text{Rings}_{\text{dg2}}$ are morphisms of DG-rings such that the map $f_0: ''B^0 \longrightarrow {}'B^0$ is an isomorphism. 2-morphisms $f \xrightarrow{z} g$ in $\text{Rings}_{\text{dg2}}$ between morphisms f and g belonging to $\text{Rings}_{\text{dg2}}$ are arbitrary 2-morphisms from f to g in the 2-category of DG-rings.

3.3 Self-Consistency Equations

Let $\widetilde{A}$ be a weak nonhomogeneous quadratic ring over its subring $R \subset \widetilde{A}$ with the R-R-subbimodule of generators $\widetilde{V} \subset \widetilde{A}$. Consider the related increasing filtration F on the ring $\widetilde{A}$, as constructed in Sect. 3.1, and let $A = \text{gr}^F \widetilde{A}$ be the associated graded ring. Then A is a nonnegatively graded ring generated by its degree-one component $A_1 = V = \widetilde{V}/R$ over the degree-zero component $A_0 = R$.

Denote by $I \subset V \otimes_R V$ the kernel of the multiplication map $A_1 \otimes_R A_1 \longrightarrow A_2$, and consider the quadratic graded ring $qA = T_R(V)/(I)$. Then we have a natural (adjunction) homomorphism of graded rings $qA \longrightarrow A$, which is an isomorphism in the degrees $n = 0$, 1, and 2, and a surjective map in the degrees $n \geq 3$.

Definition 3.11 We will say that a weak nonhomogeneous quadratic ring $\widetilde{A}$ is *3-left finitely projective* if the quadratic graded ring $qA = q\,\mathrm{gr}^F\widetilde{A}$ is 3-left finitely projective in the sense of Definition 1.4.

Let $\widetilde{A}$ be a weak nonhomogeneous quadratic ring. *In the rest of this section, we will assume that the R-R-bimodule $V = \widetilde{V}/R = A_1$ is projective as a left R-module and the R-R-bimodule $F_2\widetilde{A}/F_1\widetilde{A} = A_2$ is flat as a left R-module.*

Then, in particular, in the short exact sequence of R-R-bimodules $0 \longrightarrow R \longrightarrow \widetilde{V} \longrightarrow V \longrightarrow 0$, all the bimodules are projective as left R-modules. Therefore, this sequence splits as a short exact sequence of left R-modules, and we can choose a splitting $V \longrightarrow \widetilde{V}$. Let $V' \subset \widetilde{V}$ be the image of V under such a splitting map; so V' is a left (but not a right) R-submodule in the R-R-bimodule $\widetilde{V} \subset \widetilde{A}$ such that $\widetilde{V} = R \oplus V'$ as a left R-module. We will call V' a *submodule of strict generators* of $\widetilde{A}$.

In the rest of this section, we will identify V with V'. Let $v \in V$ and $r \in R$ be two elements. Let rv and $vr \in V$ denote the elements obtained by applying to $v \in V$ the left and right actions of the element $r \in R$ in the R-R-bimodule V, and let $r * v$ and $v * r \in \widetilde{V}$ denote the products of the elements $v \in V \simeq V' \subset \widetilde{V} \subset \widetilde{A}$ and $r \in R \subset \widetilde{A}$ in the ring $\widetilde{A}$. Then we have

$$r * v = rv \quad \text{and} \quad v * r = vr + q(v, r) \in \widetilde{V}, \tag{3.1}$$

where $q(v, r) \in R$ is a certain uniquely defined element. The map $q\colon V \times R \longrightarrow R$ is obviously biadditive, so it can be (uniquely) extended to an abelian group homomorphism $q\colon V\otimes_{\mathbb{Z}} R \longrightarrow R$. Since R is a subring in $\widetilde{A}$, we also have $r * s = rs$ (where the left-hand side denotes the product in $\widetilde{A}$ and the right-hand side is the product in R) for any pair of elements $r, s \in R$.

Furthermore, let $i \in I$ be an element. Consider the natural surjective map $V\otimes_{\mathbb{Z}}V \longrightarrow V\otimes_R V$ from the tensor product of two copies of V over the ring of integers $\mathbb{Z}$ to their tensor product over R, and denote by $\widehat{I} \subset V\otimes_{\mathbb{Z}}V$ the full preimage of the subbimodule $I \subset V\otimes_R V$ under the map $V\otimes_{\mathbb{Z}}V \longrightarrow V\otimes_R V$. Let $\hat{i} \in \widehat{I}$ denote some preimage of the element $i \in I$ under the natural surjective map $\widehat{I} \longrightarrow I$.

The element $i \in I \subset V\otimes_R V$ can be presented as a finite sum of decomposable tensors, $i = \sum_\alpha i_{1,\alpha}\otimes_R i_{2,\alpha}$, $i_{1,\alpha}, i_{2,\alpha} \in V$. We will suppress the notation for the sum over α and write simply $i = i_1\otimes i_2$. Similarly, the element $\hat{i} \in \widehat{I} \subset V\otimes_{\mathbb{Z}}V$ can be presented as a finite sum $\hat{i} = \sum_\alpha \hat{i}_{1,\alpha}\otimes_{\mathbb{Z}}\hat{i}_{2,\alpha}$, where $\hat{i}_{1,\alpha}, \hat{i}_{2,\alpha} \in V$. Once again, we will omit the notation for the sum over α and write $\hat{i} = \hat{i}_1\otimes\hat{i}_2$. All our formulas will be biadditive in $\hat{i}_1$ and $\hat{i}_2$ (or, as may be the case, appropriately R-bilinear in i_1 and i_2), so such simplification of the notation will be harmless.

For any two elements u and $v \in V$, we identify u and v with their images under the embedding $V \simeq V' \hookrightarrow \widetilde{V} \subset \widetilde{A}$ and consider their product $u * v \in \widetilde{A}$ in the ring $\widetilde{A}$. The assignment of the element $u * v$ to a pair of elements u and v can be uniquely extended to well-defined homomorphism of abelian groups (or left R-modules) $V \otimes_{\mathbb{Z}} V \longrightarrow \widetilde{A}$, which does *not*, generally speaking, factorize through $V \otimes_R V$.

In particular, for any element $i = i_1 \otimes i_2 \in I$ and any of its preimages $\hat{i} = \hat{i}_1 \otimes \hat{i}_2 \in \widehat{I}$, we consider the element $\hat{i}_1 * \hat{i}_2 \in \widetilde{A}$. We have $\hat{i}_1, \hat{i}_2 \in V' \subset \widetilde{V} = F_1 \widetilde{A}$, so $\hat{i}_1 * \hat{i}_2 \in F_2 \widetilde{A}$. Moreover, the image of the element $\hat{i}_1 * \hat{i}_2$ under the natural surjection $F_2 \widetilde{A} \longrightarrow F_2 \widetilde{A}/F_1 \widetilde{A}$ is equal to the product $i_1 i_2$ computed in the associated graded ring $A = \mathrm{gr}^F \widetilde{A}$. Now we have $i_1 i_2 = 0$ in $F_2 \widetilde{A}/F_1 \widetilde{A} = A_2$, since the composition $I \longrightarrow V \otimes_R V \longrightarrow A_2$ vanishes by construction. Thus $\hat{i}_1 * \hat{i}_2 \in F_1 \widetilde{A} = \widetilde{V} \subset \widetilde{A}$. Therefore, there exist uniquely defined elements $p(\hat{i}) \in V$ and $h(\hat{i}) \in R$ such that

$$\hat{i}_1 * \hat{i}_2 = p(\hat{i}) - h(\hat{i}) \in \widetilde{V} \subset \widetilde{A}, \qquad (3.2)$$

where the element $p(\hat{i}) \in V$ is identified with its image under the splitting $V \simeq V' \hookrightarrow \widetilde{V}$ and the element $h(\hat{i}) \in R$ is identified with its image under the subring inclusion $R \hookrightarrow \widetilde{V} \subset \widetilde{A}$. So we obtain two maps (homomorphisms of abelian groups) $p \colon \widehat{I} \longrightarrow V$ and $h \colon \widehat{I} \longrightarrow R$.

Examples 3.12

(1) Let $\widetilde{A}$ be an associative ring with a subring R. Assume that the R-R-bimodule $\widetilde{A}/R$ is projective as a left R-module. Then the filtration $F_0 \widetilde{A} = R$, $F_n \widetilde{A} = \widetilde{A}$ for $n \geq 1$ on the ring $\widetilde{A}$ satisfies the assumptions of this section. One has $V = \widetilde{A}/R$ and $I = V \otimes_R V$ (cf. Example 3.3(1)). Choosing a left R-submodule $V' \subset \widetilde{A}$ complementary to the subbimodule $R \subset \widetilde{A}$, one obtains the maps q, p, and h (3.1–3.2) providing a complete description of the multiplication map $\widetilde{A} \times \widetilde{A} \longrightarrow \widetilde{A}$ in terms of the direct sum decomposition $\widetilde{A} = R \oplus V'$.

(2) Let R be an associative ring and V be an R-R-bimodule which is projective as a left R-module. Consider the quadratic graded ring $A = T_R(V)$, and let $(\widetilde{A}, F)$ be a nonhomogeneous quadratic ring such that $\mathrm{gr}^F \widetilde{A} = A$ (cf. Example 3.3(2)). Then $I = 0$, so the maps p and h carry no essential information (more precisely, one has $h = 0$, and the map p is uniquely determined by the map q, as we will see soon in Proposition 3.14(e–f)). However, the map q can well be nontrivial. Moreover, knowing the R-R-bimodule V and the map q, one can uniquely recover the filtered ring $\widetilde{A}$; see Example 3.22 below.

Denote by $I^{(3)}$ the intersection $I \otimes_R V \cap V \otimes_R I \subset V \otimes_R V \otimes_R V$. Notice that, since the left R-modules V and $A_2 = (V \otimes_R V)/I$ are flat by assumption, the tensor products $I \otimes_R V$ and $V \otimes_R I$ are subbimodules in the triple tensor product $V \otimes_R V \otimes_R V$ (see the proofs of Propositions 1.8 and 2.1).

We are also interested in "the intersection $\widehat{I} \otimes_{\mathbb{Z}} V \cap V \otimes_{\mathbb{Z}} \widehat{I} \subset V \otimes_{\mathbb{Z}} V \otimes_{\mathbb{Z}} V$," but here we cannot claim that $\widehat{I} \otimes_{\mathbb{Z}} V$ and $V \otimes_{\mathbb{Z}} \widehat{I}$ are subgroups in $V \otimes_{\mathbb{Z}} V \otimes_{\mathbb{Z}} V$. So the intersection as

such is not well-defined. Instead, we have two abelian group homomorphisms $\widehat{I} \otimes_{\mathbb{Z}} V \longrightarrow$
$V \otimes_{\mathbb{Z}} V \otimes_{\mathbb{Z}} V$ and $V \otimes_{\mathbb{Z}} \widehat{I} \longrightarrow V \otimes_{\mathbb{Z}} V \otimes_{\mathbb{Z}} V$ induced by the inclusion $\widehat{I} \longrightarrow V \otimes_{\mathbb{Z}} V$. We
denote by $\widehat{I}^{(3)}$ the fibered product of the abelian groups $\widehat{I} \otimes_{\mathbb{Z}} V$ and $V \otimes_{\mathbb{Z}} \widehat{I}$ over the group
$V \otimes_{\mathbb{Z}} V \otimes_{\mathbb{Z}} V$.

Lemma 3.13 *The natural surjections* $\widehat{I} \otimes_{\mathbb{Z}} V \longrightarrow I \otimes_R V$, $V \otimes_{\mathbb{Z}} \widehat{I} \longrightarrow V \otimes_R I$, and
$V \otimes_{\mathbb{Z}} V \otimes_{\mathbb{Z}} V \longrightarrow V \otimes_R V \otimes_R V$ *induce a surjective map* $\widehat{I}^{(3)} \longrightarrow I^{(3)}$.

Proof Denote by $\widehat{I} \,\overline{\otimes}_{\mathbb{Z}}\, V$ and $V \,\overline{\otimes}_{\mathbb{Z}}\, \widehat{I} \subset V \otimes_{\mathbb{Z}} V \otimes_{\mathbb{Z}} V$ the images of the maps $\widehat{I} \otimes_{\mathbb{Z}} V \longrightarrow$
$V \otimes_{\mathbb{Z}} V \otimes_{\mathbb{Z}} V$ and $V \otimes_{\mathbb{Z}} \widehat{I} \longrightarrow V \otimes_{\mathbb{Z}} V \otimes_{\mathbb{Z}} V$, and let $\overline{I}^{(3)} \subset V \otimes_{\mathbb{Z}} V \otimes_{\mathbb{Z}} V$ stand for the
intersection of $\widehat{I} \,\overline{\otimes}_{\mathbb{Z}}\, V$ and $V \,\overline{\otimes}_{\mathbb{Z}}\, \widehat{I}$. Then there is a natural surjective map $\widehat{I}^{(3)} \longrightarrow \overline{I}^{(3)}$.
The surjection $V \otimes_{\mathbb{Z}} V \otimes_{\mathbb{Z}} V \longrightarrow V \otimes_R V \otimes_R V$ restricts to a map $\overline{I}^{(3)} \longrightarrow I^{(3)}$, and the
map $\widehat{I}^{(3)} \longrightarrow I^{(3)}$ is equal to the composition $\widehat{I}^{(3)} \longrightarrow \overline{I}^{(3)} \longrightarrow I^{(3)}$. It remains to prove
that the map $\overline{I}^{(3)} \longrightarrow I^{(3)}$ is surjective.

We have a short exact sequence of abelian groups $0 \longrightarrow \widehat{I} \longrightarrow V \otimes_{\mathbb{Z}} V \longrightarrow A_2 \longrightarrow 0$.
Hence the subgroups $X_1 = \widehat{I} \,\overline{\otimes}_{\mathbb{Z}}\, V$ and $X_2 = V \,\overline{\otimes}_{\mathbb{Z}}\, \widehat{I} \subset V \otimes_{\mathbb{Z}} V \otimes_{\mathbb{Z}} V$ are the
kernels of the surjective maps $V \otimes_{\mathbb{Z}} V \otimes_{\mathbb{Z}} V \longrightarrow A_2 \otimes_{\mathbb{Z}} V$ and $V \otimes_{\mathbb{Z}} V \otimes_{\mathbb{Z}} V \longrightarrow V \otimes_{\mathbb{Z}} A_2$,
respectively. Denote by Y_1 and $Y_2 \subset V \otimes_{\mathbb{Z}} V \otimes_{\mathbb{Z}} V$ the kernels of the natural surjective
maps $V \otimes_{\mathbb{Z}} V \otimes_{\mathbb{Z}} V \longrightarrow V \otimes_R V \otimes_{\mathbb{Z}} V$ and $V \otimes_{\mathbb{Z}} V \otimes_{\mathbb{Z}} V \longrightarrow V \otimes_{\mathbb{Z}} V \otimes_R V$. Then we have
$Y_1 \subset X_1$ and $Y_2 \subset X_2 \subset V \otimes_{\mathbb{Z}} V \otimes_{\mathbb{Z}} V$.

Furthermore, $Y_1 + Y_2$ is the kernel of the natural surjective map $V \otimes_{\mathbb{Z}} V \otimes_{\mathbb{Z}} V \longrightarrow$
$V \otimes_R V \otimes_R V$. Similarly, $X_1 + Y_2$ is the kernel of the surjective map $V \otimes_{\mathbb{Z}} V \otimes_{\mathbb{Z}} V \longrightarrow$
$A_2 \otimes_R V$, and $Y_1 + X_2$ is the kernel of the surjective map $V \otimes_{\mathbb{Z}} V \otimes_{\mathbb{Z}} V \longrightarrow V \otimes_R A_2$. It
follows that the subbimodule $I \otimes_R V \subset V \otimes_R V \otimes_R V$ is the image of $X_1 \subset V \otimes_{\mathbb{Z}} V \otimes_{\mathbb{Z}} V$
under the map $V \otimes_{\mathbb{Z}} V \otimes_{\mathbb{Z}} V \longrightarrow V \otimes_R V \otimes_R V$, and the subbimodule $V \otimes_R I \subset$
$V \otimes_R V \otimes_R V$ is the image of $X_2 \subset V \otimes_{\mathbb{Z}} V \otimes_{\mathbb{Z}} V$ under the same map.

The assertion that the map $\overline{I}^{(3)} \longrightarrow I^{(3)}$ is surjective is now expressed by the
distributivity equation

$$X_1 \cap X_2 + (Y_1 + Y_2) = (X_1 + Y_2) \cap (Y_1 + X_2)$$

on subgroups in the abelian group $W = V \otimes_{\mathbb{Z}} V \otimes_{\mathbb{Z}} V$. What we have here is two filtrations
$0 \subset Y_1 \subset X_1 \subset W$ and $0 \subset Y_2 \subset X_2 \subset W$ of an abelian group W. It remains to
observe that *any two filtrations of an abelian category object generate a distributive lattice
of its subobjects*, which is a particular case of [34, Theorem 5] or [53, Corollary 6.4 in
Chapter 1]. $\qquad\qquad\qquad\qquad\qquad\qquad\qquad\qquad\qquad\qquad\qquad\qquad\qquad\qquad\qquad\qquad\qquad\qquad\qquad\quad\square$

Let $j \in I^{(3)}$ be an element and $\hat{j} \in \widehat{I}^{(3)}$ be one of its preimages. The element
$j \in I^{(3)} \subset V^{\otimes_R 3}$ can be presented as a finite sum of decomposable tensors, $j =$
$\sum_\alpha j_{1,\alpha} \otimes_R j_{2,\alpha} \otimes_R j_{3,\alpha}$, $j_{1,\alpha}$, $j_{2,\alpha}$, $j_{3,\alpha} \in V$. As above, we will suppress the notation for
the sum over α and write simply $j = j_1 \otimes j_2 \otimes j_3$. Moreover, the image of the element

$\hat{j} \in \widehat{I}^{(3)}$ under the natural map $\widehat{I}^{(3)} \longrightarrow V^{\otimes_\mathbb{Z} 3}$ can be presented as a finite sum $\sum_\alpha \hat{j}_{1,\alpha} \otimes_\mathbb{Z} \hat{j}_{2,\alpha} \otimes_\mathbb{Z} \hat{j}_{3,\alpha}$, where $\hat{j}_{1,\alpha}$, $\hat{j}_{2,\alpha}$, $\hat{j}_{3,\alpha} \in V$. We will write simply $\hat{j} = \hat{j}_1 \otimes \hat{j}_2 \otimes \hat{j}_3$, omitting the notation for the sum over α and ignoring the distinction between an element of $\widehat{I}^{(3)}$ and its image in $V^{\otimes_\mathbb{Z} 3}$ in this notation.

For any three elements u, v, and $w \in V$, we will identify u, v, and w with their images under the embedding $V \simeq V' \hookrightarrow \widetilde{V} \subset \widetilde{A}$ and consider the triple product $(u * v) * w = u * v * w = u * (v * w) \in \widetilde{A}$ in the ring $\widetilde{A}$. In particular, for any element $\hat{j} \in \widehat{I}^{(3)}$, the element $\hat{j}_1 * \hat{j}_2 * \hat{j}_3 \in \widetilde{A}$ is well-defined.

Proposition 3.14 *Let $\widetilde{A}$ be a weak nonhomogeneous quadratic ring such that the left R-module $V = A_1$ is projective and the left R-module A_2 is flat. Suppose that a left R-linear splitting $V \simeq V' \hookrightarrow \widetilde{V}$ of the surjective R-R-bimodule map $\widetilde{V} \longrightarrow \widetilde{V}/R = V$ has been chosen. Then the maps $q \colon V \times R \longrightarrow R$, $p \colon \widehat{I} \longrightarrow V$, and $h \colon \widehat{I} \longrightarrow R$ defined above in (3.1–3.2) satisfy the following self-consistency equations:*

(a) $q(rv, s) = rq(v, s)$ *for all r, $s \in R$ and $v \in V$.*

(b) $q(v, rs) = q(vr, s) + q(v, r)s$ *for all r, $s \in R$ and $v \in V$.*

(c) $p(r\hat{\imath}_1 \otimes \hat{\imath}_2) = rp(\hat{\imath}_1 \otimes \hat{\imath}_2)$ *for all $r \in R$ and $\hat{\imath} \in \widehat{I}$.*

(d) $h(r\hat{\imath}_1 \otimes \hat{\imath}_2) = rh(\hat{\imath}_1 \otimes \hat{\imath}_2)$ *for all $r \in R$ and $\hat{\imath} \in \widehat{I}$.*

(e) $p(u \otimes rv - ur \otimes v) = q(u, r)v$ *for all $r \in R$ and u, $v \in V$.*

(f) $h(u \otimes rv - ur \otimes v) = 0$ *for all $r \in R$ and u, $v \in V$.*

(g) $p(\hat{\imath}_1 \otimes \hat{\imath}_2 r) = p(\hat{\imath}_1 \otimes \hat{\imath}_2)r - \hat{\imath}_1 q(\hat{\imath}_2, r)$ *for all $r \in R$ and $\hat{\imath} = \hat{\imath}_1 \otimes \hat{\imath}_2 \in \widehat{I}$.*

(h) $h(\hat{\imath}_1 \otimes \hat{\imath}_2 r) = h(\hat{\imath}_1 \otimes \hat{\imath}_2)r - q(p(\hat{\imath}_1 \otimes \hat{\imath}_2), r) + q(\hat{\imath}_1, q(\hat{\imath}_2, r))$ *for all $r \in R$ and $\hat{\imath} \in \widehat{I}$.*

(i) $p(\hat{j}_1 \otimes \hat{j}_2) \otimes \hat{j}_3 - \hat{j}_1 \otimes p(\hat{j}_2 \otimes \hat{j}_3) \in \widehat{I} \subset V \otimes_\mathbb{Z} V$ *for all $\hat{j} = \hat{j}_1 \otimes \hat{j}_2 \otimes \hat{j}_3 \in \widehat{I}^{(3)}$.*

(j) $p(p(\hat{j}_1 \otimes \hat{j}_2) \otimes \hat{j}_3 - \hat{j}_1 \otimes p(\hat{j}_2 \otimes \hat{j}_3)) = h(\hat{j}_1 \otimes \hat{j}_2)\hat{j}_3 - \hat{j}_1 h(\hat{j}_2 \otimes \hat{j}_3)$ *for all $\hat{j} \in \widehat{I}^{(3)}$.*

(k) $h(p(\hat{j}_1 \otimes \hat{j}_2) \otimes \hat{j}_3 - \hat{j}_1 \otimes p(\hat{j}_2 \otimes \hat{j}_3)) = q(\hat{j}_1, h(\hat{j}_2 \otimes \hat{j}_3))$ *for all $\hat{j} \in \widehat{I}^{(3)}$.*

Examples 3.15

(1) The Jacobi identity, defining the concept of a Lie algebra (e.g., over a field), is a classical example of a nontrivial system of self-consistency equations for the coefficients of a system of nonhomogeneous quadratic relations. More precisely, the equation (j) in the proposition is a generalization of the Jacobi identity for the Lie bracket, both for Lie algebras and for Lie algebroids. All the other identities in the definition of a Lie algebroid can also be found encoded in the equations above (with $h = 0$). We refer to Sect. 10.9 for a more detailed discussion.

(2) An example of a system of nonhomogeneous quadratic relations (over a field, with Koszul homogeneous quadratic principal part) which does *not* satisfy the self-consistency equations listed in the proposition, with the consequences that the algebra defined by such nonhomogeneous quadratic relations collapses to zero, can be found in [64, formula (7) in Section 5].

Proof of Proposition 3.14 All these equations follow, in one way or another, from the associativity of multiplication in the ring $\widetilde{A}$. The specific computations proving each of the formulas are presented below one by one.

Part (a):   compare $r * v * s = (r * v) * s = (rv) * s = rvs + q(rv, s)$ with $r * v * s = r * (v * s) = r * (vs + q(v, s)) = rvs + rq(v, s)$.

Part (b):   compare $v * r * s = v * (r * s) = v * (rs) = vrs + q(v, rs)$ with $v * r * s = (v * r) * s = (vr + q(v, r)) * s = vrs + q(vr, s) + q(v, r)s$.

Parts (c) and (d):   notice first of all that $\sum_\alpha r\hat{i}_{1,\alpha} \otimes \hat{i}_{2,\alpha} \in \widehat{I}$ whenever $r \in R$ and $\sum_\alpha \hat{i}_{1,\alpha} \otimes \hat{i}_{2,\alpha} \in \widehat{I}$. So the left-hand side of both the equations is well-defined. To deduce the equations, compare

$$r * \hat{i}_1 * \hat{i}_2 = (r * \hat{i}_1) * \hat{i}_2 = (r\hat{i}_1) * \hat{i}_2 = p(r\hat{i}_1 \otimes \hat{i}_2) - h(r\hat{i}_1 \otimes \hat{i}_2)$$

with

$$r * \hat{i}_1 * \hat{i}_2 = r * (\hat{i}_1 * \hat{i}_2) = r * (p(\hat{i}_1 \otimes \hat{i}_2) - h(\hat{i}_1 \otimes \hat{i}_2)) = rp(\hat{i}_1 \otimes \hat{i}_2) - rh(\hat{i}_1 \otimes \hat{i}_2),$$

and equate the terms belonging to $V \simeq V' \subset \widetilde{V}$ separately and the terms belonging to $R \subset \widetilde{V}$ separately.

Parts (e) and (f):   first of all, one has $u \otimes rv - ur \otimes v \in \widehat{I}$ for all $r \in R$ and $u, v \in V$. So the left-hand side of the equations is well-defined. Furthermore,

$$0 = u * (r * v) - (u * r) * v = u * (rv) - (ur) * v - q(u, r) * v$$

$$= p(u \otimes rv - ur \otimes v) - h(u \otimes rv - ur \otimes v) - q(u, r)v,$$

and it remains to equate separately the terms belonging to V' and to R.

Parts (g) and (h):   we have $\sum_\alpha \hat{i}_{1,\alpha} \otimes \hat{i}_{2,\alpha} r \in \widehat{I}$ whenever $r \in R$ and $\sum_\alpha \hat{i}_{1,\alpha} \otimes \hat{i}_{2,\alpha} \in \widehat{I}$, so the left-hand side of both the equations is well-defined. Now compare

$$\hat{i}_1 * \hat{i}_2 * r = (\hat{i}_1 * \hat{i}_2) * r = (p(\hat{i}_1 \otimes \hat{i}_2) - h(\hat{i}_1 \otimes \hat{i}_2)) * r$$

$$= p(\hat{i}_1 \otimes \hat{i}_2)r + q(p(\hat{i}_1 \otimes \hat{i}_2), r) - h(\hat{i}_1 \otimes \hat{i}_2)r$$

with

$$\hat{i}_1 * \hat{i}_2 * r = \hat{i}_1 * (\hat{i}_2 * r) = \hat{i}_1 * (\hat{i}_2 r + q(\hat{i}_2, r))$$

$$= p(\hat{i}_1 \otimes \hat{i}_2 r) - h(\hat{i}_2 \otimes \hat{i}_2 r) + \hat{i}_1 q(\hat{i}_2, r) + q(\hat{i}_1, q(\hat{i}_2, r)),$$

and equate separately the terms belonging to V' and to R.

Parts (i–k):   given an element $j \in \widehat{I}^{(3)}$, we have $(\hat{j}_1 * \hat{j}_2) * \hat{j}_3 = \hat{j}_1 * \hat{j}_2 * \hat{j}_3 = \hat{j}_1 * (\hat{j}_2 * \hat{j}_3)$ in $F_3 \widetilde{A} \subset \widetilde{A}$. By construction, the value of this triple product in $\widetilde{A}$ only depends on the image of the element j in the group $V \otimes_{\mathbb{Z}} V \otimes_{\mathbb{Z}} V$. We will compute the value of $(\hat{j}_1 * \hat{j}_2) * \hat{j}_3$ in terms of the image of j in $\widehat{I} \otimes_{\mathbb{Z}} V$ and the value of $\hat{j}_1 * (\hat{j}_2 * \hat{j}_3)$ in terms of the image of j in $V \otimes_{\mathbb{Z}} \widehat{I}$ and then equate the two expressions.

Specifically, we have

$$(\hat{j}_1 * \hat{j}_2) * \hat{j}_3 = p(\hat{j}_1 \otimes \hat{j}_2) * \hat{j}_3 - h(\hat{j}_1 \otimes \hat{j}_2) * \hat{j}_3$$

and

$$\hat{j}_1 * (\hat{j}_2 * \hat{j}_3) = \hat{j}_1 * p(\hat{j}_2 \otimes \hat{j}_3) - \hat{j}_1 * h(\hat{j}_2 \otimes \hat{j}_3),$$

and hence

$$p(\hat{j}_1 \otimes \hat{j}_2) * \hat{j}_3 - \hat{j}_1 * p(\hat{j}_2 \otimes \hat{j}_3) = h(\hat{j}_1 \otimes \hat{j}_2) * \hat{j}_3 - \hat{j}_1 * h(\hat{j}_2 \otimes \hat{j}_3). \tag{3.3}$$

Now, first of all, the right-hand side of (3.3) belongs to $\widetilde{V} \subset \widetilde{A}$, and hence so does the left-hand side. Both summands in the left-hand side belong to $F_2 \widetilde{A}$. So the image of the left-hand side in $A_2 = F_2 \widetilde{A}/F_1 \widetilde{A}$ has to vanish, which means that the expression $p(\hat{j}_1 \otimes \hat{j}_2) \otimes_R \hat{j}_3 - \hat{j}_1 \otimes_R p(\hat{j}_2 \otimes \hat{j}_3)$ belongs to $I \subset V \otimes_R V$. This proves part (i).

It remains to compute both sides of (3.3) as

$$p(\hat{j}_1 \otimes \hat{j}_2) * \hat{j}_3 - \hat{j}_1 * p(\hat{j}_2 \otimes \hat{j}_3)$$
$$= p(p(\hat{j}_1 \otimes \hat{j}_2) \otimes \hat{j}_3 - \hat{j}_1 \otimes p(\hat{j}_2 \otimes \hat{j}_3)) - h(p(\hat{j}_1 \otimes \hat{j}_2) \otimes \hat{j}_3 - \hat{j}_1 \otimes p(\hat{j}_2 \otimes \hat{j}_3))$$

and

$$h(\hat{j}_1 \otimes \hat{j}_2) * \hat{j}_3 - \hat{j}_1 * h(\hat{j}_2 \otimes \hat{j}_3) = h(\hat{j}_1 \otimes \hat{j}_2)\hat{j}_3 - \hat{j}_1 h(\hat{j}_2 \otimes \hat{j}_3) - q(\hat{j}_1, h(\hat{j}_2 \otimes \hat{j}_3)).$$

Comparing and equating separately the terms belonging to V' and to R produce the desired formulas (j–k). □

3.4 The CDG-Ring Corresponding to a Nonhomogeneous Quadratic Ring

Let R and S be associative rings, U be an R-S-bimodule, and $U^\vee = \text{Hom}_R(U, R)$ be the dual S-R-bimodule. We will use the notation

$$\langle u, f \rangle = f(u) \in R \quad \text{for any } u \in U \text{ and } f \in U^\vee.$$

Then the condition that $f : U \longrightarrow R$ is a left R-module homomorphism is expressed by the identity

$$\langle ru, f \rangle = r\langle u, f \rangle \quad \text{for all } r \in R, \ u \in U, \text{ and } f \in U^\vee,$$

while the construction of the left S-module structure on $U^\vee$ is expressed by the identity

$$\langle u, sf \rangle = \langle us, f \rangle \quad \text{for all } u \in U, \ s \in S, \text{ and } f \in U^\vee,$$

and the construction of the right R-module structure on $U^\vee$ is expressed by

$$\langle u, fr \rangle = \langle u, f \rangle r \quad \text{for all } u \in U, \ f \in U^\vee, \text{ and } r \in R.$$

Furthermore, given three rings R, S, and T, an R-S-bimodule U, and an S-T-bimodule V, the construction of the natural homomorphism of T-R-bimodules

$$\operatorname{Hom}_S(V, S) \otimes_S \operatorname{Hom}_R(U, R) \longrightarrow \operatorname{Hom}_R(U \otimes_S V, \ R)$$

from Lemma 1.1(a) can be expressed by the formula

$$\langle u \otimes v, \ g \otimes f \rangle = \langle u\langle v, g \rangle, \ f \rangle = \langle u, \ \langle v, g \rangle f \rangle$$

for all $u \in U$, $v \in V$, $g \in \operatorname{Hom}_S(V, S)$, and $f \in \operatorname{Hom}_R(U, R)$.

Proposition 3.16 *Let $\widetilde{A}$ be a 3-left finitely projective weak nonhomogeneous quadratic ring over its subring $R \subset \widetilde{A}$ with the R-R-bimodule of generators $R \subset \widetilde{V} \subset \widetilde{A}$. Denote by B the 3-right finitely projective quadratic graded ring quadratic dual to the 3-left finitely projective quadratic graded ring $\mathrm{q}A = \mathrm{q}\operatorname{gr}^F \widetilde{A}$. Suppose that a left R-linear splitting $V \simeq V' \hookrightarrow \widetilde{V}$ of the surjective R-R-bimodule map $\widetilde{V} \longrightarrow \widetilde{V}/R = V$ has been chosen. Then the formulas*

$$\langle v, d_0(r) \rangle = q(v, r) \tag{3.4}$$

and

$$\langle i, d_1(b) \rangle = \langle p(\hat{\imath}_1 \otimes \hat{\imath}_2), b \rangle - q(\hat{\imath}_1, \langle \hat{\imath}_2, b \rangle) \tag{3.5}$$

for all $r \in R$, $v \in V$, $i \in I$, and $b \in B^1$, where the maps q and p are given by (3.1–3.2), specify well-defined abelian group homomorphisms $d_0 \colon B^0 \longrightarrow B^1$ and $d_1 \colon B^1 \longrightarrow B^2$. Furthermore, the map $h \colon \widehat{I} \longrightarrow R$ descends uniquely to a well-defined left R-linear map

$I \longrightarrow R$, *providing an element* $h \in \text{Hom}_R(I, R) = B^2$. *The maps* d_0 *and* d_1 *satisfy the equations*

(a) $d_0(rs) = d_0(r)s + r d_0(s)$ *for all* $r, s \in R$.

(b) $d_1(rb) = d_0(r)b + r d_1(b)$ *for all* $r \in R$, $b \in B^1$.

(c) $d_1(br) = d_1(b)r - b d_0(r)$ *for all* $r \in R$, $b \in B^1$.

(d) $d_1(d_0(r)) = hr - rh$ *for all* $r \in R$.

(e) $\sum_\alpha d_1(e_{1,\alpha}) e_{2,\alpha} - \sum_\alpha e_{1,\alpha} d_1(e_{2,\alpha}) = 0$ *in* B^3 *for all tensors* $e = \sum_\alpha e_{1,\alpha} \otimes_R e_{2,\alpha} \in B^1 \otimes_R B^1$ *such that the image of* e *vanishes in* B^2.

The formula

$$d_2(e_1 e_2) = d_1(e_1) e_2 - e_1 d_1(e_2) \quad \text{for all } e_1, e_2 \in B^1$$

specifies a well-defined abelian group homomorphism $d_2 \colon B^2 \longrightarrow B^3$, *which satisfies the equations*

(f) $d_2(d_1(b)) = hb - bh$ *for all* $b \in B^1$.

(g) $d_2(h) = 0$.

In other words, the maps d_0 *and* d_1 *admit a unique extension to an odd derivation* $d \colon B \longrightarrow B$ *of degree* 1, *and the triple* (B, d, h) *is a CDG-ring.*

Example 3.17 Let $\tilde{A}$ be an associative ring and $R \subset \tilde{A}$ be a subring. Assume that the R-R-bimodule $\tilde{A}/R$ is finitely generated and projective as a left R-module. Then the ring $\tilde{A}$ with the subring R and the R-R-bimodule of generators $\tilde{V} = \tilde{A}$ is 3-left finitely projective nonhomogeneous quadratic (cf. Examples 3.3(1) and 3.12(1)). In fact, the filtered ring $\tilde{A}$ is left finitely projective nonhomogeneous Koszul over R in the sense of Sect. 4.6 below.

Assume further that there is a left ideal $\tilde{A}^+ \subset \tilde{A}$ such that $\tilde{A} = R \oplus \tilde{A}^+$, and choose $V = A/R \simeq \tilde{A}^+ = V' \hookrightarrow \tilde{V}$ as our left R-linear splitting of the surjective R-R-bimodule map $\tilde{V} = \tilde{A} \longrightarrow A/R = V$ (cf. Sect. 3.8 below). Then the CDG-ring produced by the construction of the proposition is a DG-ring; this DG-ring is the cobar complex (2.9) for the $\tilde{A}$-modules $L = R = M$, with the multiplicative structure described at the end of Sect. 2.2. Here the isomorphism $R \simeq \tilde{A}/\tilde{A}^+$ endows the ring R with a left $\tilde{A}$-module structure (as $\tilde{A}^+ \subset \tilde{A}$ is a left ideal).

Proof of Proposition 3.16 Recall that, by the definition of quadratic duality, we have $B^0 = R$, $B^1 = \text{Hom}_R(V, R)$, and $B^2 = \text{Hom}_R(I, R)$ (where I is the kernel of the surjective multiplication map $A_1 \otimes_R A_1 \longrightarrow A_2 \simeq (qA)_2$; the latter isomorphism holds since the graded ring A is generated by A_1 over $R = A_0$). The grading on the ring B

was denoted by lower indices in Chaps. 1–2, but we denote it by upper indices here; the convention is $B^n = B_n$ for all $n \geq 0$ (and $B^n = 0 = B_n$ for $n < 0$).

Firstly we have to check that the maps d_0 and d_1 are well-defined by the formulas (3.4–3.5). Concerning d_0, it needs to be checked that $v \longmapsto \langle v, d_0(r) \rangle$ is a left R-linear map $V \longrightarrow R$ for every $r \in R$. Indeed, we have

$$\langle sv, d_0(r) \rangle = q(sv, r) = sq(v, r) = s\langle v, d_0(r) \rangle$$

for all $r, s \in R$ and $v \in V$ by Proposition 3.14(a).

Concerning d_1, it needs to be checked that, for every element $b \in B^1$, the map $\widehat{I} \longrightarrow R$ defined by the formula $\hat{\imath} \longmapsto \langle p(\hat{\imath}_1 \otimes \hat{\imath}_2), b \rangle - q(\hat{\imath}_1, \langle \hat{\imath}_2, b \rangle)$ descends to a left R-linear map $I \longrightarrow R$. Indeed, for all $u, v \in V$ and $r \in R$ we have

$$\langle p(u \otimes rv - ur \otimes v), b \rangle - q(u, \langle rv, b \rangle) + q(ur, \langle v, b \rangle)$$

$$= \langle q(u, r)v, b \rangle - q(u, \langle rv, b \rangle) + q(ur, \langle v, b \rangle)$$

$$= q(u, r)\langle v, b \rangle - q(u, r\langle v, b \rangle) + q(ur, \langle v, b \rangle) = 0$$

by Proposition 3.14(e) and (b), the latter of which is being applied to the elements $u \in V$ and $r, \langle v, b \rangle \in R$. Since $\widehat{I} \longrightarrow I$ is a surjective map with the kernel spanned, as an abelian group, by the elements $u \otimes rv - ur \otimes v$, it follows that our map $\widehat{I} \longrightarrow R$ descends uniquely to a map $d_1(b): I \longrightarrow R$. To prove that the latter map is left R-linear, we compute

$$\langle ri, d_1(b) \rangle = \langle p(r\hat{\imath}_1 \otimes \hat{\imath}_2), b \rangle - q(r\hat{\imath}_1, \langle \hat{\imath}_2, b \rangle)$$

$$= \langle rp(\hat{\imath}_1 \otimes \hat{\imath}_2), b \rangle - rq(\hat{\imath}_1, \langle \hat{\imath}_2, b \rangle)$$

$$= r\langle p(\hat{\imath}_1 \otimes \hat{\imath}_2), b \rangle - rq(\hat{\imath}_1, \langle \hat{\imath}_2, b \rangle) = r\langle i, d_1(b) \rangle$$

using Proposition 3.14(c) and (a).

Similarly, the map $h: \widehat{I} \longrightarrow R$ descends to a well-defined map $I \longrightarrow R$ by Proposition 3.14(f), and the latter map is left R-linear by Proposition 3.14(d).

Now we have to prove the equations (a–g). Part (a): for every element $v \in V$, one has

$$\langle v, d_0(rs) \rangle = q(v, rs) = q(vr, s) + q(v, r)s = \langle vr, d_0(s) \rangle + \langle v, d_0(r) \rangle s$$

$$= \langle v, rd_0(s) \rangle + \langle v, d_0(r)s \rangle = \langle v, rd_0(s) + d_0(r)s \rangle$$

by Proposition 3.14(b).

Part (b): for every element $i \in I$ and its preimage $\hat{i} \in \widehat{I}$, one has

$$\langle i, d_1(rb) \rangle = \langle p(\hat{i}_1 \otimes \hat{i}_2), rb \rangle - q(\hat{i}_1, \langle \hat{i}_2, rb \rangle) = \langle p(\hat{i}_1 \otimes \hat{i}_2)r, b \rangle - q(\hat{i}_1, \langle \hat{i}_2 r, b \rangle)$$

$$= \langle \hat{i}_1 q(\hat{i}_2, r), b \rangle + \langle p(\hat{i}_1 \otimes \hat{i}_2 r), b \rangle - q(\hat{i}_1, \langle \hat{i}_2 r, b \rangle)$$

$$= \langle \hat{i}_1 \langle \hat{i}_2, d_0(r) \rangle, b \rangle + \langle \hat{i}r, d_1(b) \rangle = \langle i, d_0(r)b \rangle + \langle i, rd_1(b) \rangle$$

by Proposition 3.14(g).

Part (c): for every element $i \in I$ and its preimage $\hat{i} \in \widehat{I}$, one has

$$\langle i, d_1(br) \rangle = \langle p(\hat{i}_1 \otimes \hat{i}_2), br \rangle - q(\hat{i}_1, \langle \hat{i}_2, br \rangle) = \langle p(\hat{i}_1 \otimes \hat{i}_2), b \rangle r - q(\hat{i}_1, \langle \hat{i}_2, b \rangle r)$$

$$= \langle p(\hat{i}_1 \otimes \hat{i}_2), b \rangle r - q(\hat{i}_1, \langle \hat{i}_2, b \rangle)r - q(\hat{i}_1 \langle \hat{i}_2, b \rangle, r)$$

$$= \langle i, d_1(b) \rangle r - \langle \hat{i}_1 \langle \hat{i}_2, b \rangle, d_0(r) \rangle = \langle i, d_1(b)r \rangle - \langle i, bd_0(r) \rangle$$

by Proposition 3.14(b) applied to the elements $\hat{i}_1 \in V$ and $\langle \hat{i}_2, b \rangle, r \in R$.

Part (d): for every element $i \in I$ and its preimage $\hat{i} \in \widehat{I}$, one has

$$\langle i, d_1(d_0(r)) \rangle = \langle p(\hat{i}_1 \otimes \hat{i}_2), d_0(r) \rangle - q(\hat{i}_1, \langle \hat{i}_2, d_0(r) \rangle)$$

$$= q(p(\hat{i}_1 \otimes \hat{i}_2), r) - q(\hat{i}_1, q(\hat{i}_2, r)) = h(\hat{i}_1 \otimes \hat{i}_2)r - h(\hat{i}_1 \otimes \hat{i}_2 r)$$

$$= \langle i, h \rangle r - \langle ir, h \rangle = \langle i, hr \rangle - \langle i, rh \rangle$$

by Proposition 3.14(h).

In order to prove parts (e–g), we recall the natural isomorphism $B^3 \simeq \operatorname{Hom}_R(I^{(3)}, R)$, which holds for a 3-left finitely projective quadratic graded ring qA and its quadratic dual 3-right finitely projective quadratic graded ring B according to the proof of Proposition 1.8. In view of this isomorphism, in order to verify an equation in the group B^3, it suffices to evaluate it on every element $j \in I^{(3)}$ and check that the resulting equation in R holds.

Furthermore, for any tensor $e = \sum_\alpha e_{1,\alpha} \otimes e_{2,\alpha} = e_1 \otimes e_2 \in B^1 \otimes_R B^1$, and for every element $j \in I^{(3)}$ and its preimage $\hat{j} \in \widehat{I}^{(3)}$, we compute

$$\langle j, d_1(e_1)e_2 - e_1 d_1(e_2) \rangle = \langle j_1 \langle j_2 \otimes j_3, d_1(e_1) \rangle, e_2 \rangle - \langle j_1 \otimes j_2 \langle j_3, e_1 \rangle, d_1(e_2) \rangle$$

$$= \langle \hat{j}_1 \langle p(\hat{j}_2 \otimes \hat{j}_3), e_1 \rangle, e_2 \rangle - \langle \hat{j}_1 q(\hat{j}_2, \langle \hat{j}_3, e_1 \rangle), e_2 \rangle$$

$$- \langle p(\hat{j}_1 \otimes \hat{j}_2 \langle \hat{j}_3, e_1 \rangle), e_2 \rangle + q(\hat{j}_1, \langle \hat{j}_2 \langle \hat{j}_3, e_1 \rangle, e_2 \rangle)$$

$$= \langle \hat{j}_1 \langle p(\hat{j}_2 \otimes \hat{j}_3), e_1 \rangle, e_2 \rangle - \langle p(\hat{j}_1 \otimes \hat{j}_2) \langle \hat{j}_3, e_1 \rangle, e_2 \rangle + q(\hat{j}_1, \langle \hat{j}_2 \langle \hat{j}_3, e_1 \rangle, e_2 \rangle)$$

$$= \langle \hat{j}_1 \otimes p(\hat{j}_2 \otimes \hat{j}_3) - p(\hat{j}_1 \otimes \hat{j}_2) \otimes \hat{j}_3, e_1 \otimes e_2 \rangle + q(\hat{j}_1, \langle \hat{j}_2 \otimes \hat{j}_3, e_1 \otimes e_2 \rangle) \quad (3.6)$$

by Proposition 3.14(g) applied to the tensor $\hat{i}_1 \otimes \hat{i}_2 = \hat{j}_1 \otimes \hat{j}_2$ and the element $r = \langle \hat{j}_3, e_1 \rangle$.

Now, tensors $e \in B^1 \otimes_R B^1$ whose image vanishes in B^2 form the R-R-bimodule $I_B \subset B^1 \otimes_R B^1$ of quadratic relations in the quadratic graded ring B. By Proposition 1.6, we have $I_B = \mathrm{Hom}_R(A_2, R) = \mathrm{Hom}_R((V \otimes_R V)/I, R)$. To prove part (e), it remains to observe that, in view of Proposition 3.14(i), both the summands in the final expression in (3.6) involve the pairing of an element of $I \subset V \otimes_R V$ with the element e. Therefore, both the summands vanish for $e \in I_B \subset B^1 \otimes_R B^1$.

Part (f):   choose an element $e_1 \otimes e_2 = \sum_\alpha e_{1,\alpha} \otimes e_{2,\alpha} \in B^1 \otimes_R B^1$ whose image under the multiplication map $B^1 \otimes_R B^1 \longrightarrow B^2$ is equal to $d_1(b)$. Then, for every element $j \in I^{(3)}$ and its preimage $\hat{j} \in \widehat{I}^{(3)}$, we have

$$\langle j, d_2(d_1(b)) \rangle = \langle j, d_1(e_1)e_2 - e_1 d_1(e_2) \rangle$$
$$= \langle \hat{j}_1 \otimes p(\hat{j}_2 \otimes \hat{j}_3) - p(\hat{j}_1 \otimes \hat{j}_2) \otimes \hat{j}_3, \, e_1 \otimes e_2 \rangle + q(\hat{j}_1, \langle \hat{j}_2 \otimes \hat{j}_3, e_1 \otimes e_2 \rangle)$$
$$= \langle \hat{j}_1 \otimes p(\hat{j}_2 \otimes \hat{j}_3) - p(\hat{j}_1 \otimes \hat{j}_2) \otimes \hat{j}_3, \, d_1(b) \rangle + q(\hat{j}_1, \langle \hat{j}_2 \otimes \hat{j}_3, d_1(b) \rangle)$$
$$= \langle p(\hat{j}_1 \otimes p(\hat{j}_2 \otimes \hat{j}_3) - p(\hat{j}_1 \otimes \hat{j}_2) \otimes \hat{j}_3), \, b \rangle$$
$$- q(\hat{j}_1, \langle p(\hat{j}_2 \otimes \hat{j}_3), b \rangle) + q(p(\hat{j}_1 \otimes \hat{j}_2), \langle \hat{j}_3, b \rangle)$$
$$+ q(\hat{j}_1, \langle p(\hat{j}_2 \otimes \hat{j}_3), b \rangle) - q(\hat{j}_1, q(\hat{j}_2, \langle \hat{j}_3, b \rangle))$$
$$= \langle \hat{j}_1 h(\hat{j}_2 \otimes \hat{j}_3), b \rangle - \langle h(\hat{j}_1 \otimes \hat{j}_2)\hat{j}_3, b \rangle + q(p(\hat{j}_1 \otimes \hat{j}_2), \langle \hat{j}_3, b \rangle) - q(\hat{j}_1, q(\hat{j}_2, \langle \hat{j}_3, b \rangle))$$
$$= \langle \hat{j}_1 h(\hat{j}_2 \otimes \hat{j}_3), b \rangle - h(\hat{j}_1 \otimes \hat{j}_2)\langle \hat{j}_3, b \rangle + q(p(\hat{j}_1 \otimes \hat{j}_2), \langle \hat{j}_3, b \rangle) - q(\hat{j}_1, q(\hat{j}_2, \langle \hat{j}_3, b \rangle))$$
$$= \langle \hat{j}_1 h(\hat{j}_2 \otimes \hat{j}_3), b \rangle - h(\hat{j}_1 \otimes \hat{j}_2 \langle \hat{j}_3, b \rangle)$$
$$= \langle j_1 \langle \hat{j}_2 \otimes j_3, h \rangle, b \rangle - \langle \hat{j}_1 \otimes j_2 \langle j_3, b \rangle, h \rangle = \langle j, hb - bh \rangle$$

by (3.6), (3.5), and Proposition 3.14(j) and (h), the latter of which is being applied to the tensor $\hat{i}_1 \otimes \hat{i}_2 = \hat{j}_1 \otimes \hat{j}_2$ and the element $r = \langle \hat{j}_3, b \rangle$.

Part (g):   choose an element $h_1 \otimes h_2 = \sum_\alpha h_{1,\alpha} \otimes h_{2,\alpha} \in B^1 \otimes_R B^1$ whose image under the multiplication map $B^1 \otimes_R B^1 \longrightarrow B^2$ is equal to h. Then, for every element $j \in I^{(3)}$ and its preimage $\hat{j} \in \widehat{I}^{(3)}$, we have

$$\langle j, d_2(h) \rangle = \langle j, d_1(h_1)h_2 - h_1 d_1(h_2) \rangle$$
$$= \langle \hat{j}_1 \otimes p(\hat{j}_2 \otimes \hat{j}_3) - p(\hat{j}_1 \otimes \hat{j}_2) \otimes \hat{j}_3, \, h_1 \otimes h_2 \rangle + q(\hat{j}_1, \langle \hat{j}_2 \otimes \hat{j}_3, h_1 \otimes h_2 \rangle)$$
$$= h(\hat{j}_1 \otimes p(\hat{j}_2 \otimes \hat{j}_3) - p(\hat{j}_1 \otimes \hat{j}_2) \otimes \hat{j}_3) + q(\hat{j}_1, h(\hat{j}_2 \otimes \hat{j}_3)) = 0$$

by (3.6) and Proposition 3.14(k).

Finally, for any quadratic graded ring B, any pair of maps $d_0 \colon B^0 \longrightarrow B^1$ and $d_1 \colon B^1 \longrightarrow B^2$ satisfying (a–c) and (e) can be extended to an odd derivation $d \colon B \longrightarrow B$ of degree 1 in a unique way. This assertion is provable, e.g., using the next lemma (where one takes $W = B^0 \oplus B^1$). The equations (d) and (f) imply that $d(d(x)) = [h, x]$ for all $x \in B$, since B is generated by B^1 over B^0.   □

Lemma 3.18

(a) *Let* $W = W_{\bar{0}} \oplus W_{\bar{1}}$ *be a* $\mathbb{Z}/2\mathbb{Z}$-*graded abelian group (where* $\mathbb{Z}/2\mathbb{Z} = \{\bar{0}, \bar{1}\}$ *is the group of order 2), and let* $T_{\mathbb{Z}}(W) = \bigoplus_{n=0}^{\infty} W^{\otimes_{\mathbb{Z}} n}$ *denote the free associative ring spanned by* W. *Endow* $T_{\mathbb{Z}}(W)$ *with the* $\mathbb{Z}/2\mathbb{Z}$-*grading induced by that of* W. *Let* $d_W \colon W \longrightarrow T_{\mathbb{Z}}(W)$ *be an arbitrary odd homomorphism of* $\mathbb{Z}/2\mathbb{Z}$-*graded abelian groups. Then there exists a unique odd derivation* $d_T \colon T_{\mathbb{Z}}(W) \longrightarrow T_{\mathbb{Z}}(W)$ *of the free ring* $T_{\mathbb{Z}}(W)$ *extending the map* d_W *from* $W \subset T_{\mathbb{Z}}(W)$.

(b) *Let* $L \subset T_{\mathbb{Z}}(W)$ *be a* $\mathbb{Z}/2\mathbb{Z}$-*homogeneous subgroup and* $K \subset T_{\mathbb{Z}}(W)$ *be the two-sided* $\mathbb{Z}/2\mathbb{Z}$-*homogeneous ideal generated by* L *in* $T_{\mathbb{Z}}(W)$, *so* $K = (L)$. *Suppose that* $d_T(L) \subset K$. *Then* $d_T(K) \subset K$, *and* d_T *descends to a well-defined odd derivation* d *of the quotient ring* $T_{\mathbb{Z}}(W)/K$.

Proof Part (a): put $d_T(w_1 \otimes \cdots \otimes w_n) = d_W(w_1) \otimes w_2 \otimes \cdots \otimes w_n + (-1)^{|w_1|} w_1 \otimes d_W$ $(w_2) \otimes w_3 \otimes \cdots \otimes w_n + \cdots + (-1)^{|w_1| + \cdots + |w_{n-1}|} w_1 \otimes w_2 \otimes \cdots \otimes w_{n-1} \otimes d_W(w_n)$ for all $w_i \in W_{|w_i|}$, $1 \le i \le n$, $n \ge 0$. Part (b) is obvious.   □

3.5   Change of Strict Generators

Let $\widetilde{A}$ be a weak nonhomogeneous quadratic ring over a subring $R \subset \widetilde{A}$ with the R-R-subbimodule of generators $\widetilde{V} \subset \widetilde{A}$. Assume that the R-R-bimodule $V = \widetilde{V}/R$ is projective as a left R-module and that the R-R-bimodule $F_2\widetilde{A}/F_1\widetilde{A}$ is flat as a left R-module.

Furthermore, assume that we are given two left R-linear splittings $V \simeq V'' \subset \widetilde{V}$ and $V \simeq V' \subset \widetilde{V}$ of the surjective R-R-bimodule morphism $\widetilde{V} \longrightarrow V$. Given an element $v \in V$, we will denote by $v'' \in V''$ and $v' \in V'$ its images under the two splittings. Then $v \longmapsto v'' - v'$ is a left R-linear map $V \longrightarrow R$, which we will denote by a. Conversely, given a left R-linear splitting $V \simeq V' \subset \widetilde{V}$ and a left R-linear map $a \colon V \longrightarrow R$, one can construct a second splitting $V \simeq V'' \subset \widetilde{V}$ by the rule

$$V'' = \{ v' + a(v) \mid v \in V \} \subset \widetilde{V}. \tag{3.7}$$

Denote the maps q, p, and h defined by the formulas (3.1–3.2) using the splitting $V' \subset \widetilde{V}$ by

$$q' : V \times R \longrightarrow R, \quad p' : \widehat{I} \longrightarrow V, \text{ and } h' : \widehat{I} \longrightarrow R$$

and the similar maps constructed using the splitting $V'' \subset \widetilde{V}$ by

$$q'' : V \times R \longrightarrow R, \quad p'' : \widehat{I} \longrightarrow V, \text{ and } h'' : \widehat{I} \longrightarrow R.$$

Proposition 3.19 *The maps q'', p'', and h'' can be obtained from the maps q', p', and h' and the map $a : V \longrightarrow R$ by the formulas*

(a) $q''(v, r) = q'(v, r) + a(v)r - a(vr)$ *for all $r \in R$ and $v \in V$.*
(b) $p''(\hat{i}_1 \otimes \hat{i}_2) = p'(\hat{i}_1 \otimes \hat{i}_2) + a(\hat{i}_1)\hat{i}_2 + \hat{i}_1 a(\hat{i}_2)$ *for all tensors $\hat{i} = \hat{i}_1 \otimes \hat{i}_2 \in \widehat{I}$.*
(c) $h''(\hat{i}_1 \otimes \hat{i}_2) = h'(\hat{i}_1 \otimes \hat{i}_2) + a(p'(\hat{i}_1 \otimes \hat{i}_2)) - q'(\hat{i}_1, a(\hat{i}_2)) + a(\hat{i}_1 a(\hat{i}_2))$ *for all $\hat{i} \in \widehat{I}$.*

Proof In our new notation, the formulas (3.1–3.2) take the form

$$r * v' = (rv)' \quad \text{and} \quad r * v'' = (rv)'',$$
$$v' * r = (vr)' + q'(v, r) \quad \text{and} \quad v'' * r = (vr)'' + q''(v, r),$$

and

$$\hat{i}'_1 * \hat{i}'_2 = p'(\hat{i}_1 \otimes \hat{i}_2)' - h'(\hat{i}_1 \otimes \hat{i}_2) \quad \text{and} \quad \hat{i}''_1 * \hat{i}''_2 = p''(\hat{i}_1 \otimes \hat{i}_2)'' - h''(\hat{i}_1 \otimes \hat{i}_2)$$

for all $r \in R$, $v \in V$, and $\hat{i} \in \widehat{I}$. Here the elements rv, $vr \in V$ correspond to the elements $(rv)'$, $(vr)' \in V'$ and $(rv)''$, $(vr)'' \in V''$. Similarly, $p'(\hat{i}_1 \otimes \hat{i}_2)$, $p''(\hat{i}_1 \otimes \hat{i}_2) \in V$, while $p'(\hat{i}_1 \otimes \hat{i}_2)' \in V'$ and $p''(\hat{i}_1 \otimes \hat{i}_2)'' \in V''$.

Part (a): one has, on the one hand,

$$v' * r = (vr)' + q'(v, r) = (vr)'' - a(vr) + q'(v, r)$$

and, on the other hand,

$$v' * r = (v'' - a(v)) * r = (vr)'' + q''(v, r) - a(v)r,$$

since $v' = v'' - a(v)$ for all $v \in V$.

Parts (b–c): we have, on the one hand,

$$\hat{i}'_1 * \hat{i}'_2 = p'(\hat{i}_1 \otimes \hat{i}_2)' - h'(\hat{i}_1 \otimes \hat{i}_2) = p'(\hat{i}_1 \otimes \hat{i}_2)'' - a(p'(\hat{i}_1 \otimes \hat{i}_2)) - h'(\hat{i}_1 \otimes \hat{i}_2) \qquad (3.8)$$

and, on the other hand,

$$\hat{i}'_1 * \hat{i}'_2 = (\hat{i}''_1 - a(\hat{i}_1)) * (\hat{i}''_2 - a(\hat{i}_2))$$

$$= \hat{i}''_1 * \hat{i}''_2 - a(\hat{i}_1) * \hat{i}''_2 - \hat{i}''_1 * a(\hat{i}_2) + a(\hat{i}_1) * a(\hat{i}_2)$$

$$= p''(\hat{i}_1 \otimes \hat{i}_2)'' - h''(\hat{i}_1 \otimes \hat{i}_2) - (a(\hat{i}_1)\hat{i}_2)'' - (\hat{i}_1 a(\hat{i}_2))'' - q''(\hat{i}_1, a(\hat{i}_2)) + a(\hat{i}_1)a(\hat{i}_2).$$
(3.9)

Comparing (3.8) with (3.9) and equating separately the terms belonging to $V'' \subset \tilde{V}$ and to $R \subset \tilde{V}$, we obtain part (b) as well as the equation

$$h''(\hat{i}_1 \otimes \hat{i}_2) = h'(\hat{i}_1 \otimes \hat{i}_2) + a(p'(\hat{i}_1 \otimes \hat{i}_2)) - q''(\hat{i}_1, a(\hat{i}_2)) + a(\hat{i}_1)a(\hat{i}_2).$$

In order to deduce part (c), it remains to take into account the equation

$$-q''(\hat{i}_1, a(\hat{i}_2)) + a(\hat{i}_1)a(\hat{i}_2) = -q'(\hat{i}_1, a(\hat{i}_2)) + a(\hat{i}_1 a(\hat{i}_2))$$

obtained by substituting $v = \hat{i}_1$ and $r = a(\hat{i}_2)$ into part (a).   □

Proposition 3.20 *Let $\tilde{A}$ be a 3-left finitely projective weak nonhomogeneous quadratic ring over its subring $R \subset \tilde{A}$ with the R-R-bimodule of generators $R \subset \tilde{V} \subset \tilde{A}$. Let $V \simeq V' \hookrightarrow \tilde{V}$ and $V \simeq V'' \hookrightarrow \tilde{V}$ be two left R-linear splittings of the surjective R-R-bimodule map $\tilde{V} \longrightarrow \tilde{V}/R = V$. Denote by $'B = (B, d', h')$ and $''B = (B, d'', h'')$ the two related CDG-ring structures on the 3-right finitely projective graded ring B quadratic dual to qA, as constructed in Proposition 3.16. Let $a \in \operatorname{Hom}_R(V, R) = B^1$ be the element for which the two splittings $V' \subset \tilde{V}$ and $V'' \subset \tilde{V}$ are related by the rule $V' = \{v' \mid v \in V\}$ and $V'' = \{v'' \mid v \in V\}$ with $v', v'' \longmapsto v$ under the map $\tilde{V} \longrightarrow V$ and $v'' = v' + a(v)$ (3.7). Then the equations*

(a) $d''_0(r) = d'_0(r) + ar - ra$ *for all* $r \in R$,
(b) $d''_1(b) = d'_1(b) + ab + ba$ *for all* $b \in B^1$, *and*
(c) $h'' = h' + d'(a) + a^2$

hold in B, showing that $(\operatorname{id}, a)\colon\ ''B \longrightarrow {}'B$ *is a CDG-ring isomorphism.*

Proof In our new notation, the formulas (3.4–3.5) take the form

$$\langle v, d'_0(r) \rangle = q'(v, r) \quad \text{and} \quad \langle v, d''_0(r) \rangle = q''(v, r),$$

$$\langle i, d'_1(b) \rangle = \langle p'(\hat{i}_1 \otimes \hat{i}_2), b \rangle - q'(\hat{i}_1, \langle \hat{i}_2, b \rangle),$$

$$\langle i, d''_1(b) \rangle = \langle p''(\hat{i}_1 \otimes \hat{i}_2), b \rangle - q''(\hat{i}_1, \langle \hat{i}_2, b \rangle).$$

Part (a):   for every element $v \in V$, one has

$$\langle v, d_0''(r) \rangle = q''(v, r) = q'(v, r) + a(v)r - a(vr)$$

$$= \langle v, d_1'(r) \rangle + \langle v, a \rangle r - \langle vr, a \rangle = \langle v, d_1'(r) \rangle + \langle v, ar \rangle - \langle v, ra \rangle$$

by Proposition 3.19(a).

Part (b):   for every element $i \in I$ and its preimage $\hat{\imath} \in \widehat{I}$, one has

$$\langle i, d_1''(b) \rangle = \langle p''(\hat{\imath}_1 \otimes \hat{\imath}_2), b \rangle - q''(\hat{\imath}_1, \langle \hat{\imath}_2, b \rangle)$$

$$= \langle p'(\hat{\imath}_1 \otimes \hat{\imath}_2), b \rangle + \langle a(\hat{\imath}_1)\hat{\imath}_2, b \rangle + \langle \hat{\imath}_1 a(\hat{\imath}_2), b \rangle$$

$$- q'(\hat{\imath}_1, \langle \hat{\imath}_2, b \rangle) - a(\hat{\imath}_1)\langle \hat{\imath}_2, b \rangle + a(\hat{\imath}_1 \langle \hat{\imath}_2, b \rangle)$$

$$= \langle i, d_1'(b) \rangle + \langle \hat{\imath}_1 \langle \hat{\imath}_2, a \rangle, b \rangle + \langle \hat{\imath}_1 \langle \hat{\imath}_2, b \rangle, a \rangle = \langle i, d_1'(b) \rangle + \langle i, ab \rangle + \langle i, ba \rangle$$

by Proposition 3.19(b) and (a), the latter of which is being applied to the elements $v = \hat{\imath}_1$ and $r = \langle \hat{\imath}_2, b \rangle$.

Part (c):   for every element $i \in I$ and its preimage $\hat{\imath} \in \widehat{I}$, one has

$$\langle i, h'' \rangle = \langle i, h' \rangle + \langle p'(\hat{\imath}_1 \otimes \hat{\imath}_2), a \rangle - q'(\hat{\imath}_1, \langle \hat{\imath}_2, a \rangle) + \langle \hat{\imath}_1 \langle \hat{\imath}_2, a \rangle, a \rangle$$

$$= \langle i, h' \rangle + \langle i, d_1'(a) \rangle + \langle i, a^2 \rangle$$

by Proposition 3.19(c).

Finally, the equations (a) and (b) imply that $d''(x) = d'(x) + [a, x]$ for all $x \in B$, since B is generated by B^1 over $B^0 = R$.   □

3.6   The Nonhomogeneous Quadratic Duality Functor

Let R be an associative ring. We denote by $R\text{-rings}_{\mathsf{fil}}$ the category of filtered rings $(\widetilde{A}, F)$ with increasing filtrations F and the filtration component $F_0 \widetilde{A}$ identified with R.

So the objects of $R\text{-rings}_{\mathsf{fil}}$ are associative rings $\widetilde{A}$ endowed with a filtration $0 = F_{-1}\widetilde{A} \subset F_0\widetilde{A} \subset F_1\widetilde{A} \subset F_2\widetilde{A} \subset \cdots$ such that $\widetilde{A} = \bigcup_{n=0}^{\infty} F_n\widetilde{A}$, the filtration F is compatible with the multiplication in $\widetilde{A}$, and an associative ring isomorphism $R \simeq F_0\widetilde{A}$ has been chosen. Morphisms $('\widetilde{A}, F) \longrightarrow (''\widetilde{A}, F)$ in $R\text{-rings}_{\mathsf{fil}}$ are ring homomorphisms $f : '\widetilde{A} \longrightarrow ''\widetilde{A}$ such that $f(F_n\,'\widetilde{A}) \subset F_n\,''\widetilde{A}$ for all $n \geq 0$ and the ring homomorphism $F_0 f : F_0\,'\widetilde{A} \longrightarrow F_0\,''\widetilde{A}$ forms a commutative triangle diagram with the fixed isomorphisms $R \simeq F_0\,'\widetilde{A}$ and $R \simeq F_0\,''\widetilde{A}$.

The *category of 3-left finitely projective weak nonhomogeneous quadratic rings over* R, denoted by $R\text{-rings}_{\mathsf{wnlq}}$, is defined as the full subcategory in $R\text{-rings}_{\mathsf{fil}}$ whose objects are the 3-left finitely projective weak nonhomogeneous quadratic rings $R \subset \widetilde{V} \subset \widetilde{A}$ endowed with the filtration F generated by $F_1\widetilde{A} = \widetilde{V}$ over $F_0\widetilde{A} = R$. In other words, this means that a morphism $(\widetilde{A}, '\widetilde{V}) \longrightarrow ("\widetilde{A}, "\widetilde{V})$ in $R\text{–rings}_\text{wnlq}$ is a ring homomorphism $f \colon '\widetilde{A} \longrightarrow "\widetilde{A}$ forming a commutative triangle diagram with the inclusions $R \simeq F_0'\widetilde{A} \hookrightarrow '\widetilde{A}$ and $R \simeq F_0"\widetilde{A} \hookrightarrow "\widetilde{A}$ and satisfying the condition that $f('\widetilde{V}) \subset "\widetilde{V}$ (cf. the discussion in Sect. 3.1).

Furthermore, the *category of 3-right finitely projective quadratic CDG-rings over R*, denoted by $R\text{–rings}_\text{cdg,rq}$, is the full subcategory in the category $R\text{–rings}_\text{cdg}$ (as defined in Sect. 3.2) consisting of all the CDG-rings (B, d, h) whose underlying nonnegatively graded ring B is 3-right finitely projective quadratic over R.

Theorem 3.21 *The constructions of Propositions 3.16 and 3.20 define a fully faithful contravariant functor*

$$(R\text{–rings}_\text{wnlq})^\text{op} \longrightarrow R\text{–rings}_\text{cdg,rq} \tag{3.10}$$

from the category of 3-left finitely projective weak nonhomogeneous quadratic rings to the category of 3-right finitely projective quadratic CDG-rings over R.

Proof Let $R\text{–rings}_\text{wnlq}^\text{sg}$ be the category whose objects are 3-left finitely projective weak nonhomogeneous quadratic rings $R \subset \widetilde{V} \subset \widetilde{A}$ with a chosen submodule of strict generators $V' \subset \widetilde{V}$. So V' is a left R-submodule in $\widetilde{V}$ such that $\widetilde{V} = R \oplus V'$ as a left R-module. Morphisms $(\widetilde{A}, '\widetilde{V}, 'V) \longrightarrow ("\widetilde{A}, "\widetilde{V}, "V)$ in $R\text{–rings}_\text{wnlq}^\text{sg}$ are the same as morphisms $(\widetilde{A}, '\widetilde{V}) \longrightarrow ("\widetilde{A}, "\widetilde{V})$ in $R\text{–rings}_\text{wnlq}^\text{sg}$, so a morphism in $R\text{–rings}_\text{wnlq}^\text{sg}$ has to take $'\widetilde{V}$ into $"\widetilde{V}$, but it does not need to respect the chosen submodules of strict generators $'V \subset '\widetilde{V}$ and $"V \subset "\widetilde{V}$ in any way. Then the functor $R\text{–rings}_\text{wnlq}^\text{sg} \longrightarrow R\text{–rings}_\text{wnlq}$ forgetting the choice of the submodule of strict generators $V' \subset \widetilde{V}$ is fully faithful and surjective on objects, so it is an equivalence of categories.

Furthermore, let $R\text{–rings}_\text{wnlq}^\text{sgsm}$ be the subcategory in $R\text{–rings}_\text{wnlq}^\text{sg}$ whose objects are all the objects of $R\text{–rings}_\text{wnlq}^\text{sg}$ and whose morphisms are the morphisms $f \colon (\widetilde{A}, '\widetilde{V}, 'V) \longrightarrow ("\widetilde{A}, "\widetilde{V}, "V)$ in $R\text{–rings}_\text{wnlq}^\text{sg}$ such that $f('V) \subset "V$. The category $R\text{–rings}_\text{wnlq}^\text{sg}$ can be called the category of 3-left finitely projective weak nonhomogeneous quadratic rings *with strict generators chosen*, while the category $R\text{–rings}_\text{wnlq}^\text{sgsm}$ is the category of 3-left finitely projective weak nonhomogeneous quadratic rings *with strict generators and strict morphisms*.

Finally, let $R\text{–rings}_\text{cdg}^\text{sm}$ denote the subcategory in $R\text{–rings}_\text{cdg}$ whose objects are all the objects of $R\text{–rings}_\text{cdg}$ and whose morphisms are the strict morphisms only, i.e., all morphisms of the form $(g, 0) \colon ("B, d", h") \longrightarrow ('B, d', h')$ in $R\text{–rings}_\text{cdg}$. We denote by $R\text{–rings}_\text{cdg,rq}^\text{sm}$ the intersection $R\text{–rings}_\text{cdg}^\text{sm} \cap R\text{–rings}_\text{cdg,rq} \subset R\text{–rings}_\text{cdg}$, that is, the category of 3-right finitely projective quadratic CDG-rings over R and strict morphisms between them.

Then the construction of Proposition 3.16 assigns an object $(B, d, h) \in R\text{-rings}_{\text{cdg,rq}}$ to every object $(\widetilde{A}, \widetilde{V}, V') \in R\text{-rings}_{\text{wnlq}}^{\text{sg}}$ in a natural way. Given a morphism $f : (\widetilde{A}, {}'\widetilde{V}, {}'V) \longrightarrow ({}''\widetilde{A}, {}''\widetilde{V}, {}''V)$ in $R\text{-rings}_{\text{wnlq}}^{\text{sg}}$ such that $f({}'V) \subset {}''V$, we consider the induced morphism of 3-left finitely projective quadratic graded rings $\operatorname{q} \operatorname{gr}^F f : \operatorname{q} \operatorname{gr}^F {}'\widetilde{A} \longrightarrow \operatorname{q} \operatorname{gr}^F {}''\widetilde{A}$. According to Propositions 1.6 and 1.8, the morphism $\operatorname{q} \operatorname{gr}^F f$ induces a morphism in the opposite direction between the quadratic dual 3-right finitely projective quadratic graded rings, $g : {}''B \longrightarrow {}'B$.

Let $({}'B, d', h')$ and $({}''B, d'', h'')$ denote the 3-right finitely projective quadratic CDG-rings assigned to the 3-left finitely projective weak nonhomogeneous quadratic rings $({}'\widetilde{A}, {}'\widetilde{V})$ and $({}''\widetilde{A}, {}''\widetilde{V})$ with the submodules of strict generators ${}'V \subset {}'\widetilde{V}$ and ${}''V \subset {}''\widetilde{V}$ by the construction of Proposition 3.16. Assigning the morphism $(g, 0) : ({}''B, d'', h'') \longrightarrow ({}'B, d', h')$ to the morphism $f : ({}'\widetilde{A}, {}'\widetilde{V}, {}'V) \longrightarrow ({}''\widetilde{A}, {}''\widetilde{V}, {}''V)$, we obtain a contravariant functor

$$(R\text{-rings}_{\text{wnlq}}^{\text{sgsm}})^{\text{op}} \longrightarrow R\text{-rings}_{\text{cdg,rq}}^{\text{sm}}. \tag{3.11}$$

We still have to check that the functor (3.11) is well-defined, i.e., that $(g, 0) : ({}''B, d'', h'') \longrightarrow ({}'B, d', h')$ is indeed a morphism of CDG-rings. Simultaneously we will see that the functor (3.11) is fully faithful.

Indeed, specifying a morphism $f : ({}'\widetilde{A}, {}'\widetilde{V}, {}'V) \longrightarrow ({}''\widetilde{A}, {}''\widetilde{V}, {}''V)$ in $R\text{-rings}_{\text{wnlq}}^{\text{sgsm}}$ means specifying an R-R-bimodule map $f_1 : {}'\widetilde{V}/R \longrightarrow {}''\widetilde{V}/R$, which, interpreted as a map ${}'V \longrightarrow {}''V$ and taken together with the identity map $R \longrightarrow R$, extends (necessarily uniquely) to a ring homomorphism ${}'\widetilde{A} \longrightarrow {}''\widetilde{A}$. The latter condition is equivalent to the following two:

(i) The map $\tilde{f}_1 = \operatorname{id}_R \oplus f_1 : {}''\widetilde{V} = R \oplus {}''V \longrightarrow R \oplus {}'V = {}'\widetilde{V}$ agrees with the right R-module structures on ${}''\widetilde{V}$ and ${}'\widetilde{V}$.

(ii) The tensor ring homomorphism $T_R(\tilde{f}_1) : T_R({}'\widetilde{V}) \longrightarrow T_R({}''\widetilde{V})$ induced by the map $\tilde{f}_1 : {}'\widetilde{V} \longrightarrow {}''\widetilde{V}$ takes the R-R-subbimodule $\widetilde{I}_{{}'\widetilde{A}} \subset R \oplus {}'\widetilde{V} \oplus {}'\widetilde{V}^{\otimes_R 2}$ of nonhomogeneous quadratic relations in the ring ${}'\widetilde{A}$ into the R-R-subbimodule $\widetilde{I}_{{}''\widetilde{A}} \subset R \oplus {}''\widetilde{V} \oplus {}''\widetilde{V}^{\otimes_R 2}$ of nonhomogeneous quadratic relations in the ring ${}''\widetilde{A}$ (see Sect. 3.1 for the notation).

Denote the maps defined by the formulas (3.1–3.2) for $({}'\widetilde{A}, {}'\widetilde{V}, {}'V)$ and $({}''\widetilde{A}, {}''\widetilde{V}, {}''V)$ by

$$q' : {}'V \times R \longrightarrow R, \quad p' : {}'\widehat{I} \longrightarrow {}'V, \quad h' : {}'\widehat{I} \longrightarrow R,$$
$$q'' : {}''V \times R \longrightarrow R, \quad p'' : {}''\widehat{I} \longrightarrow {}''V, \quad h'' : {}''\widehat{I} \longrightarrow R,$$

where the notation $'V = '\widetilde{V}/R$ and $''V = ''\widetilde{V}/R$ is presumed, while $'\widehat{I} \subset 'V \otimes_{\mathbb{Z}} 'V$ and $''\widehat{I} \subset ''V \otimes_{\mathbb{Z}} ''V$ are the full preimages of the R-R-subbimodules $'I \subset 'V \otimes_R 'V$ and $''I \subset ''V \otimes_R ''V$ of quadratic relations in the graded rings $'A = \mathrm{gr}^F {}'\widetilde{A}$ and $''A = \mathrm{gr}^F {}''\widetilde{A}$, respectively. Then condition (i) is equivalent to the equation

$$q'(v, r) = q''(f_1(v), r) \quad \text{for all } v \in {}'V, \ r \in R. \tag{3.12}$$

Assuming (i) or (3.12), condition (ii) is equivalent to the combination of the inclusion

$$(f_1 \otimes f_1)('I) \subset {}''I \tag{3.13}$$

with the equations

$$f_1(p'(\hat{i}_1 \otimes \hat{i}_2)) = p''(f_1(\hat{i}_1) \otimes f_1(\hat{i}_2)) \quad \text{for all } \hat{i}_1 \otimes \hat{i}_2 \in {}'\widehat{I}$$
$$h'(i_1' \otimes i_2') = h''(f_1(i_1) \otimes f_1(i_2)) \quad \text{for all } i_1 \otimes i_2 \in {}'I. \tag{3.14}$$

Finally, the inclusion (3.13) holds if and only if the R-R-bimodule morphism $g_1 = \mathrm{Hom}_R(f_1, R)\colon {}''B^1 = \mathrm{Hom}_R(''V, R) \longrightarrow \mathrm{Hom}_R('V, R) = {}'B^1$ together with the identity map $''B^0 = R \longrightarrow R = {}'B^0$ can be extended to a graded ring homomorphism $g\colon {}''B \longrightarrow {}'B$. Assuming (3.13) or (equivalently) the existence of $g = (g_n)_{n=0}^\infty$, Eqs. (3.12) and (3.14) are equivalent to the equations

$$d_0'(r) = g_1(d_0''(r)) \quad \text{for all } r \in R,$$
$$d_1'(g_1(b)) = g_2(d_1''(b)) \quad \text{for all } b \in {}''B^1, \tag{3.15}$$
$$h' = g_2(h''),$$

which mean that $(g, 0)\colon (''B, d'', h'') \longrightarrow ('B, d', h')$ is a CDG-ring morphism.

Conversely, given a strict morphism $(g, 0)\colon (''B, d'', h'') \longrightarrow ('B, d', h')$ between 3-right finitely projective quadratic CDG-rings coming from 3-left finitely projective nonhomogeneous quadratic rings $('\widetilde{A}, '\widetilde{V}, 'V)$ and $(''\widetilde{A}, ''\widetilde{V}, ''V)$, we put $f_1 = \mathrm{Hom}_{R^{\mathrm{op}}}(g_1, R)\colon {}'V = \mathrm{Hom}_{R^{\mathrm{op}}}('B^1, R) \longrightarrow \mathrm{Hom}_{R^{\mathrm{op}}}(''B^1, R) = {}''V$ and extend the map f_1 together with the identity map $'\widetilde{A} \supset R \longrightarrow R \subset {}''\widetilde{A}$ to a ring homomorphism $f\colon {}'\widetilde{A} \longrightarrow {}''\widetilde{A}$. This can be done, since the combination of conditions (i) and (ii) is equivalent to the existence of a (necessarily unique) graded ring homomorphism $g\colon {}''B \longrightarrow {}'B$ extending the given map $g_1\colon {}''B^1 \longrightarrow {}'B^1$ together with the identity map $''B^0 = R \longrightarrow R = {}'B^0$ and satisfying Eq. (3.15).

Now we will construct a fully faithful functor

$$(R\text{--rings}_{\mathrm{wnlq}}^{\mathrm{sg}})^{\mathrm{op}} \longrightarrow R\text{--rings}_{\mathrm{cdg, rq}} \tag{3.16}$$

extending the functor (3.11) to nonstrict morphisms. For a 3-left finitely projective weak nonhomogeneous quadratic ring $(\widetilde{A}, \widetilde{V}) \in R\text{–rings}_{\text{wnlq}}$ and any two choices of a submodule of strict generators V', $V'' \subset \widetilde{V}$, we have two objects $(\widetilde{A}, \widetilde{V}, V')$ and $(\widetilde{A}, \widetilde{V}, V'') \in R\text{–rings}_{\text{wnlq}}^{\text{sg}}$ connected by an isomorphism $(\widetilde{A}, \widetilde{V}, V') \longrightarrow (\widetilde{A}, \widetilde{V}, V'')$ corresponding to the identity map $\text{id} : \widetilde{A} \longrightarrow \widetilde{A}$. Let us call such isomorphisms in $R\text{–rings}_{\text{wnlq}}^{\text{sg}}$ the *change-of-strict-generators isomorphisms*.

Let $f : ('\widetilde{A}, '\widetilde{V}, 'V) \longrightarrow (''\widetilde{A}, ''\widetilde{V}, ''V)$ be an arbitrary morphism in $R\text{–rings}_{\text{wnlq}}^{\text{sg}}$. Denote by $'V' \subset '\widetilde{V}$ the full preimage of the left R-submodule $''V \subset ''\widetilde{V}$ under the R-R-bimodule morphism $F_1 f : '\widetilde{V} \longrightarrow ''\widetilde{V}$. Then one has $'\widetilde{V} = R \oplus 'V'$, so $'V' \subset '\widetilde{V}$ is another choice of a submodule of strict generators in $'\widetilde{A}$, alternative to $'V \subset '\widetilde{V}$. Any morphism $f : (\widetilde{A}, '\widetilde{V}, 'V) \longrightarrow (''\widetilde{A}, ''\widetilde{V}, ''V)$ in $R\text{–rings}_{\text{wnlq}}^{\text{sg}}$ decomposes uniquely into a change-of-strict-generators isomorphism $('\widetilde{A}, '\widetilde{V}, 'V) \longrightarrow ('\widetilde{A}, '\widetilde{V}, 'V')$ followed by a strict morphism $('\widetilde{A}, '\widetilde{V}, 'V') \longrightarrow (''\widetilde{A}, ''\widetilde{V}, ''V)$.

The construction of Proposition 3.20 assigns a change-of-connection isomorphism in $R\text{–rings}_{\text{cdg,rq}}$ to every change-of-strict-generators isomorphism in $R\text{–rings}_{\text{wnlq}}^{\text{sg}}$. Decomposing any morphism in $R\text{–rings}_{\text{wnlq}}^{\text{sg}}$ into a change-of-strict-generators isomorphism followed by a strict morphism, one extends the functor (3.11) to a contravariant functor (3.16). We omit further details, which are straightforward.

We still have to show that the functor (3.16) is fully faithful. It is clear from the construction of Proposition 3.20 that this functor restricts to a fully faithful functor from the subcategory of change-of-strict-generators isomorphisms in $R\text{–rings}_{\text{wnlq}}^{\text{sg}}$ to the subcategory of change-of-connection isomorphisms in $R\text{–rings}_{\text{cdg,rq}}$. Since the functor (3.11) is fully faithful as well, it follows that so is the functor (3.16).

In order to construct the desired fully faithful functor (3.10), it remains to choose any quasi-inverse functor to the category equivalence $R\text{–rings}_{\text{wnlq}}^{\text{sg}} \longrightarrow R\text{–rings}_{\text{wnlq}}$ and compose it with the fully faithful functor (3.16).   □

Example 3.22 Let R be an associative ring and V be an R-R-bimodule which is finitely generated and projective as a left R-module. Let $(\widetilde{A}, F)$ be a nonhomogenous quadratic ring such that the quadratic graded ring $A = \text{gr}^F \widetilde{A}$ is isomorphic to the tensor ring $T_R(V)$, as in Examples 3.3(2) and 3.12(2).

Then A is a 3-left finitely projective quadratic graded ring (so $\widetilde{A}$ is a 3-left finitely projective nonhomogeneous quadratic ring), and the 3-right finitely projective quadratic graded ring quadratic dual to A is $B = R \oplus V^{\vee}$, as per Examples 1.5(1) and 1.7(1). What is the CDG-ring structure (B, d, h) on the graded ring B corresponding to the 3-left finitely projective nonhomogeneous quadratic ring $\widetilde{A}$ under the construction of Proposition 3.16?

We have $B^n = 0$ for $n \geq 0$, and hence $h = 0$ and $d_n = 0$ for $n \geq 1$. The only possibly nontrivial component of the CDG-ring structure on B is thus the differential $d_0 : R \longrightarrow B^1 = V^{\vee}$, which is defined in terms of the map $q : V \times R \longrightarrow R$. The map d_0 must be a derivation of the ring R taking values in the R-R-bimodule B^1, that is, the equation $d_0(rs) = d_0(r)s + r d_0(s)$ has to be satisfied.

According to Proposition 3.20, the filtered ring $\widetilde{A}$ determines the CDG-ring (B, d, h) uniquely up to change-of-connection isomorphisms, which transform a derivation $d_0' : R \longrightarrow B^1$ into a derivation $d_0'' : R \longrightarrow B^1$ given by the formula $d_0''(r) = d_0'(r) + ar - ra$ for all $r \in R$, where $a \in B^1$ is some chosen element.

Theorem 3.21 tells that two nonhomogeneous quadratic rings $'\widetilde{A}$ and $''\widetilde{A}$ with $\mathrm{gr}^F \,'\widetilde{A} = A = \mathrm{gr}^F \,''\widetilde{A}$ are isomorphic (as filtered rings with a fixed subring $F_0 \,'\widetilde{A} = R = F_0 \,''\widetilde{A}$) if and only if the corresponding two derivations d_0' and d_0'' are connected by a change-of-connection isomorphism as above. Furthermore, as we will see in Theorem 4.25 below, *any* derivation $d_0 : R \longrightarrow B^1$ corresponds to a nonhomogeneous quadratic ring $\widetilde{A}$ with $\mathrm{gr}^F \widetilde{A} = A$ (since the graded ring B is right finitely projective Koszul by Examples 2.31(1) and 2.39(1)).

To give a concrete nontrivial example, it is instructive to consider the case when the ring R is commutative and the left and right actions of R in V coincide. In this case, one has $ar = ra$ for all $r \in R$ and $a \in B^1$, so the change-of-connection transformations leave the derivations $d_0 : R \longrightarrow B^1$ unchanged. Thus any nonzero derivation $d : R \longrightarrow B^1 = V^\vee$ defines a nonhomogeneous quadratic ring $\widetilde{A}$ with $\mathrm{gr}^F \widetilde{A} = A$ such that $\widetilde{A}$ is *not* isomorphic to A as a filtered ring. The ring $\widetilde{A}$ can be viewed as a nonhomogeneous deformation of the tensor ring $A = T_R(V)$.

Remark 3.23 The counterexamples in [56, Section 3.4] or [53, Section 2 in Chapter 5] show that the fully faithful contravariant functor in Theorem 3.21 is *not* an anti-equivalence of categories (even when $R = k$ is the ground field). We will see below in Sect. 4.6 that this functor becomes an anti-equivalence when restricted to the full subcategories of, respectively, left and right finitely projective Koszul rings in $R\text{–rings}_{\mathsf{wnlq}}$ and $R\text{–rings}_{\mathsf{cdg,rq}}$.

3.7 Nonhomogeneous Duality 2-Functor

We define the 2-*category of filtered rings* $\mathsf{Rings}_{\mathsf{fil2}}$ as follows. The objects of $\mathsf{Rings}_{\mathsf{fil2}}$ are associative rings $\widetilde{A}$ endowed with an exhaustive increasing filtration $0 = F_{-1}\widetilde{A} \subset F_0\widetilde{A} \subset F_1\widetilde{A} \subset F_2\widetilde{A} \subset \cdots$ compatible with the multiplication on $\widetilde{A}$. Morphisms $('\widetilde{A}, F) \longrightarrow (''\widetilde{A}, F)$ in $\mathsf{Rings}_{\mathsf{fil2}}$ are ring homomorphisms $f : \,'\widetilde{A} \longrightarrow ''\widetilde{A}$ such that $f(F_n \,'\widetilde{A}) \subset F_n \,''\widetilde{A}$ for all $n \geq 0$ and the map $F_0 f : F_0 \,'\widetilde{A} \longrightarrow F_0 \,''\widetilde{A}$ is an isomorphism. 2-morphisms $f \overset{z}{\longrightarrow} g$ between a pair of parallel morphisms $f, g : ('\widetilde{A}, F) \longrightarrow (''\widetilde{A}, F)$ are invertible elements $z \in F_0 \,''\widetilde{A}$ such that $g(c) = zf(c)z^{-1}$ for all $c \in \,'\widetilde{A}$.

The vertical composition of two 2-morphisms $f' \overset{w}{\longrightarrow} f'' \overset{z}{\longrightarrow} f'''$ is the 2-morphism $f' \overset{zw}{\longrightarrow} f'''$. The identity 2-morphism is the 2-morphism $f \overset{1}{\longrightarrow} f$. The horizontal composition of two 2-morphisms $g' \overset{w}{\longrightarrow} g'' : (\widetilde{A}, F) \longrightarrow (\widetilde{B}, F)$ and $f' \overset{z}{\longrightarrow} f'' : (\widetilde{B}, F) \longrightarrow (\widetilde{C}, F)$ is the 2-morphism $f'g' \overset{z \circ w}{\longrightarrow} f''g'' : (\widetilde{A}, F) \longrightarrow (\widetilde{C}, F)$ with the element $z \circ w = zf'(w) = f''(w)z$.

All the 2-morphisms of filtered rings are invertible. If $f: (\,'\widetilde{A}, F) \longrightarrow (\,''\widetilde{A}, F)$ is a morphism of filtered rings and $z \in F_0 \,''\widetilde{A}$ is an invertible element, then $g: c \longmapsto zf(c)z^{-1}$ is also a morphism of filtered rings $g: (\,'\widetilde{A}, F) \longrightarrow (\,''\widetilde{A}, F)$. The morphisms f and g are connected by the 2-isomorphism $f \xrightarrow{z} g$.

The condition about the map $F_0 f: F_0 \,'\widetilde{A} \longrightarrow F_0 \,''\widetilde{A}$ being an isomorphism could be harmlessly dropped from the above definition (which makes sense without this condition just as well), but we need it for the purposes of the next definition. The 2-*category of 3-left finitely projective weak nonhomogeneous quadratic rings*, denoted by $\mathsf{Rings}_{\mathsf{wnlq2}}$, is defined as the following 2-subcategory in $\mathsf{Rings}_{\mathsf{fil2}}$. The objects of $\mathsf{Rings}_{\mathsf{wnlq2}}$ are the 3-left finitely projective weak nonhomogeneous quadratic rings $R \subset \widetilde{V} \subset \widetilde{A}$ with the filtration F generated by $F_1 \widetilde{A} = \widetilde{V}$ over $F_0 \widetilde{A} = R$. All morphisms in $\mathsf{Rings}_{\mathsf{fil2}}$ between objects of $\mathsf{Rings}_{\mathsf{wnlq2}}$ are morphisms in $\mathsf{Rings}_{\mathsf{wnlq2}}$, and all 2-morphisms in $\mathsf{Rings}_{\mathsf{fil2}}$ between morphisms of $\mathsf{Rings}_{\mathsf{wnlq2}}$ are 2-morphisms in $\mathsf{Rings}_{\mathsf{wnlq2}}$.

Furthermore, the 2-*category of 3-right finitely projective quadratic CDG-rings*, denoted by $\mathsf{Rings}_{\mathsf{cdg2,rq}}$, is the following 2-subcategory in the 2-category $\mathsf{Rings}_{\mathsf{cdg2}}$ (as defined in Sect. 3.2). The objects of $\mathsf{Rings}_{\mathsf{cdg2,rq}}$ are all the CDG-rings (B, d, h) whose underlying nonnegatively graded ring B is 3-right finitely projective quadratic (over its degree-zero component B^0). All morphisms in $\mathsf{Rings}_{\mathsf{cdg2}}$ between objects of $\mathsf{Rings}_{\mathsf{cdg2,rq}}$ are morphisms in $\mathsf{Rings}_{\mathsf{cdg2,rq}}$, and all 2-morphisms in $\mathsf{Rings}_{\mathsf{cdg2}}$ between morphisms of $\mathsf{Rings}_{\mathsf{cdg2,rq}}$ are 2-morphisms in $\mathsf{Rings}_{\mathsf{cdg2,rq}}$.

Lemma 3.24 *Let $R \subset \widetilde{V} \subset \widetilde{A}$ be a 3-left finitely projective weak nonhomogeneous quadratic ring, and let $'V \subset \widetilde{V}$ be a submodule of strict generators of $\widetilde{A}$. Let (B, d, h) be the 3-right finitely projective quadratic CDG-ring corresponding to $(\widetilde{A}, \widetilde{V}, 'V)$ under the construction of Proposition 3.16. Let $z \in R$ be an invertible element. Consider the conjugation morphism $f_{z^{-1}}: \widetilde{A} \longrightarrow \widetilde{A}$ taking any element $c \in \widetilde{A}$ to the element $f_{z^{-1}}(c) = z^{-1}cz$. Then the CDG-ring morphism $(B, d, h) \longrightarrow (B, d, h)$ corresponding to the morphism $f_{z^{-1}}$ under the duality functor of Theorem 3.21 is equal to*

$$(g_z, a_z): (B, d, h) \longrightarrow (B, d, h),$$

where $g_z: B \longrightarrow B$ is the conjugation map taking any element $b \in B$ to the element $g_z(b) = zbz^{-1}$, and the element $a_z \in B^1$ is given by the formula $a_z = -d(z)z^{-1}$.

Proof Strictly speaking, the assertion of the lemma does not literally make sense as stated, and we need to make it more precise before proving it. The problem is that $f_{z^{-1}}: \widetilde{A} \longrightarrow \widetilde{A}$ is *not* a morphism in the category $R\text{–rings}_{\mathsf{wnlq}}$ or $R\text{–rings}_{\mathsf{fil}}$, as defined in Sect. 3.6, because it does not restrict to the identity map $\widetilde{A} \supset R \longrightarrow R \subset \widetilde{A}$, but rather to the map $R \longrightarrow R$ of conjugation with z^{-1}. For the same reason, $g_z: B \longrightarrow B$ is not a morphism in $R\text{–rings}_{\mathsf{gr}}$, and consequently $(g_z, a_z): (B, d, h) \longrightarrow (B, d, h)$ is not a morphism in $R\text{–rings}_{\mathsf{cdg}}$ or $R\text{–rings}_{\mathsf{cdg,rq}}$. So some preparatory work is needed before the functor (3.10) could be applied to the map $f_{z^{-1}}$.

Denote by $t_z \colon R \longrightarrow R$ the conjugation map $r \longmapsto zrz^{-1}$, $r \in R$. Let $(\widetilde{A}^{(z)}, \widetilde{V}^{(z)})$ denote the following 3-left finitely projective weak nonhomogeneous quadratic ring over R. As an associative ring, $\widetilde{A}^{(z)}$ coincides with $\widetilde{A}$. Denoting by $\iota \colon R \longrightarrow \widetilde{A}$ the embedding of the ring R as a subring of the ring $\widetilde{A}$, the map $\iota^{(z)} = \iota t_z \colon R \longrightarrow \widetilde{A} = \widetilde{A}^{(z)}$ makes R a subring of $\widetilde{A}^{(z)}$. For any element $c \in \widetilde{A}$, we will denote by $c^{(z)} \in \widetilde{A}^{(z)}$ the corresponding element of the ring $\widetilde{A}^{(z)}$. So in particular, we have $\iota^{(z)}(r) = (z\iota(r)z^{-1})^{(z)}$ for all $r \in R$. Furthermore, as a subgroup of $\widetilde{A}^{(z)} = \widetilde{A}$, the group $\widetilde{V}^{(z)}$ coincides with $\widetilde{V}$. Then the map $f_{z^{-1}} \colon \widetilde{A}^{(z)} \longrightarrow \widetilde{A}$ taking an element $c^{(z)} \in \widetilde{A}^{(z)}$ to the element $z^{-1}cz \in \widetilde{A}$ is a morphism in the category $R\text{–rings}_{\mathsf{wnlq}}$.

Let $(B^{(z)}, d^{(z)}, h^{(z)})$ denote the following 3-right finitely projective quadratic CDG-ring over R. As a graded associative ring, $B^{(z)}$ coincides with B, and both the differential $d^{(z)} \colon B^{(z)} \longrightarrow B^{(z)}$ and the curvature element $h \in B^{(z),2}$ coincide with the differential d and the curvature element h in B. However, denoting by $\iota \colon R \longrightarrow B^0$ the identification of the ring R with the degree-zero component of the graded ring B, the identification of the ring R with the degree-zero component of the graded ring $B^{(z)}$ is provided by the map $\iota^{(z)} = \iota t_z \colon R \longrightarrow B^{(z),0} = B^0$.

For any element $b \in B$, we denote by $b^{(z)} \in B^{(z)}$ the corresponding element of the ring $B^{(z)}$. So, in particular, we have $d^{(z)}(b^{(z)}) = (d(b))^{(z)}$ for all $b \in B$, and our notation for the element $h^{(z)}$ is consistent: $h^{(z)} \in B^{(z),2}$ is the element corresponding to $h \in B^2$. Then the map $g_z \colon B \longrightarrow B^{(z)}$ taking an element $b \in B$ to the element $(zbz^{-1})^{(z)} \in B^{(z)}$ is a morphism in the category $R\text{–rings}_{\mathsf{gr}}$. Furthermore, let $a_z^{(z)} \in B^{(z),1}$ be the element corresponding to the element $a_z = -d(z)z^{-1} \in B^1$ under the identity isomorphism $B = B^{(z)}$. Then the pair $(g_z, a_z^{(z)})$ is a morphism of CDG-rings $(B, d, h) \longrightarrow (B^{(z)}, d^{(z)}, h^{(z)})$, as one can readily check. Therefore, it is also a morphism in the category $R\text{–rings}_{\mathsf{cdg}}$ and in the category $R\text{–rings}_{\mathsf{cdg,rq}}$.

Now the promised precise formulation of the lemma claims that the duality functor (3.10) takes the morphism $f_{z^{-1}} \colon \widetilde{A}^{(z)} \longrightarrow \widetilde{A}$ in the category $R\text{–rings}_{\mathsf{wnlq}}$ to the morphism $(g_z, a_z^{(z)}) \colon (B, d, h) \longrightarrow (B^{(z)}, d^{(z)}, h^{(z)})$ in the category $R\text{–rings}_{\mathsf{cdg,rq}}$.

To be even more precise, we need to establish first that the functor (3.10) takes the object $\widetilde{A}^{(z)} \in R\text{–rings}_{\mathsf{wnlq}}$ to the object $(B^{(z)}, d^{(z)}, h^{(z)}) \in R\text{–rings}_{\mathsf{cdg,rq}}$. Put $A^{(z)} = \operatorname{gr}^F \widetilde{A}^{(z)}$; then the identity isomorphism of rings $A_0^{(z)} = B^{(z),0}$ commuting with the identifications $A_0^{(z)} \simeq R \simeq B^{(z),0}$ together with the $A_0^{(z)}\text{-}A_0^{(z)}$-bimodule isomorphism $\operatorname{Hom}_{A_0^{(z)}}(A_1^{(z)}, A_0^{(z)}) = B^{(z),1}$ allows to consider $B^{(z)}$ as the quadratic dual ring to $A^{(z)}$. Let the subgroup $'V^{(z)} \subset \widetilde{V}^{(z)}$ coincide with the subgroup $'V \subset \widetilde{V}$; then the construction of Proposition 3.16 takes the 3-left finitely projective weak nonhomogeneous quadratic ring $(\widetilde{A}^{(z)}, \widetilde{V}^{(z)})$ with the submodule of strict generators $'V^{(z)} \subset \widetilde{V}^{(z)}$ to the 3-right finitely projective quadratic CDG-ring $(B^{(z)}, d^{(z)}, h^{(z)})$.

At last, we can perform the computation proving the lemma. We have to check that the functor (3.16) takes the morphism $f_{z^{-1}} \colon (\widetilde{A}^{(z)}, \widetilde{V}^{(z)}, 'V^{(z)}) \longrightarrow (\widetilde{A}, \widetilde{V}, 'V)$ to the morphism $(g_z, a_z^{(z)}) \colon (B, d, h) \longrightarrow (B^{(z)}, d^{(z)}, h^{(z)})$.

Set $"V^{(z)} = z'V^{(z)}z^{-1} \subset \widetilde{V}^{(z)}$. Then our morphism $f_{z^{-1}} \colon (\widetilde{A}^{(z)}, \widetilde{V}^{(z)}, 'V^{(z)}) \longrightarrow$ $(\widetilde{A}, \widetilde{V}, 'V)$ decomposes into a change-of-strict-generators morphism $(\widetilde{A}^{(z)}, \widetilde{V}^{(z)}, 'V^{(z)})$ $\longrightarrow (\widetilde{A}^{(z)}, \widetilde{V}^{(z)}, "V^{(z)})$, which acts by the identity map on the underlying associative ring $\widetilde{A}^{(z)}$, followed by the strict morphism $f_{z^{-1}} \colon (\widetilde{A}^{(z)}, \widetilde{V}^{(z)}, "V^{(z)}) \longrightarrow (\widetilde{A}, \widetilde{V}, 'V)$. Denote by $(B^{(z)}, d_{(z)}, h_{(z)})$ the CDG-ring assigned to the 3-left finitely projective weak nonhomogeneous quadratic ring $(\widetilde{A}^{(z)}, \widetilde{V}^{(z)})$ with the submodule of strict generators $"V^{(z)} \subset \widetilde{V}^{(z)}$ by the construction of Proposition 3.16.

Let $q' \colon V^{(z)} \times R \longrightarrow R$ be the map defined by the formula (3.1) using the splitting $'V^{(z)} \subset \widetilde{V}^{(z)}$ of the bimodule of generators of the 3-left finitely projective weak nonhomogeneous quadratic ring $\widetilde{A}^{(z)}$. Denoting by $u' \in 'V^{(z)}$ and $u'' \in "V^{(z)}$ the elements corresponding to an element $u = v^{(z)} \in V^{(z)} = \widetilde{V}^{(z)}/R$, we have

$$u'' = z * (z^{-1}uz)' * z^{-1} = (uz)' * z^{-1} = u' + q'(uz, z^{-1}).$$

Furthermore, by Proposition 3.14(b),

$$0 = q'(u, 1) = q'(u, zz^{-1}) = q'(uz, z^{-1}) + q'(u, z)z^{-1},$$

and hence

$$u'' = u' - q'(u, z)z^{-1} = u' - \langle u, d_0(z) \rangle z^{-1} = u' - \langle u, d_0(z)z^{-1} \rangle = u' + a_z^{(z)}(u).$$

By Proposition 3.20, it follows that $(\mathrm{id}, a_z^{(z)}) \colon (B^{(z)}, d_{(z)}, h_{(z)}) \longrightarrow (B^{(z)}, d^{(z)}, h^{(z)})$ is a change-of-connection morphism of CDG-rings over R. Hence one can compute that $d_{(z)}(zb^{(z)}z^{-1}) = zd^{(z)}(b^{(z)})z^{-1} = z(d(b))^{(z)}z^{-1}$ for all $b \in B$, and $h_{(z)} = zh^{(z)}z^{-1}$. By construction, $(\mathrm{id}, a_z^{(z)}) \colon (B^{(z)}, d_{(z)}, h_{(z)}) \longrightarrow (B^{(z)}, d^{(z)}, h^{(z)})$ is *the* change-of-connection morphism of CDG-rings over R assigned to the change-of-strict-generators morphism $(\widetilde{A}^{(z)}, \widetilde{V}^{(z)}, 'V^{(z)}) \longrightarrow (\widetilde{A}^{(z)}, \widetilde{V}^{(z)}, "V^{(z)})$ by the functor (3.16).

Finally, we need to show that the functor (3.11) takes the strict morphism $f_{z^{-1}} \colon (\widetilde{A}^{(z)}, \widetilde{V}^{(z)}, "V^{(z)}) \longrightarrow (\widetilde{A}, \widetilde{V}, 'V)$ to the strict morphism $(g_z, 0) \colon (B, d, h) \longrightarrow (B^{(z)}, d_{(z)}, h_{(z)})$. For this purpose, it suffices to check that the morphism $\bar{f}_{z^{-1}} = \mathrm{gr}^F f_{z^{-1}} \colon A^{(z)} = \mathrm{gr}^F \widetilde{A}^{(z)} \longrightarrow \mathrm{gr}^F \widetilde{A} = A$ corresponds to the morphism $g_z \colon B \longrightarrow B^{(z)}$ under the homogeneous quadratic duality of Propositions 1.6 and 1.8. All we need to do is to observe that

$$\langle v^{(z)}, g_z(b) \rangle = \langle v^{(z)}, zb^{(z)}z^{-1} \rangle = z\langle z^{-1}v^{(z)}z, b^{(z)} \rangle z^{-1}$$

$$= t_z(\langle z^{-1}v^{(z)}z, b^{(z)} \rangle) = \langle z^{-1}vz, b \rangle = \langle \bar{f}_{z^{-1}}(v^{(z)}), b \rangle$$

for all $v^{(z)} \in V^{(z)} = A_1^{(z)}$ and $b \in B^1$.

It remains to compute the image of the composition of our morphisms in the category $R\text{--rings}_{\text{wnlq}}^{\text{sg}}$

$$(\widetilde{A}^{(z)}, \widetilde{V}^{(z)}, 'V^{(z)}) \longrightarrow (\widetilde{A}^{(z)}, \widetilde{V}^{(z)}, ''V^{(z)}) \xrightarrow{f_z^{-1}} (\widetilde{A}, \widetilde{V}, 'V)$$

under the functor (3.16). This is equal, by construction, to the composition of morphisms of CDG-rings

$$(\text{id}, a_z^{(z)}) \circ (g_z, 0) = (g_z, a_z^{(z)}),$$

as desired.

In the context of the functor (3.18) below instead of the functor (3.10), the assertion of the lemma becomes literally true as stated, without the additional discussion in the first half of the above proof.   □

Theorem 3.25 *The constructions of Theorem 3.21 and Lemma 3.24 define a fully faithful strict contravariant 2-functor*

$$(\text{Rings}_{\text{wnlq2}})^{\text{op}} \longrightarrow \text{Rings}_{\text{cdg2,rq}} \tag{3.17}$$

from the 2-category of 3-left finitely projective weak nonhomogeneous quadratic rings to the 2-category of 3-right finitely projective quadratic CDG-rings.

Proof The 2-functor (3.17) is fully faithful in the strict sense: for any two objects $('\widetilde{A}, '\widetilde{V})$ and $(''\widetilde{A}, ''\widetilde{V}) \in \text{Rings}_{\text{wnlq2}}$ and the corresponding CDG-rings $('B, d', h')$ and $(''B, d'', h'') \in \text{Rings}_{\text{cdg2,rq}}$, the 2-functor (3.17) induces a bijection between morphisms $('\widetilde{A}, '\widetilde{V}) \longrightarrow (''\widetilde{A}, ''\widetilde{V})$ in $\text{Rings}_{\text{wnlq2}}$ and morphisms $(''B, d'', h'') \longrightarrow ('B, d', h')$ in $\text{Rings}_{\text{cdg2,rq}}$. Furthermore, for any pair of parallel morphisms f', $f'': ('\widetilde{A}, '\widetilde{V}) \longrightarrow (''\widetilde{A}, ''\widetilde{V})$ in $\text{Rings}_{\text{wnlq2}}$ and the corresponding pair of parallel morphisms g', $g'': (''B, d'', h'') \longrightarrow ('B, d', h')$ in $\text{Rings}_{\text{cdg2,rq}}$, the 2-functor (3.17) induces a bijection between 2-morphisms $f' \xrightarrow{z} f''$ in $\text{Rings}_{\text{wnlq2}}$ and 2-morphisms $g'' \xrightarrow{w} g'$ in $\text{Rings}_{\text{cdg2,rq}}$.

To construct the desired 2-functor, denote by $\text{Rings}_{\text{wnlq}} \subset \text{Rings}_{\text{wnlq2}}$ the category whose objects are the objects of the 2-category $\text{Rings}_{\text{wnlq2}}$ and whose morphisms are the morphisms of the 2-category $\text{Rings}_{\text{wnlq2}}$ (but there are no 2-morphisms in $\text{Rings}_{\text{wnlq}}$). Similarly, denote by $\text{Rings}_{\text{cdg,rq}}$ the category whose objects are the objects of the 2-category $\text{Rings}_{\text{cdg2,rq}}$ and whose morphisms are the morphisms of the 2-category $\text{Rings}_{\text{cdg2,rq}}$ (but there are no 2-morphisms in $\text{Rings}_{\text{cdg,rq}}$). Then essentially the same construction that was used to define the functor (3.10) in Theorem 3.21 provides a fully faithful contravariant functor

$$(\mathsf{Rings_{wnlq}})^{\mathrm{op}} \longrightarrow \mathsf{Rings_{cdg,rq}}. \qquad (3.18)$$

In order to extend the functor (3.18) to a 2-functor (3.17), we notice that, in the notation of the first paragraph of this proof, for every 2-morphism $f' \xrightarrow{z} f''$ in $\mathsf{Rings_{wnlq2}}$, both the morphisms $F_0 f' : F_0 \,'\widetilde{A} \longrightarrow F_0 \,''\widetilde{A}$ and $F_0 f'' : F_0 \,'\widetilde{A} \longrightarrow F_0 \,''\widetilde{A}$ are ring isomorphisms and the preimages of the element $z \in F_0 \,''\widetilde{A}$ under these two isomorphisms coincide, $(F_0 f')^{-1}(z) = (F_0 f'')^{-1}(z)$, because the element z is preserved by the conjugation with z. Similarly, for any 2-morphism $g'' \xrightarrow{w} g'$ in $\mathsf{Rings_{cdg2,rq}}$, both the morphisms $g_0' : \,''B^0 \longrightarrow \,'B^0$ and $g_0'' : \,''B^0 \longrightarrow \,'B^0$ are ring isomorphisms and the preimages of the element $w \in \,'B^0$ under these two isomorphisms coincide, $g_0'^{-1}(w) = g_0''^{-1}(w)$. By construction, we have $'B^0 = F_0 \,'\widetilde{A}$ and $''B^0 = F_0 \,''\widetilde{A}$. The maps $F_0 f' : F_0 \,'\widetilde{A} \longrightarrow F_0 \,''\widetilde{A}$ and $g_0' : \,''B^0 \longrightarrow \,'B^0$ are mutually inverse under this identification and so are the two maps $F_0 f'' : F_0 \,'\widetilde{A} \longrightarrow F_0 \,''\widetilde{A}$ and $g_0'' : \,''B^0 \longrightarrow \,'B^0$, that is, $g_0' = (F_0 f')^{-1}$ and $g_0'' = (F_0 f'')^{-1}$.

We assign a 2-morphism $g'' \xrightarrow{w} g'$ in $\mathsf{Rings_{cdg2,rq}}$ to a 2-morphism $f' \xrightarrow{z} f''$ in $\mathsf{Rings_{wnlq2}}$ if $g_0'(z) = w = g_0''(z)$, or equivalently, if $(F_0 f')(w) = z = (F_0 f'')(w)$. It only needs to be checked that $g'' \xrightarrow{w} g'$ is a 2-morphism in $\mathsf{Rings_{cdg2,rq}}$ if and only if $f' \xrightarrow{z} f''$ is a 2-morphism in $\mathsf{Rings_{wnlq2}}$. Then the compatibility with the vertical and horizontal compositions of 2-morphisms will be clear from the construction of such compositions in the beginning of this section and in Sect. 3.2.

Let us define *basic* 2-*morphisms* in the 2-category $\mathsf{Rings_{wnlq2}}$ as 2-morphisms of the form $f_{z^{-1}} \xrightarrow{z} \mathrm{id}_{(\widetilde{A}, \widetilde{V})}$, where $(\widetilde{A}, \widetilde{V})$ is an object of $\mathsf{Rings_{wnlq}}$, $z \in F_0 \widetilde{A}$ is an invertible element, $f_{z^{-1}} : (\widetilde{A}, \widetilde{V}) \longrightarrow (\widetilde{A}, \widetilde{V})$ is the morphism taking an element $c \in \widetilde{A}$ to the element $f_{z^{-1}}(c) = z^{-1} c z \in \widetilde{A}$, and $\mathrm{id}_{(\widetilde{A}, \widetilde{V})} : (\widetilde{A}, \widetilde{V}) \longrightarrow (\widetilde{A}, \widetilde{V})$ is the identity morphism. Then any 2-morphism in $\mathsf{Rings_{wnlq2}}$ decomposes uniquely as a morphism followed by a basic 2-morphism and also as a basic 2-morphism followed by a morphism. Specifically, a 2-morphism $f' \xrightarrow{z} f''$ as above is the composition of the morphism f'' followed by the basic 2-morphism $f_{z^{-1}} \xrightarrow{z} \mathrm{id}_{(''\widetilde{A}, ''\widetilde{V})}$, and it is also the composition of the basic 2-morphism $f_{w^{-1}} \xrightarrow{w} \mathrm{id}_{('\widetilde{A}, '\widetilde{V})}$ followed by the 2-morphism f''.

Similarly, we define *basic* 2-*morphisms* in the 2-category $\mathsf{Rings_{cdg2,rq}}$ as 2-morphisms of the form $(\mathrm{id}_B, 0) \xrightarrow{z} (g_z, a_z) : (B, d, h) \longrightarrow (B, d, h)$, where (B, d, h) is an object of $\mathsf{Rings_{cdg,rq}}$, $z \in B^0$ is an invertible element, $g_z : B \longrightarrow B$ is the graded ring homomorphism taking an element $b \in B$ to the element $g_z(b) = zbz^{-1} \in B$, and $a_z = -d(z)z^{-1} \in B^1$. Then any 2-morphism in $\mathsf{Rings_{wnlq2}}$ decomposes uniquely as a morphism followed by a basic 2-morphism and also as a basic 2-morphism followed by a morphism. Specifically, a 2-morphism $g'' \xrightarrow{w} g'$ as above is the composition of the morphism g'' followed by the basic 2-morphism $(\mathrm{id}_{'B}, 0) \xrightarrow{w} (g_w, a_w) : ('B, d', h') \longrightarrow$

$('B, d', h')$, and it is also the composition of the basic 2-morphism $(\mathrm{id}_{''B}, 0) \xrightarrow{z}$ $(g_z, a_z)\colon (''B, d'', h'') \longrightarrow (''B, d'', h'')$ followed by the morphism g''.

In other words, $f' \xrightarrow{z} f''$ is a 2-morphism in $\mathsf{Rings}_{\mathsf{wnlq2}}$ if and only if $f' = f_{z^{-1}} f''$, or equivalently, $f' = f'' f_{w^{-1}}$. Similarly, $g' \xrightarrow{w} g''$ is a 2-morphism in $\mathsf{Rings}_{\mathsf{cdg2,rq}}$ if and only if $g' = (g_w, a_w) \circ g''$, or equivalently, $g' = g'' \circ (g_z, a_z)$.

It remains to refer to Lemma 3.24 for the assertion that, for any object $(\widetilde{A}, \widetilde{V}) \in$ $\mathsf{Rings}_{\mathsf{wnlq2}}$ and the corresponding object $(B, d, h) \in \mathsf{Rings}_{\mathsf{cdg2,rq}}$, basic 2-morphisms $f_{z^{-1}} \xrightarrow{z} \mathrm{id}_{(\widetilde{A}, \widetilde{V})}$ in $\mathsf{Rings}_{\mathsf{wnlq2}}$ correspond to basic 2-morphisms $(\mathrm{id}_B, 0) \xrightarrow{z}$ $(g_z, a_z)\colon (B, d, h) \longrightarrow (B, d, h)$ in $\mathsf{Rings}_{\mathsf{cdg2,rq}}$. $\square$

3.8 Augmented Nonhomogeneous Quadratic Rings

We start with the definition of a one-sided augmentation.

Definition 3.26 Let $\widetilde{A}$ be an associative ring and $R \subset \widetilde{A}$ be a subring. A *left augmentation of $\widetilde{A}$ over R* is a left ideal $\widetilde{A}^+ \subset \widetilde{A}$ such that $\widetilde{A} = R \oplus \widetilde{A}^+$. Equivalently, a left augmentation is a left action of $\widetilde{A}$ in R extending the regular left action of R in itself.

Given a left augmentation ideal $\widetilde{A}^+ \subset \widetilde{A}$, such a left action of $\widetilde{A}$ in R is obtained by identifying R with the quotient left $\widetilde{A}$-module $\widetilde{A}/\widetilde{A}^+ = R$. Conversely, given a left augmentation action of $\widetilde{A}$ in R, the left augmentation ideal $\widetilde{A}^+ \subset \widetilde{A}$ is recovered as the annihilator of the element $1 \in R$.

The *category of left augmented rings over R*, denoted by $R\text{–}\mathsf{rings}^{\mathsf{laug}}$, is defined as follows. The objects of $R\text{–}\mathsf{rings}^{\mathsf{laug}}$ are associative rings $\widetilde{A}$ endowed with a subring identified with R and a left augmentation ideal $\widetilde{A}^+ \subset \widetilde{A}$. Morphisms $('\widetilde{A}, '\widetilde{A}^+) \longrightarrow$ $(''\widetilde{A}, ''\widetilde{A}^+)$ in $R\text{–}\mathsf{rings}^{\mathsf{laug}}$ are ring homomorphisms $f\colon '\widetilde{A} \longrightarrow ''\widetilde{A}$ forming a commutative triangle diagram with the embeddings $R \longrightarrow '\widetilde{A}$ and $R \longrightarrow ''\widetilde{A}$ and satisfying the condition of compatibility with the augmentations, namely, that $f('\widetilde{A}^+) \subset ''\widetilde{A}^+$. Equivalently, both the conditions on f can be expressed by saying that the left action of $'\widetilde{A}$ in R coincides with the action obtained from the left action of $''\widetilde{A}$ in R by the restriction of scalars via f.

Moreover, one can define the *2-category of left augmented rings*, denoted by $\mathsf{Rings}^{\mathsf{laug2}}$, in the following way. The objects of $\mathsf{Rings}^{\mathsf{laug2}}$ are associative rings $\widetilde{A}$ endowed with a subring $F_0\widetilde{A} \subset \widetilde{A}$ and a left ideal $\widetilde{A}^+ \subset \widetilde{A}$ such that $\widetilde{A} = F_0\widetilde{A} \oplus \widetilde{A}^+$. Morphisms $(\widetilde{A}, F_0'\widetilde{A}, '\widetilde{A}^+) \longrightarrow (''\widetilde{A}, F_0''\widetilde{A}, ''\widetilde{A}^+)$ in $\mathsf{Rings}^{\mathsf{laug2}}$ are ring homomorphisms $f\colon '\widetilde{A} \longrightarrow ''\widetilde{A}$ such that f restricts to an isomorphism $F_0 f\colon F_0'\widetilde{A} \longrightarrow F_0''\widetilde{A}$ and $f('\widetilde{A}^+) \subset ''\widetilde{A}^+$. 2-morphisms $f \xrightarrow{z} g$ between a pair of parallel morphisms f, $g\colon (\widetilde{A}, F_0'\widetilde{A}, '\widetilde{A}^+) \longrightarrow (''\widetilde{A}, F_0''\widetilde{A}, ''\widetilde{A}^+)$, are invertible elements $z \in F_0'\widetilde{A}$ such that $g(c) = f(zcz^{-1})$ for all $c \in '\widetilde{A}$.

Notice that it follows from the latter condition that $z\,'\widetilde{A}^+z^{-1} = \,'\widetilde{A}^+$, or equivalently, $'\widetilde{A}^+z = \,'\widetilde{A}^+$. Indeed, $zcz^{-1} = r + a$ for $c, a \in \,'\widetilde{A}^+$ and $r \in F_0\,'\widetilde{A}$ implies $g(c) = f(zcz^{-1}) = f(r) + f(a) \in \,''\widetilde{A}$ with $f(r) \in F_0\,''\widetilde{A}$ and $f(a) \in \,''\widetilde{A}^+$, and hence $f(r) = 0$ and $r = 0$. Similarly, $z^{-1}cz = r + a$ for $c, a \in \,'\widetilde{A}^+$ and $r \in F_0\,'\widetilde{A}$ implies $f(c) = g(z^{-1}cz) = g(r) + g(a) \in \,''\widetilde{A}$, and hence $r = 0$.

The vertical composition of two 2-morphisms $f' \xrightarrow{w} f'' \xrightarrow{z} f'''$ is the 2-morphism $f' \xrightarrow{wz} f'''$. The identity 2-morphism is the 2-morphism $f \xrightarrow{1} f$. The horizontal composition of two 2-morphisms $g' \xrightarrow{w} g'' \colon (\widetilde{A}, F_0\widetilde{A}, \widetilde{A}^+) \longrightarrow (\widetilde{B}, F_0\widetilde{B}, \widetilde{B}^+)$ and $f' \xrightarrow{z} f'' \colon (\widetilde{B}, F_0\widetilde{B}, \widetilde{B}^+) \longrightarrow (\widetilde{C}, F_0\widetilde{C}, \widetilde{C}^+)$ is the 2-morphism $f'g' \xrightarrow{z \circ w} f''g'' \colon (\widetilde{A}, F_0\widetilde{A}, \widetilde{A}^+) \longrightarrow (\widetilde{C}, F_0\widetilde{C}, \widetilde{C}^+)$ with the element $z \circ w = (F_0g')^{-1}(z)w = w(F_0g'')^{-1}(z)$, where $(F_0g')^{-1}$, $(F_0g'')^{-1} \colon F_0\widetilde{B} \longrightarrow F_0\widetilde{A}$ are the inverse maps to the ring isomorphisms F_0g', $F_0g'' \colon F_0\widetilde{A} \longrightarrow F_0\widetilde{B}$.

All the 2-morphisms of left augmented rings are invertible. If $f \colon (\widetilde{A}, F_0\,'\widetilde{A}, \,'\widetilde{A}^+) \longrightarrow (\,''\widetilde{A}, F_0\,''\widetilde{A}, \,''\widetilde{A}^+)$ is a morphism of left augmented rings and $z \in F_0\,'\widetilde{A}$ is an invertible element such that $'\widetilde{A}^+z = \,'\widetilde{A}^+$, then $g \colon c \longmapsto f(zcz^{-1})$ is also a morphism of left augmented rings $g \colon (\widetilde{A}, F_0\,'\widetilde{A}, \,'\widetilde{A}^+) \longrightarrow (\,''\widetilde{A}, F_0\,''\widetilde{A}, \,''\widetilde{A}^+)$. The morphisms f and g are connected by the 2-isomorphism $f \xrightarrow{z} g$.

Let $(\widetilde{A}, F)$ be a filtered ring with an increasing filtration $0 = F_{-1}\widetilde{A} \subset F_0\widetilde{A} \subset F_1\widetilde{A} \subset F_2\widetilde{A} \subset \cdots$ (which, as above, is presumed to be exhaustive and compatible with the multiplication in $\widetilde{A}$). The filtered ring $(\widetilde{A}, F)$ is said to be *left augmented* if the ring $\widetilde{A}$ is left augmented over its subring $F_0\widetilde{A}$. In other words, this means that a left ideal $\widetilde{A}^+ \subset \widetilde{A}$ is chosen such that $\widetilde{A} = F_0\widetilde{A} \oplus \widetilde{A}^+$.

We denote by $R\text{–}\mathsf{rings}^{\mathsf{laug}}_{\mathsf{fil}}$ the category of left augmented filtered rings with the filtration component $F_0\widetilde{A}$ identified with R. So the objects of $R\text{–}\mathsf{rings}^{\mathsf{laug}}_{\mathsf{fil}}$ are left augmented filtered rings $(\widetilde{A}, F, \widetilde{A}^+)$ for which a ring isomorphism $R \simeq F_0\widetilde{A}$ has been chosen. Morphisms $(\widetilde{A}, F, \,'\widetilde{A}^+) \longrightarrow (\,''\widetilde{A}, F, \,''\widetilde{A}^+)$ in $R\text{–}\mathsf{rings}^{\mathsf{laug}}_{\mathsf{fil}}$ are ring homomorphisms $f \colon \,'\widetilde{A} \longrightarrow \,''\widetilde{A}$ such that $f(F_n\,'\widetilde{A}) \subset F_n\,''\widetilde{A}$ for all $n \geq 0$, $f(\,'\widetilde{A}^+) \subset \,''\widetilde{A}^+$, and the ring homomorphism $F_0f \colon F_0\,'\widetilde{A} \longrightarrow F_0\,''\widetilde{A}$ forms a commutative triangle diagram with the fixed isomorphisms $R \simeq F_0\,'\widetilde{A}$ and $R \simeq F_0\,''\widetilde{A}$.

The definition of the 2-*category of left augmented filtered rings*, denoted by $\mathsf{Rings}^{\mathsf{laug2}}_{\mathsf{fil}}$, is similar to the above. The objects of $\mathsf{Rings}^{\mathsf{laug2}}_{\mathsf{fil}}$ are left augmented filtered rings $(\widetilde{A}, F, \widetilde{A}^+)$. Morphisms $(\widetilde{A}, F, \,'\widetilde{A}^+) \longrightarrow (\,''\widetilde{A}, F, \,''\widetilde{A}^+)$ are ring homomorphisms $f \colon \,'\widetilde{A} \longrightarrow \,''\widetilde{A}$ such that $f(F_n\,'\widetilde{A}) \subset F_n\,''\widetilde{A}$ for all $n \geq 0$, the map $F_0f \colon F_0\,'\widetilde{A} \longrightarrow F_0\,''\widetilde{A}$ is an isomorphism, and $f(\,'\widetilde{A}^+) \subset \,''\widetilde{A}^+$. 2-morphisms $f \xrightarrow{z} g$ between a pair of parallel morphisms $f, g \colon (\widetilde{A}, F, \,'\widetilde{A}^+) \longrightarrow (\,''\widetilde{A}, F, \,''\widetilde{A}^+)$ are invertible elements $z \in F_0\,'\widetilde{A}$ such that $g(c) = f(zcz^{-1})$ for all $c \in \,'\widetilde{A}$. The composition of 2-morphisms in $\mathsf{Rings}^{\mathsf{laug2}}_{\mathsf{fil}}$ is defined in the same way as in the category $\mathsf{Rings}^{\mathsf{laug2}}$, so there is an obvious forgetful strict 2-functor $\mathsf{Rings}^{\mathsf{laug2}}_{\mathsf{fil}} \longrightarrow \mathsf{Rings}^{\mathsf{laug2}}$.

A weak nonhomogeneous quadratic ring $R \subset \widetilde{V} \subset \widetilde{A}$ is said to be *left augmented* if the ring $\widetilde{A}$ is endowed with a left augmentation over its subring R. The *category of 3-left finitely projective left augmented weak nonhomogeneous quadratic rings over R*, denoted by R–$\mathsf{rings}^{\mathsf{laug}}_{\mathsf{wnlq}}$, is defined as the full subcategory in R–$\mathsf{rings}^{\mathsf{laug}}_{\mathsf{fil}}$ whose objects are the left augmented weak nonhomogeneous quadratic rings over R that are 3-left finitely projective as weak nonhomogeneous quadratic rings.

The 2-*category of* 3-*left finitely projective left augmented weak nonhomogeneous quadratic rings*, denoted by $\mathsf{Rings}^{\mathsf{laug2}}_{\mathsf{wnlq}}$, is defined as the following 2-subcategory in $\mathsf{Rings}^{\mathsf{laug2}}_{\mathsf{fil}}$. The objects of $\mathsf{Rings}^{\mathsf{laug2}}_{\mathsf{fil}}$ are the 3-left finitely projective left augmented weak nonhomogeneous quadratic rings $R \subset \widetilde{V} \subset \widetilde{A} \supset \widetilde{A}^+$ with the filtration F generated by $F_1\widetilde{A}$ over $F_0\widetilde{A}$. All morphisms in $\mathsf{Rings}^{\mathsf{laug2}}_{\mathsf{fil}}$ between objects of $\mathsf{Rings}^{\mathsf{laug2}}_{\mathsf{wnlq}}$ are morphisms in $\mathsf{Rings}^{\mathsf{laug2}}_{\mathsf{wnlq}}$, and all 2-morphisms in $\mathsf{Rings}^{\mathsf{laug2}}_{\mathsf{fil}}$ between morphisms of $\mathsf{Rings}^{\mathsf{laug2}}_{\mathsf{wnlq}}$ are 2-morphisms in $\mathsf{Rings}^{\mathsf{laug2}}_{\mathsf{wnlq}}$.

Furthermore, the *category of* 3-*right finitely projective quadratic DG-rings over R*, denoted by R–$\mathsf{rings}_{\mathsf{dg,rq}}$, is the full subcategory in the category R–$\mathsf{rings}_{\mathsf{dg}}$ (as defined in Sect. 3.2) consisting of all the DG-rings (B, d) whose underlying nonnegatively graded ring B is 3-right finitely projective quadratic over R.

The 2-*category of* 3-*right finitely projective quadratic DG-rings*, denoted by $\mathsf{Rings}_{\mathsf{dg2,rq}}$, is the similar 2-subcategory in the 2-category $\mathsf{Rings}_{\mathsf{dg2}}$. The objects of $\mathsf{Rings}_{\mathsf{dg2,rq}}$ are all the DG-rings (B, d) whose underlying nonnegatively graded ring B is 3-right finitely projective quadratic over B^0. All morphisms in $\mathsf{Rings}_{\mathsf{dg2}}$ between objects of $\mathsf{Rings}_{\mathsf{dg2,rq}}$ are morphisms in $\mathsf{Rings}_{\mathsf{dg2,rq}}$, and all 2-morphisms in $\mathsf{Rings}_{\mathsf{dg2}}$ between morphisms of $\mathsf{Rings}_{\mathsf{dg2,rq}}$ are 2-morphisms in $\mathsf{Rings}_{\mathsf{dg2,rq}}$.

Theorem 3.27 *The nonhomogeneous quadratic duality functor of Theorem 3.21 restricts to a fully faithful contravariant functor*

$$(R\text{–}\mathsf{rings}^{\mathsf{laug}}_{\mathsf{wnlq}})^{\mathsf{op}} \longrightarrow R\text{–}\mathsf{rings}_{\mathsf{dg,rq}} \tag{3.19}$$

from the category of 3-*left finitely projective left augmented weak nonhomogeneous quadratic rings to the category of* 3-*right finitely projective quadratic DG-rings over R. Moreover, a* 3-*right finitely projective quadratic DG-ring* (B, d) *over R belongs to the essential image of the functor (3.19) if and only if, viewed as a CDG-ring* $(B, d, 0)$, *it belongs to the essential image of the functor (3.10) from Theorem 3.21.*

Examples 3.28

(1) One class of examples of left augmented nonhomogeneous quadratic rings and the quadratic DG-rings corresponding to them under the fully faithful functor described in the theorem was described in Example 3.17 above.

(2) Let R be an associative ring and V be an R-R-bimodule which is finitely generated and projective as a left R-module. Let $A = T_R(V)$ be the tensor ring, and let $(\widetilde{A}, F)$ be a nonhomogeneous quadratic ring such that $\mathrm{gr}^F \widetilde{A} = A$. Then it follows from

Theorem 3.27 and the discussion in Examples 3.12(2) and 3.22 that any choice of a left R-linear splitting of the surjective R-R-bimodule map $\widetilde{V} = F_1\widetilde{A} \longrightarrow A_1 = V$ defines a left augmentation of the ring $\widetilde{A}$ over its subring R. So a left augmentation of $\widetilde{A}$ over R exists. Similarly, if V is a finitely generated projective right R-module, then a right augmentation of $\widetilde{A}$ over R exists, and such right augmentations correspond bijectively to right R-linear splittings of the surjective R-R-bimodule map $\widetilde{V} \longrightarrow V$. However, a *two-sided* augmentation ideal in $\widetilde{A}$ over R usually does *not* exist; in fact, it exists if and only if $\widetilde{A}$ is isomorphic to A as a filtered ring.

Proof of Theorem 3.27 It is clear from the above discussion that R–$\mathrm{rings}_{\mathrm{dg,rq}}$ is a subcategory in R–$\mathrm{rings}_{\mathrm{cdg,rq}}$. Moreover, the category of 3-right finitely projective quadratic DG-rings R–$\mathrm{rings}_{\mathrm{dg,rq}}$ is a full subcategory in the category R–$\mathrm{rings}^{\mathrm{sm}}_{\mathrm{cdg,rq}}$ of 3-right finitely projective quadratic CDG-rings over R and strict morphisms between them (which was introduced in the proof of Theorem 3.21).

Similarly, we observe that the category of 3-left finitely projective left augmented weak nonhomogeneous quadratic rings R–$\mathrm{rings}^{\mathrm{laug}}_{\mathrm{wnlq}}$ is a full subcategory in the category R–$\mathrm{rings}^{\mathrm{sgsm}}_{\mathrm{wnlq}}$ of 3-left finitely projective weak nonhomogeneous quadratic rings $(\widetilde{A}, \widetilde{V})$ with a fixed submodule of strict generators $V' \subset \widetilde{V}$ and morphisms $f\colon ('\widetilde{A}, '\widetilde{V}, 'V) \longrightarrow (''\widetilde{A}, ''\widetilde{V}, ''V)$ preserving the submodule of strict generators. Indeed, given a 3-left finitely projective left augmented weak nonhomogeneous quadratic ring $(\widetilde{A}, \widetilde{V}, \widetilde{A}^+)$ over R, we choose the left R-submodule $V' = \widetilde{A}^+ \cap \widetilde{V} \subset \widetilde{V}$ as the submodule of strict generators of $\widetilde{A}$. The left augmentation ideal $\widetilde{A}^+ \subset \widetilde{A}$ can then be recovered as the left ideal (equivalently, the subring without unit) generated by V' in $\widetilde{A}$.

Moreover, the essential image of the fully faithful functor R–$\mathrm{rings}^{\mathrm{laug}}_{\mathrm{wnlq}} \longrightarrow R$–$\mathrm{rings}^{\mathrm{sgsm}}_{\mathrm{wnlq}}$ can be explicitly described as follows. Given an object $(\widetilde{A}, \widetilde{V}, V') \in R$–$\mathrm{rings}^{\mathrm{sgsm}}_{\mathrm{wnlq}}$, consider the related maps $q\colon V \times R \longrightarrow R$, $p\colon \widehat{I} \longrightarrow R$, and $h\colon \widehat{I} \longrightarrow R$ defined by the formulas (3.1–3.2). Then the object $(\widetilde{A}, \widetilde{V}, V')$ corresponds to a (3-left finitely projective) left augmented weak nonhomogeneous quadratic ring if and only if one has $\hat{\imath}_1 * \hat{\imath}_2 \in V' \subset \widetilde{V}$ for all $\hat{\imath} \in \widehat{I}$, that is, $h = 0$.

Indeed, the "only if" assertion is obvious. To prove the "if," one observes that, for any weak nonhomogeneous quadratic ring $(\widetilde{A}, \widetilde{V})$ satisfying the assumptions of Sect. 3.3 and any chosen submodule of strict generators $V' \subset \widetilde{V}$, the ring $\widetilde{A}$ is generated by the ring R and the abelian group V' with the defining relations (3.1–3.2). It is clear from the form of these relations that the left ideal generated by V' in $\widetilde{A}$ does not intersect R whenever $h = 0$.

The latter condition means exactly that the quadratic CDG-ring (B, d, h) assigned to $(\widetilde{A}, \widetilde{V}, V')$ by the functor (3.11) is a DG-ring. The desired fully faithful contravariant functor (3.19) can be now obtained as a restriction of the fully faithful contravariant functor (3.11) to the full subcategory R–$\mathrm{rings}^{\mathrm{laug}}_{\mathrm{wnlq}} \subset R$–$\mathrm{rings}^{\mathrm{sgsm}}_{\mathrm{wnlq}}$. $\qquad\square$

Theorem 3.29 *The nonhomogeneous quadratic duality 2-functor of Theorem 3.25 restricts to a fully faithful strict contravariant 2-functor*

$$(\mathsf{Rings}_{\mathsf{wnlq}}^{\mathsf{laug2}})^{\mathrm{op}} \longrightarrow \mathsf{Rings}_{\mathsf{dg2,rq}} \tag{3.20}$$

from the 2-category of 3-left finitely projective left augmented weak nonhomogeneous quadratic rings to the 2-category of 3-right finitely projective quadratic DG-rings.

Proof Similarly to the functor (3.17), the functor (3.20) is fully faithful in the strict sense. For any two objects $('\widetilde{A}, '\widetilde{V}, '\widetilde{A}^+)$ and $(''\widetilde{A}, ''\widetilde{V}, ''\widetilde{A}^+) \in \mathsf{Rings}_{\mathsf{wnlq}}^{\mathsf{laug2}}$ and the corresponding DG-rings $('B, d')$ and $(''B, d'') \in \mathsf{Rings}_{\mathsf{dg2,rq}}$, the 2-functor (3.20) induces a bijection between morphisms $('\widetilde{A}, '\widetilde{V}, '\widetilde{A}^+) \longrightarrow (''\widetilde{A}, ''\widetilde{V}, ''\widetilde{A}^+)$ in $\mathsf{Rings}_{\mathsf{wnlq}}^{\mathsf{laug2}}$ and morphisms $(''B, d'') \longrightarrow ('B, d')$ in $\mathsf{Rings}_{\mathsf{dg2,rq}}$. Furthermore, for any pair of parallel morphisms f', $f'' : ('\widetilde{A}, '\widetilde{V}, '\widetilde{A}^+) \longrightarrow (''\widetilde{A}, ''\widetilde{V}, ''\widetilde{A}^+)$ in $\mathsf{Rings}_{\mathsf{wnlq}}^{\mathsf{laug2}}$ and the corresponding pair of parallel morphisms g', $g'' : (''B, d'') \longrightarrow ('B, d')$ in $\mathsf{Rings}_{\mathsf{dg2,rq}}$, the 2-functor (3.20) induces a bijection between 2-morphisms $f' \xrightarrow{z} f''$ in $\mathsf{Rings}_{\mathsf{wnlq}}^{\mathsf{laug2}}$ and 2-morphisms $g'' \xrightarrow{z} g'$ in $\mathsf{Rings}_{\mathsf{dg2,rq}}$.

It is clear from the discussion above in this section that $\mathsf{Rings}_{\mathsf{dg2,rq}}$ is a 2-subcategory in $\mathsf{Rings}_{\mathsf{cdg2,rq}}$. Similarly, the 2-category $\mathsf{Rings}_{\mathsf{wnlq}}^{\mathsf{laug2}}$ can be viewed as a 2-subcategory of the 2-category $\mathsf{Rings}_{\mathsf{wnlq2}}$ in the following way. To any object $(\widetilde{A}, \widetilde{V}, \widetilde{A}^+) \in \mathsf{Rings}_{\mathsf{wnlq}}^{\mathsf{laug2}}$, one assigns the object $(\widetilde{A}, \widetilde{V}) \in \mathsf{Rings}_{\mathsf{wnlq2}}$, and to any morphism $f : (\widetilde{A}, '\widetilde{V}, '\widetilde{A}^+) \longrightarrow (''\widetilde{A}, ''\widetilde{V}, ''\widetilde{A}^+)$ in $\mathsf{Rings}_{\mathsf{wnlq}}^{\mathsf{laug2}}$, one assigns the morphism $f : (\widetilde{A}, '\widetilde{V}) \longrightarrow (''\widetilde{A}, ''\widetilde{V})$ in $\mathsf{Rings}_{\mathsf{wnlq2}}$. Finally, to any 2-morphism $f \xrightarrow{z} g : (\widetilde{A}, '\widetilde{V}, '\widetilde{A}^+) \longrightarrow (''\widetilde{A}, ''\widetilde{V}, ''\widetilde{A}^+)$ in $\mathsf{Rings}_{\mathsf{wnlq}}^{\mathsf{laug2}}$, one assigns the 2-morphism $f \xrightarrow{w} g : (\widetilde{A}, '\widetilde{V}) \longrightarrow (''\widetilde{A}, ''\widetilde{V})$ in $\mathsf{Rings}_{\mathsf{wnlq2}}$ with the element $w = (F_0 f)(z) = (F_0 g)(z)$. Here $z \in F_0\,'\widetilde{A}$ is an invertible element such that $'\widetilde{A}^+ z = '\widetilde{A}^+$ and $w \in F_0\,''\widetilde{A}$ is an invertible element, while $F_0 f$ and $F_0 g : F_0\,'\widetilde{A} \longrightarrow F_0\,''\widetilde{A}$ are two ring isomorphisms whose values coincide on the element z.

Denote by $\mathsf{Rings}_{\mathsf{wnlq}}^{\mathsf{laug}} \subset \mathsf{Rings}_{\mathsf{wnlq}}^{\mathsf{laug2}}$ the category whose objects are the objects of the 2-category $\mathsf{Rings}_{\mathsf{wnlq}}^{\mathsf{laug2}}$ and whose morphisms are the morphisms of the 2-category $\mathsf{Rings}_{\mathsf{wnlq}}^{\mathsf{laug2}}$ (but there are no 2-morphisms in $\mathsf{Rings}_{\mathsf{wnlq}}^{\mathsf{laug}}$). Similarly, denote by $\mathsf{Rings}_{\mathsf{dg,rq}}$ the category whose objects are the objects of the 2-category $\mathsf{Rings}_{\mathsf{dg2,rq}}$ and whose morphisms are the morphisms of the 2-category $\mathsf{Rings}_{\mathsf{dg2,rq}}$ (but there are no 2-morphisms in $\mathsf{Rings}_{\mathsf{dg,rq}}$). Then essentially the same argument that was used to restrict the functor (3.10) to the functor (3.19) in Theorem 3.27 shows that the fully faithful functor (3.18) restricts to a fully faithful functor

$$(\mathsf{Rings}_{\mathsf{wnlq}}^{\mathsf{laug}})^{\mathrm{op}} \longrightarrow \mathsf{Rings}_{\mathsf{dg,rq}}. \tag{3.21}$$

To deduce the existence of a fully faithful functor (3.20) from the existence of the fully faithful functors (3.17) and (3.21), one can observe that both the embeddings of 2-categories $\mathsf{Rings}_{\mathsf{wnlq}}^{\mathsf{laug2}} \longrightarrow \mathsf{Rings}_{\mathsf{wnlq2}}$ and $\mathsf{Rings}_{\mathsf{dg2,rq}} \longrightarrow \mathsf{Rings}_{\mathsf{cdg2,rq}}$ are *fully faithful on the level of 2-morphisms*. In other words, this means that any 2-morphism in $\mathsf{Rings}_{\mathsf{wnlq2}}$ between a pair of parallel morphisms in $\mathsf{Rings}_{\mathsf{wnlq}}^{\mathsf{laug2}}$ belongs to $\mathsf{Rings}_{\mathsf{wnlq}}^{\mathsf{laug2}}$, and similarly, any 2-morphism in $\mathsf{Rings}_{\mathsf{cdg2,rq}}$ between a pair of parallel morphisms in $\mathsf{Rings}_{\mathsf{dg2,rq}}$ belongs to $\mathsf{Rings}_{\mathsf{dg2,rq}}$. (More generally, $\mathsf{Rings}_{\mathsf{fil}}^{\mathsf{laug2}}$ is a 2-subcategory in $\mathsf{Rings}_{\mathsf{fil2}}$ such that any 2-morphism in $\mathsf{Rings}_{\mathsf{fil2}}$ between a pair of parallel morphisms in $\mathsf{Rings}_{\mathsf{fil}}^{\mathsf{laug2}}$ belongs to $\mathsf{Rings}_{\mathsf{fil}}^{\mathsf{laug2}}$, and $\mathsf{Rings}_{\mathsf{dg2}}$ is a 2-subcategory in $\mathsf{Rings}_{\mathsf{cdg2}}$ such that any 2-morphism in $\mathsf{Rings}_{\mathsf{cdg2}}$ between a pair of parallel morphisms in $\mathsf{Rings}_{\mathsf{dg2}}$ belongs to $\mathsf{Rings}_{\mathsf{dg2}}$.) This suffices to prove the theorem.

Alternatively, one can construct the fully faithful strict 2-functor (3.20) in the way similar to the construction of the fully faithful strict 2-functor (3.17) in the proof of Theorem 3.25. For this purpose, one defines the "basic 2-morphisms" in the 2-categories $\mathsf{Rings}_{\mathsf{wnlq}}^{\mathsf{laug2}}$ and $\mathsf{Rings}_{\mathsf{dg2,rq}}$ and observes that any 2-morphism in $\mathsf{Rings}_{\mathsf{wnlq}}^{\mathsf{laug2}}$ decomposes uniquely as a basic 2-morphism followed by a morphism (but *not* in the other order), while any 2-morphism in $\mathsf{Rings}_{\mathsf{dg2,rq}}$ decomposes uniquely as a morphism followed by a basic 2-morphism (but *not* in the other order).

The reason is, essentially, that the cocycle equation $d'(z) = 0$ in the definition of a 2-morphism of DG-rings $f \xrightarrow{z} g\colon ({}''B, d'') \longrightarrow ({}'B, d')$ does *not* imply the equation $d''(w) = 0$ for the element $w = f_0^{-1}(z) = g_0^{-1}(z)$. The latter equation is stronger than the former one and does not need to hold. Similarly, the condition ${}'\widetilde{A}^+ z = {}'\widetilde{A}^+$ related to the definition a 2-morphism of left augmented rings $f \xrightarrow{z} g\colon ({}'\widetilde{A}, F_0{}'\widetilde{A}, {}'\widetilde{A}^+) \longrightarrow ({}''\widetilde{A}, F_0{}''\widetilde{A}, {}''\widetilde{A}^+)$ does *not* imply the condition ${}''\widetilde{A}^+ w = {}''\widetilde{A}^+$ for the element $w = (F_0 f)(z) = (F_0 g)(z)$. The latter condition is stronger than the former one and does not need to hold. $\square$

The Poincaré–Birkhoff–Witt Theorem

4

4.1 Central Element Theorem

Let $\widehat{A} = \bigoplus_{n=0}^{\infty} \widehat{A}_n$ be a nonnegatively graded ring with the degree-zero component $R = \widehat{A}_0$, and let $t \in \widehat{A}_1$ be a central element. Let $A = \widehat{A}/\widehat{A}t$ denote the quotient ring of $\widehat{A}$ by the homogeneous ideal generated by t. So $A = \bigoplus_{n=0}^{\infty} A_n$ is also a nonnegatively graded ring with the degree-zero component $A_0 = R$, the degree-one component $A_1 = \widehat{A}_1/Rt$, and the degree n component $A_n = \widehat{A}_n/\widehat{A}_{n-1}t$ for all $n \geq 1$. We will say that t is a *nonzero-divisor* in $\widehat{A}$ if $at = 0$ implies $a = 0$ for any $a \in \widehat{A}_n$, $n \geq 0$.

Proposition 4.1 *Let $\widehat{A}$ be a nonnegatively graded ring and $t \in \widehat{A}_1$ be a central element. Then*

(a) *The graded ring $\widehat{A}$ is generated by $\widehat{A}_1$ over $\widehat{A}_0$ if and only if the graded ring $A = \widehat{A}/\widehat{A}t$ is generated by A_1 over A_0.*

(b) *Assuming that t is a nonzero-divisor in $\widehat{A}$, the graded ring $\widehat{A}$ is quadratic if and only if the graded ring $A = \widehat{A}/\widehat{A}t$ is quadratic.*

Examples 4.2

(1) It is easy to give a counterexample showing that the "if" implication in part (b) does *not* hold without the assumption that t is a nonzero-divisor. Let k be a field; consider the graded ring $\widehat{A} = k[t]/(t^n)$, where $k[t]$ is the ring of polynomials in one variable t over k, graded so that t is a homogeneous element of degree 1, and (t^n) is the principal ideal generated by the element $t^n \in k[t]$, where $n \geq 3$. Then $\widehat{A}/\widehat{A}t = k$ is a quadratic graded ring (over k), but the graded ring $\widehat{A}$ is not quadratic.

© The Author(s), under exclusive license to Springer Nature Switzerland AG 2021
L. Positselski, *Relative Nonhomogeneous Koszul Duality*, Frontiers in Mathematics,
https://doi.org/10.1007/978-3-030-89540-2_4

(2) Another example to the same effect is provided by the graded ring $\widehat{A} = k[t, x]/(t^n x^m)$,
where $k[t, x]$ is the (commutative) ring of polynomials in two variables t and x over k,
graded so that both t and x are homogeneous elements of degree 1, and $(t^n x^m)$ is the
principal ideal generated by the element $t^n x^m \in k[t, x]$, where $n \geq 1$, $n + m \geq 3$.
Then $\widehat{A}/\widehat{A}t = k[x]$ is a quadratic graded ring (over k), but the graded ring $\widehat{A}$ is not
quadratic.

The same examples also show that the condition that t is a nonzero-divisor cannot be
dropped in Theorem 4.4 below. Indeed, in both cases, the graded ring $\widehat{A}/\widehat{A}t$ is (left and
right flat) Koszul, but the graded ring $\widehat{A}$ is not.

Proof of Proposition 4.1

Part (a):   for any nonnegatively graded ring $C = \bigoplus_{n=0}^{\infty} C_n$ generated by C_1 over C_0, and
for any homogeneous ideal $H \subset C$, the quotient ring $A = C/H$ is generated by A_1
over A_0. This proves the implication "only if."

To prove the "if," suppose that we are given a nonnegatively graded ring C and a
homogeneous ideal $H \subset C$ which is generated, as a two-sided ideal, by its degree-one
component H_1. Suppose further that the quotient ring $A = C/H$ is generated by A_1
over A_0. Let $C' \subset C$ denote the subring in C generated by $C'_1 = C_1$ over $C'_0 = C_0$. Then
the composition $C' \longrightarrow C \longrightarrow A$ is surjective, so we have $C = C' + H$. Arguing by
induction, we will prove that $C'_n = C_n$ for every $n \geq 2$. Indeed, assume that $C'_k = C_k$ for
all $k \leq n - 1$. Then $H_n = \sum_{k=1}^{n} C_{k-1} H_1 C_{n-k} = \sum_{k=1}^{n} C'_{k-1} H_1 C'_{n-k} \subset C'_n$, and hence
$C_n = C'_n + H_n = C'_n$. Thus, $C' = C$, so C is generated by C_1 over C_0.

Part (b):   for any quadratic graded ring $C = \bigoplus_{n=0}^{\infty} C_n$ and any homogeneous ideal $H \subset
C$ that is generated, as a two-sided ideal, by its components H_1 and H_2, the quotient
ring $A = C/H$ is quadratic. This proves the implication "only if" (which does not
depend on the assumption that t is a nonzero-divisor).

To prove the "if," suppose that $t \in \widehat{A}_1$ is a central nonzero-divisor and the graded ring
$A = \widehat{A}/\widehat{A}t$ is quadratic. Then, by part (a), the graded ring $\widehat{A}$ is generated by $\widehat{A}_1$ over
$R = \widehat{A}_0$.

Let $\widehat{I} \subset \widehat{A}_1 \otimes_R \widehat{A}_1$ be the kernel of the multiplication map $\widehat{A}_1 \otimes_R \widehat{A}_1 \longrightarrow \widehat{A}_2$ and
$q\widehat{A} = T_R(\widehat{A}_1)/(\widehat{I})$ be the quadratic graded ring generated by $\widehat{A}_1$ with the relations $\widehat{I}$
over R. Then we have a unique surjective homomorphism of graded rings $q\widehat{A} \longrightarrow \widehat{A}$
acting by the identity maps on the components of degrees 0 and 1. By construction, the
graded ring map $q\widehat{A} \longrightarrow \widehat{A}$ is also an isomorphism in degree 2.

The isomorphism $q\widehat{A}_1 \simeq \widehat{A}_1$ allows to consider t as an element of the ring $q\widehat{A}$.
Moreover, $t \in q\widehat{A}_1$ is a central element, since $q\widehat{A}$ is generated by $q\widehat{A}_1$ over R and the
relations of commutativity of t with the elements of R and $q\widehat{A}_1$ have degree ≤ 2, so they
hold in $q\widehat{A}$ whenever they hold in $\widehat{A}$.

Furthermore, by the "only if" assertion (which we have already explained), the quotient ring $A' = q\widehat{A}/(q\widehat{A})t$ is quadratic. We have the induced homomorphism of graded rings $A' = q\widehat{A}/(q\widehat{A})t \longrightarrow \widehat{A}/\widehat{A}t = A$. Since the map $q\widehat{A} \longrightarrow \widehat{A}$ is an isomorphism in degree $\leq$ 2, so is the map $A' \longrightarrow A$. Since the graded ring A is quadratic by assumption, it follows by virtue of Lemma 3.5 that the map $A' \longrightarrow A$ is an isomorphism of graded rings.

It follows that the kernel $H \subset q\widehat{A}$ of the graded ring homomorphism $q\widehat{A} \longrightarrow \widehat{A}$ is contained in $(q\widehat{A})t \subset q\widehat{A}$. Now we will prove by induction in $n \geq 3$ that $H_n = 0$. Indeed, let $h \in H_n$ be an element. Then $h = h't$ for some $h' \in q\widehat{A}_{n-1}$. The image of h under the ring homomorphism $q\widehat{A} \longrightarrow \widehat{A}$ vanishes, and since t is a nonzero-divisor in $\widehat{A}$, it follows that the image of h' under the same homomorphism vanishes as well. Hence $h' \in H_{n-1} = 0$ by the induction assumption and $h = h't = 0$.

We have shown that $q\widehat{A} \longrightarrow \widehat{A}$ is an isomorphism of graded rings, and it follows that the graded ring $\widehat{A}$ is quadratic. $\qquad\square$

Lemma 4.3 *Let $\widehat{A}$ be a nonnegatively graded ring, $t \in \widehat{A}_1$ be a central nonzero-divisor, and $A = \widehat{A}/\widehat{A}t$ be the quotient ring. Let $n \geq 0$ be an integer. Assume that A_j is a finitely generated projective (projective, or flat) left module over the ring $R = \widehat{A}_0 = A_0$ for all $0 \leq j \leq n$. Then $\widehat{A}_j$ is a finitely generated projective (respectively, projective or flat) left R-module for all $0 \leq j \leq n$.*

Proof It is provable by induction in n using the short exact sequences of R-R-bimodules
$$0 \longrightarrow \widehat{A}_{n-1} \xrightarrow{t} \widehat{A}_n \longrightarrow A_n \longrightarrow 0. \qquad\square$$

The following theorem extends to the relative context a very specific particular case of the result of [57, second assertion of Theorem 6.1].

Theorem 4.4 *Let $\widehat{A}$ be a nonnegatively graded ring and $t \in \widehat{A}_1$ be a central nonzero-divisor. Assume that $A_n = \widehat{A}_n/\widehat{A}_{n-1}t$ is a flat left R-module for every $n \geq 1$. Then the graded ring $\widehat{A}$ is left flat Koszul if and only if the graded ring A is left flat Koszul.*

Proof For any homomorphism of (graded) rings $C \longrightarrow A$, any (graded) right C-module N, and any (graded) left A-module M, the isomorphism of left derived functors of tensor product
$$(N\otimes_C^{\mathbb{L}}A)\otimes_A^{\mathbb{L}}M \simeq N\otimes_C^{\mathbb{L}}M$$
on the derived categories of modules leads to a spectral sequence of (internally graded) abelian groups
$$E_{p,q}^2 = \mathrm{Tor}_p^A(\mathrm{Tor}_q^C(N, A), M) \implies E_{p,q}^{\infty} = \mathrm{gr}_p^F \mathrm{Tor}_{p+q}^C(N, M)$$
with the differentials $d_{p,q}^r : E_{p,q}^r \longrightarrow E_{p-r,q+r-1}^r$.

In particular, for any homomorphism of nonnegatively graded rings $C \longrightarrow A$ acting by the identity map on their degree-zero components $C_0 = R = A_0$, we have a spectral sequence of internally graded R-R-bimodules

$$E^2_{p,q} = \mathrm{Tor}^A_p(\mathrm{Tor}^C_q(R, A), R) \implies E^\infty_{p,q} = \mathrm{gr}^F_p \mathrm{Tor}^C_{p+q}(R, R). \tag{4.1}$$

In the situation at hand with $C = \widehat{A}$ and $A = \widehat{A}/t$, where $t \in \widehat{A}_1$ is a central nonzero-divisor, we have

$$\mathrm{Tor}^{\widehat{A}}_q(R, A) = \begin{cases} R & \text{for } q = 0, \\ Rt & \text{for } q = 1, \\ 0 & \text{for } q \geq 2, \end{cases}$$

where the R-R-bimodule $\mathrm{Tor}^{\widehat{A}}_0(R, A) = R$ is situated in the internal degree 0 and the R-R-bimodule $\mathrm{Tor}^{\widehat{A}}_1(R, A) = Rt$ is situated in the internal degree 1.

Furthermore, the assumption that A_n is a flat left R-module for every $n \geq 1$ implies that $\widehat{A}_n$ is a flat left R-module as well, as one can show arguing by induction in n and using the short exact sequences of R-R-bimodules $0 \longrightarrow \widehat{A}_{n-1} \overset{t}{\longrightarrow} \widehat{A}_n \longrightarrow A_n \longrightarrow 0$. Now if $\mathrm{Tor}^A_{i,j}(R, R) = 0$ for all $i \neq j$, then every term $E^2_{p,q}$ of the spectral sequence (4.1) is concentrated in the internal degree $p+q$, hence so is the term $E^\infty_{p,q}$. It follows immediately that $\mathrm{Tor}^{\widehat{A}}_{i,j}(R, R) = 0$ for all $i \neq j$. So the conditions of Theorem 2.30(a) hold for $\widehat{A}$ whenever they hold for A. This proves the implication "if."

To prove the "only if," one can proceed by induction in i. Assume that the graded R-R-bimodule $\mathrm{Tor}^A_p(R, R)$ is concentrated in the internal degree $j = p$ for all $p \leq i - 1$. Then the terms $E^2_{p,q}$ are concentrated in the internal degree $p + q$ for all $p \leq i - 1$. Furthermore, if the graded R-R-bimodule $\mathrm{Tor}^{\widehat{A}}_i(R, R)$ is concentrated in the internal degree i, then so are the R-R-bimodules $E^\infty_{p,q}$ for all $p+q = i$. In particular, the term $E^\infty_{i,0}$ is concentrated in the internal degree i.

The only possibly nontrivial differentials passing through $E^r_{i,0}$ with $r \geq 2$ are $d^2_{i,0} \colon E^2_{i,0} \longrightarrow E^2_{i-2,1}$. As the term $E^2_{i-2,1}$ is concentrated in the internal degree $i - 1$ by the induction assumption and the above discussion, and the term $E^\infty_{i,0}$ is concentrated in the internal degree i, it follows that the term $E^2_{i,0} = \mathrm{Tor}^A_i(R, R)$ can only have nonzero components in the internal degrees $j = i - 1$ and i. It remains to recall that $\mathrm{Tor}^A_{i,j}(R, R) = 0$ for $j < i$ by Proposition 2.1(a). Thus the graded R-R-bimodule $\mathrm{Tor}^A_i(R, R)$ is concentrated in the internal degree $j = i$. $\qquad\square$

The next result is a kind of Poincaré–Birkhoff–Witt theorem (see the first proof of Theorem 4.25 in Sect. 4.6).

Theorem 4.5 *Let $\widehat{A}$ be a quadratic graded ring and $t \in \widehat{A}_1$ be a central element. Assume that the left R-modules $\widehat{A}_n$ are flat for all $n \geq 1$ and the graded ring $A = \widehat{A}/\widehat{A}t$ is left flat Koszul. Assume further that the three maps $R \xrightarrow{t} \widehat{A}_1 \xrightarrow{t} \widehat{A}_2 \xrightarrow{t} \widehat{A}_3$ are injective. Then the central element t is a nonzero-divisor in $\widehat{A}$.*

Proof For any graded module M over a graded ring C, let us denote by $M(1)$ the same module with the shifted grading, $M(1)_n = M_{n-1}$. Denote by $H \subset \widehat{A}$ the kernel of the multiplication map $\widehat{A} \xrightarrow{t} \widehat{A}$. Then we have a four-term exact sequence of graded $\widehat{A}$-$\widehat{A}$-bimodules

$$0 \longrightarrow H(1) \longrightarrow \widehat{A}(1) \xrightarrow{t} \widehat{A} \longrightarrow A \longrightarrow 0.$$

It follows that $\mathrm{Tor}_2^{\widehat{A}}(R, A) \simeq R \otimes_{\widehat{A}} H(1)$ and $\mathrm{Tor}_0^{\widehat{A}}(\mathrm{Tor}_2^{\widehat{A}}(R, A), R) \simeq R \otimes_{\widehat{A}} H(1) \otimes_{\widehat{A}} R$ as an internally graded R-R-bimodule. We also have $\mathrm{Tor}_1^{\widehat{A}}(R, A) = Rt \simeq R(1)$ and $\mathrm{Tor}_0^{\widehat{A}}(R, A) = R$.

By assumption, we have $H_n = 0$ for $n \leq 2$. Arguing by induction in $n \geq 3$, we will prove that $H_n = 0$ for all n. Assuming that $H_{n-1} = 0$, the R-R-bimodule H_n is isomorphic to the degree n component of the R-R-bimodule $R \otimes_{\widehat{A}} H \otimes_{\widehat{A}} R$, or which is the same, the degree $n + 1$ component of the R-R-bimodule $R \otimes_{\widehat{A}} H(1) \otimes_{\widehat{A}} R$.

We will make use of the spectral sequence (4.1) for the graded ring homomorphism $C = \widehat{A} \longrightarrow A$. In the induction assumption as above, we need to check that the degree $n + 1$ component of the term $E_{0,2}^2$ vanishes. Indeed, we have $\mathrm{Tor}_{2,j}^{\widehat{A}}(R, R) = 0$ for $j \geq 3$ by Proposition 2.4(b) (since the graded ring $\widehat{A}$ is quadratic and its grading components are flat left R-modules by assumption), and hence the term $E_{p,q}^\infty$ has no grading components in the internal degrees $j \geq 3$ when $p + q = 2$. In particular, this applies to the term $E_{0,2}^\infty$.

The only possibly nontrivial differentials passing through $E_{0,2}^r$ with $r \geq 2$ are $d_{2,1}^2 : E_{2,1}^2 \longrightarrow E_{0,2}^2$ and $d_{3,0}^3 : E_{3,0}^3 \longrightarrow E_{0,2}^3$. Since the graded ring A is left flat Koszul by assumption, we have $\mathrm{Tor}_{2,j}^A(R, R) = 0$ for $j \geq 3$ and $\mathrm{Tor}_{3,j}^A(R, R) = 0$ for $j \geq 4$ by Theorem 2.30(a). Therefore, both the terms $E_{2,1}^2$ and $E_{3,0}^2$ have no grading components in the internal degrees $j \geq 4$. It follows that the term $E_{0,2}^2$ cannot have a nonzero grading component in the degree $n + 1 \geq 4$, and we are done. $\square$

4.2 Quasi-Differential Graded Rings and CDG-Rings

The following definition plays an important role in our theory.

Definition 4.6 A *quasi-differential graded ring* $(\widehat{B}, \partial)$ is a graded associative ring $\widehat{B} = \bigoplus_{n \in \mathbb{Z}} \widehat{B}^n$ endowed with an odd derivation $\partial : \widehat{B} \longrightarrow \widehat{B}$ of degree -1 such that $\partial^2 = 0$ and the homology ring $H_\partial(\widehat{B}) = \ker \partial / \mathrm{im} \, \partial$ vanishes. The latter condition holds (for an odd derivation ∂ of degree -1 satisfying $\partial^2 = 0$) if and only if the homology class of the unit element $1 \in \widehat{B}^0$ vanishes, that is, $1 \in \partial(\widehat{B}^1)$.

The *underlying graded ring* B of a quasi-differential graded ring $(\widehat{B}, \partial)$ is defined as the kernel of the differential $\partial \colon \widehat{B} \longrightarrow \widehat{B}$, that is, $B = \ker \partial \subset \widehat{B}$. A *quasi-differential structure* on a graded ring B is the datum of a quasi-differential graded ring $(\widehat{B}, \partial)$ together with a graded ring isomorphism $B \simeq \ker \partial \subset \widehat{B}$.

A *morphism of quasi-differential graded rings* $\hat{f} \colon (''\widehat{B}, \partial'') \longrightarrow ('\widehat{B}, \partial')$ is a morphism of graded rings $\hat{f} \colon ''\widehat{B} \longrightarrow '\widehat{B}$ such that $\hat{f}\partial'' = \partial'\hat{f}$. The composition of morphisms of quasi-differential graded rings is defined in the obvious way. These rules define the *category of quasi-differential graded rings*.

A quasi-differential graded ring $(\widehat{B}, \partial)$ is said to be *nonnegatively graded* if its underlying graded ring B is nonnegatively graded, $B = \bigoplus_{n=0}^{\infty} B^n$, or equivalently, the graded ring $\widehat{B}$ is nonnegatively graded, $\widehat{B} = \bigoplus_{n=0}^{\infty} \widehat{B}^n$. For a nonnegatively graded quasi-differential ring $(\widehat{B}, \partial)$, one has $B^0 = \widehat{B}^0$.

We will denote the category of nonnegatively graded quasi-differential rings $(\widehat{B}, \partial)$ with the fixed degree-zero component $\widehat{B}^0 = B^0 = R$ by $R\text{-rings}_{\mathsf{qdg}}$. Morphisms $\hat{f} \colon (''\widehat{B}, \partial'') \longrightarrow ('\widehat{B}, \partial')$ in $R\text{-rings}_{\mathsf{qdg}}$ are morphisms of quasi-differential rings such that the graded ring homomorphism $\hat{f} \colon ''\widehat{B} \longrightarrow '\widehat{B}$ forms a commutative triangle diagram with the fixed isomorphisms $R \simeq ''\widehat{B}^0$ and $R \simeq '\widehat{B}^0$, or equivalently, the induced homomorphism $f \colon ''B \longrightarrow 'B$ between the graded rings $''B = \ker \partial'' \subset ''\widehat{B}$ and $'B = \ker \partial' \subset '\widehat{B}$ forms a commutative diagram with the fixed isomorphisms $R \simeq ''B^0$ and $R \simeq 'B^0$.

Theorem 4.7 *The category of quasi-differential graded rings is equivalent to the category of CDG-rings. In particular, for any fixed ring R, the category $R\text{-rings}_{\mathsf{qdg}}$ is equivalent to the category $R\text{-rings}_{\mathsf{cdg}}$.*

Proof To a CDG-ring (B, d, h), one assigns the following quasi-differential graded ring $(\widehat{B}, \partial)$. The graded ring $\widehat{B} = B[\delta]$ is obtained by adjoining to the graded ring B a new element $\delta \in \widehat{B}^1$ satisfying the relations

$$[\delta, b] = \delta b - (-1)^{|b|} b\delta = d(b) \quad \text{for all } b \in B \tag{4.2}$$

and

$$\delta^2 = h. \tag{4.3}$$

In other words, the elements of the grading component $\widehat{B}^n$ are all the formal expressions $b + c\delta$ with $b \in B^n$ and $c \in B^{n-1}$. The unit element in $\widehat{B}$ is $1_{\widehat{B}} = 1_B + 0\delta$. The multiplication in $\widehat{B}$ is given by the formula

$$(b' + c'\delta)(b'' + c''\delta) = (b'b'' + c'd(b'') + (-1)^{|c''|}c'c''h) + (b'c'' + (-1)^{|b''|}c'b'' + c'd(c''))\delta.$$

The odd derivation ∂ on $\widehat{B}$ can be described informally as the partial derivative $\partial = \partial/\partial\delta$. Explicitly, we put $\partial(b + c\delta) = (-1)^{|c|}c + 0\delta$. So the kernel $\ker \partial \subset \widehat{B}$ clearly coincides with the subring $B \subset \widehat{B}$ embedded by the obvious rule $b \longmapsto b + 0\delta$.

Let $(\mathrm{id}, a) \colon (B, d'', h'') \longrightarrow (B, d, h)$ be a change-of-connection isomorphism of CDG-rings, so $d''(b) = d'(b) + [a, b]$ for all $b \in B$ and $h'' = h' + d'(a) + a^2$. Let $''\widehat{B} = B[\delta'']$ and $'\widehat{B} = B[\delta']$ denote the quasi-differential graded rings corresponding to (B, d'', h'') and (B, d', h'), respectively. Then the rules $b \longmapsto b$ for all $b \in B$ and $\delta'' \longmapsto \delta' + a$ define an isomorphism of graded rings $''\widehat{B} \simeq {}'\widehat{B}$ forming a commutative square diagram with the odd derivations $\partial'' \colon {}''\widehat{B} \longrightarrow {}''\widehat{B}$ and $\partial' \colon {}'\widehat{B} \longrightarrow {}'\widehat{B}$.

Generally, to a morphism of CDG-rings $(f, a) \colon (''B, d'', h'') \longrightarrow ('B, d', h')$, one assigns the morphism of quasi-differential graded rings $\hat{f} \colon (''\widehat{B}, \partial'') \longrightarrow ('\widehat{B}, \partial')$, where the graded ring homomorphism $\hat{f} \colon {}''\widehat{B} = {}''B[\delta''] \longrightarrow {}'B[\delta'] = {}'\widehat{B}$ takes any element $b \in {}''B \subset {}''\widehat{B}$ to the element $\hat{f}(b) = f(b) \in {}'B \subset {}'\widehat{B}$ and the element $\delta'' \in {}''\widehat{B}$ to the element $\hat{f}(\delta'') = \delta' + a \in {}'\widehat{B}$. This construction defines a functor from the category of CDG-rings to the category of quasi-differential graded rings, and one can easily see that this functor is fully faithful.

To construct the inverse functor, one needs to choose, for each quasi-differential graded ring $(\widehat{B}, \partial)$, an element $\delta \in \widehat{B}^1$ such that $\partial(\delta) = 1$. Then the CDG-ring (B, d, h) is recovered by the rules $B = \ker \partial \subset \widehat{B}$, $d(b) = [\delta, b]$ for all $b \in B \subset \widehat{B}$, and $h = \delta^2$. One has $\partial([\delta, b]) = [\partial(\delta), b] - [\delta, \partial(b)] = [1, b] - [\delta, 0] = 0$ for all $b \in B$ and $\partial(\delta^2) = [\partial(\delta), \delta] = [1, \delta] = 0$, hence $[\delta, b] \in B$ and $\delta^2 \in B$, as desired. The construction of the morphism of CDG-rings assigned to a morphism of quasi-differential graded rings is obvious from the above. □

One can define the 2-*category of quasi-differential graded rings* in the way similar to the definition of the 2-category of DG-rings in Sect. 3.2. Let $\hat{f}, \hat{g} \colon (''\widehat{B}, \partial'') \longrightarrow ('\widehat{B}, \partial')$ be a pair of parallel morphisms of quasi-differential graded rings. A 2-morphism $\hat{f} \xrightarrow{z} \hat{g}$ is an invertible element $z \in {}'\widehat{B}^0 = \ker(\partial_0 \colon {}'\widehat{B}^0 \to {}'\widehat{B}^{-1})$ such that $\hat{g}(c) = z\hat{f}(c)z^{-1}$ for all $c \in {}''\widehat{B}$. The composition of 2-morphisms is defined in the same way as in Sect. 3.2.

The 2-category $\mathsf{Rings_{qdg2}}$ of nonnegatively graded quasi-differential rings is defined as the following subcategory of the 2-category of quasi-differential graded rings. The objects of $\mathsf{Rings_{qdg2}}$ are nonnegatively graded quasi-differential rings $(\widehat{B}, \partial)$. Morphisms $\hat{f} \colon (''\widehat{B}, \partial'') \longrightarrow ('\widehat{B}, \partial')$ in $\mathsf{Rings_{qdg2}}$ are morphisms of quasi-differential graded rings such that the map $\hat{f}_0 \colon {}''\widehat{B}^0 \longrightarrow {}'\widehat{B}^0$ is an isomorphism. 2-morphisms $\hat{f} \xrightarrow{z} \hat{g}$ in

Rings$_{\mathsf{qdg2}}$ between morphisms $\hat{f}$ and $\hat{g}$ belonging to Rings$_{\mathsf{qdg2}}$ are arbitrary 2-morphisms from $\hat{f}$ to $\hat{g}$ in the 2-category of quasi-differential graded rings.

Theorem 4.8 *The equivalence of categories from Theorem 4.7 can be extended to a strict equivalence between the 2-category of CDG-rings and the 2-category of quasi-differential graded rings. In particular, the 2-category of nonnegatively graded CDG-rings* Rings$_{\mathsf{cdg2}}$ *is strictly equivalent to the 2-category of quasi-differential graded rings* Rings$_{\mathsf{qdg2}}$.

Proof Let (f, a) and $(g, b)\colon ({''}B, d'', h'') \longrightarrow ({'}B, d', h')$ be a pair of parallel morphisms of CDG-rings and $\hat{f}$, $\hat{g}\colon {''}\widehat{B} = {''}B[\delta''] \longrightarrow {'}B[\delta'] = {'}\widehat{B}$ be the corresponding pair of parallel morphisms of quasi-differential graded rings. Let $z \in {'}B^0$ be an invertible element. We leave it to the reader to check that $(f, a) \xrightarrow{\;z\;} (g, b)$ is a 2-morphism in the 2-category of CDG-rings if and only if $\hat{f} \xrightarrow{\;z\;} \hat{g}$ is a 2-morphism in the 2-category of quasi-differential graded rings. □

4.3 Quadratic Quasi-Differential Graded Rings

We will say that a nonnegatively graded quasi-differential ring $(\widehat{B}, \partial)$ is *quadratic* if its underlying graded ring $B = \ker \partial \subset \widehat{B}$ is quadratic.

Lemma 4.9 *Let $(\widehat{B}, \partial)$ be a nonnegatively graded quasi-differential ring and $B = \ker \partial \subset \widehat{B}$ be its underlying graded ring. Then*

(a) *The graded ring $\widehat{B}$ is generated by $\widehat{B}^1$ over $\widehat{B}^0$ whenever the graded ring B is generated by B^1 over B^0.*
(b) *The graded ring $\widehat{B}$ is quadratic whenever the graded ring B is quadratic.*

Proof

Part (a):   following the proof of Theorem 4.7, we have $\widehat{B} = B[\delta]$, where $\delta \in \widehat{B}^1$. So the ring $\widehat{B}$ is generated by B and δ, and if B is generated by B^1 over B^0, then it follows that $\widehat{B}$ is generated by $\widehat{B}^1$ over $\widehat{B}^0 = B^0$.

Part (b):   the ring $\widehat{B} = B[\delta]$ is generated by the ring B and the adjoined element δ with the defining relations (4.2–4.3). The relation (4.3) has degree 2, and it remains to observe that, whenever the ring B is generated by B^1 over B^0, it suffices to impose the relation (4.2) for elements $b \in B^0$ and B^1 only. This makes the ring $\widehat{B}$ defined by relations of degree ≤ 2 between generators of degree ≤ 1 provided that the ring B can be so defined.

□

Examples 4.10 The following examples show that the converse implications in the context of Lemma 4.9 are *not* true.

(1) Let $\widehat{B} = k[\delta]$ be the polynomial ring in one variable δ over a field k, endowed with the grading in which $\deg \delta = 1$. Let $\partial \colon \widehat{B} \longrightarrow \widehat{B}$ be the unique odd derivation on $\widehat{B}$ such that $\partial(\delta) = 1$, so $\partial(\delta^{2n}) = 0$ and $\partial(\delta^{2n+1}) = \delta^{2n}$ for all the integers $n \geq 0$. Then $\partial^2 = 0$, and the graded ring $\widehat{B}$ is generated by $\widehat{B}^1$ over $\widehat{B}^0 = k$ (in fact, $\widehat{B}$ is quadratic), but the subring $B = \ker \partial \subset \widehat{B}$ is *not* generated by B^1 over B^0.

(2) To give another counterexample to the converse assertion in part (a) of the lemma, let $B = k[x, y]/(x^2)$ be the free graded commutative k-algebra with the generators x and y of degrees $\deg x = 1$ and $\deg y = 2$. The rules $d(x) = y$ and $d(y) = 0$ define an odd derivation $d \colon B \longrightarrow B$ of degree 1 with $d^2 = 0$, making (B, d) a DG-ring. Let $\widehat{B} = B[\delta]$ be the quasi-differential ring corresponding to the CDG-ring $(B, d, 0)$ under the equivalence of categories from Theorem 4.7. Then the ring $\widehat{B}$ is generated by the elements x and $\delta \in B^1$, because the element y can be expressed as $y = \delta x + x\delta$. So the graded ring $\widehat{B}$ is generated by $\widehat{B}^1$ over $\widehat{B}^0 = k$, but the graded ring B is *not* generated by B^1 over B^0.

Alternatively, instead of the free graded commutative k-algebra $B = k[x, y]/(x^2)$, one can use the free associative algebra $B' = k\{x, y\}$ with two generators x and y (with $\deg x = 1$, $\deg y = 2$, and $d(x) = y$) in the same construction.

(3) Let $E = \Lambda_k(x, y, z)$ be the exterior algebra in three variables x, y, and z over a field k (so the relations $x^2 = y^2 = z^2 = 0 = xy + yx = xz + zx = yz + zy$ hold in E). Define the grading on E by the standard rule $\deg x = \deg y = \deg z = 1$, so E is a quadratic graded ring. Let $d \colon E \longrightarrow E$ be the odd derivation of degree 1 defined by the formulas $d(x) = xy$, $d(y) = d(z) = 0$; then $d^2 = 0$ on E. Put $h = xz \in E^2$. Then (E, d, h) is *not* a CDG-ring: the equation $d^2(e) = 0 = [h, e]$ holds for all $e \in E$, but $d^2(h) = xyz \neq 0$ in E^3.

Put $B = E/(xyz)$, so the cubic relation $xyz = 0$ is imposed in the graded ring B, which is consequently *not* quadratic. Then the differential $d \colon E \longrightarrow E$ induces an odd derivation $d \colon B \longrightarrow B$ with $d^2 = 0$. Denote for simplicity the image of the element h in B^2 also by h. Then (B, d, h) is a CDG-ring.

Now let $\widehat{B} = B[\delta]$ be the quasi-differential ring corresponding to the CDG-ring (B, d, h). Then $\widehat{B}$ is a quadratic graded ring. Indeed, the graded k-algebra $\widehat{B}$ is generated by the elements x, y, z, δ of degree 1 with the quadratic relations listed above as the defining relations of the exterior algebra E together with the quadratic relations involving δ, namely, $\delta x + x\delta = xy$, $\delta y + y\delta = 0$, $\delta z + z\delta = 0$, and $\delta^2 = xz$. The relation $xyz = 0$ in $\widehat{B}^3$ follows from these quadratic relations, as one has $0 = \delta^3 - \delta^3 = \delta xz - xz\delta = (\delta x + x\delta)z - x(\delta z + z\delta) = xyz$.

(4) Here is another counterexample to the converse assertion to part (b) of the lemma. Let $E = k\{x, y\}$ be the free associative algebra with two generators x and y over a field k. Define the grading on E by the standard rule $\deg x = \deg y = 1$, so E is a quadratic graded ring. Let $d \colon E \longrightarrow E$ be the odd derivation of degree 1 defined by the formulas

$d(x) = xy$ and $d(y) = 0$. Then $d^2 \neq 0$ on E; in fact, one has $d^2(x) = xy^2 \neq 0$. So (E, d) is *not* a DG-ring and $(E, d, 0)$ is not a CDG-ring.

Put $B = E/(xy^2)$, where (xy^2) is the two-sided ideal generated by the element $xy^2 \in E^3$, so the cubic relation $xy^2 = 0$ is imposed in the graded ring B, which is therefore *not* quadratic. Then the differential $d \colon E \longrightarrow E$ induces an odd derivation $d \colon B \longrightarrow B$ with $d^2 = 0$ on B. Hence (B, d) is a DG-ring and $(B, d, 0)$ is a CDG-ring.

Let $\widehat{B} = B[\delta]$ be the quasi-differential ring corresponding to the CDG-ring $(B, d, 0)$. Then $\widehat{B}$ is a quadratic graded ring. Indeed, the graded k-algebra $\widehat{B}$ is generated by the elements x, y, δ of degree 1 with the quadratic relations $\delta x + x\delta = xy$, $\delta y + y\delta = 0$, and $\delta^2 = 0$. The relation $xy^2 = 0$ in $\widehat{B}^3$ follows from these quadratic relations, as one has $0 = \delta^2 x - x\delta^2 = \delta(\delta x + x\delta) - (\delta x + x\delta)\delta = \delta xy - xy\delta = (\delta x + x\delta)y - x(\delta y + y\delta) = xy^2$.

Lemma 4.11 *Let $(\widehat{B}, \partial)$ be a nonnegatively graded quasi-differential ring and $B = \ker \partial \subset \widehat{B}$ be its underlying graded ring.*

(a) *Assume that the graded ring $\widehat{B}$ is generated by $\widehat{B}^1$ over $\widehat{B}^0$. Then the graded ring B is generated by B^1 over B^0 if and only if the component B^2 is generated by B^1, that is, the multiplication map $B^1 \otimes_{B^0} B^1 \longrightarrow B^2$ is surjective.*

(b) *Assume that the graded ring B is generated by B^1 over B^0 and the graded ring $\widehat{B}$ is quadratic. Then the graded ring B is quadratic if and only if it has no relations of degree 3, that is, in other words, the natural homomorphism of graded rings $qB \longrightarrow B$ is injective in degree 3.*

Proof The "only if" assertion in both (a) and (b) is obvious. To prove the "if," choose an element $\delta \in \widehat{B}^1$ such that $\partial(\delta) = 1$ and consider the CDG-ring (B, d, h) corresponding to $(\widehat{B}, \partial)$ under the equivalence of categories from Theorem 4.7.

Part (a):   let $'B \subset B$ denote the graded subring in B generated by $'B^1 = B^1$ over $'B^0 = B^0$. By assumption, we have $'B^2 = B^2$. Hence the curvature element $h \in B^2$ belongs to $'B$. Moreover, we have $d('B^1) \subset 'B^2$, and it follows that $d('B) \subset 'B \subset B$. Thus $('B, d|_{'B}, h)$ is a CDG-ring. Let $'\widehat{B} = 'B[\delta]$ be the quasi-differential graded ring corresponding to the CDG-ring $('B, d|_{'B}, h)$ under the construction of Theorem 4.7. Then $'\widehat{B}$ is naturally a graded subring in $\widehat{B}$. Furthermore, we have $'\widehat{B}^0 = \widehat{B}^0$ and $'\widehat{B}^1 = \widehat{B}^1$, since $'B^0 = B^0$ and $'B^1 = B^1$. Since the graded ring $\widehat{B}$ is generated by $\widehat{B}^1$ over $\widehat{B}^0$ by assumption, it follows that $'\widehat{B} = \widehat{B}$. In view of the equivalence of categories from Theorem 4.7, we can conclude that $'B = B$.

Part (b):   the natural homomorphism of graded rings $qB \longrightarrow B$, which we will denote by f, is surjective in our assumptions. If it is injective in degree 3, this means that it is an isomorphism in degrees 2 and 3. So we can consider the curvature element $h \in B^2$ as an element of qB^2.

Furthermore, the components $d_0 \colon B^0 \longrightarrow B^1$ and $d_1 \colon B^1 \longrightarrow B^2$ of the odd derivation $d \colon B \longrightarrow B$ can be considered as maps $d_0' \colon qB^0 \longrightarrow qB^1$ and $d_1' \colon qB^1 \longrightarrow qB^2$. The possibility of extending these maps to an odd derivation $d' \colon qB \longrightarrow qB$ presents itself if and only if they respect the defining relations of the graded ring qB, which all live in the degrees ≤ 2. The same relations hold in the graded ring B. Applying the maps d_0' and d_1' to the defining relations of qB, one obtains equations for elements of the components qB^1, qB^2, and qB^3. These are isomorphic to B^1, B^2, and B^3, respectively, via the graded ring homomorphism $f \colon qB \longrightarrow B$. Since the odd derivation $d \colon B \longrightarrow B$ exists, it follows that one can indeed extend the maps d_0' and d_1' to an odd derivation $d' \colon qB \longrightarrow qB$.

Finally, the equation $d'(h) = 0$ holds in qB^3 since the equation $d(h) = 0$ holds in B^3. Concerning the equation $d'(d'(b)) = [h, b]$ for all $b \in qB$, it suffices to check it for $b \in qB^0$ and qB^1, since the graded ring qB is generated by qB^1 over qB^0. Once again, the equations need to hold in the components qB^2 and qB^3, which are isomorphic to B^2 and B^3 via the graded ring homomorphism f. Since (B, d, h) is a CDG-ring, it follows that (qB, d', h) is a CDG-ring as well. We have a strict morphism of CDG-rings $(f, 0) \colon (qB, d', h) \longrightarrow (B, d, h)$.

Let $'\widehat{B} = (qB)[\delta]$ denote the quasi-differential graded ring corresponding to the CDG-ring (qB, d', h) under the construction of Theorem 4.7, and let $\hat{f} \colon {'\widehat{B}} \longrightarrow \widehat{B}$ denote the morphism of quasi-differential graded rings corresponding to the CDG-ring morphism $(f, 0)$. Then the graded ring homomorphism $\hat{f}$ is an isomorphism in degrees 0, 1, and 2, since so is the graded ring homomorphism f. Since the graded ring $\widehat{B}$ is quadratic by assumption and the graded ring $'\widehat{B}$ is quadratic by Lemma 4.9, we can conclude that the graded ring homomorphism $\hat{f} \colon {'\widehat{B}} \longrightarrow \widehat{B}$ is an isomorphism (in all the degrees) by Lemma 3.5. In view of the equivalence of categories from Theorem 4.7, it follows that the graded ring homomorphism $f \colon qB \longrightarrow B$ is an isomorphism, too. Thus the graded ring B is quadratic.   □

Lemma 4.12 *Let $(\widehat{B}, \partial)$ be a nonnegatively graded quasi-differential ring and $B = \ker \partial \subset \widehat{B}$ be its underlying graded ring. Let $n \geq 0$ be an integer. Then $\widehat{B}^j$ is a finitely generated projective (projective, or flat) right module over the ring $R = \widehat{B}^0 = B^0$ for all $0 \leq j \leq n$ if and only if B^j is a finitely generated projective (respectively, projective or flat) right R-module for all $0 \leq j \leq n$.*

Proof First we observe that, for any odd derivation ∂ of degree -1 on a nonnegatively graded ring $\widehat{B}$, the component $\partial_0 \colon \widehat{B}^0 \longrightarrow \widehat{B}^{-1}$ vanishes, since $\widehat{B}^{-1} = 0$. In other words, we have $\partial(R) = 0$; hence ∂ is an R-R-bimodule map.

Returning to the situation at hand, the case $n = 0$ is obvious. Arguing by induction in n, it suffices to assume that B^{n-1} is a finitely generated projective (respectively, projective or flat) right R-module and prove that $\widehat{B}^n$ is a finitely generated projective (respectively, projective or flat) right R-module if and only if B^n is. This is clear from the short exact sequence of R-R-bimodules $0 \longrightarrow B^n \longrightarrow \widehat{B}^n \overset{\partial}{\longrightarrow} B^{n-1} \longrightarrow 0$.   □

The following theorem is the relative version of a very specific particular case of [57, Corollary 6.2(b)].

Theorem 4.13 *Let $(\widehat{B}, \partial)$ be a nonnegatively graded quasi-differential ring and $B = \ker \partial \subset \widehat{B}$ be its underlying graded ring. Then*

(a) *The graded ring $\widehat{B}$ is right flat Koszul whenever the graded ring B is.*
(b) *The graded ring $\widehat{B}$ is right finitely projective Koszul whenever the graded ring B is.*

Proof In view of Lemma 4.12, it suffices to prove part (a). Then, by the same lemma, we know that $\widehat{B}^n$ is a flat right R-module for every $n \geq 0$.

In the notation of the proof of Theorem 4.5, we have a short exact sequence of R-B-bimodules $0 \longrightarrow B \longrightarrow \widehat{B} \xrightarrow{\partial} B(1) \longrightarrow 0$. So the graded right B-module $\widehat{B}$ is projective (in fact, free). Hence the spectral sequence (4.1) for the morphism of graded rings $B \longrightarrow \widehat{B}$ (with the roles of the left and right sides switched) degenerates to an isomorphism

$$\mathrm{Tor}_i^B(R, R) \simeq \mathrm{Tor}_i^{\widehat{B}}(R, \mathrm{Tor}_0^B(\widehat{B}, R)) \quad \text{for all } i \geq 0.$$

Furthermore, the internally graded R-R-bimodule $\mathrm{Tor}_0^B(\widehat{B}, R)$ can be computed as

$$\widehat{B}/\widehat{B}B^+ = \mathrm{Tor}_0^B(\widehat{B}, R) = R \oplus R(1),$$

where $B^+ = \bigoplus_{n=1}^{\infty} B^n$. The R-R-bimodule direct summand $R(1) \subset \mathrm{Tor}_0^B(\widehat{B}, R)$ is a left $\widehat{B}$-subbimodule, since the ring $\widehat{B}$ is nonnegatively graded. So we have a short exact sequence of graded $\widehat{B}$-R-bimodules

$$0 \longrightarrow R(1) \longrightarrow \mathrm{Tor}_0^B(\widehat{B}, R) \longrightarrow R \longrightarrow 0,$$

where $\widehat{B}$ acts in $R(1)$ and in R via the augmentation map $B \longrightarrow B/B^+ = R$. We are arriving to a long exact sequence of R-R-bimodules

$$\cdots \longrightarrow \mathrm{Tor}_{i,j-1}^{\widehat{B}}(R, R) \longrightarrow \mathrm{Tor}_{i,j}^B(R, R)$$

$$\longrightarrow \mathrm{Tor}_{i,j}^{\widehat{B}}(R, R) \longrightarrow \mathrm{Tor}_{i-1,j-1}^{\widehat{B}}(R, R) \longrightarrow \cdots,$$

where the index $i \geq 0$ denotes the homological grading, while the index $j \geq 0$ denotes the internal grading of the Tor induced by the grading of B and $\widehat{B}$.

Now if $\mathrm{Tor}_{i,j}^B(R, R) = 0$ for $i \neq j$ and $\mathrm{Tor}_{i-1,j-1}^{\widehat{B}}(R, R) = 0$, then it follows that $\mathrm{Tor}_{i,j}^{\widehat{B}}(R, R) = 0$. Arguing by induction in $i \geq 0$ (or in $j \geq 0$), one proves that $\mathrm{Tor}_{i,j}^{\widehat{B}}(R, R) = 0$ for $i \neq j$. $\qquad\square$

4.4   Central Elements and Acyclic Derivations

Let $\widehat{A} = \bigoplus_{n=0}^{\infty} \widehat{A}_n$ be a 2-left finitely projective quadratic graded ring with the degree-zero component $R = \widehat{A}_0$, and let $\widehat{B} = \bigoplus_{n=0}^{\infty} \widehat{B}^n$, $\widehat{B}^0 = R$, be the 2-right finitely projective quadratic graded ring quadratic dual to $\widehat{A}$. The aim of this section is to establish and study a correspondence between central elements $t \in \widehat{A}_1$ and odd derivations $\partial : \widehat{B} \longrightarrow \widehat{B}$ of degree -1.

First of all, to every element $t \in A_1 \simeq \mathrm{Hom}_{R^{\mathrm{op}}}(\widehat{B}^1, R)$, we assign a right R-module morphism $\partial_1 : \widehat{B}^1 \longrightarrow R$ and vice versa. This is done in the obvious way except for a sign rule. In the pairing notation of Sect. 3.4, we put

$$\partial_1(b) = -\langle t, b \rangle \in R \qquad \text{for all } b \in \widehat{B}^1.$$

Lemma 4.14 *For any element $t \in \widehat{A}_1$, consider the related right R-module morphism $\partial_1 : \widehat{B}^1 \longrightarrow R$. Then*

(a) *The element t commutes with all the elements of $\widehat{A}_0 = R$ in $\widehat{A}$ if and only if the map ∂_1 is an R-R-bimodule morphism.*

(b) *The element t is central in $\widehat{A}$ if and only if the map ∂_1 can be (uniquely) extended to an odd derivation $\partial : \widehat{B} \longrightarrow \widehat{B}$ of degree -1.*

Examples 4.15

(1) Let V be a finite-dimensional vector space over a field k and $\widehat{A} = \mathrm{Sym}_k(V)$ be the symmetric algebra of V. Let V^* be the dual vector space and $\widehat{B} = \Lambda_k(V^*)$ be its exterior algebra. Then $\widehat{A}$ and $\widehat{B}$ are quadratic graded rings (by Examples 1.3(4–5)) which are quadratic dual to each other (see Example 1.7(3)).

Let $x_1, \ldots, x_m$ be a basis in V and $\xi_1, \ldots, \xi_m$ be the dual basis in V^*. Then any element $a_1 x_1 + \cdots + a_m x_m \in V$, where $a_1, \ldots, a_m \in k$, is central in $\widehat{A} = \mathrm{Sym}_k(V)$. The corresponding odd derivation on $\widehat{B}$ is $a_1 \frac{\partial}{\partial \xi_1} + \cdots + a_m \frac{\partial}{\partial \xi_m} : \Lambda_k(V^*) \longrightarrow \Lambda_k(V^*)$.

(2) Let R be a commutative ring and V be a finitely generated projective R-module. Put $V^{\vee} = \mathrm{Hom}_R(V, R)$. Let $\widehat{A} = \mathrm{Sym}_R(V)$ be the symmetric algebra of V and $\widehat{B} = \Lambda_R(V^{\vee})$ be the exterior algebra of $V^{\vee}$ over R. Then $\widehat{A}$ and $\widehat{B}$ are 2-left and right finitely projective quadratic graded rings which are quadratic dual to each other (see Examples 1.5(2) and 1.7(3)).

Any element $t \in V$ is central in $\widehat{A} = \mathrm{Sym}_R(V)$, and any R-linear map $V^{\vee} \longrightarrow R$ can be (uniquely) extended to an odd derivation $\partial : \widehat{B} \longrightarrow \widehat{B}$ of degree -1.

(3) Let R be a commutative ring and V be a free R-module with one generator. Let $\widehat{A} = \Lambda_R(V)$ be the exterior algebra of V and $\widehat{B} = \mathrm{Sym}_R(V^{\vee})$ be the symmetric algebra of $V^{\vee}$ over R. In this very special particular case, the exterior algebra $\widehat{A}$ is not only graded commutative but also commutative, so all the elements of $\widehat{A}_1$ are central

in $\widehat{A}$. Accordingly, any R-linear map $\widehat{B}_1 \longrightarrow R$ can be extended to an odd derivation $\partial : \widehat{B} \longrightarrow \widehat{B}$ (cf. Example 4.10(1)).

(4) Let R be a commutative algebra over a field of characteristic 2 and V be a projective R-module. Put $\widehat{A} = \Lambda_R(V)$ and $\widehat{B} = \mathrm{Sym}_R(V^{\vee})$. In this special case, any graded commutative ring is commutative, so in particular the exterior algebra $\widehat{A}$ is commutative and all the elements of $\widehat{A}_1$ are central. Hence any R-linear map $\widehat{B}_1 \longrightarrow R$ can be extended to an odd derivation $\partial : \widehat{B} \longrightarrow \widehat{B}$.

Notice that there is no difference between odd and even derivations in characteristic 2. For example, if V is a free R-module with a set of free generators $\xi_1, \ldots, \xi_m$, while $x_1, \ldots, x_m$ is the dual set of free generators in the free R-module $V^{\vee}$, then $a_1 \frac{\partial}{\partial x_1} + \cdots + a_m \frac{\partial}{\partial x_m} : \mathrm{Sym}(V^{\vee}) \longrightarrow \mathrm{Sym}(V^{\vee})$ is the odd derivation of $\widehat{B}$ corresponding to an element $a_1 \xi_1 + \cdots + a_m \xi_m \in \widehat{A}_1 = V$, where $a_1, \ldots, a_m \in R$.

(5) Let V be a vector space of dimension ≥ 2 over a field k of characteristic different from 2. Put $\widehat{A} = \Lambda_k(V)$ and $\widehat{B} = \mathrm{Sym}_k(V^*)$. Then there are no nonzero central elements in $\widehat{A}_1$, and accordingly there are no nonzero odd derivations of degree -1 on $\widehat{B}$. For example, if $\dim_k V = 2$ and x, y is a basis in V^*, so $\widehat{B} = k[x, y]$ is the polynomial ring with the standard grading $\deg x = \deg y = 1$, then there does *not* exist an odd derivation of degree -1 on $\widehat{B}$ taking x to 1 and y to 0. Indeed, for such a derivation ∂ one would have $\partial(xy) = \partial(x)y - x\partial(y) = y$ but $\partial(yx) = \partial(y)x - y\partial(x) = -y$, a contradiction.

(6) Similarly one can take $\widehat{A} = T_k(V)$ and $\widehat{B} = k \oplus V^*$, where V is a vector space of dimension ≥ 2 over a field k (see Example 1.7(1)). Then there are no nonzero central elements in $\widehat{A}_1$ and no nonzero odd derivations of degree -1 on $\widehat{B}$.

(7) On the other hand, if $\widehat{A} = k \oplus V$ and $\widehat{B} = T_k(V^*)$, then all the elements of $\widehat{A}_1$ are central in $\widehat{A}$, and any k-linear map $V^* \longrightarrow k$ can be extended to an odd derivation of degree -1 on $\widehat{B}$.

Proof of Lemma 4.14

Part (a):   we have $\partial_1(b) = -\langle t, b \rangle$ for all $b \in \widehat{B}^1$. Furthermore,

$$\langle rt, b \rangle = r\langle t, b \rangle \quad \text{and} \quad \langle tr, b \rangle = \langle t, rb \rangle \qquad \text{for all } r \in R \text{ and } b \in \widehat{B}^1.$$

Hence one has $rt = tr$ for all $r \in R$ if and only if $r\partial_1(b) = \partial_1(rb)$ for all $r \in R$ and $b \in \widehat{B}^1$, i.e., if and only if ∂_1 is a left R-module map.

Part (b):   any odd derivation $\partial : \widehat{B} \longrightarrow \widehat{B}$ of degree -1 is an R-R-bimodule map, as it was explained in the proof of Lemma 4.12. Moreover, the odd derivation ∂ is uniquely determined by its component ∂_1, since the graded ring $\widehat{B}$ is generated by $\widehat{B}^1$ over $\widehat{B}^0$. Similarly, the element t is central in $\widehat{A}$ if and only if it commutes with all the elements of $\widehat{A}_0$ and $\widehat{A}_1$, since the graded ring $\widehat{A}$ is generated by $\widehat{A}_1$ over $\widehat{A}_0$.

So we can assume that both the equivalent conditions of part (a) hold, and it remains to prove that the element t commutes with all the elements of $\widehat{A}_1$ if and only if the

R-R-bimodule map $\partial_1\colon \widehat{B}^1 \longrightarrow \widehat{B}^0$ can be extended to an odd derivation of degree -1. The latter condition can be equivalently restated as follows. Consider the R-R-bimodule morphism

$$\tilde{\partial}_2\colon \widehat{B}^1\otimes_R\widehat{B}^1 \longrightarrow \widehat{B}^1$$

defined by the formula

$$\tilde{\partial}_2(b_1\otimes b_2) = \partial_1(b_1)b_2 - b_1\partial_1(b_2) \quad \text{for all } b_1, b_2 \in \widehat{B}^1.$$

For any R-R-bimodule map $\partial_1\colon \widehat{B}^1 \longrightarrow R$, the corresponding map $\tilde{\partial}_2$ is well-defined. Since the graded ring $\widehat{B}$ is quadratic, the map ∂_1 extends to an odd derivation of $\widehat{B}$ if and only if the map $\tilde{\partial}_2$ annihilates the kernel $\widehat{J} \subset \widehat{B}^1\otimes_R\widehat{B}^1$ of the surjective multiplication map $\widehat{B}^1\otimes_R\widehat{B}^1 \longrightarrow \widehat{B}^2$. This is provable using Lemma 3.18.

On the other hand, the condition that the element t commutes with all the elements of $\widehat{A}_1$ can be restated as follows. Consider the R-R-bimodule morphism

$$\tilde{t}\colon \widehat{A}_1 \longrightarrow \widehat{A}_1\otimes_R\widehat{A}_1$$

defined by the formula

$$\tilde{t}(a) = a\otimes t - t\otimes a.$$

Then the element $t \in \widehat{A}_1$ commutes with all the elements of $\widehat{A}_1$ if and only if the composition of the map $\tilde{t}$ with the surjective multiplication map $\widehat{A}_1\otimes_R\widehat{A}_1 \longrightarrow \widehat{A}_2$ vanishes. In order to prove the desired equivalence, it remains to recall and observe that the contravariant functor $\mathrm{Hom}_R(-, R)$ transforms the R-R-bimodule $\widehat{A}_1$ into the R-R-bimodule $\widehat{B}^1$, the R-R-bimodule $\widehat{A}_1\otimes_R\widehat{A}_1$ into the R-R-bimodule $\widehat{B}^1\otimes_R\widehat{B}^1$, the map $\tilde{t}$ into the map $-\tilde{\partial}_2$, the R-R-bimodule $\widehat{A}_2$ into the R-R-bimodule $\widehat{J}$, and the surjection $\widehat{A}_1\otimes_R\widehat{A}_1 \longrightarrow \widehat{A}_2$ into the inclusion $\widehat{J} \longrightarrow \widehat{B}^1\otimes_R\widehat{B}^1$. Besides, the R-R-bimodules $\widehat{A}_1$, $\widehat{A}_1\otimes_R\widehat{A}_1$, and $\widehat{A}_2$ are finitely generated and projective as left R-modules, while the R-R-bimodules $\widehat{B}^1$, $\widehat{B}^1\otimes_R\widehat{B}^1$, and $\widehat{J}$ are finitely generated and projective as right R-modules, so the contravariant functor $\mathrm{Hom}_{R^{\mathrm{op}}}(-, R)$ performs the inverse transformation. □

Lemma 4.16 *Let $\partial\colon \widehat{B} \longrightarrow \widehat{B}$ be an odd derivation of degree -1. Then*

(a) $\partial^2 = 0$.

(b) *The homology ring $H_\partial(\widehat{B})$ vanishes if and only if, for the central element $t \in \widehat{A}_1$ corresponding to ∂, the multiplication map $\widehat{A}_0 \xrightarrow{t} \widehat{A}_1$ is injective and its cokernel $A_1 = \widehat{A}_1/\widehat{A}_0 t$ is a projective left R-module.*

Examples 4.17

(1) Let V be a finite-dimensional k-vector space, and let $\widehat{A} = \mathrm{Sym}_k(V)$ and $\widehat{B} = \Lambda_k(V^*)$ be the pair of quadratic dual quadratic algebras as in Example 4.15(1). Let $t \in \widehat{A}_1$ be an element and $\partial \colon \widehat{B} \longrightarrow \widehat{B}$ be the corresponding odd derivation of degree -1. Then the homology ring $H_\partial(\widehat{B})$ vanishes if and only if $\partial \neq 0$, or equivalently, $t \neq 0$.

(2) Let R be a commutative ring and V be a finitely generated free R-module. Let $\widehat{A} = \mathrm{Sym}_R(V)$ and $\widehat{B} = \Lambda_R(V^\vee)$ be the pair of quadratic dual 2-left and right finitely projective quadratic algebras over R as in Example 4.15(2). Let $x_1, \ldots, x_m$ be a set of free generators of the R-module V, and $\xi_1, \ldots, \xi_m$ be the dual set of free generators of the R-module $V^\vee$. Let $t = a_1 x_1 + \cdots + a_m x_m$, where $a_1, \ldots, a_m \in R$, be an arbitrary element in $V = \widehat{A}_1$. Then t is a central element in $\widehat{A}_1$ and $\partial = a_1 \frac{\partial}{\partial \xi_1} + \cdots + a_m \frac{\partial}{\partial \xi_m}$ is the corresponding odd derivation of $\widehat{B}$. The homology ring $H_\partial(\widehat{B})$ vanishes if and only if the multiplication map $R \xrightarrow{t} V$ is injective and its cokernel V/Rt is a projective R-module and if and only if the elements $a_1, \ldots, a_m$ generate the unit ideal in the commutative ring R.

Proof of Lemma 4.16 Part (a): the compositions $\partial_{-1}\partial_0 \colon \widehat{B}^0 \longrightarrow \widehat{B}^{-1} \longrightarrow \widehat{B}^{-2}$ and $\partial_0\partial_1 \colon \widehat{B}^1 \longrightarrow \widehat{B}^0 \longrightarrow \widehat{B}^{-1}$ vanish, since $\widehat{B}^{-1} = 0$. Since the square of an odd derivation is a derivation and the graded ring $\widehat{B}$ is generated by $\widehat{B}^1$ over $\widehat{B}^0$, it follows that $\partial^2 = 0$ on the whole graded ring $\widehat{B}$. Part (b) is a particular case of the assertion that a morphism of finitely generated projective left R-modules $t \colon \widehat{A}_0 \longrightarrow \widehat{A}_1$ is injective with a projective cokernel if and only if the dual morphism of finitely generated projective right R-modules $-\partial_1 \colon \widehat{B}^1 \longrightarrow \widehat{B}^0$ is surjective.   □

Proposition 4.18 *Let $\widehat{A}$ be a 2-left finitely projective quadratic graded ring and $\widehat{B}$ be the quadratic dual 2-right finitely projective quadratic graded ring. Let $t \in \widehat{A}_1$ be a central element and $\partial \colon \widehat{B} \longrightarrow \widehat{B}$ be the corresponding odd derivation of degree -1 (see Lemma 4.14(b)). Assume that the equivalent conditions of Lemma 4.16(b) are satisfied. Let $B = \ker \partial \subset \widehat{B}$ be the underlying graded ring of the quasi-differential graded ring $(\widehat{B}, \partial)$, and let $A = \widehat{A}/\widehat{A}t$ be the quotient ring. Then*

(a) *The graded ring B is generated by B^1 over B^0 if and only if the multiplication map $\widehat{A}_1 \xrightarrow{t} \widehat{A}_2$ is injective and its cokernel $A_2 = \widehat{A}_2/\widehat{A}_1 t$ is a projective left R-module.*

(b) *If the equivalent conditions of part (a) hold, then the 2-right finitely projective quadratic graded ring $\mathrm{q}B$ is quadratic dual to the 2-left finitely projective quadratic graded ring A. The composition of graded ring homomorphisms $\mathrm{q}B \longrightarrow B \longrightarrow \widehat{B}$ is the morphism of 2-right finitely projective quadratic graded rings corresponding to the surjective morphism of 2-left finitely projective quadratic graded rings $\widehat{A} \longrightarrow A$ under the equivalence of categories from Proposition 1.6.*

Proof Part (a): according to Lemma 4.11(a), the graded ring B is generated by B^1 over B^0 if and only if the multiplication map $B^1 \otimes_R B^1 \longrightarrow B^2$ is surjective. In the notation of the proof of Lemma 4.14(b), we have

$$B^1 \otimes_R B^1 \subset \ker \tilde{\partial}_2 \subset \widehat{B}^1 \otimes_R \widehat{B}^1.$$

Here the map $B^1 \otimes_R B^1 \longrightarrow \widehat{B}^1 \otimes_R \widehat{B}^1$ induced by the inclusion of the R-R-subbimodule B^1 into the R-R-bimodule $\widehat{B}^1$ is injective, because the right R-module $\widehat{B}^1$ is projective and the quotient bimodule $\widehat{B}^1/B^1$, being naturally isomorphic to R via the surjective R-R-bimodule map $\partial_1 \colon \widehat{B}^1 \longrightarrow \widehat{B}^0$, is also projective as a right R-module. The inclusion $B^1 \otimes_R B^1 \subset \ker \tilde{\partial}_2$ is clear from the construction of the map $\tilde{\partial}_2$.

Furthermore, following the proof of Lemma 4.14(b), we have $\widehat{J} \subset \ker \tilde{\partial}_2$, since ∂ is an odd derivation of $\widehat{B}$ by assumption. The two R-R-bimodule maps

$$\widehat{B}^1 \otimes_R \widehat{B}^1 \rightrightarrows \widehat{B}^1, \qquad b_1 \otimes_R b_2 \longmapsto \partial(b_1)b_2 \text{ and } b_1 \partial(b_2)$$

restrict to one and the same map $u \colon \ker \tilde{\partial}_2 \longrightarrow \widehat{B}_1$. The subbimodule $B^1 \otimes_R B^1 \subset \ker \tilde{\partial}_2$ is the kernel of the map u.

The map $\tilde{\partial}_2$ is equal to the composition of the surjective multiplication map $\widehat{B}^1 \otimes_R \widehat{B}^1 \longrightarrow \widehat{B}^2$ with the map $\partial_2 \colon \widehat{B}^2 \longrightarrow \widehat{B}^1$. By the definition, $B^2 \subset \widehat{B}^2$ is the kernel of the latter map. It follows that the multiplication map $B^1 \otimes_R B^1 \longrightarrow B^2$ is surjective if and only if one has $\ker \tilde{\partial}_2 = \widehat{J} + B^1 \otimes_R B^1$. The latter condition holds if and only if the map

$$\bar{u} = u|_{\widehat{J}} \colon \widehat{J} \longrightarrow \widehat{B}^1$$

is surjective.

It remains to observe that the functor $\mathrm{Hom}_R(-, R)$ transforms the map $\widehat{A}_1 \overset{t}{\longrightarrow} \widehat{A}_2$ into the map $-\bar{u}$. Thus the map $\bar{u}$ is surjective if and only if the map t is injective and its cokernel A_2 is a projective left R-module.

In part (b), the graded ring A is quadratic by (the proof of) Proposition 4.1(b) and 2-left finitely projective by assumptions, while the quadratic graded ring qB is 2-right finitely projective by Lemma 4.12. Furthermore, whenever the equivalent conditions of Lemma 4.16(b) hold, the functor $\mathrm{Hom}_R(-, R)$ transforms the R-R-bimodule A_1 into the R-R-bimodule B^1. This is clear from the proof of Lemma 4.16(b). Finally, whenever the equivalent conditions of part (a) of the present proposition hold, the functor $\mathrm{Hom}_R(-, R)$ transforms the R-R-bimodule A_2 into the R-R-bimodule

$$\ker \bar{u} = \widehat{J} \cap (B^1 \otimes_R B^1) \subset \widehat{B}^1 \otimes_R \widehat{B}^1.$$

The latter R-R-bimodule is the kernel of the surjective multiplication map $B^1 \otimes_R B^1 \longrightarrow$ B^2. Then one has to check that the same functor transforms the surjective multiplication map $A_1 \otimes_R A_1 \longrightarrow A_2$ into the inclusion map $\ker \bar{u} \longrightarrow B^1 \otimes_R B^1$.   □

Proposition 4.19 *Let $\widehat{A}$ be a 3-left finitely projective quadratic graded ring and $\widehat{B}$ be the quadratic dual 3-right finitely projective quadratic graded ring. Let $t \in \widehat{A}_1$ be a central element and $\partial \colon \widehat{B} \longrightarrow \widehat{B}$ be the corresponding odd derivation of degree -1. Assume that the equivalent conditions of Lemma 4.16(b) are satisfied and the equivalent conditions of Proposition 4.18(a) are satisfied as well. Then the graded ring B is quadratic if and only if the multiplication map $\widehat{A}_2 \xrightarrow{t} \widehat{A}_3$ is injective and its cokernel $A_3 = \widehat{A}_3/\widehat{A}_2 t$ is a projective left R-module.*

Proof By Lemma 4.12, the right R-module B^3 is projective in our assumptions. So, if the graded ring B is quadratic, then it is 3-right finitely projective quadratic. Since qB is quadratic dual to A by Proposition 4.18(b), it then follows by virtue of Proposition 1.8 that the quadratic graded ring A is 3-left finitely projective. Thus we can assume that A_3 is a projective left R-module in all cases.

Then, again by Propositions 4.18(b) and 1.8, it follows that the quadratic graded ring qB is 3-right finitely projective. According to Lemma 4.11(b), the graded ring B is quadratic if and only if the surjective homomorphism of graded rings q$B \longrightarrow B$ is an isomorphism in degree 3. Equivalently, this means that the composition q$B^3 \longrightarrow B^3 \longrightarrow \widehat{B}^3$ is an injective map. Furthermore, since both qB^3 and B^3 are projective right R-modules, and since the right R-module $\widehat{B}^3/B^3 \simeq B^2$ is projective as well, we can conclude that the map q$B^3 \longrightarrow B^3$ is an isomorphism if and only if applying the functor $\mathrm{Hom}_{R^{\mathrm{op}}}(-, R)$ to the composition q$B^3 \longrightarrow B^3 \longrightarrow \widehat{B}^3$ produces a surjective map $f \colon \mathrm{Hom}_{R^{\mathrm{op}}}(\widehat{B}^3, R) \longrightarrow \mathrm{Hom}_{R^{\mathrm{op}}}(\mathrm{q}B^3, R)$.

Denote by $I \subset A_1 \otimes_R A_1$ and $\widehat{I} \subset \widehat{A}_1 \otimes_R \widehat{A}_1$ the kernels of the multiplication maps $A_1 \otimes_R A_1 \longrightarrow A_1$ and $\widehat{A}_1 \otimes_R \widehat{A}_1 \longrightarrow \widehat{A}_1$. Both the graded rings A and $\widehat{A}$ are 3-left finitely projective quadratic in our assumptions; qB and $\widehat{B}$, respectively, are their quadratic dual 3-right finitely projective quadratic rings. Put

$$I^{(3)} = (I \otimes_R A_1) \cap (A_1 \otimes_R I) \subset A_1 \otimes_R A_1 \otimes_R A_1$$

and

$$\widehat{I}^{(3)} = (\widehat{I} \otimes_R \widehat{A}_1) \cap (\widehat{A}_1 \otimes_R \widehat{I}) \subset \widehat{A}_1 \otimes_R \widehat{A}_1 \otimes_R \widehat{A}_1.$$

Following the proof of Proposition 1.8, we have natural isomorphisms of R-R-bimodules $\mathrm{Hom}_{R^{\mathrm{op}}}(\mathrm{q}B^3, R) \simeq I^{(3)}$ and $\mathrm{Hom}_{R^{\mathrm{op}}}(\widehat{B}^3, R) \simeq \widehat{I}^{(3)}$. The map

$$f \colon \widehat{I}^{(3)} = \mathrm{Hom}_{R^{\mathrm{op}}}(\widehat{B}^3, R) \longrightarrow \mathrm{Hom}_{R^{\mathrm{op}}}(\mathrm{q}B^3, R) = I^{(3)}$$

that we are interested in is induced by the surjective morphism of quadratic graded rings $\widehat{A} \longrightarrow A$ (which is quadratic dual to the morphism of quadratic graded rings $qB \longrightarrow \widehat{B}$ according to Proposition 4.18(b)).

We have shown that the graded ring B is quadratic if and only if the natural map $f : \widehat{I}^{(3)} \longrightarrow I^{(3)}$ is surjective. Let us show that the latter condition holds if and only if the multiplication map $\widehat{A}_2 \overset{t}{\longrightarrow} \widehat{A}_3$ is injective.

Consider the three-term complex

$$\widehat{A}_1 \otimes_R \widehat{A}_1 \otimes_R \widehat{A}_1 \longrightarrow (\widehat{A}_1 \otimes_R \widehat{A}_2) \oplus (\widehat{A}_2 \otimes_R \widehat{A}_1) \longrightarrow \widehat{A}_3 \longrightarrow 0$$

and denote it by $\widehat{C}_\bullet$. For the sake of certainty of notation, let us place the complex $\widehat{C}_\bullet$ in the homological degrees 1, 2, and 3, so that $\widehat{C}_3 = \widehat{A}_1^{\otimes_R 3}$ and $\widehat{C}_1 = \widehat{A}_3$. Endow the graded ring $\widehat{A}$ with a multiplicative decreasing filtration by homogeneous ideals $\widehat{A} = G^0 \widehat{A} \supset G^1 \widehat{A} \supset G^2 \widehat{A} \supset \cdots$ defined by the rule $G^p \widehat{A} = t^p \widehat{A}$, and endow the complex $\widehat{C}_\bullet$ with the induced decreasing filtration. So, in particular, one has $\widehat{A}/G^1 \widehat{A} = A$, and the complex $\widehat{C}_\bullet/G^1 \widehat{C}_\bullet$ is isomorphic to the complex

$$A_1 \otimes_R A_1 \otimes_R A_1 \longrightarrow (A_1 \otimes_R A_2) \oplus (A_2 \otimes_R A_1) \longrightarrow A_3 \longrightarrow 0,$$

which we denote by $C_\bullet = C_\bullet(A)$.

Since the graded rings $\widehat{A}$ and A are quadratic, we have $H_i(\widehat{C}_\bullet) = 0 = H_i(C_\bullet)$ for all $i \neq 3$, while $H_3(\widehat{C}_\bullet) = \widehat{I}^{(3)}$ and $H_3(C_\bullet) = I^{(3)}$. It is clear from the homological long exact sequence related to the short exact sequence of complexes

$$0 \longrightarrow G^1 \widehat{C}_\bullet \longrightarrow \widehat{C}_\bullet \longrightarrow C_\bullet \longrightarrow 0 \tag{4.4}$$

that the natural map $f : \widehat{I}^{(3)} \longrightarrow I^{(3)}$ is surjective if and only if $H_2(G^1 \widehat{C}_\bullet) = 0$.

Consider the nonnegatively graded ring $\mathbb{Z}[\bar{t}]$ of polynomials in one variable $\bar{t}$ of degree 1 with integer coefficients. Let $A[\bar{t}] = A \otimes_{\mathbb{Z}} \mathbb{Z}[\bar{t}]$ be the tensor product of the graded rings A and $\mathbb{Z}[\bar{t}]$, taken over the ring $\mathbb{Z}$ and endowed with the induced grading. Obviously, the graded ring $A[\bar{t}]$ is quadratic. Furthermore, there is a natural morphism of graded rings

$$A[\bar{t}] \longrightarrow \mathrm{gr}_G \widehat{A} = \bigoplus_{p=0}^{\infty} G^p \widehat{A}/G^{p+1} \widehat{A} \tag{4.5}$$

whose restriction to $A \subset A[\bar{t}]$ is equal to the inclusion $A \simeq \widehat{A}/G^1 \widehat{A} \hookrightarrow \mathrm{gr}_G \widehat{A}$ and which takes the element $\bar{t} \in A[\bar{t}]$ to the coset $t + G^2 \widehat{A} \in G^1 \widehat{A}/G^2 \widehat{A}$.

One easily observes that the map (4.5) is surjective. Furthermore, the map $A[\bar{t}]_n \longrightarrow (\mathrm{gr}_G \widehat{A})_n$ is injective (equivalently, an isomorphism) if and only if the map $\widehat{A}_{j-1} \overset{t}{\longrightarrow} \widehat{A}_j$ is injective for all $j \leq n$. In our assumptions, we know that this injectivity holds in the internal degrees $j = 1$ and 2, and we are interested in knowing whether it holds for $j = 3$. So the map $\widehat{A}_2 \overset{t}{\longrightarrow} \widehat{A}_3$ is injective if and only if the map $A[\bar{t}]_3 \longrightarrow \mathrm{gr}_G \widehat{A}_3$ is.

Put $\widetilde{A} = A[\bar{t}]$ and consider the three-term complex

$$\widetilde{A}_1 \otimes_R \widetilde{A}_1 \otimes_R \widetilde{A}_1 \longrightarrow (\widetilde{A}_1 \otimes_R \widetilde{A}_2) \oplus (\widetilde{A}_2 \otimes_R \widetilde{A}_1) \longrightarrow \widetilde{A}_3 \longrightarrow 0,$$

denoted by $\widetilde{C}_\bullet = C_\bullet(\widetilde{A})$. Since the graded ring A is 2-left finitely projective, the complex $\mathrm{gr}_G \widehat{C}_\bullet$ is isomorphic to the complex $C_\bullet(\mathrm{gr}_G \widehat{A})$. The surjective morphism of graded rings $\widetilde{A} \longrightarrow \mathrm{gr}_G \widehat{A}$ induces a surjective morphism of complexes $\widetilde{C}_\bullet \longrightarrow \mathrm{gr}_G \widehat{C}_\bullet$.

Moreover, the maps $\widetilde{C}_i \longrightarrow \mathrm{gr}_G \widehat{C}_i$ are isomorphisms for $i \neq 1$, since the map $\widetilde{A}_j \longrightarrow \mathrm{gr}_G \widehat{A}_j$ is an isomorphism for $j < 3$. Furthermore, the complex $\widetilde{C}_i$ is acyclic in the homological degrees $i \neq 3$, since the graded ring $\widetilde{A} = A[\bar{t}]$ is quadratic. Since the map $\widetilde{C}_3 \longrightarrow \mathrm{gr}_G \widehat{C}_3$ is surjective, it follows that the map $A[\bar{t}]_3 \longrightarrow \mathrm{gr}_G \widehat{A}_3$ is an isomorphism if and only if $H_2(\mathrm{gr}_G \widehat{C}_\bullet) = 0$.

It remains to show that $H_2(G^1 \widehat{C}_\bullet) = 0$ if and only if $H_2(\mathrm{gr}_G \widehat{C}_\bullet) = 0$. The implication "if" is obvious. To prove the "only if," we assume that $H_2(G^1 \widehat{C}_\bullet) = 0$, the map $f \colon \widehat{I}^{(3)} \longrightarrow I^{(3)}$ is surjective, and the graded ring B is quadratic.

Consider the additional grading p on the ring $\mathrm{gr}_G \widehat{A}$ and the complex $\mathrm{gr}_G \widehat{C}_\bullet$ induced by the indexing of the filtration G. The related grading p on the ring $\widetilde{A} = A[\bar{t}]$ and the complex $\widetilde{C}_\bullet = C_\bullet(\widetilde{A})$ is induced by the grading of the ring $\mathbb{Z}[\bar{t}]$. The additional grading p on the rings $\mathrm{gr}_G \widehat{A}$ and $\widetilde{A}$ takes values in the monoid of nonnegative integers $p \geq 0$. On the complexes $\mathrm{gr}_G \widehat{C}_\bullet$ and $\widetilde{C}_\bullet$, the additional grading takes values $0 \leq p \leq 3$.

In the additional grading $p = 0$, the morphism of graded rings $A[\bar{t}] \longrightarrow \mathrm{gr}_G \widehat{A}$ is an isomorphism, since $A = \widehat{A}/\widehat{A}t$. Therefore, the morphism of complexes $\widetilde{C}_\bullet \longrightarrow \mathrm{gr}_G \widehat{C}_\bullet$ is an isomorphism in the additional grading $p = 0$. Any possible homology classes in $H_2(\mathrm{gr}_G \widehat{C}_\bullet)$ would occur in the additional grading $p = 1$, 2, or 3.

We have $H_1(\mathrm{gr}_G \widehat{C}_\bullet) = 0$, as the graded ring $\mathrm{gr}_G \widehat{A}$ is generated by its component $A \oplus R[\bar{t}]$ of degree $n = 1$ over its component R of degree $n = 0$. Let us compute the p-graded abelian group (R-R-bimodule) $\widehat{I}^{(3)} = H_3(\widetilde{C}_\bullet) = H_3(\mathrm{gr}_G \widehat{C}_\bullet)$.

The quadratic graded ring $\widetilde{A} = A[\bar{t}]$ (in the grading n) is 3-left finitely projective, and its quadratic dual 3-right finitely projective graded ring $\widetilde{B}$ can be computed as the graded ring $\widetilde{B} = B[\bar{\delta}] = B \otimes_{\mathbb{Z}}^{-1} \mathbb{Z}[\bar{\delta}]$ obtained by adjoining to B a generator $\bar{\delta}$ of degree $n = 1$ (and $p = 1$) with the relations $\bar{\delta}b + b\bar{\delta} = 0$ for all $b \in B^1$ and $\bar{\delta}^2 = 0$. Here the notation $B \otimes_{\mathbb{Z}}^q \mathbb{Z}[\bar{\delta}]$ stands for the relations $\bar{\delta}b = qb\bar{\delta}$, $b \in B^1$, with $q = -1$ [53, Section 1 of Chapter 3]. Consequently, $\widetilde{B}^3 = B^3 \oplus B^2\bar{\delta}$ and $\widetilde{I}^{(3)} \simeq \mathrm{Hom}_{R^{\mathrm{op}}}(\widetilde{B}^3, R) = I^{(3)} \oplus \bar{\epsilon}I$, where $\bar{\epsilon}$ is a variable dual to $\bar{\delta}$. The direct summand $I^{(3)}$ of the p-graded group $\widetilde{I}^{(3)}$ sits in the additional grading $p = 0$ and the direct summand $\bar{\epsilon}I \simeq I$ sits in the additional grading $p = 1$.

Thus we have

$$H_3(G^p \widehat{C}_\bullet / G^{p+1} \widehat{C}_\bullet) = \begin{cases} I^{(3)} & \text{for } p = 0 \\ \bar{\epsilon}I & \text{for } p = 1 \\ 0 & \text{for } p \geq 2. \end{cases}$$

In view of the spectral sequence connecting the homology groups of the complexes $G^1 \widehat{C}_\bullet$ and $\mathrm{gr}_G \widehat{C}^\bullet$, it remains to check that the map

$$H_3(G^1\widehat{C}_\bullet) \longrightarrow H_3(G^1\widehat{C}_\bullet/G^2\widehat{C}_\bullet) = \bar{\epsilon}I \tag{4.6}$$

is surjective (or equivalently, an isomorphism). Here the group $H_3(G^1\widehat{C}_\bullet)$ is the kernel of the natural map $f \colon \widehat{I}^{(3)} \longrightarrow I^{(3)}$ (see short exact sequence (4.4)).

Following the discussion in the beginning of this proof, the map $f \colon \widehat{I}^{(3)} \longrightarrow I^{(3)}$ can be obtained by applying the functor $\mathrm{Hom}_{R^{\mathrm{op}}}(-, R)$ to the composition of maps $\mathrm{q}B^3 \longrightarrow B^3 \longrightarrow \widehat{B}^3$. In our present assumptions, the map f is surjective and the graded ring B is quadratic. The kernel of the map f is $\ker(f) \simeq \mathrm{Hom}_{R^{\mathrm{op}}}(\widehat{B}^3/B^3, R) = \mathrm{Hom}_{R^{\mathrm{op}}}(B^2\bar{\delta}, R) \simeq \bar{\epsilon}I$, as desired. Notice that the existence of a well-defined odd derivation $\partial = \partial/\partial\delta$ on $\widehat{B}$ ensures injectivity (equivalently, bijectivity) of the multiplication map $B^2 \xrightarrow{\bar{\delta}} \widehat{B}^3/B^3$. □

4.5   Nonhomogeneous Quadratic Duality via Quasi-Differential Graded Rings

The construction of the nonhomogeneous quadratic duality functors

$$(R\text{–rings}_{\mathrm{wnlq}})^{\mathrm{op}} \longrightarrow R\text{–rings}_{\mathrm{cdg,rq}} \quad \text{and} \quad (\mathrm{Rings}_{\mathrm{wnlq2}})^{\mathrm{op}} \longrightarrow \mathrm{Rings}_{\mathrm{cdg2,rq}}$$

in Theorems 3.21 and 3.25 is based on the computations in Sects. 3.3–3.5, which are beautiful, but quite involved. The definitions and results above in Chap. 4 allow to produce the nonhomogeneous duality functors in a more conceptual fashion.

A nonnegatively graded quasi-differential ring $(\widehat{B}, \partial)$ is said to be *3-right finitely projective quadratic* if its underlying graded ring $B = \ker \partial \subset \widehat{B}$ is 3-right finitely projective quadratic. The results of Sect. 4.3 explain how such properties of the graded ring B are related to the similar properties of the graded ring $\widehat{B}$.

Let R be an associative ring. The *category of 3-right finitely projective quadratic quasi-differential graded rings over R*, denoted by $R\text{–rings}_{\mathrm{qdg,rq}}$, is the full subcategory in the category $R\text{–rings}_{\mathrm{qdg}}$ (as defined in Sect. 4.2) consisting of all the 3-right finitely projective quadratic quasi-differential graded rings.

Furthermore, the *2-category of 3-right finitely projective quadratic quasi-differential graded rings*, denoted by $\mathrm{Rings}_{\mathrm{qdg2,rq}}$, is the following 2-subcategory in the 2-category $\mathrm{Rings}_{\mathrm{qdg2}}$ (which was also defined in Sect. 4.2). The objects of $\mathrm{Rings}_{\mathrm{qdg2,rq}}$ are all the 3-right finitely projective quadratic quasi-differential graded rings. All morphisms in $\mathrm{Rings}_{\mathrm{qdg2}}$ between objects of $\mathrm{Rings}_{\mathrm{qdg2,rq}}$ are morphisms in $\mathrm{Rings}_{\mathrm{qdg2,rq}}$, and all 2-morphisms in $\mathrm{Rings}_{\mathrm{qdg2}}$ between morphisms of $\mathrm{Rings}_{\mathrm{qdg2,rq}}$ are 2-morphisms in $\mathrm{Rings}_{\mathrm{qdg2,rq}}$.

Restricting the equivalence of categories $R\text{-rings}_{\mathsf{qdg}} \simeq R\text{-rings}_{\mathsf{cdg}}$ provided by Theorem 4.7 to the full subcategories of 3-right finitely projective quadratic CDG-rings and quasi-differential graded rings over R, we obtain an equivalence of categories

$$R\text{-rings}_{\mathsf{qdg,rq}} \simeq R\text{-rings}_{\mathsf{cdg,rq}}. \tag{4.7}$$

Similarly, a restriction of the strict equivalence of 2-categories $\mathsf{Rings}_{\mathsf{qdg2}} \simeq \mathsf{Rings}_{\mathsf{cdg2}}$ provided by Theorem 4.8 produces a strict equivalence between the 2-categories of 3-right finitely projective quadratic CDG-rings and quasi-differential graded rings,

$$\mathsf{Rings}_{\mathsf{qdg2,rq}} \simeq \mathsf{Rings}_{\mathsf{cdg2,rq}}. \tag{4.8}$$

Let $(\widetilde{A}, F)$ be a filtered ring with an increasing filtration $0 = F_{-1}\widetilde{A} \subset F_0\widetilde{A} \subset F_1\widetilde{A} \subset F_2\widetilde{A} \subset \cdots$. Consider two graded rings related to such a filtration: the associated graded ring $A = \mathrm{gr}^F \widetilde{A}$ and the Rees ring $\widehat{A} = \bigoplus_{n=0}^{\infty} F_n\widetilde{A}$.

The unit element $1 \in F_0\widetilde{A}$, viewed as an element of $F_1\widetilde{A}$, represents a central nonzero-divisor $t \in \widehat{A}_1$. The quotient ring $\widehat{A}/\widehat{A}t$ is naturally isomorphic to the associated graded ring A of the filtration F on the ring $\widetilde{A}$. By Proposition 4.1, the graded ring $\widehat{A}$ is generated by $\widehat{A}_1$ over $\widehat{A}_0$ if and only if the graded ring A is generated by A_1 over A_0, and moreover, the graded ring $\widehat{A}$ is quadratic if and only if the graded ring A is.

Let $\widetilde{A}$ be a weak nonhomogeneous quadratic ring over a subring $R \subset \widetilde{A}$ with the R-R-bimodule of generators $\widetilde{V} \subset \widetilde{A}$. Let F be the related increasing filtration on the ring $\widetilde{A}$, as defined in Sect. 3.1. Then the graded ring $\widehat{A}$ is generated by $\widehat{A}_1$ over $\widehat{A}_0$, and the graded ring A is generated by A_1 over A_0. The graded rings $\widehat{A}$ and A do *not* need to be quadratic (as we only assume $\widetilde{A}$ be a *weak* nonhomogeneous quadratic ring), so let us consider the quadratic graded rings $\mathrm{q}\widehat{A}$ and $\mathrm{q}A$.

Lemma 4.20 *Let $\widehat{A}$ be a nonnegatively graded ring, and let $t \in \widehat{A}$ be a central element. Let $A = \widehat{A}/\widehat{A}t$ be the quotient ring. Then*

(a) *$t \in \mathrm{q}\widehat{A}_1 = \widehat{A}_1$ is a central element in the quadratic graded ring $\mathrm{q}\widehat{A}$.*
(b) *For each $n = 1$, 2, or 3, the multiplication map $\mathrm{q}\widehat{A}_{n-1} \xrightarrow{t} \mathrm{q}\widehat{A}_n$ is injective whenever the multiplication map $\widehat{A}_{n-1} \xrightarrow{t} \widehat{A}_n$ is.*
(c) *There is a natural isomorphism of quadratic graded rings $\mathrm{q}\widehat{A}/(\mathrm{q}\widehat{A})t \simeq \mathrm{q}A$.*

Proof In part (a), the element $t \in \mathrm{q}\widehat{A}_1$ commutes with all the elements of $\mathrm{q}\widehat{A}_0$ and $\mathrm{q}\widehat{A}_1$, since the graded ring homomorphism $\mathrm{q}\widehat{A} \longrightarrow \widehat{A}$ is an isomorphism in degree $n \leq 1$ and a monomorphism in degree $n = 2$, and the element $t \in \widehat{A}_1$ is central in $\widehat{A}$. Since the graded ring $\mathrm{q}\widehat{A}$ is generated by $\mathrm{q}\widehat{A}_1$ over $\mathrm{q}\widehat{A}_0$, it follows that the element $t \in \mathrm{q}\widehat{A}_1$ is central in $\mathrm{q}\widehat{A}$.

Part (b) follows from injectivity of the map $q\widehat{A}_{n-1} \longrightarrow \widehat{A}_{n-1}$ for $n \leq 3$ and commutativity of the square diagram $q\widehat{A}_{n-1} \xrightarrow{t} q\widehat{A}_n \longrightarrow \widehat{A}_n$, $q\widehat{A}_{n-1} \longrightarrow \widehat{A}_{n-1} \xrightarrow{t} \widehat{A}_n$.

In part (c), the graded ring homomorphism $q\widehat{A} \longrightarrow \widehat{A}$ induces a graded ring homomorphism between the quotient rings $q\widehat{A}/(q\widehat{A})t \longrightarrow \widehat{A}/\widehat{A}t = A$. Since the graded ring $q\widehat{A}/(q\widehat{A})t$ is quadratic by Proposition 4.1, the latter morphism, in turn, induces the desired graded ring homomorphism $f \colon q\widehat{A}/(q\widehat{A})t \longrightarrow qA$. The map f is an isomorphism in the degrees $n \leq 2$ by construction, and hence it is an isomorphism of quadratic graded rings by Lemma 3.5.   □

Proposition 4.21 *Let $R \subset \widetilde{V} \subset \widetilde{A}$ be a 3-left finitely projective weak nonhomogeneous quadratic ring. Consider the corresponding graded rings $\widehat{A} = \bigoplus_{n=0}^{\infty} F_n \widetilde{A}$ and $A = \mathrm{gr}^F \widetilde{A}$. Then the quadratic graded rings $q\widehat{A}$ and qA are 3-left finitely projective.*

Let $\widehat{B}$ be the 3-right finitely projective quadratic graded ring quadratic dual to $q\widehat{A}$, and let $\partial \colon \widehat{B} \longrightarrow \widehat{B}$ be the odd derivation of degree -1 corresponding to the central element $t \in \widehat{A}_1$, as in Lemma 4.14(b). Then $(\widehat{B}, \partial)$ is a quasi-differential graded ring. The underlying graded ring $B = \ker \partial \subset \widehat{B}$ is 3-right finitely projective quadratic and quadratic dual to qA.

Proof The quadratic graded ring qA is 3-left finitely projective by the definition of what it means for a weak nonhomogeneous quadratic ring $\widetilde{A}$ to be 3-left finitely projective (see Sect. 3.3). To prove that the left R-module $q\widehat{A}_n$ is finitely generated projective for $n \leq 3$, one argues by induction in $0 \leq n \leq 3$, using the short exact sequences of R-R-bimodules

$$0 \longrightarrow q\widehat{A}_{n-1} \xrightarrow{t} q\widehat{A}_n \longrightarrow qA_n \longrightarrow 0$$ provided by Lemma 4.20. Here the multiplication maps $q\widehat{A}_{n-1} \xrightarrow{t} q\widehat{A}_n$ are injective for $n \leq 3$ by Lemma 4.20(b), since t is a nonzero-divisor in $\widehat{A}$.

The pair $(\widehat{B}, \partial)$ is a quasi-differential graded ring by Lemma 4.16 (applied to the quadratic graded ring $q\widehat{A}$ with its central element t). The graded ring $B = \ker \partial \subset \widehat{B}$ is generated by B^1 over B^0 by Proposition 4.18(a) and quadratic by Proposition 4.19. The quadratic graded ring B is 3-right finitely projective by Lemma 4.12. It is quadratic dual to qA by Proposition 4.18(b).   □

Theorem 4.22 *The construction of Proposition 4.21 defines a fully faithful contravariant functor*

$$(R\text{–rings}_{\mathrm{wnlq}})^{\mathrm{op}} \longrightarrow R\text{–rings}_{\mathrm{qdg,rq}} \tag{4.9}$$

from the category of 3-left finitely projective weak nonhomogeneous quadratic rings to the category of 3-right finitely projective quasi-differential quadratic graded rings over R. The functor (4.9) forms a commutative triangle diagram with the fully faithful contravariant functor (3.10) and the equivalence of categories (4.7).

The same construction also defines a fully faithful strict contravariant 2-functor

$$(\text{Rings}_{\text{wnlq2}})^{\text{op}} \longrightarrow \text{Rings}_{\text{qdg2,rq}} \tag{4.10}$$

from the 2-category of 3-left finitely projective weak nonhomogeneous quadratic rings to the 2-category of 3-right finitely projective quasi-differential quadratic graded rings. The strict 2-functor (4.10) forms a commutative triangle diagram with the fully faithful strict contravariant 2-functor (3.17) and the strict equivalence of 2-categories (4.8).

Proof The proof is straightforward.   □

4.6   PBW Theorem

The Poincaré–Birkhoff–Witt theorem in nonhomogeneous quadratic duality tells that, when restricted to the left/right finitely projective Koszul rings on both sides, the fully faithful contravariant nonhomogeneous quadratic duality functors from Theorems 3.21, 3.25, 3.27, 3.29, and 4.22 become anti-equivalences of categories (or strict anti-equivalences of 2-categories). In other words, *every right finitely projective Koszul CDG-ring (or quasi-differential graded ring) arises from a left finitely projective nonhomogeneous Koszul ring.*

Here are the relevant definitions.

Definition 4.23 A (weak) nonhomogeneous quadratic ring $R \subset \widetilde{V} \subset \widetilde{A}$ is said to be *left finitely projective Koszul* if the quadratic graded ring $qA = \text{qgr}^F \widetilde{A}$ is left finitely projective Koszul (in the sense of the definition in Sect. 2.10).

A nonnegatively graded CDG-ring (B, d, h) is said to be *right finitely projective Koszul* if the nonnegatively graded ring B is right finitely projective Koszul.

A nonnegatively graded quasi-differential ring $(\widehat{B}, \partial)$ is said to be *right finitely projective Koszul* if its underlying graded ring $B = \ker \partial \subset \widehat{B}$ is right finitely projective Koszul.

The following result was mentioned in Remark 3.6 in Sect. 3.1.

Theorem 4.24 *If a weak nonhomogeneous quadratic ring is left finitely projective Koszul, then it is nonhomogeneous quadratic.*

So we will call such filtered rings as in Theorem 4.24 *left finitely projective nonhomogeneous Koszul rings.* Notice that a filtered ring $(\widetilde{A}, F)$ with an increasing filtration F such that $A_n = \text{gr}_n^F \widetilde{A}$ is a projective left A_0-module for all $n \geq 0$ is left finitely projective Koszul (i.e., the graded ring $A = \text{gr}^F \widetilde{A}$ is left finitely projective Koszul) if and only if the

Rees ring $\widehat{A} = \bigoplus_{n=0}^{\infty} F_n \widetilde{A}$ is left finitely projective Koszul. This is clear from Lemma 4.3 and Theorem 4.4.

The proof of Theorem 4.24 will be given at the end of Sect. 4.6.

Theorem 4.25 *Every right finitely projective Koszul CDG-ring arises from a left finitely projective nonhomogeneous Koszul ring via the construction of Proposition 3.16. Equivalently, every right finitely projective Koszul quasi-differential graded ring arises from a left finitely projective nonhomogeneous Koszul ring via the construction of Proposition 4.21.*

The two assertions in Theorem 4.25 are equivalent in view of Theorems 4.7 and 4.22. We will prove the second assertion.

First Proof of Theorem 4.25 Let $(\widehat{B}, \partial)$ be a right finitely projective Koszul quasi-differential graded ring. Then the graded ring B is right finitely projective Koszul by definition and the graded ring $\widehat{B}$ is right finitely projective Koszul by Theorem 4.13(b). Let $\widehat{A}$ be the left finitely projective Koszul ring quadratic dual to $\widehat{B}$ (see Proposition 2.37), and let $t \in \widehat{A}_1$ be the central element corresponding to the odd derivation $\partial \colon \widehat{B} \longrightarrow \widehat{B}$ (see Lemma 4.14). Let $A = \widehat{A}/\widehat{A}t$ be the quotient ring.

By Lemma 4.16, the multiplication map $\widehat{A}_0 \xrightarrow{t} \widehat{A}_1$ is injective and A_1 is a projective left R-module. By Proposition 4.18(a), the multiplication map $\widehat{A}_1 \xrightarrow{t} \widehat{A}_2$ is injective and A_2 is a projective left R-module. By Proposition 4.18(b), the right finitely projective Koszul ring B is quadratic dual to the 2-left finitely projective quadratic ring A. Therefore, the quadratic graded ring A is left finitely projective Koszul. By Proposition 4.19, the multiplication map $\widehat{A}_2 \xrightarrow{t} \widehat{A}_3$ is injective.

Applying Theorem 4.5, we conclude that t is a nonzero-divisor in $\widehat{A}$. It follows that $\widehat{A}$ is the Rees ring of the filtered ring $\widetilde{A} = \widehat{A}/\widehat{A}(t-1)$ with the filtration $F_n \widetilde{A} = \widehat{A}_n + \widehat{A}(t-1)$, and A is the associated graded ring, $A \simeq \mathrm{gr}^F \widetilde{A}$. We have obtained the desired left finitely projective nonhomogeneous Koszul ring $\widetilde{A}$. $\qquad\square$

Second Proof of Theorem 4.25 This is a particular case of the argument in [58, proof of Theorem 11.6]. Let $(\widehat{B}, \partial)$ be a right finitely projective Koszul quasi-differential graded ring and $B = \ker \partial \subset \widehat{B}$ be its underlying right finitely projective Koszul graded ring. Then the right R-modules B^n are finitely generated projective by definition and the right R-modules $\widehat{B}^n$ are finitely generated projective by Lemma 4.12.

The (essentially) spectral sequence argument below is to be compared with, and distinguished from, a quite different (and simpler) spectral sequence argument proving the Poincaré–Birkhoff–Witt theorem for nonhomogeneous quadratic algebras *over the ground field* in [56, Section 3.3] and [53, Proposition 7.2 in Chapter 5].

Put $D_n = \mathrm{Hom}_{R^{\mathrm{op}}}(B^n, R)$. Then $D = \bigoplus_{n=0}^{\infty} D_n$ is a graded coring over the ring R with the counit map $\varepsilon \colon D \longrightarrow D_0 \simeq R$ dual to the inclusion map $R \simeq B^0 \longrightarrow B$ and the comultiplication maps $\mu_{i,j} \colon D_{i+j} \longrightarrow D_i \otimes_R D_j$ obtained by dualizing the multiplication maps $B^j \otimes_R B^i \longrightarrow B^{i+j}$ in the graded ring B,

$$D_{i+j} = \mathrm{Hom}_{R^{\mathrm{op}}}(B^{i+j}, R) \longrightarrow \mathrm{Hom}_{R^{\mathrm{op}}}(B^j \otimes_R B^i, R)$$

$$\simeq \mathrm{Hom}_{R^{\mathrm{op}}}(B^i, R) \otimes_R \mathrm{Hom}_{R^{\mathrm{op}}}(B^j, R) = D_i \otimes_R D_j.$$

In the pairing notation of Sect. 3.4, we have $\langle \mu(f), b_1 \otimes b_2 \rangle = \langle f, b_1 b_2 \rangle$ for all $f \in D$ and $b_1, b_2 \in B$.

Similarly, we set $\widehat{D}_n = \mathrm{Hom}_{R^{\mathrm{op}}}(\widehat{B}^n, R)$, so $\widehat{D} = \bigoplus_{n=0}^{\infty} \widehat{D}_n$ is also a graded coring over R. The odd derivation $\partial \colon \widehat{B} \longrightarrow \widehat{B}$ is an R-R-bilinear map, so it dualizes to an R-R-bilinear map $\mathrm{Hom}_{R^{\mathrm{op}}}(\partial, R) \colon \widehat{D} \longrightarrow \widehat{D}$, which we denote for brevity also by ∂. There is a sign rule involved: in the pairing notation, we put $\langle \partial(f), b \rangle = (-1)^{|f|+1} \langle f, \partial(b) \rangle$ for $f \in \widehat{D}$ and $b \in \widehat{B}$. The map $\partial \colon \widehat{D} \longrightarrow \widehat{D}$ is an odd coderivation of degree 1 on the coring $\widehat{D}$, in the sense that its components act as $\partial_n \colon \widehat{D}_n \longrightarrow \widehat{D}_{n+1}$ and, in the symbolic notation $\mu(f) = \mu_1(f) \otimes \mu_2(f)$ for the comultiplication, one has

$$\partial(\mu(f)) = \partial \mu_1(f) \otimes \mu_2(f) + (-1)^{|\mu_1(f)|} \mu_1(f) \otimes \partial \mu_2(f) \quad \text{for all } f \in \widehat{D}.$$

The odd coderivation ∂ on the graded coring $\widehat{D}$ has zero square, $\partial^2 = 0$ (so one can say that $\widehat{D}$ is a *DG-coring over* R). Furthermore, the odd coderivation ∂ on $\widehat{D}$ has vanishing cohomology, $H_\partial(\widehat{D}) = 0$. The cokernel of ∂ is the quotient coring D of the coring $\widehat{D}$, that is, $\widehat{D} \twoheadrightarrow \mathrm{coker}(\partial) \simeq D$.

Consider the bigraded R-R-bimodule K with the components $K^{p,q} = \widehat{D}_{q-p}$ for $p \le 0$, $q \le 0$, and $K^{p,q} = 0$ otherwise. The R-R-bimodule K can be viewed as a bigraded coring over R with the comultiplication inherited from the comultiplication of $\widehat{D}$. Considered as a graded coring in the total grading $p + q$, the coring K has an odd coderivation ∂_K of degree 1 with the components $\partial_K^{p,q} \colon K^{p,q} \longrightarrow K^{p,q+1}$ given by the rule $\partial_K^{p,q} = \partial_{q-p} \colon \widehat{D}_{q-p} \longrightarrow \widehat{D}_{q-p+1}$. So for every fixed $p = -n$, $n \ge 0$, the components $K^{p,q}$ with varying q form a complex of R-R-bimodules

$$0 \longrightarrow K^{p,p} \xrightarrow{\partial_K^{p,p}} K^{p,p+1} \xrightarrow{\partial_K^{p,p+1}} \cdots \xrightarrow{\partial_K^{p,-2}} K^{p,-1} \xrightarrow{\partial_K^{p,-1}} K^{p,0} \longrightarrow 0$$

isomorphic to

$$0 \longrightarrow \widehat{D}_0 \xrightarrow{\partial_0} \widehat{D}_1 \xrightarrow{\partial_1} \cdots \xrightarrow{\partial_{n-2}} \widehat{D}_{n-1} \xrightarrow{\partial_{n-1}} \widehat{D}_n \longrightarrow 0.$$

The only nontrivial cohomology bimodule of this complex occurs at its rightmost term ($q = 0$) and is isomorphic to D_n. So there is a surjective morphism of bigraded corings $K \longrightarrow D$ inducing an isomorphism of the bimodules/corings of cohomology, where the coring D is placed in the bigrading $D^{p,0} = D_n = D_{-p}$ and endowed with the zero differential.

Denote by K_+ the cokernel of the inclusion $R \simeq K^{0,0} \longrightarrow K$. Consider the tensor ring $T_R(K_+)$ of the bigraded R-R-bimodule K_+. By the definition, $T_R(K_+)$ is a trigraded ring with the gradings $p \le 0$ and $q \le 0$ inherited from the bigrading of K_+ and the additional grading $r \ge 0$ by the number of tensor factors. We will consider $T_R(K_+)$ as a graded ring in the total grading $p + q + r$. The graded ring $T_R(K_+)$ is endowed with three odd derivations of total degree 1, which we will now introduce.

For any (graded) R-R-bimodule V, derivations of the tensor ring $T(V)$ (say, odd derivations with respect to the total parity on $T(V)$) annihilating the subring $R \subset T(V)$ are uniquely determined by their restriction to the subbimodule $V \subset T(V)$, which can be an arbitrary R-R-bilinear map $V \longrightarrow T(V)$ (that is homogeneous of the prescribed degrees). In the situation at hand, we let ∂_T be the only odd derivation of $T_R(K_+)$ which preserves the subbimodule $K_+ \subset T_R(K_+)$ and whose restriction to K_+ is equal to $-\partial_K$. Let d_T be the only odd derivation of $T_R(K_+)$ which maps K_+ into $K_+ \otimes_R K_+$ by the comultiplication map μ with the sign rule

$$d_T(k) = (-1)^{p_1+q_1} \mu_1(k) \otimes \mu_2(k), \quad k \in K_+, \ \mu_1(k) \in K^{p_1,q_1}.$$

Finally, let δ_T be the only odd derivation of $T_R(K_+)$ whose restriction to K_+ is the identity map of the component $K^{-1,-1} = R$ to the unit component $T_R(K_+)^{0,0,0} = R$ and zero on all the remaining components of K_+.

All the three derivations are constructed to be odd derivations in the parity $p+q+r$. The derivations ∂_T, d_T, and δ_T have tridegrees $(0, 1, 0)$, $(0, 0, 1)$, and $(1, 1, -1)$, respectively, in the trigrading (p, q, r) of the tensor ring $T_R(K_+)$. All the three derivations have zero squares, and they pairwise anti-commute.

There is an increasing filtration F on the graded ring $T_R(K_+)$ whose component $F_n T_R(K_+)$ is the direct sum of all the trigrading components $T_R(K_+)^{p,q,r}$ with $-p \le n$. This filtration is compatible with the differentials ∂_T, d_T, and δ_T; the graded ring $\mathrm{gr}^F T_R(K_+) = \bigoplus_{n=0}^{\infty} F_n T_R(K_+)/F_{n-1} T_R(K_+)$ with the differential induced by $\partial_T + d_T + \delta_T$ is naturally isomorphic to $T_R(K_+)$ with the differential $\partial_T + d_T$.

Let us also consider the tensor ring $T_R(D_+)$ of the R-R-bimodule $D_+ = D/R$. Similarly to the above, we endow $T_R(D_+)$ with the grading p coming from the grading $D^p = D_n = D_{-p}$ of D_+ and the grading r by the number of tensor factors. As to the grading q, we set it to be identically zero on $T_R(D_+)$. We will consider $T_R(D_+)$ as a graded ring in the total grading $p+r$. Let d'_T be the only odd derivation of $T_R(D_+)$ which maps the subbimodule $D_+ \subset T_R(D_+)$ to $D_+ \otimes_R D_+$ by the comultiplication map μ with the sign rule similar to the above.

The DG-ring $T_R(D_+)$ is naturally isomorphic to the cobar complex (2.13) (with the roles of the left and right sides switched) computing the bigraded $\mathrm{Ext}_{B^{\mathrm{op}}}(R, R)$. There is some sign rule involved in this isomorphism, which is discussed in [58, proof of Theorem 11.6]. By Proposition 2.37, it follows that the bigraded ring of cohomology of the DG-ring $T_R(D_+)$ is isomorphic to the left finitely projective Koszul graded ring

$A = \bigoplus_{n=0}^{\infty} A_n$ quadratic dual to B, placed in the diagonal bigrading $-p = r = n$. In the total grading $p + r$, the cohomology ring of the DG-ring $T_R(D_+)$ is the whole ring A placed in the degree $p + r = 0$.

Consider the morphism of DG-rings $(T_R(K_+), \partial_T + d_T) \longrightarrow (T_R(D_+), d'_T)$ induced by the surjective morphism of graded corings $K \longrightarrow D$. The morphism of DG-corings $K \longrightarrow D$ is a quasi-isomorphism, so the induced morphism of DG-rings is a quasi-isomorphism, too, due to the right flatness/projectivity conditions imposed on the R-R-bimodule D and the presence of a nonpositive (essentially, negative) internal grading p. The fact that the components of fixed grading p in the DG-coring K are finite complexes (of projective right R-modules, with projective right R-modules of cohomology) is relevant here. Thus we have $H^0_{\partial_T + d_T}(T_R(K_+)) \simeq A$ and $H^i_{\partial_T + d_T}(T_R(K_+)) = 0$ for $i \neq 0$.

Finally, we put $\widetilde{A} = H^0_{\partial_T + d_T + \delta_T}(T_R(K_+))$. Then the ring $\widetilde{A}$ is endowed with an increasing filtration F induced by the filtration F of the DG-ring $(T_R(K_+), \partial_T + d_T + \delta_T)$. Since $H^i_{\partial_T + d_T}(T_R(K_+)) = 0$ for $i \neq 0$, we can conclude that the associated graded ring of the ring $\widetilde{A}$ is naturally isomorphic to the graded ring A, that is, $\operatorname{gr}^F \widetilde{A} \simeq A$, while $H^i_{\partial_T + d_T + \delta_T}(T_R(K_+)) = 0$ for $i \neq 0$.

Since the graded ring $A = \operatorname{gr}^F \widetilde{A}$ is left finitely projective Koszul, the graded ring $\widehat{A} = \bigoplus_{n=0}^{\infty} F_n \widetilde{A}$ is left finitely projective Koszul as well (by Lemma 4.3 and Theorem 4.4). Let $\widehat{B}'$ be the right finitely projective Koszul graded ring quadratic dual to $\widehat{A}$. By Proposition 4.21, the graded ring $\widehat{B}'$ is endowed with an odd derivation $\partial' \colon \widehat{B}' \longrightarrow \widehat{B}'$ of degree -1, making it a right finitely projective Koszul quasi-differential graded ring. The underlying graded ring $B' = \ker \partial' \subset \widehat{B}'$ of the quasi-differential graded ring $(\widehat{B}', \partial')$ is quadratic dual to A, so we have $B' \simeq B$. It remains to construct a natural isomorphism of quasi-differential graded rings $(\widehat{B}', \partial') \simeq (\widehat{B}, \partial)$.

For this purpose, let us consider the dual graded corings $D' = \operatorname{Hom}_{R^{\mathrm{op}}}(B', R)$ and $\widehat{D}' = \operatorname{Hom}_{R^{\mathrm{op}}}(\widehat{B}', R)$. As in the beginning of this proof, we have an odd coderivation $\partial' \colon \widehat{D}' \longrightarrow \widehat{D}'$ of degree 1 dual to the odd derivation $\partial' \colon \widehat{B}' \longrightarrow \widehat{B}'$. It suffices to construct a natural isomorphism $(\widehat{D}, \partial) \longrightarrow (\widehat{D}', \partial')$ of DG-corings over R.

The embedding of the component $\widehat{D}_1 = T_R(K_+)^{-1,0,1} \longrightarrow T_R(K_+)$ induces an isomorphism of R-R-bimodules $\widehat{D}_1 \simeq F_1 \widetilde{A}$. The composition $\widehat{D}_2 \longrightarrow \widehat{D}_1 \otimes_R \widehat{D}_1 \simeq F_1 \widetilde{A} \otimes_R F_1 \widetilde{A} \longrightarrow F_2 \widetilde{A}$ of the comultiplication and multiplication maps vanishes, being killed by the differential $(\partial_T + d_T + \delta_T)^{-2,0,1} = d_T^{-2,0,1} \colon T_R(K_+)^{-2,0,1} \longrightarrow T_R(K_+)^{-2,0,2}$. So there is a natural morphism of graded corings $\widehat{D} \longrightarrow \widehat{D}'$. Since the embedding $R = F_0 \widetilde{A} \longrightarrow F_1 \widetilde{A}$ corresponds to the map $\partial_0 \colon R = \widehat{D}_0 \longrightarrow \widehat{D}_1$ under the isomorphisms $F_0 \widetilde{A} = R = \widehat{D}_0$ and $F_1 \widetilde{A} \simeq \widehat{D}_1$, the graded coring morphism $\widehat{D} \longrightarrow \widehat{D}'$ forms a commutative square diagram with the differentials ∂ on $\widehat{D}$ and ∂' on $\widehat{D}'$.

The induced morphism $\operatorname{coker}(\partial) \longrightarrow \operatorname{coker}(\partial')$ coincides with the natural isomorphism $D \longrightarrow D'$ on the components of degree 1 and hence on the other components as well. Therefore, the morphism of corings $\widehat{D} \longrightarrow \widehat{D}'$ is also an isomorphism.   □

Proof of Theorem 4.24 Let $R \subset \widetilde{V} \subset \widetilde{A}$ be a weak nonhomogeneous quadratic ring and $A = \mathrm{gr}^F \widetilde{A}$ be its associated graded ring with respect to the filtration F generated by $F_1\widetilde{A} = \widetilde{V}$ over $F_0\widetilde{A} = R$. Assume that the quadratic graded ring $qA = q\mathrm{gr}^F \widetilde{A}$ is left finitely projective Koszul. Let (B, d, h) be the CDG-ring corresponding to $\widetilde{A}$ under the construction of Proposition 3.16. Equivalently, one can consider the quasi-differential graded ring $(\widehat{B}, \partial)$ with the underlying graded ring $B = \ker \partial \subset \widehat{B}$ corresponding to $\widetilde{A}$ under the construction of Proposition 4.21.

Whichever one of these two points of view one takes, the graded ring B is quadratic and quadratic dual to the quadratic graded ring qA by construction. Hence the graded ring B is right finitely projective Koszul by Proposition 2.37. Applying Theorem 4.25, we see that the CDG-ring (B, d, h) or the quasi-differential ring $(\widehat{B}, \partial)$ comes from a left finitely projective nonhomogeneous Koszul ring $(\widetilde{A}', F)$. Following either one of the two proofs of Theorem 4.25, the graded ring $\mathrm{gr}^F \widetilde{A}'$ is quadratic and left finitely projective Koszul.

It remains to observe that the nonhomogeneous quadratic duality functor assigns the same CDG-ring (B, d, h) (or the same quasi-differential graded ring $(\widehat{B}, \partial)$) to the two 3-left finitely projective (weak) nonhomogeneous quadratic rings $\widetilde{A}$ and $\widetilde{A}'$. Since the nonhomogeneous quadratic duality functor is fully faithful by Theorem 3.21 or 4.22, it follows that the two (weak) nonhomogeneous quadratic rings $(\widetilde{A}, F)$ and $(\widetilde{A}', F)$ are isomorphic. Hence the graded ring $A = \mathrm{gr}^F \widetilde{A} \simeq \mathrm{gr}^F \widetilde{A}'$ is quadratic (and left finitely projective Koszul). In other words, the weak nonhomogeneous quadratic ring $R \subset \widetilde{V} \subset \widetilde{A}$ is actually nonhomogeneous quadratic. $\square$

4.7 Anti-Equivalences of Koszul Ring Categories

Let R be an associative ring. The *category of left finitely projective nonhomogeneous Koszul rings over R*, denoted by $R\text{–rings}_{\mathsf{nlk}}$, is defined as the full subcategory of the category of filtered rings $R\text{–rings}_{\mathsf{fil}}$ (see Sect. 3.6) whose objects are the left finitely projective nonhomogeneous Koszul rings $(\widetilde{A}, F)$. The *category of right finitely projective Koszul CDG-rings over R*, denoted by $R\text{–rings}_{\mathsf{cdg,rk}}$, is the full subcategory in the category of nonnegatively graded CDG-rings $R\text{–rings}_{\mathsf{cdg}}$ whose objects are the right finitely projective Koszul CDG-rings. The *category of right finitely projective Koszul quasi-differential rings over R*, denoted by $R\text{–rings}_{\mathsf{qdg,rk}}$, is the full subcategory in the category of nonnegatively graded quasi-differential rings $R\text{–rings}_{\mathsf{qdg}}$ whose objects are the right finitely projective Koszul quasi-differential graded rings.

Corollary 4.26 *The constructions of Theorems 3.21, 4.7, and 4.22 define natural (anti)-equivalences*

$$(R\text{–rings}_{\mathsf{nlk}})^{\mathsf{op}} \simeq R\text{–rings}_{\mathsf{qdg,rk}} \simeq R\text{–rings}_{\mathsf{cdg,rk}}$$

between the categories of left finitely projective nonhomogeneous Koszul rings, right finitely projective Koszul quasi-differential rings, and right finitely projective Koszul CDG-rings over R.

Proof It follows from the mentioned theorems and Theorem 4.25.   □

The 2-*category of left finitely projective nonhomogeneous Koszul rings*, denoted by $\mathsf{Rings}_{\mathsf{nlk2}}$, is defined as the following 2-subcategory in the 2-category of filtered rings $\mathsf{Rings}_{\mathsf{fil2}}$ (see Sect. 3.7). The objects of $\mathsf{Rings}_{\mathsf{nlk2}}$ are the left finitely projective nonhomogeneous Koszul rings $(\widetilde{A}, F)$. All morphisms in $\mathsf{Rings}_{\mathsf{fil2}}$ between objects of $\mathsf{Rings}_{\mathsf{nlk2}}$ are morphisms in $\mathsf{Rings}_{\mathsf{nlk2}}$, and all 2-morphisms in $\mathsf{Rings}_{\mathsf{fil2}}$ between morphisms of $\mathsf{Rings}_{\mathsf{nlk2}}$ are 2-morphisms in $\mathsf{Rings}_{\mathsf{nlk2}}$.

The 2-*category of right finitely projective Koszul CDG-rings*, denoted by $\mathsf{Rings}_{\mathsf{cdg2,rk}}$, is the following 2-subcategory in the 2-category of nonnegatively graded CDG-rings $\mathsf{Rings}_{\mathsf{cdg2}}$. The objects of $\mathsf{Rings}_{\mathsf{cdg2,rk}}$ are all the right finitely projective Koszul CDG-rings (B, d, h). All morphisms in $\mathsf{Rings}_{\mathsf{cdg2}}$ between objects of $\mathsf{Rings}_{\mathsf{cdg2,rk}}$ are morphisms in $\mathsf{Rings}_{\mathsf{cdg2,rk}}$, and all 2-morphisms in $\mathsf{Rings}_{\mathsf{cdg2}}$ between morphisms of $\mathsf{Rings}_{\mathsf{cdg2,rk}}$ are 2-morphisms in $\mathsf{Rings}_{\mathsf{cdg2,rk}}$.

The 2-*category of right finitely projective Koszul quasi-differential graded rings*, denoted by $\mathsf{Rings}_{\mathsf{qdg2,rk}}$, is the following 2-subcategory in the 2-category of nonnegatively graded quasi-differential rings $\mathsf{Rings}_{\mathsf{qdg2}}$ (see Sect. 4.2). The objects of $\mathsf{Rings}_{\mathsf{qdg2,rk}}$ are all the right finitely projective Koszul quasi-differential graded rings $(\widehat{B}, \partial)$. All morphisms in $\mathsf{Rings}_{\mathsf{qdg2}}$ between objects of $\mathsf{Rings}_{\mathsf{qdg2,rk}}$ are morphisms in $\mathsf{Rings}_{\mathsf{qdg2,rk}}$, and all 2-morphisms in $\mathsf{Rings}_{\mathsf{qdg2}}$ between morphisms of $\mathsf{Rings}_{\mathsf{qdg2,rk}}$ are 2-morphisms in $\mathsf{Rings}_{\mathsf{qdg2,rk}}$.

Corollary 4.27 *The constructions of Theorems 3.25, 4.8, and 4.22 define natural strict (anti)-equivalences*

$$(\mathsf{Rings}_{\mathsf{nlk2}})^{\mathsf{op}} \simeq \mathsf{Rings}_{\mathsf{qdg2,rk}} \simeq \mathsf{Rings}_{\mathsf{cdg2,rk}}$$

between the 2-categories of left finitely projective nonhomogeneous Koszul rings, right finitely projective Koszul quasi-differential rings, and right finitely projective Koszul CDG-rings.

Proof It follows from the mentioned theorems and Theorem 4.25.   □

A left finitely projective nonhomogeneous Koszul ring $(\widetilde{A}, F)$ is said to be *left augmented* if the ring $\widetilde{A}$ is left augmented over its subring $F_0\widetilde{A}$. In other words, this means that a left ideal $\widetilde{A}^+ \subset \widetilde{A}$ is chosen such that $\widetilde{A} = F_0\widetilde{A} \oplus \widetilde{A}^+$ (see Sect. 3.8). The *category of left augmented left finitely projective nonhomogeneous Koszul rings over R*, denoted by $R\text{--}\mathsf{rings}_{\mathsf{nlk}}^{\mathsf{laug}}$, is defined as the full subcategory in the category of left augmented filtered rings $R\text{–rings}_{\mathrm{fil}}^{\mathrm{laug}}$ whose objects are the left augmented left finitely projective nonhomogeneous Koszul rings.

A nonnegatively graded DG-ring (B, d) (in the sense of Sect. 3.2) is said to be *right finitely projective Koszul* if the graded ring B is right finitely projective Koszul. The *category of right finitely projective Koszul DG-rings over* R, denoted by $R\text{–rings}_{\mathrm{dg,rk}}$ is defined as the full subcategory in the category of nonnegatively graded DG-rings $R\text{–rings}_{\mathrm{dg}}$ whose objects are the right finitely projective Koszul DG-rings.

Corollary 4.28 *The construction of Theorem 3.27 defines a natural anti-equivalence*

$$(R\text{–rings}_{\mathrm{nlk}}^{\mathrm{laug}})^{\mathrm{op}} \simeq R\text{–rings}_{\mathrm{dg,rk}}$$

between the category of left augmented left finitely projective nonhomogeneous Koszul rings and the category of right finitely projective Koszul DG-rings.

Proof It follows from Theorems 4.25 and 3.27.   □

The 2-*category of left augmented left finitely projective nonhomogeneous Koszul rings*, denoted by $\mathsf{Rings}_{\mathrm{nlk}}^{\mathrm{laug2}}$, is defined as the following 2-subcategory in the 2-category of left augmented filtered rings $\mathsf{Rings}_{\mathrm{fil}}^{\mathrm{laug2}}$ (see Sect. 3.8). The objects of $\mathsf{Rings}_{\mathrm{nlk}}^{\mathrm{laug2}}$ are the left augmented left finitely projective nonhomogeneous Koszul rings $(\widetilde{A}, F, A^+)$. All morphisms in $\mathsf{Rings}_{\mathrm{fil}}^{\mathrm{laug2}}$ between objects of $\mathsf{Rings}_{\mathrm{nlk}}^{\mathrm{laug2}}$ are morphisms in $\mathsf{Rings}_{\mathrm{nlk}}^{\mathrm{laug2}}$, and all 2-morphisms in $\mathsf{Rings}_{\mathrm{fil}}^{\mathrm{laug2}}$ between morphisms of $\mathsf{Rings}_{\mathrm{nlk}}^{\mathrm{laug2}}$ are 2-morphisms in $\mathsf{Rings}_{\mathrm{nlk}}^{\mathrm{laug2}}$.

The 2-*category of right finitely projective Koszul DG-rings*, denoted by $\mathsf{Rings}_{\mathrm{dg2,rk}}$, is defined as the following 2-subcategory in the 2-category of nonnegatively graded DG-rings $\mathsf{Rings}_{\mathrm{dg2}}$. The objects of $\mathsf{Rings}_{\mathrm{dg2,rk}}$ are the right finitely projective Koszul DG-rings (B, d). All morphisms in $\mathsf{Rings}_{\mathrm{dg2}}$ between objects of $\mathsf{Rings}_{\mathrm{dg2,rk}}$ are morphisms in $\mathsf{Rings}_{\mathrm{dg2,rk}}$, and all 2-morphisms in $\mathsf{Rings}_{\mathrm{dg2}}$ between morphisms of $\mathsf{Rings}_{\mathrm{dg2,rk}}$ are 2-morphisms in $\mathsf{Rings}_{\mathrm{dg2,rk}}$.

Corollary 4.29 *The construction of Theorem 3.29 defines a natural strict anti-equivalence*

$$(\mathsf{Rings}_{\mathrm{nlk}}^{\mathrm{laug2}})^{\mathrm{op}} \simeq \mathsf{Rings}_{\mathrm{dg2,rk}}$$

between the 2-category of left augmented left finitely projective nonhomogeneous Koszul rings and the 2-category of right finitely projective Koszul DG-rings.   □

Examples of left finitely projective nonhomogeneous Koszul rings and right finitely projective Koszul (C)DG-rings corresponding to each other under the equivalences of categories from Corollaries 4.26 and 4.28 will be presented and discussed in detail in Sects. 10.2–10.10 below.

Some such examples have been already mentioned in Examples 3.12, 3.17, 3.22, and 3.28 in Chap. 3. In particular, the 3-left finitely projective nonhomogeneous quadratic rings in Examples 3.17, 3.22, and 3.28 are left finitely projective Koszul.

Comodules and Contramodules Over Graded Rings

5

5.1 Ungraded Comodules Over Nonnegatively Graded Rings

Let $B = \bigoplus_{n=0}^{\infty} B_n$ be a nonnegatively graded ring. We denote the underlying ungraded ring of B by the same letter $B = \Sigma B$ (see Sect. 2.1) and consider ungraded right B-modules.

Definition 5.1 An ungraded right B-module M is said to be a *B-comodule* (or an *ungraded right B-comodule*) if for every element $x \in M$ there exists an integer $m \geq 0$ such that $xb = 0$ in M for all $b \in B_n$, $n > m$. Notice that any right B-comodule has a natural structure of right module over the ring $\Pi B = \prod_{n=0}^{\infty} B_n$.

We will denote the full subcategory of right B-comodules by $\mathsf{comod}{-}B \subset \mathsf{mod}{-}B$. The following examples explain the terminology.

Examples 5.2

(1) Let B be a nonnegatively graded associative algebra over a field $k = B_0$ with finite-dimensional grading components, $\dim_k B_n < \infty$ for all $n \geq 0$. Then the graded dual vector space $C = \bigoplus_{n=0}^{\infty} B_n^*$ to B has a natural structure of graded coassociative coalgebra over k. An ungraded right B-comodule in the sense of the above definition is the same thing as an ungraded right C-comodule.

(2) More generally, let B be a nonnegatively graded ring with the degree-zero grading component $R = B_0$. Assume that B_n is a finitely generated projective right R-module for every $n \geq 0$, and put $C = \bigoplus_{n=0}^{\infty} C_n$, where $C_n = \mathrm{Hom}_{R^{\mathrm{op}}}(B_n, R)$. Then, in view of Lemma 1.1(b), applying the functor $\mathrm{Hom}_{R^{\mathrm{op}}}(-, R)$ to the multiplication maps $B_i \otimes_R B_j \longrightarrow B_{i+j}$ produces comultiplication maps $C_{i+j} \longrightarrow C_j \otimes_R C_i$ endowing

L. Positselski, *Relative Nonhomogeneous Koszul Duality*, Frontiers in Mathematics,
https://doi.org/10.1007/978-3-030-89540-2_5

the graded R-R-bimodule C with a natural structure of graded coring over R. An ungraded right B-comodule M in the sense of the above definition is the same thing as an ungraded right C-comodule, i.e., a right R-module M endowed with a coassociative, counital comultiplication map $M \longrightarrow M \otimes_R C$ (see Example 5.10(3) below for a discussion of corings and comodules).

(3) In the context of (1), one can also say that ungraded left B-comodules are the same thing as ungraded left C-comodules. But let us *warn* the reader that, in the context of (2), left B-comodules are, generally speaking, entirely *unrelated* to left C-comodules. Rather, in order to describe left C-comodules, one needs to consider the graded ring $^\#B$ defined in Remark 2.13. Then, assuming that C_n is a finitely generated projective right R-module for every $n \geq 0$, an ungraded left $^\#B$-comodule is the same thing as an ungraded left C-comodule (and the same holds for graded comodules as defined in Sect. 5.5 below; cf. Example 5.23).

Example 5.3 Let $B = k[x]$ be the ring of polynomials in one variable x over a field k, endowed with the usual grading with $\deg x = 1$. Let $k[x, x^{-1}]$ be the B-module of Laurent polynomials in x over k. Then the quotient module $k[x, x^{-1}]/k[x]$ (known as the *Prüfer module*) is a B-comodule. The B-module $k[x]/(x^n)$ is also a B-comodule for every $n \geq 1$. On the other hand, the B-modules $k[x]$, $k[x, x^{-1}]$, and $k[x]/(x - a)^n$, where $a \in k \setminus \{0\}$ and $n \geq 1$, are *not* B-comodules.

Obviously, the full subcategory $\mathsf{comod}{-}B \subset \mathsf{mod}{-}B$ is closed under subobjects, quotients, and infinite direct sums. In fact, $\mathsf{comod}{-}B$ is a hereditary pretorsion class in $\mathsf{mod}{-}B$ corresponding to the filter/topology of right ideals in B in which the two-sided ideals $B_{\geq m} = \bigoplus_{n \geq m} B_n \subset B$, $m \geq 1$, form a base [78, Section VI.4] and [68, Sections 2.3–2.4]. So $\mathsf{comod}{-}B$ is a Grothendieck abelian category and its inclusion $\mathsf{comod}{-}B \longrightarrow \mathsf{mod}{-}B$ is an exact functor preserving infinite direct sums.

Let us show that, under certain assumptions, the full subcategory $\mathsf{comod}{-}B$ is also closed under extensions in $\mathsf{mod}{-}B$ (in other words, it is a *hereditary torsion class*).

Lemma 5.4 *Let $B = \bigoplus_{n=0}^{\infty} B_n$ be a nonnegatively graded ring and $m \geq 1$ be an integer. Then the following five conditions are equivalent:*

(a) *There exists an integer $k \geq 1$ such that the graded right B-module $B_{\geq 1} = \bigoplus_{n=1}^{\infty} B_n$ is generated by elements of degree $\leq k$.*

(b) *There exists an integer $l \geq m$ such that the graded right B-module $B_{\geq m} = \bigoplus_{n=m}^{\infty} B_n$ is generated by elements of degree $\leq l$.*

(c) *There exists an integer $k \geq 1$ such that for every $n > k$ the multiplication map $\bigoplus_{i+j=n}^{i,j \geq 1} B_i \otimes_R B_j \longrightarrow B_n$ is surjective.*

(d) *There exists an integer $l \geq 2m - 1$ such that for every $n > l$ the multiplication map $\bigoplus_{i+j=n}^{i,j\geq m} B_i \otimes_R B_j \longrightarrow B_n$ is surjective.*

(e) *There exists an integer $k \geq 1$ such that the ring B is generated by its subgroup $\bigoplus_{n=0}^{k} B_k \subset B$.*

Proof The equivalences (a) $\Longleftrightarrow$ (c) $\Longleftrightarrow$ (e) (for the same k) are easy, so are the implications (d) $\Longrightarrow$ (b) $\Longrightarrow$ (c) (for $k = l$). The implication (e) $\Longrightarrow$ (d) holds for $l = 2m - 2 + k$. □

Notice that it follows from Lemma 5.4 that the conditions (a–b) are left-right symmetric (because the conditions (c–e) are). It also follows that the conditions (b) and (d) do not depend on m (because the conditions (a), (c), and (e) do not).

Lemma 5.5 *Let $B = \bigoplus_{n=0}^{\infty} B_n$ be a nonnegatively graded ring and $m \geq 1$ be an integer. Then, for $m = 1$, the following two conditions are equivalent:*

(i) *$B_{\geq m} = \bigoplus_{n=m}^{\infty} B_n$ is a finitely generated (graded) right B-module.*
(ii) *B_n is a finitely generated right B_0-module for every $n \geq m$, and any one of the equivalent conditions of Lemma 5.4 holds.*

For any $m \geq 1$, the implication (ii) $\Longrightarrow$ (i) is true.

Proof (i) $\Longrightarrow$ (ii) Clearly, (i) $\Longrightarrow$ (b) (for the same m), so it remains to show that (i) implies the first part of (ii). Indeed, let $\{b_{s,n} \in B_n \mid n \geq m, 1 \leq s \leq t_n\}$ be a finite set of homogeneous generators of the right B-module $B_{\geq m}$. Put $R = B_0$. Then, for every $n \geq m$, the cokernel of the multiplication map $\bigoplus_{i+j=n}^{i\geq m, j\geq 1} B_i \otimes_R B_j \longrightarrow B_n$ is generated by the cosets of the elements $b_{s,n}$, $1 \leq s \leq t_n$ as a right R-module. For any R-R-bimodules U and V, if the right R-modules U and V are finitely generated, then so is the right R-module $U \otimes_R V$. This allows to prove by induction in n that the right R-module B_n is finitely generated for $n \geq m$.

(ii) $\Longrightarrow$ (i) Choose l as in (b), and for every $m \leq n \leq l$, choose a finite set of generators $b_{s,n}$, $1 \leq s \leq t_n$ of the right R-module B_n. Then the right B-module $B_{\geq m}$ is generated by the elements $\{b_{s,n} \in B_n \mid m \leq n \leq l, 1 \leq s \leq t_n\}$. □

Lemma 5.6 *Let $B = \bigoplus_{n=0}^{\infty} B_n$ be a nonnegatively graded ring such that the augmentation ideal $B_{\geq 1}$ is finitely generated as a right ideal in B. Then, for every $m \geq 1$, the right ideal $B_{\geq m} \subset B$ is also finitely generated.*

Proof It follows from Lemmas 5.4 and 5.5. □

The following particular case is important for our purposes. Let B be a nonnegatively graded ring generated by B_1 over B_0. Assume the right B_0-module B_1 is finitely generated.

Then it is clear from Lemma 5.4 (take $k = 1$) and Lemma 5.5 that the augmentation ideal $B_{\geq 1} \subset B$ is finitely generated as a right ideal in B. In particular, this holds for any 2-right finitely projective quadratic graded ring B.

Proposition 5.7 *Assume that the augmentation ideal* $B_{\geq 1} = \bigoplus_{n=1}^{\infty} B_n$ *of a nonnegatively graded ring* B *is finitely generated as a right ideal in* B. *Then the full subcategory of ungraded right* B-comodules comod$-B$ *is closed under extensions in the category of right* B-modules mod$-B$.

Proof Let $0 \longrightarrow K \longrightarrow L \longrightarrow M \longrightarrow 0$ be a short exact sequence of (ungraded) right B-modules. Assume that K and M are B-comodules. Let $x \in L$ be an element, and let $y \in M$ be the image of x under the surjective B-module morphism $L \longrightarrow M$. By assumption, there exists $m \geq 1$ such that $y B_{\geq m} = 0$ in M.

By Lemma 5.6, the right ideal $B_{\geq m} \subset B$ is finitely generated. Let $\{b_s \mid 1 \leq s \leq t\}$ be a finite set of its homogeneous generators, so $B_{\geq m} = \sum_{s=1}^{t} b_s B$. Let k_s be the homogeneous degree of the element $b_s \in B$, and let $k \geq m$ be an integer such that $k \geq k_s$ for all $1 \leq s \leq t$. Then we have $B_n = \sum_{s=1}^{t} b_s B_{n-k_s}$ for all $n \geq k$.

For every $1 \leq s \leq t$, we have $z_s = x b_s \in K \subset L$. By assumption, there exists $j_s \geq 1$ such that $z_s B_{\geq j_s} = 0$ in K. Let $j \geq 1$ be an integer such that $j \geq j_s$ for all $1 \leq s \leq t$. Then, for every $n \geq k + j$, we have $x B_n = \sum_{s=1}^{t} x b_s B_{n-k_s} = \sum_{s=1}^{t} z_s B_{n-k_s} = 0$ in L, since $n - k_s \geq j_s$. Thus $x B_{\geq k+j} = 0$, as desired. $\qquad\qquad\square$

5.2 Right Exact Monads on Abelian Categories

Let A be a category. A *monad* on A is a monoid object in the monoidal category of endofunctors of A (with respect to the composition of functors). In other words, this means that a monad $\mathbb{M} \colon$ A $\longrightarrow$ A is a covariant functor endowed with two natural transformations of *monad unit* $e \colon \mathrm{Id}_A \longrightarrow \mathbb{M}$ and *monad multiplication* $m \colon \mathbb{M} \circ \mathbb{M} \longrightarrow \mathbb{M}$ satisfying the associativity and unitality axioms. We refer to [43, Chapter VI] for the definitions of monads and algebras over monads.

Since we are interested in abelian categories arising from monads, we prefer to use the term "modules over a monad" for what are usually called algebras over a monad. Given a monad $\mathbb{M} \colon$ A $\longrightarrow$ A, we denote the category of algebras/modules over $\mathbb{M}$ (known also as the "Eilenberg–Moore category of $\mathbb{M}$") by $\mathbb{M}-$mod. The category $\mathbb{M}-$mod comes together with a faithful forgetful functor $\mathbb{M}-$mod $\longrightarrow$ A, which has a left adjoint functor A $\longrightarrow \mathbb{M}-$mod. The latter functor takes any object $A \in$ A to the object $\mathbb{M}(A) \in$ A, which is endowed with a natural structure of an $\mathbb{M}$-module, making it an object of $\mathbb{M}-$mod. The object $\mathbb{M}(A) \in \mathbb{M}-$mod is called the *free* $\mathbb{M}$-*module* spanned by A.

Lemma 5.8 *Let* A *be an additive category and* $\mathbb{M}\colon A \longrightarrow A$ *be a monad. Then the following three conditions are equivalent:*

(a) *The category* $\mathbb{M}-\mathsf{mod}$ *is additive.*
(b) *The category* $\mathbb{M}-\mathsf{mod}$ *is additive, and both the forgetful functor* $\mathbb{M}-\mathsf{mod} \longrightarrow A$ *and the free module functor* $A \longrightarrow \mathbb{M}-\mathsf{mod}$ *are additive.*
(c) *The functor* $\mathbb{M}\colon A \longrightarrow A$ *is additive.*

Proof (a) $\Longleftrightarrow$ (b) $\Longrightarrow$ (c) The functor $\mathbb{M}\colon A \longrightarrow A$ is the composition of two adjoints $A \longrightarrow \mathbb{M}-\mathsf{mod} \longrightarrow A$. Any functor between additive categories having an adjoint functor is additive, and the composition of two additive functors is additive.

(c) $\Longrightarrow$ (b) If the functor $\mathbb{M}\colon A \longrightarrow A$ is additive, then for any two $\mathbb{M}$-modules P and Q, the set $\mathrm{Hom}_\mathbb{M}(P, Q)$ of morphisms $P \longrightarrow Q$ in $\mathbb{M}-\mathsf{mod}$ is a subgroup in the group $\mathrm{Hom}_A(P, Q)$ of morphisms $P \longrightarrow Q$ in the category A. This shows that the category $\mathbb{M}-\mathsf{mod}$ is preadditive and the faithful forgetful functor $\mathbb{M}-\mathsf{mod} \longrightarrow A$ is additive. Finally, for any finite collection of objects $P_i \in \mathbb{M}-\mathsf{mod}$, $i = 1, \ldots, n$, the direct sum ${}^A\bigoplus_{i=1}^n P_i$ taken in the category A is naturally endowed with an $\mathbb{M}$-module structure, which makes it the direct sum of the objects P_i in the category $\mathbb{M}-\mathsf{mod}$. Thus finite (co)products exist in $\mathbb{M}-\mathsf{mod}$. $\qquad\square$

Lemma 5.9 *Let* A *be an abelian category and* $\mathbb{M}\colon A \longrightarrow A$ *be a monad. Then the following two conditions are equivalent:*

(a) *The category* $\mathbb{M}-\mathsf{mod}$ *is abelian and* *the forgetful functor* $\mathbb{M}-\mathsf{mod} \longrightarrow A$ *is exact.*
(b) *The functor* $\mathbb{M}\colon A \longrightarrow A$ *is right exact.*

Proof Notice first of all that any exact or one-sided exact functor is additive by definition, so Lemma 5.8 is applicable. This allows us to presume that all the categories and functors involved are additive.

(a) $\Longrightarrow$ (b) Any left adjoint functor preserves colimits; hence any left adjoint functor between abelian categories is right exact. So the free module functor $A \longrightarrow \mathbb{M}-\mathsf{mod}$ is right exact. If the forgetful functor $\mathbb{M}-\mathsf{mod} \longrightarrow A$ is exact, it follows that the composition $\mathbb{M}\colon A \longrightarrow A$ of two adjoints $A \longrightarrow \mathbb{M}-\mathsf{mod} \longrightarrow A$ is right exact.

(b) $\Longrightarrow$ (a) For any monad $\mathbb{M}\colon A \longrightarrow A$ and any morphism $f\colon P \longrightarrow Q$ in $\mathbb{M}-\mathsf{mod}$, the kernel $\ker^A(f)$ of the morphism f taken in the category A is naturally endowed with an $\mathbb{M}$-module structure, which makes it the kernel of f in the category $\mathbb{M}-\mathsf{mod}$. When the functor $\mathbb{M}\colon A \longrightarrow A$ is right exact, the same holds true for the cokernel of f. Both the properties in (a) follow easily. $\qquad\square$

The following examples are illuminating.

Examples 5.10

(1) Let $g: R \longrightarrow S$ be an associative ring homomorphism. Put $\mathsf{A} = R-\mathsf{mod}$, and consider the functor $\mathbb{M} = S \otimes_R -: R-\mathsf{mod} \longrightarrow R-\mathsf{mod}$. The map g is an R-R-bimodule morphism, so it induces a natural transformation $\mathrm{Id}_\mathsf{A} \longrightarrow \mathbb{M}$. The multiplication map $S \otimes_R S \longrightarrow S$ is an R-R-bimodule morphism, too, and it induces a natural transformation $\mathbb{M} \circ \mathbb{M} \longrightarrow \mathbb{M}$. These two morphisms of functors make $\mathbb{M}$ a monad on $\mathsf{A} = R-\mathsf{mod}$. The category of $\mathbb{M}$-modules $\mathbb{M}-\mathsf{mod}$ is equivalent to $S-\mathsf{mod}$. The forgetful functor $\mathbb{M}-\mathsf{mod} \longrightarrow \mathsf{A}$ is the functor of restriction of scalars $S-\mathsf{mod} \longrightarrow R-\mathsf{mod}$, while the free module functor $\mathsf{A} \longrightarrow \mathbb{M}-\mathsf{mod}$ is the functor of extension of scalars $S \otimes_R -: R-\mathsf{mod} \longrightarrow S-\mathsf{mod}$.

(2) The notion of a *comonad* on a category A is dual to that of a monad. Specifically, a comonad $\mathbb{C}$ on A is a covariant endofunctor $\mathbb{C}: \mathsf{A} \longrightarrow \mathsf{A}$ endowed with two natural transformations of *counit* $\mathbb{C} \longrightarrow \mathrm{Id}_\mathsf{A}$ and *comultiplication* $\mathbb{C} \longrightarrow \mathbb{C} \circ \mathbb{C}$ satisfying the coassociativity and counitality axioms. One can say that a comonad on A is the same thing as a monad on A^{op}. Denote by $\mathbb{C}-\mathsf{comod}$ the category of comodules (usually called "coalgebras") over a comonad $\mathbb{C}$ on A.

   Let A be an abelian category and $\mathbb{C}$ be a comonad on A. Then the dual version of Lemma 5.9 tells that the category $\mathbb{C}-\mathsf{comod}$ is abelian with exact forgetful functor $\mathbb{C}-\mathsf{comod} \longrightarrow \mathsf{A}$ if and only if the functor $\mathbb{C}: \mathsf{A} \longrightarrow \mathsf{A}$ is left exact.

(3) Let R be a ring and C be a (coaccosiative, counital) coring over R, that is, a comonoid object in the monoidal category of R-R-bimodules. Then the functor $\mathbb{C} = - \otimes_R C: \mathsf{mod}-R \longrightarrow \mathsf{mod}-R$ is a comonad on the category of right R-modules $\mathsf{mod}-R$. The counit map $C \longrightarrow R$ of the coring C induces the counit morphism $\mathbb{C} \longrightarrow \mathrm{Id}_{\mathsf{mod}-R}$, and the comultiplication map $C \longrightarrow C \otimes_R C$ induces the comultiplication morphism $\mathbb{C} \longrightarrow \mathbb{C} \circ \mathbb{C}$. Comodules over the comonad $\mathbb{C}$ are known as *right comodules* over the coring C. In other words, a right C-comodule N is a right R-module endowed with a *right coaction* map $N \longrightarrow N \otimes_R C$, which must be a right R-module morphism satisfying the coassociativity and counitality equations involving the comultiplication and counit maps of the coring C. Similarly one defines *left C-comodules* (using the comonad induced by C on the category of left R-modules).

   Let $\mathsf{comod}-C = \mathbb{C}-\mathsf{comod}$ denote the category of right C-comodules. Then, as a particular case of (2), one obtains the assertion that the category $\mathsf{comod}-C$ is abelian with exact forgetful functor $\mathsf{comod}-C \longrightarrow \mathsf{mod}-R$ if and only if C is a flat left R-module [58, Section 1.1.2], [61, Section 2.5].

(4) Let R be a ring and C be a coring over R, as in (3). Then the functor $\mathbb{M} = \mathrm{Hom}_R(C, -): R-\mathsf{mod} \longrightarrow R-\mathsf{mod}$ is a monad on the category of left R-modules. The counit map of the coring C induces the unit morphism $\mathrm{Id}_{R-\mathsf{mod}} \longrightarrow \mathbb{M}$, and the comultiplication map of the coring C induces the multiplication morphism $\mathbb{M} \circ \mathbb{M} \longrightarrow \mathbb{M}$. Modules over the monad $\mathbb{M}$ are known as *left contramodules* over

the coring C. In other words, a left C-contramodule P is a left R-module endowed with a *left contraaction* map $\mathrm{Hom}_R(C, P) \longrightarrow P$, which must be a left R-module morphism satisfying (contra)associativity and (contra)unitality equations involving the comultiplication and counit of the coring C. We refer to [58, Section 3.1] or [61, Section 2.5] for details on contramodules over corings.

Let $C-\mathsf{contra} = \mathbb{M}-\mathsf{mod}$ denote the category of left C-contramodules. Then it follows from Lemma 5.9 that the category $C-\mathsf{contra}$ is abelian with exact forgetful functor $C-\mathsf{contra} \longrightarrow R-\mathsf{mod}$ if and only if C is a projective left R-module [58, Section 3.1.2], [61, Proposition 2.5].

5.3 Ungraded Contramodules Over Nonnegatively Graded Rings

Let $B = \bigoplus_{n=0}^{\infty} B_n$ be a nonnegatively graded ring, and let K be an associative ring endowed with a ring homomorphism $K \longrightarrow B_0$. Consider the following monad $\mathbb{M} = \mathbb{M}_K : K-\mathsf{mod} \longrightarrow K-\mathsf{mod}$ on the category of left K-modules.

To any left K-module L, the monad $\mathbb{M}$ assigns the left K-module

$$\mathbb{M}_K(L) = \prod_{n=0}^{\infty} (B_n \otimes_K L).$$

The monad unit map

$$e_L : L \longrightarrow \prod_{n=0}^{\infty} B_n \otimes_K L$$

is the composition

$$L \longrightarrow B_0 \otimes_K L \longrightarrow \prod_{n=0}^{\infty} B_n \otimes_K L$$

of the map $L \longrightarrow B_0 \otimes_K L$ induced by the ring homomorphism $K \longrightarrow B_0$ with the inclusion of the $(n = 0)$-indexed component $B_0 \otimes_K L \longrightarrow \prod_{n=0}^{\infty} B_n \otimes_K L$. The monad multiplication map

$$m_L : \prod_{p=0}^{\infty} B_p \otimes_K \left(\prod_{q=0}^{\infty} B_q \otimes_K L \right) \longrightarrow \prod_{n=0}^{\infty} B_n \otimes_K L$$

is the composition

$$\prod_{p=0}^{\infty} B_p \otimes_K \left(\prod_{q=0}^{\infty} B_q \otimes_K L \right) \longrightarrow \prod_{p,q=0}^{\infty} B_p \otimes_K B_q \otimes_K L \longrightarrow \prod_{n=0}^{\infty} B_n \otimes_K L.$$

Here the leftmost arrow is the product over $p \geq 0$ of the maps $B_p \otimes_K \left(\prod_{q=0}^{\infty} B_q \otimes_K L \right)$ $\longrightarrow$ $\prod_{q=0}^{\infty} B_p \otimes_K B_q \otimes_K L$ whose components are the direct summand projections $B_p \otimes_K \left(\prod_{k=0}^{\infty} B_k \otimes_K L \right) \longrightarrow B_p \otimes_K B_q \otimes_K L$. The rightmost arrow is induced by the multiplication maps $\prod_{p+q=n}^{p,q \geq 0} B_p \otimes_K B_q \longrightarrow B_n$.

Definition 5.11 A *left B-contramodule* (or an *ungraded left B-contramodule*) is a module over the monad $\mathbb{M}_K : K-\text{mod} \longrightarrow K-\text{mod}$. In other words, a left B-contramodule P is a left K-module endowed with a *left B-contraaction map*

$$\pi_P : \mathbb{M}_K(P) = \prod_{n=0}^{\infty} B_n \otimes_K P \longrightarrow P,$$

which must be a morphism of left K-modules satisfying the (contra)associativity and (contra)unitality equations involving the multiplication and unit maps of the monad $\mathbb{M}_K$. This means that the compositions

$$\prod_{i=0}^{\infty} B_i \otimes_K \left(\prod_{j=0}^{\infty} B_j \otimes_K P \right) \rightrightarrows \prod_{n=0}^{\infty} B_n \otimes_K P \longrightarrow P$$

of the monad multiplication map m_P and the map $\mathbb{M}_K(\pi_P)$ with the contraaction map π_P are equal to each other, while the composition

$$P \longrightarrow \prod_{n=0}^{\infty} B_n \otimes_K P \longrightarrow P$$

of the monad unit map e_P and the contraaction map π_P is equal to the identity map id_P.

We denote the category of ungraded left B-contramodules by $B-\text{contra} = \mathbb{M}_K-\text{mod}$. Given two ungraded left B-contramodules P and Q, the group of morphisms $P \longrightarrow Q$ in $B-\text{contra}$ is denoted by $\text{Hom}^B(P, Q)$.

The next proposition shows that the notion of an ungraded left B-contramodule does not depend on the choice of a ring K. One can consider two polar cases. On the one hand, one can take $K = \mathbb{Z}$ and the unique ring homomorphism $\mathbb{Z} \longrightarrow B_0$. On the other hand, one can take $K = B_0$ and the identity map $K \longrightarrow B_0$.

Proposition 5.12 *There are natural equivalences of categories*

$$\mathbb{M}_{\mathbb{Z}}-\text{mod} \simeq \mathbb{M}_K-\text{mod} \simeq \mathbb{M}_{B_0}-\text{mod},$$

making the notation $B-\text{contra}$ unambiguous. The category $B-\text{contra}$ of ungraded left B-contramodules is abelian with enough projective objects. The forgetful functor $B-\text{contra} \longrightarrow K-\text{mod}$ can be naturally lifted to a forgetful functor $B-\text{contra} \longrightarrow$

$B-$mod *taking values in the category of ungraded left B-modules. The forgetful functor*
$B-$contra $\longrightarrow B-$mod *is exact and preserves infinite products.*

Proof To lift the forgetful functor $\mathbb{M}_K-$mod $\longrightarrow K-$mod to a functor $\mathbb{M}_K-$mod $\longrightarrow$
$B-$mod, consider an $\mathbb{M}_K$-module P and restrict the contraaction map π_P to the left
K-submodule

$$B\otimes_K P = \bigoplus_{n=0}^{\infty} B_n\otimes_K P \subset \prod_{n=0}^{\infty} B_n\otimes_K P.$$

The resulting map $B\otimes_K P \longrightarrow P$ extends the left K-module structure on P to a left
B-module structure, as explained in Example 5.10(1).

The functor $\mathbb{M}_K \colon K-$mod $\longrightarrow K-$mod is right exact, since the infinite product
functor is exact in $K-$mod and the tensor product functors $B_n\otimes_K-$ are right exact. By
Lemma 5.9, it follows that the category $\mathbb{M}_K-$mod is abelian and the forgetful functor
$\mathbb{M}_K-$mod $\longrightarrow K-$mod is exact. The latter functor, being a right adjoint, also preserves
all limits. Since the forgetful functor $B-$mod $\longrightarrow K-$mod is faithful and exact and
preserves all limits, it follows that the forgetful functor $\mathbb{M}_K-$mod $\longrightarrow B-$mod is also
exact and preserves all limits.

The free object functor $K-$mod $\longrightarrow \mathbb{M}_K-$mod, being left adjoint to an exact
functor, takes projective objects of $K-$mod to projective objects of $\mathbb{M}_K-$mod. One
can easily check that any object $P \in \mathbb{M}_K-$mod is a quotient object of an $\mathbb{M}_K$-module
$\mathbb{M}(K^{(X)})$ obtained by applying the free $\mathbb{M}_K$-module functor to a free left K-module with
X generators, for some set X. Indeed, if the underlying left K-module of P is a quotient
of a free left K-module $K^{(X)}$, then the $\mathbb{M}_K$-module P is a quotient of the $\mathbb{M}_K$-module
$\mathbb{M}_K(K^{(X)})$.

It remains to prove the first assertion. For this purpose, we will show that, for any
$\mathbb{M}_K$-module P, the contraaction map π_P factorizes through the natural surjective map
$\prod_{n=0}^{\infty} B_n\otimes_K P \longrightarrow \prod_{n=0}^{\infty} B_n\otimes_{B_0} P$. Indeed, passing to the product over $n \geq 0$ of
the right exact sequences (bar-complex fragments) $B_n\otimes_K B_0\otimes_K P \longrightarrow B_n\otimes_K P \longrightarrow$
$B_n\otimes_{B_0} P \longrightarrow 0$, one obtains a right exact sequence

$$\prod_{n=0}^{\infty} B_n\otimes_K B_0\otimes_K P \longrightarrow \prod_{n=0}^{\infty} B_n\otimes_K P \longrightarrow \prod_{n=0}^{\infty} B_n\otimes_{B_0} P \longrightarrow 0.$$

The map $\prod_{n=0}^{\infty} B_n\otimes_K B_0\otimes_K P \longrightarrow \prod_{n=0}^{\infty} B_n\otimes_K P$ is constructed as the difference of two
maps, one of which is induced by the multiplication maps $B_n\otimes_K B_0 \longrightarrow B_n$ and the other
one by the action maps $B_0\otimes_K P \longrightarrow P$.

This pair of maps can be obtained by restricting the pair of maps m_P, $\mathbb{M}_K(\pi_P)$:
$\prod_{i=0}^{\infty}\left(B_i\otimes_K\left(\prod_{j=0}^{\infty} B_j\otimes_K P\right)\right)$ $\rightrightarrows$ $\prod_{n=0}^{\infty} B_n\otimes_K P$ to the direct summand
$\prod_{i=0}^{\infty} B_i\otimes_K B_0\otimes_K P \subset \prod_{i=0}^{\infty}\left(B_i\otimes_K\left(\prod_{j=0}^{\infty} B_j\otimes_K P\right)\right)$ corresponding to the value
of the index $j = 0$. Thus the contraassociativity equation for the $\mathbb{M}_K$-module P implies

the desired factorization. Now it is straightforward to check that the natural functors $\mathbb{M}_{B_0}-\mathrm{mod} \longrightarrow \mathbb{M}_K-\mathrm{mod} \longrightarrow \mathbb{M}_\mathbb{Z}-\mathrm{mod}$ are category equivalences.   □

Explicitly, the $\mathbb{M}_K$-module $\mathbb{M}_K(K^{(X)})$ mentioned in the above proof has the form

$$\mathbb{M}_K(K^{(X)}) = \prod_{n=0}^{\infty} B_n \otimes_K K^{(X)} = \prod_{n=0}^{\infty} B_n^{(X)} = \Pi(B^{(X)}),$$

where the notation $A^{(X)}$ stands for the direct sum of X copies of a (graded) abelian group A. The left B-contramodule $\mathbb{M}_K(K^{(X)})$ does not depend on the choice of a ring K, but only on the nonnegatively graded ring B and a set X. The $\mathbb{M}_K$-modules $\mathbb{M}_K(K^{(X)})$ are called the *free ungraded left B-contramodules*. Since there are enough free ungraded left B-contramodules, it follows that every projective ungraded left B-contramodule is a direct summand of a free one.

The forgetful functor lifting assertion in Proposition 5.12 can be strengthened as follows. In fact, the underlying left B-module structure of any ungraded left B-contramodule can be extended naturally to a structure of left module over the ring $\Pi B = \prod_{n=0}^{\infty} B_n$. To define the underlying left ΠB-module structure on a left B-contramodule $P \in \mathbb{M}_K-\mathrm{mod}$, it suffices to compose the contraaction map π_P with the natural map

$$\Pi B \otimes_K P = \left(\prod_{n=0}^{\infty} B_n\right) \otimes_K P \longrightarrow \prod_{n=0}^{\infty} (B_n \otimes_K P) = \mathbb{M}_K(P)$$

and use Example 5.10(1).

Examples 5.13

(1) Let B be a nonnegatively graded algebra over a field $k = B_0$ with finite-dimensional components B_n, as in Example 5.2(1), and let $C = \bigoplus_{n=0}^{\infty} B_n^*$ be the graded dual coalgebra over k. Then an ungraded left B-contramodule in the sense of the above definition is the same thing as an ungraded left C-contramodule in the sense of [61, Section 1.1] and [58, Appendix A].

(2) More generally, let B be a nonnegatively graded ring with the degree-zero component $R = B_0$ such that B_n is a finitely generated projective right R-module for every n, as in Example 5.2(2). Let $C = \bigoplus_{n=0}^{\infty} C_n$, where $C_n = \mathrm{Hom}_{R^{op}}(B_n, R)$, be the graded (right) dual coring. Then an ungraded left B-contramodule in the sense of the above definition is the same thing as an ungraded left C-contramodule in the sense of [58, Section 3.1.1] and [61, Section 2.5] (see Example 5.10(4)).

Example 5.14 Let E be an associative ring, N be an E-B-bimodule, and U be a left E-module. Assume that the right B-module N is a right B-comodule. Then the left B-module structure on the Hom group $\mathrm{Hom}_E(N, U)$ can be extended to a left B-contramodule structure in a natural way.

To construct the contraaction map

$$\pi : \prod_{n=0}^{\infty} B_n \otimes_K \mathrm{Hom}_E(N, U) \longrightarrow \mathrm{Hom}_E(N, U),$$

consider an element $y \in N$, and choose an integer $m \geq 0$ such that $y B_{\geq m+1} = 0$. Let $w \in \prod_{n=0}^{\infty} B_n \otimes_K \mathrm{Hom}_E(N, U)$ be an arbitrary element. In order to evaluate the element $\pi(w) \in \mathrm{Hom}_E(N, U)$ on the element $y \in N$, consider the composition

$$\prod_{n=0}^{\infty} B_n \otimes_K \mathrm{Hom}_E(N, U) \longrightarrow \bigoplus_{n=0}^{m} B_n \otimes_K \mathrm{Hom}_E(N, U) \longrightarrow \mathrm{Hom}_E(N, U)$$

of the direct summand projection $\prod_{n=0}^{\infty} B_n \otimes_K \mathrm{Hom}_E(N, U) \longrightarrow \bigoplus_{n=0}^{m} B_n \otimes_K \mathrm{Hom}_E(N, U)$ and the map $\bigoplus_{n=0}^{m} B_n \otimes_K \mathrm{Hom}_E(N, U) \longrightarrow \mathrm{Hom}_E(N, U)$ provided by the left action of B in $\mathrm{Hom}_E(N, U)$. This composition of maps needs to be applied to the element w, and the resulting element in $\mathrm{Hom}_E(N, U)$ needs to be evaluated on the element $y \in N$, producing the desired element $\pi(w)(y) \in U$.

Remark 5.15 Following the discussion in Sect. 5.1, the hereditary pretorsion class comod$-B \subset$ mod$-B$ corresponds to a right linear (in fact, two-sided linear) topology on the ring $B = \bigoplus_{n=0}^{\infty} B_n$ with a base of neighborhoods of zero formed by the ideals $B_{\geq m} = \bigoplus_{n \geq m} B_n$. The completion of B with respect to this topology is the topological ring $\Pi B = \prod_{n=0}^{\infty} B_n$ with the product topology. The abelian category $B-$contra of ungraded left B-contramodules, as defined above, is in fact equivalent to the abelian category of left contramodules over the topological ring ΠB, as defined in [60, Section 1.2], [61, Section 2.1], [69, Sections 1.2 and 5], [67, Example 1.3(2)], [71, Section 6.2], [68, Section 2.7], etc.

Let us explain the connection between the two definitions. The difference is that in the above references, unlike in the present chapter, we were working with *monads on the category of sets*. This allowed for an extra flexibility, in that one could easily define contramodules over topological rings of arithmetic flavor, like the p-adic integers [61, Section 1.4]. In the context of this book, such additional flexibility or extra generality is not needed or relevant, so we are mostly restricting ourselves to the more straightforward definition above.

Still it is interesting and important for our purposes to establish a comparison between the two definitions, i.e., show that they are indeed equivalent for the topological ring ΠB. This is provable along the lines of the argument in [60, Section 1.10] and [61, Section 2.3]. Let us spell out some details.

The category of left contramodules over the topological ring ΠB is, by the definition, the category of modules over the monad $X \longmapsto \Pi B[[X]]$ on the category of sets, assigning to a set X the set of all infinite formal linear combinations of elements of X with the coefficients in ΠB, where the family of coefficients converges to zero in the topology of

ΠB. For any set X, we have a natural isomorphism of left K-modules

$$\Pi B[[X]] \simeq \mathbb{M}_K(K[X]).$$

For any left K-module A, there is an obvious surjective map of sets (or left K-modules)

$$\gamma_A \colon \Pi B[[A]] \longrightarrow \mathbb{M}_K(A).$$

Composing the map γ_P with the contraaction map $\pi_P \colon \mathbb{M}_K(P) \longrightarrow P$, one defines a left ΠB-contramodule structure on every ungraded left B-contramodule P. This construction defines a functor $B-\mathsf{contra} \longrightarrow \Pi B-\mathsf{contra}$, where $\Pi B-\mathsf{contra}$ denotes the category of left contramodules over the topological ring ΠB. This functor is fully faithful, because the map γ_P is surjective. In order to show that this is an equivalence of categories, it remains to check that, for any left ΠB-contramodule C, the contraaction map $\Pi B[[C]] \longrightarrow C$ factorizes through the surjection γ_C.

First of all, we recall that the underlying set of any contramodule over the topological ring ΠB has a natural left ΠB-module structure, hence also a natural left K-module structure. The two forgetful functors $B-\mathsf{contra} \longrightarrow \Pi B-\mathsf{mod}$ and $\Pi B-\mathsf{contra} \longrightarrow \Pi B-\mathsf{mod}$ form a commutative triangle diagram with the above comparison functor $B-\mathsf{contra} \longrightarrow \Pi B-\mathsf{contra}$.

For any set X, we denote by $K[X] = K^{(X)}$ the free left K-module with X generators. Then, for any left K-module A, one can construct the monadic bar resolution related to the forgetful functor from $K-\mathsf{mod}$ to the category of sets. We are interested in the fragment

$$K[K[A]] \rightrightarrows K[A] \longrightarrow A. \tag{5.1}$$

Here, for any left K-module A, there is a natural surjective map $p_A \colon K[A] \longrightarrow A$; this is the rightmost map in (5.1). The leftmost pair of maps is formed by the maps $p_{K[A]}$ and $K[p_A]$. Passing to the difference of the leftmost pair of maps, one obtains a right exact sequence of left K-modules

$$K[K[A]] \longrightarrow K[A] \longrightarrow A \longrightarrow 0. \tag{5.2}$$

Applying the right exact functor $\mathbb{M}_K$ to (5.1) and (5.2), we obtain a right exact sequence of left K-modules

$$\Pi B[[K[A]]] \longrightarrow \Pi B[[A]] \longrightarrow \mathbb{M}_K(A) \longrightarrow 0, \tag{5.3}$$

where the rightmost map in (5.3) is γ_A. We have obtained a description of the kernel of γ_A as the image of the leftmost map in (5.3), which is the difference of two maps $\mathbb{M}_K(p_{K[A]})$ and $\mathbb{M}_K(K[p_A]) = \Pi B[[p_A]]$.

Returning to a left ΠB-contramodule C, we recall that, by the definition, its contraaction map $\Pi B[[C]] \longrightarrow C$ has to satisfy the (contra)associativity equation of a module over the monad $X \longmapsto \Pi B[[X]]$ on the category of sets. This equation, claiming that two compositions of maps of sets

$$\Pi B[[\Pi B[[C]]]] \rightrightarrows \Pi B[[C]] \longrightarrow C$$

are equal to each other, can also be expressed by the vanishing of the composition of two maps in the sequence of left K-module morphisms

$$\Pi B[[\Pi B[[C]]]] \longrightarrow \Pi B[[C]] \longrightarrow C \longrightarrow 0. \tag{5.4}$$

It remains to observe that there is a natural commutative square of a morphism from the leftmost map in (5.3) for $A = C$ to the leftmost map in (5.4). Hence we get the induced map on the rightmost terms $\mathbb{M}_K(C) \longrightarrow C$, as desired.

Let P be an ungraded left B-contramodule. One can define a natural decreasing filtration $P = G^0 P \supset G^1 P \supset G^2 P \supset \cdots$ on P by the rule

$$G^m P = \mathrm{im}\left(\left(\prod\nolimits_{n \geq m} B_n \otimes_K P \right) \xrightarrow{\;\pi_P\;} P \right).$$

Here the infinite product $\prod_{n \geq m} B_n \otimes_K P$ is viewed as a K-submodule in $\mathbb{M}_K(P) = \prod_{n \geq 0} B_n \otimes_K P$, the contraaction map $\pi_P \colon \mathbb{M}_K(P) \longrightarrow P$ is restricted onto this submodule, and the image of the restriction is taken.

It follows from the contraassociativity axiom for π_P that $G^m P$ is a B-subcontramodule in P for every $m \geq 0$ and the successive quotient contramodules $G^m P / G^{m+1} P$ have *trivial* ungraded left B-contramodule structures. Here an ungraded left B-contramodule Q is said to be trivial if the contraaction map π_Q vanishes on the submodule $\prod_{n=1}^{\infty} B_n \otimes_K Q \subset \prod_{n=0}^{\infty} B_n \otimes_K Q$ (in other words, this means that $G^1 Q = 0$). The full subcategory of trivial ungraded left B-contramodules in B–contra is equivalent to the category of left B_0-modules.

An ungraded left B-contramodule P is said to be *separated* if $\bigcap_{m \geq 0} G^m P = 0$, that is, in other words, the natural B-contramodule map $\lambda_{B,P} \colon P \longrightarrow \varprojlim_{m \geq 0} P / G^m P$ is injective. A B-contramodule P is said to be *complete* if the map $\lambda_{B,P}$ is surjective.

The following result is essentially well-known for contramodules over a topological ring with a countable base of neighborhoods of zero.

Proposition 5.16 *All ungraded left B-contramodules are complete (but not necessarily separated). All projective ungraded left B-contramodules are separated. Every ungraded left B-contramodule is the cokernel of an injective morphism of separated ungraded left B-contramodules.*

Proof By Remark 5.15, we can consider left contramodules over the topological ring ΠB in lieu of the ungraded left B-contramodules. Then the first assertion is [69, Lemma 6.3(b)] (see also [68, Theorem 5.3] or [58, Lemma A.2.3 and Remark A.3]). The second assertion is clear from the explicit description of projective ungraded left B-contramodules above (cf. [69, Lemma 6.9]). The third assertion follows from the second one together with the existence of sufficiently many projective contramodules, as all subcontramodules of separated contramodules are separated (in particular, any subcontramodule of a projective ungraded left B-contramodule is separated). □

Theorem 5.17 *Assume that the augmentation ideal $B_{\geq 1} = \bigoplus_{n=1}^{\infty} B_n$ of a nonnegatively graded ring B is finitely generated as a right ideal in B. Then the forgetful functor $B-\mathsf{contra} \longrightarrow B-\mathsf{mod}$ from the category of ungraded left B-contramodules to the category of ungraded left B-modules is fully faithful.*

Proof By Proposition 5.7, the full subcategory $\mathsf{comod}-B \subset \mathsf{mod}-B$ is closed under extensions under our assumptions. In other words, $\mathsf{comod}-B$ is a hereditary torsion class in $\mathsf{mod}-B$. In the terminology of [78, Section VI.5] and [68, Section 2.4], this means that the topology on B with a base formed by the ideals $B_{\geq m}$, $m \geq 1$, is a *right Gabriel topology*. By Lemma 5.6, the right ideals $B_{\geq m} \subset B$ are finitely generated.

Now we can apply [68, Corollary 6.7] in order to conclude that the forgetful functor $\Pi B-\mathsf{contra} \longrightarrow B-\mathsf{mod}$ is fully faithful. Alternatively, it is straightforward to see that the right ideal $B_{\geq m} \subset B$ is strongly (finitely) generated in the sense of [68, Section 6], so [68, Theorem 6.2 (ii) $\Rightarrow$ (iii)] (or even [67, Theorem 3.1]) is directly applicable. Finally, Remark 5.15 identifies the category $\Pi B-\mathsf{contra}$ of left contramodules over the topological ring ΠB with the category $B-\mathsf{contra}$ which we are interested in. □

Example 5.18 Let $B = k[x]$ be the graded ring of polynomials in one variable x over a field k, with $\deg x = 1$, as in Example 5.3. By Theorem 5.17, the forgetful functor $B-\mathsf{contra} \longrightarrow B-\mathsf{mod}$ is fully faithful, so a B-contramodule structure on a given ungraded B-module is unique if it exists.

For example, the ungraded B-modules $k[x]$, $k[x, x^{-1}]$, and $k[x, x^{-1}]/k[x]$ are *not* B-contramodules. The B-modules $k[x]/(x - a)^n$, where $a \in k \setminus \{0\}$ and $n \geq 1$, are not B-contramodules, either. On the other hand, the B-module $k[x]/(x^n)$ is a B-contramodule for every $n \geq 1$. The ungraded B-module of formal Taylor power series $k[[x]]$ in the variable x over k is a B-contramodule; this is the free ungraded B-contramodule with one generator.

5.4   Weak Koszulity and the Ext Comparison

Let $B = \bigoplus_{n=0}^{\infty} B_n$ be a nonnegatively graded ring with the degree-zero component $R = B_0$. We will say that B is *weakly right finitely projective Koszul* if the graded right B-module R has a graded B-module resolution of the form

$$\cdots \longrightarrow V_2 \otimes_R B \longrightarrow V_1 \otimes_R B \longrightarrow B \longrightarrow R \longrightarrow 0, \qquad (5.5)$$

where V_i, $i \geq 1$, are finitely generated projective graded right R-modules. This means that $V_i = \bigoplus_{j \in \mathbb{Z}} V_{i,j}$ is a graded right R-module concentrated in a finite set of internal degrees j, and for every $j \in \mathbb{Z}$, the right R-module $V_{i,j}$ is finitely generated and projective.

By the opposite version of Theorem 2.35(d) (with the roles of the rings A and B switched), any right finitely projective Koszul graded ring is weakly right finitely projective Koszul. For any weakly right finitely projective Koszul graded ring B, the augmentation ideal $B_{\geq 1} = \bigoplus_{n=1}^{\infty} B_n$ is finitely generated as a right ideal in B. In fact, any finite set of homogeneous generators of the graded right R-module V_1 gives rise to a finite set of homogeneous generators of the right ideal $B_{\geq 1} \subset B$.

Lemma 5.19

(a) *For any graded right R-module V and graded left R-module H, there is a natural map of ungraded abelian groups*

$$\Sigma V \otimes_R \Pi H \longrightarrow \Pi(V \otimes_R H).$$

This map is an isomorphism whenever V is a finitely generated projective graded right R-module.

(b) *For any graded right R-modules V and J, there is a natural map of ungraded abelian groups*

$$\Sigma \operatorname{Hom}_{R^{\mathrm{op}}}(V, J) \longrightarrow \operatorname{Hom}_{R^{\mathrm{op}}}(\Sigma V, \Sigma J).$$

This map is an isomorphism whenever V is a finitely generated projective graded right R-module. □

Theorem 5.20 *For any weakly right finitely projective Koszul graded ring $B = \bigoplus_{n=0}^{\infty} B_n$, the exact, fully faithful forgetful functor of abelian categories B–contra $\longrightarrow$ B–mod induces isomorphisms on all the Ext groups.*

Proof According to the discussion above, the augmentation ideal $B_{\geq 1} \subset B$ is finitely generated as a right ideal in B. By Theorem 5.17, it follows that the forgetful functor $B-\mathsf{contra} \longrightarrow B-\mathsf{mod}$ is indeed fully faithful. We have to show that it also induces isomorphisms of the groups Ext^i for $i > 0$.

There are enough projective objects in the abelian category $B-\mathsf{contra}$. Hence it suffices to prove that $\mathrm{Ext}^i_B(P, Q) = 0$ for all projective objects $P \in B-\mathsf{contra}$, all objects $Q \in B-\mathsf{contra}$, and all $i > 0$ (where, as usually, Ext^*_B denotes the Ext functor in the category $B-\mathsf{mod}$). By Proposition 5.16, any object of $B-\mathsf{contra}$ is the cokernel of an injective morphism of separated contramodules. Hence we can assume that Q is a separated ungraded left B-contramodule.

Any separated ungraded left B-contramodule Q is an infinitely iterated extension, in the sense of projective limit, of trivial ungraded left B-contramodules $G^m Q / G^{m+1} Q$ (because, by the same proposition, all ungraded left B-contramodules are complete). In view of the dual Eklof lemma [24, Proposition 18], we can assume that Q is a trivial ungraded left B-contramodule (that is, just a left R-module in which the augmentation ideal $B_{\geq 1} \subset B$ acts by zero).

Now the left R-module Q has a resolution by injective left R-modules of the form $F^+ = \mathrm{Hom}_{\mathbb{Z}}(F, \mathbb{Q}/\mathbb{Z})$, where F ranges over free right R-modules. This reduces the question to the case of a left R-module of the form $Q = F^+$. Furthermore, we have $\mathrm{Ext}^i_B(P, F^+) \simeq \mathrm{Tor}^B_i(F, P)^+$ (where $B_{\geq 1}$ acts by zero in F). Thus it suffices to show that $\mathrm{Tor}^B_i(R, P) = 0$ for all projective objects $P \in B-\mathsf{contra}$ and all integers $i > 0$ (where R is viewed as a right B-module with $B_{\geq 1}$ acting by zero).

We have done a chain of reductions for an object $Q \in B-\mathsf{contra}$; now we need to deal with a projective object $P \in B-\mathsf{contra}$. According to the discussion after Proposition 5.12 in Sect. 5.3, the underlying left B-module of P is a direct summand of an ungraded left B-module ΠH for some free graded left B-module $H = B^{(X)}$ with generators in degree zero.

Finally, we compute the groups $\mathrm{Tor}^B_i(R, \Pi H)$ using the projective resolution (5.5) of the right B-module R. More precisely, we use the ungraded projective resolution obtained by applying Σ to (5.5). We obtain the complex of ungraded abelian groups

$$\cdots \longrightarrow \Sigma V_2 \otimes_R \Pi H \longrightarrow \Sigma V_1 \otimes_R \Pi H \longrightarrow \Pi H \longrightarrow 0 \qquad (5.6)$$

computing $\mathrm{Tor}^B_*(R, \Pi H)$. According to Lemma 5.19(a), the complex (5.6) is isomorphic to

$$\cdots \longrightarrow \Pi(V_2 \otimes_R H) \longrightarrow \Pi(V_1 \otimes_R H) \longrightarrow \Pi(H) \longrightarrow 0. \qquad (5.7)$$

The complex (5.7) can be obtained by applying Π to the complex

$$\cdots \longrightarrow V_2 \otimes_R H \longrightarrow V_1 \otimes_R H \longrightarrow H \longrightarrow 0 \qquad (5.8)$$

computing the graded abelian groups $\text{Tor}^B_*(R, H)$. It remains to observe that $\text{Tor}^B_i(R, H) = 0$ for $i > 0$ and the functor Π is exact. Thus $\text{Tor}^B_i(R, \Pi H) \simeq \Pi \text{Tor}^B_i(R, H) = 0$ for $i > 0$ under our assumptions. $\qquad\square$

The assertion of the next theorem can be equivalently restated by saying that, under its assumptions, the full subcategory of right B-comodules $\mathsf{comod}{-}B$ is a *weakly stable torsion class* in $\mathsf{mod}{-}B$ in the sense of the paper [80].

Theorem 5.21 *For any weakly right finitely projective Koszul graded ring $B = \bigoplus_{n=0}^{\infty} B_n$, the exact, fully faithful inclusion functor of abelian categories $\mathsf{comod}{-}B \longrightarrow \mathsf{mod}{-}B$ induces isomorphisms on all the Ext groups.*

Proof There are enough injective objects in a Grothendieck abelian category of ungraded right B-comodules $\mathsf{comod}{-}B$. Hence it suffices to prove that $\text{Ext}^i_{B^{op}}(M, L) = 0$ for all objects $M \in \mathsf{comod}{-}B$, all injective objects $L \in \mathsf{comod}{-}B$, and all $i > 0$ (where $\text{Ext}^*_{B^{op}}$ denotes the Ext functor in the category $\mathsf{mod}{-}B$).

Any object $M \in \mathsf{comod}{-}B$ has a natural increasing filtration $0 = G_{-1}M \subset G_0 M \subset G_1 M \subset G_2 M \subset \cdots$ by right B-submodules $G_m M \subset M$, where $G_m M$ is the subset of all elements $x \in M$ such that $x B_{\geq m+1} = 0$. By the definition, we have $M = \bigcup_{m \geq 0} G_m M$. The successive quotient modules $G_m M / G_{m-1} M$ are *trivial* ungraded right B-comodules, i.e., right B-modules in which the augmentation ideal $B_{\geq 1}$ acts by zero. The category of trivial ungraded right B-comodules is equivalent to the category of right R-modules $\mathsf{mod}{-}R$.

In view of the Eklof lemma [24, Lemma 1], we can assume that M is a trivial ungraded right B-comodule. Now the right R-module M has a resolution by free right R-modules, which reduces the question to the case of a free right R-module $M = R^{(X)}$ (with $B_{\geq 1}$ acting by zero in M). Thus it suffices to show that $\text{Ext}^i_{B^{op}}(R, L) = 0$ for all injective objects $L \in \mathsf{comod}{-}B$ and all integers $i > 0$ (where R is viewed as a right B-module with $B_{\geq 1}$ acting by zero).

Injective objects of the category $\mathsf{comod}{-}B$ can be described as follows. The inclusion functor $\mathsf{comod}{-}B \longrightarrow \mathsf{mod}{-}B$ has a right adjoint functor $\Gamma_B : \mathsf{mod}{-}B \longrightarrow \mathsf{comod}{-}B$ assigning to an ungraded right B-module N its submodule $\Gamma_B(N) \subset N$ consisting of all the elements $x \in N$ for which there exists $m \geq 1$ such that $x B_{\geq m} = 0$. The injective objects of $\mathsf{comod}{-}B$ are precisely the direct summands of the objects $\Gamma_B(N)$, where N ranges over the injective objects of $\mathsf{mod}{-}B$.

Injective ungraded right B-modules can be further described as follows. Let $H = B^{(X)}$ be a free graded left B-module with generators in degree zero, and let $H^+ = \text{Hom}_{\mathbb{Z}}(B^{(X)}, \mathbb{Q}/\mathbb{Z})$ be the graded character module of H. So H^+ is a graded right B-module concentrated in the nonpositive degrees. Then $\Pi(H^+)$ is an injective ungraded right B-module, and all injective right B-modules are direct summands of B-modules of this form. One can easily see that $\Gamma_B(\Pi(H^+)) = \Sigma(H^+)$.

Finally, we compute the groups $\text{Ext}^i_{B^{op}}(R, \Sigma(H^+))$ using the projective resolution (5.5) of the right B-module R. More precisely, we use the projective resolution of ungraded right B-module R obtained by applying Σ to (5.5). We obtain a complex of ungraded abelian groups

$$0 \longrightarrow \Sigma(H^+) \longrightarrow \text{Hom}_{R^{op}}(\Sigma V_1, \Sigma(H^+)) \longrightarrow \text{Hom}_{R^{op}}(\Sigma V_2, \Sigma(H^+)) \longrightarrow \cdots$$
$$(5.9)$$

computing $\text{Ext}^*_{B^{op}}(R, \Sigma(H^+))$. According to Lemma 5.19(b), the complex (5.9) is isomorphic to

$$0 \longrightarrow \Sigma(H^+) \longrightarrow \Sigma \text{Hom}_{R^{op}}(V_1, H^+) \longrightarrow \Sigma \text{Hom}_{R^{op}}(V_2, H^+) \longrightarrow \cdots \quad (5.10)$$

The complex (5.10) can be obtained by applying Σ to the complex of graded abelian groups

$$0 \longrightarrow H^+ \longrightarrow \text{Hom}_{R^{op}}(V_1, H^+) \longrightarrow \text{Hom}_{R^{op}}(V_2, H^+) \longrightarrow \cdots \quad (5.11)$$

computing the graded $\text{Ext}^*_{B^{op}}(R, H^+)$ (that is, the derived functor of graded Hom of graded right B-modules). It remains to observe that H^+ is an injective object of the category of graded right B-modules and the functor Σ is exact. Thus $\text{Ext}^i_{B^{op}}(R, \Sigma(H^+)) \simeq \Sigma \text{Ext}^i_{B^{op}}(R, H^+) = 0$ for $i > 0$. $\square$

5.5   Graded Comodules Over Nonnegatively Graded Rings

Let $B = \bigoplus_{n=0}^{\infty} B_n$ be a nonnegatively graded ring. We consider graded right B-modules $M = \bigoplus_{n \in \mathbb{Z}} M_n$.

Definition 5.22 A graded right B-module M is said to be a B-*comodule* (or a *graded right B-comodule*) if its underlying ungraded right B-module ΣM is an ungraded right B-comodule. Equivalently, this means that for every (homogeneous) element $x \in M$, there exists an integer $m \geq 0$ such that $xb = 0$ in M for all $b \in B_n$, $n > m$.

Clearly, any *nonpositively* graded right B-module $M = \bigoplus_{n=-\infty}^{0} M_n$ is a B-comodule. But the free graded right B-module B is usually *not* a B-comodule (in fact, B is a B-comodule if and only if $B_n = 0$ for $n \gg 0$).

We denote the category of graded right B-modules by $\text{mod}_{gr}-B$ and the full subcategory of graded right B-comodules by $\text{comod}_{gr}-B \subset \text{mod}_{gr}-B$. Obviously, the full subcategory $\text{comod}_{gr}-B \subset \text{mod}_{gr}-B$ is closed under subobjects, quotients, and infinite direct sums (in other words, $\text{comod}_{gr}-B$ is a hereditary pretorsion class in $\mathsf{mod}_{\mathrm{gr}}-B$). Hence $\mathsf{comod}_{\mathrm{gr}}-B$ is a Grothendieck abelian category and its inclusion $\mathsf{comod}_{\mathrm{gr}}-B \longrightarrow \mathsf{mod}_{\mathrm{gr}}-B$ is an exact functor preserving infinite direct sums.

Example 5.23 Let B be a nonnegatively graded ring with the degree-zero grading component $R = B_0$. Assume that B_n is a finitely generated projective right R-module for every $n \geq 0$, and consider the graded right dual coring C as in Example 5.2(2). Then a graded right B-comodule in the sense of the above definition is the same thing as a graded right C-comodule.

Proposition 5.24 *Assume that the augmentation ideal $B_{\geq 1} = \bigoplus_{n=1}^{\infty} B_n$ of a nonnegatively graded ring B is finitely generated as a right ideal in B. Then the full subcategory of graded right B-comodules $\mathsf{comod}_{\mathrm{gr}}-B$ is closed under extensions in the category of graded right B-modules $\mathsf{mod}_{\mathrm{gr}}-B$.*

Proof This is a graded version of Proposition 5.7. It can be either proved by the same computation or deduced formally from the ungraded version.   □

Example 5.25 Consider the graded ring $B = k[x]$, as in Example 5.3. Then the graded B-modules $k[x, x^{-1}]/k[x]$ and $k[x]/(x^n)$, $n \geq 1$, are graded B-comodules. The graded B-modules $k[x]$ and $k[x, x^{-1}]$ are not B-comodules.

5.6   Graded Contramodules Over Nonnegatively Graded Rings

Let $B = \bigoplus_{n=0}^{\infty} B_n$ be a nonnegatively graded ring, and let K be an associative ring endowed with a ring homomorphism $K \longrightarrow B_0$. Consider the following monad $\mathbb{M} = \mathbb{M}_K^{\mathrm{gr}} \colon K-\mathsf{mod}_{\mathrm{gr}} \longrightarrow K-\mathsf{mod}_{\mathrm{gr}}$ on the category of graded left K-modules $K-\mathsf{mod}_{\mathrm{gr}}$ (where K is viewed as a graded ring concentrated in the grading 0).

To any graded left K-module $L = (L_j)_{j \in \mathbb{Z}}$, the monad $\mathbb{M}$ assigns the graded left K-module with the components

$$\mathbb{M}_K^{\mathrm{gr}}(L)_j = \prod_{n=0}^{\infty} \left(B_n \otimes_K L_{j-n} \right).$$

The monad unit map $e_L \colon L \longrightarrow \mathbb{M}_K^{\mathrm{gr}}(L)$ is the graded K-module morphism whose degree j component is the composition

$$L_j \longrightarrow B_0 \otimes_K L_j \longrightarrow \prod_{n=0}^{\infty} B_n \otimes_K L_{j-n}$$

of the map $L_j \longrightarrow B_0 \otimes_K L_j$ induced by the ring homomorphism $K \longrightarrow B_0$ with the inclusion of the $(n = 0)$-indexed summand $B_0 \otimes_K L_j \longrightarrow \prod_{n=0}^{\infty} B_n \otimes_K L_{j-n}$. The monad multiplication map $m_L \colon \mathbb{M}_K^{\mathrm{gr}}(\mathbb{M}_K^{\mathrm{gr}}(L)) \longrightarrow \mathbb{M}_K^{\mathrm{gr}}(L)$ is the graded K-module morphism whose degree j component is the composition

$$\prod_{p=0}^{\infty} B_p \otimes_K \left(\prod_{q=0}^{\infty} B_q \otimes_K L_{j-p-q} \right) \longrightarrow \prod_{p,q=0}^{\infty} B_p \otimes_K B_q \otimes_K L_{j-p-q} \longrightarrow \prod_{n=0}^{\infty} B_n \otimes_K L_{j-n}.$$

Here the leftmost arrow is the product over $p \geq 0$ of the maps

$$B_p \otimes_K \left(\prod_{q=0}^{\infty} B_q \otimes_K L_{j-p-q} \right) \longrightarrow \prod_{q=0}^{\infty} B_p \otimes_K B_q \otimes_K L_{j-p-q}$$

whose q-components are the direct summand projections $B_p \otimes_K \left(\prod_{k=0}^{\infty} B_k \otimes_K L_{j-p-k} \right)$ $\longrightarrow B_p \otimes_K B_q \otimes_K L_{j-p-q}$. The rightmost arrow is induced by the multiplication maps $\prod_{p+q=n}^{p,q \geq 0} B_p \otimes_K B_q \longrightarrow B_n$.

Definition 5.26 A *graded left B-contramodule* is a module over the monad $\mathbb{M}_K^{\mathrm{gr}}$: $K-\mathrm{mod}_{\mathrm{gr}} \longrightarrow K-\mathrm{mod}_{\mathrm{gr}}$. In other words, a graded left B-contramodule P is a graded left K-module endowed with a *left B-contraaction map*

$$\pi_P : \mathbb{M}_K^{\mathrm{gr}}(P) \longrightarrow P,$$

which must be a morphism of graded left K-modules satisfying the (contra)associativity and (contra)unitality equations involving the multiplication and unit maps of the monad $\mathbb{M}_K^{\mathrm{gr}}$. This means that, in every degree $j \in \mathbb{Z}$, the two compositions

$$\prod_{p=0}^{\infty} B_p \otimes_K \left(\prod_{q=0}^{\infty} B_q \otimes_K P_{j-p-q} \right) \rightrightarrows \prod_{n=0}^{\infty} B_n \otimes_K P_{j-n} \longrightarrow P_j$$

of the degree j components of the monad multiplication map m_P and the map $\mathbb{M}_K^{\mathrm{gr}}(\pi_P)$ with the degree j component of the contraaction map π_P are equal to each other, while the composition

$$P_j \longrightarrow \prod_{n=0}^{\infty} B_n \otimes_K P_{j-n} \longrightarrow P_j$$

of the degree j components of the monad unit map e_P and the contraaction map π_P is equal to the identity map id_{P_j}.

We denote the category of graded left B-contramodules by $B-\mathrm{contra}_{\mathrm{gr}} = \mathbb{M}_K^{\mathrm{gr}}-\mathrm{mod}$. The category of graded left B-modules $M = \bigoplus_{j \in \mathbb{Z}} M_j$ is denoted by $B-\mathrm{mod}_{\mathrm{gr}}$.

For any graded left B-contramodule P, the ungraded left K-module $\Pi P = \prod_{j \in \mathbb{Z}} P_j$ has a natural structure of ungraded left B-contramodule. Indeed, the functors $\mathbb{M}_K^{\mathrm{gr}} : K-\mathrm{mod}_{\mathrm{gr}} \longrightarrow K-\mathrm{mod}_{\mathrm{gr}}$ and $\mathbb{M}_K : K-\mathrm{mod} \longrightarrow K-\mathrm{mod}$ form a commutative square diagram with the functor $\Pi : K-\mathrm{mod}_{\mathrm{gr}} \longrightarrow K-\mathrm{mod}$, and the monad multiplication and unit maps of the two monads agree. So we obtain a faithful functor of forgetting the grading

$$\Pi: B-\mathsf{contra}_{\mathsf{gr}} \longrightarrow B-\mathsf{contra}.$$

It is clear from the next proposition that the functor $\Pi: B-\mathsf{contra}_{\mathsf{gr}} \longrightarrow B-\mathsf{contra}$ is exact and preserves infinite products (because the functor $\Pi: K-\mathsf{mod}_{\mathsf{gr}} \longrightarrow K-\mathsf{mod}$ has such properties).

Proposition 5.27 *There are natural equivalences of categories*

$$\mathbb{M}^{\mathsf{gr}}_{\mathbb{Z}}-\mathsf{mod} \simeq \mathbb{M}^{\mathsf{gr}}_{K}-\mathsf{mod} \simeq \mathbb{M}^{\mathsf{gr}}_{B_0}-\mathsf{mod},$$

making the notation $B-\mathsf{contra}_{\mathsf{gr}}$ unambiguous. The category $B-\mathsf{contra}_{\mathsf{gr}}$ of graded left B-contramodules is abelian with enough projective objects. The forgetful functor $B-\mathsf{contra}_{\mathsf{gr}} \longrightarrow K-\mathsf{mod}_{\mathsf{gr}}$ can be naturally lifted to a forgetful functor $B-\mathsf{contra}_{\mathsf{gr}} \longrightarrow B-\mathsf{mod}_{\mathsf{gr}}$ taking values in the category of graded left B-modules. The forgetful functor $B-\mathsf{contra}_{\mathsf{gr}} \longrightarrow B-\mathsf{mod}_{\mathsf{gr}}$ is exact and preserves infinite products.

Proof This is a graded version of Proposition 5.12. To prove the first assertion, notice that for every graded left B_0-module L there is a natural surjective map of graded left K-modules $\mathbb{M}^{\mathsf{gr}}_{K}(L) \longrightarrow \mathbb{M}^{\mathsf{gr}}_{B_0}(L)$, and for every graded left K-module L there is a natural surjective map of graded abelian groups $\mathbb{M}^{\mathsf{gr}}_{\mathbb{Z}}(L) \longrightarrow \mathbb{M}^{\mathsf{gr}}_{K}(L)$. Using these maps, one constructs a natural $\mathbb{M}^{\mathsf{gr}}_{K}$-module structure on the underlying graded left K-module of every $\mathbb{M}^{\mathsf{gr}}_{B_0}$-module and a natural $\mathbb{M}^{\mathsf{gr}}_{\mathbb{Z}}$-module structure on the underlying graded abelian group of every $\mathbb{M}^{\mathsf{gr}}_{K}$-module. The resulting functors $\mathbb{M}^{\mathsf{gr}}_{B_0}-\mathsf{mod} \longrightarrow \mathbb{M}^{\mathsf{gr}}_{K}-\mathsf{mod} \longrightarrow \mathbb{M}^{\mathsf{gr}}_{\mathbb{Z}}-\mathsf{mod}$ are fully faithful, since the above maps of graded modules are surjective.

To prove these functors are essentially surjective, it remains to show that, for every $\mathbb{M}^{\mathsf{gr}}_{K}$-module P, the contraaction map $\pi_P: \mathbb{M}^{\mathsf{gr}}_{K}(P) \longrightarrow P$ factorizes through the natural surjection $\mathbb{M}^{\mathsf{gr}}_{K}(P) \longrightarrow \mathbb{M}^{\mathsf{gr}}_{B_0}(P)$. This can be either checked directly in the way similar to the proof of the ungraded version in Proposition 5.12 or deduced formally from the ungraded version. The latter approach works as follows: the map $\Pi\pi_P: \Pi\mathbb{M}^{\mathsf{gr}}_{K}(P) \longrightarrow \Pi P$ is naturally identified with the contraaction map $\pi_{\Pi P}: \mathbb{M}_K(\Pi P) \longrightarrow \Pi P$. The latter map factorizes though $\mathbb{M}_{B_0}(\Pi P)$, as explained in the proof of Proposition 5.12; this means that the map $\Pi\pi_P$ factorizes though $\Pi\mathbb{M}^{\mathsf{gr}}_{B_0}(P)$. Hence the map π_P factorizes through $\mathbb{M}^{\mathsf{gr}}_{B_0}(P)$.

The remaining assertions are provable in the way similar to Proposition 5.12. It is important that the functor $\mathbb{M}^{\mathsf{gr}}_{K}: K-\mathsf{mod}_{\mathsf{gr}} \longrightarrow K-\mathsf{mod}_{\mathsf{gr}}$ is right exact (as a combination of tensor products and direct products). We skip the obvious details and will only explain what the projective objects of the category $B-\mathsf{contra}_{\mathsf{gr}}$ are. Enough of these can be obtained by applying the free $\mathbb{M}_K^{\mathrm{gr}}$-module functor $K-\mathsf{mod}_{\mathrm{gr}} \longrightarrow \mathbb{M}_K^{\mathrm{gr}}-\mathsf{mod}$ to free graded left K-modules (with generators possibly sitting in all the degrees).

Explicitly, let X be a *graded set*, which means a sorted set with the sorts indexed by the integers, or equivalently, a set presented as a disjoint union $X = \coprod_{i \in \mathbb{Z}} X_i$. Let $K[X]$ denote the graded left K-module with the components $K[X]_i = K^{(X_i)}$. Then, for every $j \in \mathbb{Z}$, the degree j component of the graded left B-contramodule $\mathbb{M}_K^{\mathrm{gr}}(K[X])$ has the form

$$\mathbb{M}_K^{\mathrm{gr}}(K[X])_j = \prod_{n=0}^{\infty} B_n \otimes_K K[X]_{j-n} = \prod_{n=0}^{\infty} B_n^{(X_{j-n})}.$$

The graded left B-contramodule $\mathbb{M}_K(K[X])$ does not depend on the choice of a ring K, but only on the nonnegatively graded ring B and a graded set X. The $\mathbb{M}_K^{\mathrm{gr}}$-module $\mathbb{M}_K(K[X])$ is called the *free graded left B-contramodule* spanned by a graded set X. Since there are enough free graded left B-contramodules, it follows that every projective graded left B-contramodule is a direct summand of a free one.   □

Example 5.28 Once again, let us take $B = k[x]$, as in Examples 5.18 and 5.25. As we will see below in Theorem 5.34, the forgetful functor $B-\mathsf{contra}_{\mathrm{gr}} \longrightarrow B-\mathsf{mod}_{\mathrm{gr}}$ is fully faithful in this case, so a graded B-contramodule structure on a given graded B-module is unique if it exists.

For example, the graded B-modules $k[x]$ and $k[x, x^{-1}]/k[x]$ are *not* B-contramodules. On the other hand, the graded B-module $k[x]/(x^n)$, $n \geq 1$, is a graded B-contramodule. The graded B-module $k[x]$ is also a graded B-contramodule; it is the free graded B-contramodule with one generator in degree 0. The functor $\Pi\colon B-\mathsf{contra}_{\mathrm{gr}} \longrightarrow B-\mathsf{contra}$ takes the graded B-contramodule $k[x]$ to the ungraded B-contramodule $k[[x]]$ mentioned in Example 5.18.

Examples 5.29

(1) Let B be a nonnegatively graded algebra over a field $k = B_0$ with finite-dimensional components B_n, as in Examples 5.2(1) and 5.13(1), and let $C = \bigoplus_{n=0}^{\infty} B_n^*$ be the graded dual coalgebra over k. Then a graded left B-contramodule in the sense of the above definition is the same thing as a graded left C-contramodule in the sense of [59, Section 2.2]. (We ignore the issue of the sign rule, which only becomes important when one considers differentials on coalgebras, comodules, and contramodules.)

(2) More generally, let B be a nonnegatively graded ring with the degree-zero component $R = B_0$ such that B_n is a finitely generated projective right R-module for every n, and let $C = \bigoplus_{n=0}^{\infty} \mathrm{Hom}_{R^{\mathrm{op}}}(B_n, R)$ be the graded (right) dual coring, as in Examples 5.2(2) and 5.13(2). Then a graded left B-contramodule in the sense of the above definition is the same thing as a graded left C-contramodule in the sense of [58, Section 11.1.1].

Examples 5.30

(1) Any *nonnegatively* graded left B-module $P = \bigoplus_{j=0}^{\infty} P_j$ has a unique graded left B-contramodule structure compatible with its B-module structure. The point is that the relevant direct products are essentially finite in this case and therefore coincide with the similar direct sums,

$$\prod_{n=0}^{\infty} B_n \otimes_K P_{j-n} = \bigoplus_{n=0}^{\infty} B_n \otimes_K P_{j-n} \qquad \text{for all } j \in \mathbb{Z}.$$

(2) For the same reason, there is no difference between graded left B-modules and graded left B-contramodules when $B_n = 0$ for $n \gg 0$. (There is also no difference between ungraded left B-modules and ungraded left B-contramodules in this case and similarly for graded or ungraded right B-comodules.)

(3) However, outside of the degenerate case (2), the cofree graded left B-module $B^+ = \operatorname{Hom}_{\mathbb{Z}}(B, \mathbb{Q}/\mathbb{Z})$ usually does not admit an extension of its graded B-module structure to a graded B-contramodule structure. In order to prove this for a concrete graded ring B, it suffices to show that the cofree ungraded left B-module $\Pi(B^+)$ does not admit an extension of its B-module structure to an ungraded B-contramodule structure. Let us give a specific example.

Choose a field k, and let B be the polynomial algebra $B = k[x]$ in one variable x of the degree $\deg x = 1$. Then the (graded or ungraded) B-contramodules are the same thing as the (respectively, graded or ungraded) contramodules over the coalgebra C over k graded dual to B (as in Examples 5.13(1) and 5.29(1)). Ungraded C-contramodules are described in [58, Remark A.1.1], [61, Sections 1.3 and 1.6], [60, Theorem B.1.1], and [65, Theorem 3.3] as $k[x]$-modules P such that $\operatorname{Hom}_{k[x]}(k[x, x^{-1}], P) = 0 = \operatorname{Ext}^1_{k[x]}(k[x, x^{-1}], P)$. In particular, it follows that P has no nonzero x-divisible $k[x]$-submodules [65, Lemma 3.2]. Therefore, the $k[x]$-module $\Pi(B^+)$ is not a C-contramodule because it is nonzero and x-divisible (i.e., the operator $x \colon \Pi(B^+) \longrightarrow \Pi(B^+)$ is surjective).

Example 5.31 Let $E = \bigoplus_{n \in \mathbb{Z}} E_n$ be a graded associative ring, $N = \bigoplus_{n \in \mathbb{Z}} N_n$ be a graded E-B-bimodule, and $U = \bigoplus_{n \in \mathbb{Z}} U_n$ be a graded left E-module. Assume that the graded right B-module N is a graded right B-comodule. Then the graded left B-module structure on the graded Hom group $\operatorname{Hom}_E(N, U)$ can be extended to a graded left B-contramodule structure in a natural way. The construction of the contraaction is similar to that in Example 5.14.

Specifically, for every $j \in \mathbb{Z}$, we need to construct the degree j component of the contraaction map

$$\pi_j \colon \prod_{n=0}^{\infty} B_n \otimes_K \operatorname{Hom}_E(N, U)_{j-n} \longrightarrow \operatorname{Hom}_E(N, U)_j.$$

Let $w \in \prod_{n=0}^{\infty} B_n \otimes_K \mathrm{Hom}_E(N, U)_{j-n}$ be an arbitrary element. Consider an element $y \in N_i$, $i \in \mathbb{Z}$, and choose an integer $m \geq 0$ such that $y B_{\geq m+1} = 0$. In order to evaluate element $\pi_j(w) \in \mathrm{Hom}_E(N, U)_j$ on the element $y \in N_i$, consider the composition

$$\prod_{n=0}^{\infty} B_n \otimes_K \mathrm{Hom}_E(N, U)_{j-n} \longrightarrow \bigoplus_{n=0}^{m} B_n \otimes_K \mathrm{Hom}_E(N, U)_{j-n} \longrightarrow \mathrm{Hom}_E(N, U)_j$$

of the direct summand projection $\prod_{n=0}^{\infty} B_n \otimes_K \mathrm{Hom}_E(N, U)_{j-n} \longrightarrow \bigoplus_{n=0}^{m} B_n \otimes_K \mathrm{Hom}_E(N, U)_{j-n}$ and the map $\bigoplus_{n=0}^{m} B_n \otimes_K \mathrm{Hom}_E(N, U)_{j-n} \longrightarrow \mathrm{Hom}_E(N, U)_j$ provided by the left action of B in $\mathrm{Hom}_E(N, U)$. This composition of maps has to be applied to the element w, and the resulting element in $\mathrm{Hom}_E(N, U)_j$ has to be evaluated on the element $y \in N_i$, producing the desired element $\pi_j(w)(y) \in U_{i+j}$.

Let P be a graded left B-contramodule. The natural decreasing filtration $P = G^0 P \supset G^1 P \supset G^2 P \supset \cdots$ on P is defined in the following way. For every $m \geq 0$, the graded left K-submodule $G^m P \subset P$ has the grading components

$$G^m P_j = \mathrm{im}\left(\left(\prod_{n \geq m} B_n \otimes_K P_{j-n} \right) \xrightarrow{\pi_P} P_j \right).$$

Here the degree j component of the contraaction map $\pi_{P,j} \colon \mathbb{M}_K^{\mathrm{gr}}(P)_j \longrightarrow P_j$ is restricted onto the K-submodule (in fact, a direct summand) $\prod_{n \geq m} B_n \otimes_K P_{j-n} \subset \prod_{n=0}^{\infty} B_n \otimes_K P_{j-n}$, and the image of the restriction is taken.

The key observation is that $\bigoplus_{n \geq m} B_m$ is a homogeneous two-sided ideal in B. Hence it follows from the contraassociativity axiom for π_P that $G^m P$ is a graded B-subcontramodule in P for every $m \geq 0$ and the successive quotient graded contramodules $G^m P / G^{m+1} P$ have *trivial* graded left B-contramodule structures. Here a graded left B-contramodule Q is said to be trivial if $G^1 Q = 0$. The full subcategory of trivial graded left B-contramodules in $B-\mathsf{contra}$ is equivalent to the category of graded left B_0-modules $B_0-\mathsf{mod}_{\mathrm{gr}}$.

A graded left B-contramodule P is said to be *separated* if $\bigcap_{m \geq 0} G^m P = 0$, that is, in other words, the natural graded B-contramodule map $\lambda_{B,P} \colon P \longrightarrow \varprojlim_{m \geq 0} P / G^m P$ is injective. Here the projective limit is taken in the category of graded left K-modules, which agrees with the projective limit in the category of graded left B-contramodules. A graded left B-contramodule P is said to be *complete* if the map $\lambda_{B,P}$ is surjective.

The functor $\Pi \colon B-\mathsf{contra}_{\mathrm{gr}} \longrightarrow B-\mathsf{contra}$ takes the canonical decreasing filtration G on a graded left B-contramodule P to the canonical decreasing filtration G on the ungraded left B-contramodule ΠP (which was constructed in Sect. 5.3). Consequently, the functor $\Pi \colon B-\mathsf{contra}_{\mathrm{gr}} \longrightarrow B-\mathsf{contra}$ takes the morphism $\lambda_{B,P}$ to the morphism $\lambda_{B,\Pi P}$ constructed in Sect. 5.3. Hence a graded left B-contramodule P is separated (respectively, complete) if and only if the ungraded left B-contramodule ΠP is separated (respectively, complete).

Proposition 5.32 *All graded left B-contramodules are complete (but not necessarily separated). All projective graded left B-contramodules are separated. Every graded left B-contramodule is the cokernel of an injective morphism of separated graded left B-contramodules.*

Proof This is a graded version of Proposition 5.16. The first assertion follows from the ungraded version in view of the discussion above. The second assertion is clear from the explicit description of projective graded contramodules as direct summands of the free ones (similarly to the ungraded version). The third assertion is provable similarly to the ungraded version explained in Proposition 5.16. □

Example 5.33 Here is an example of a nonseparated graded contramodule. It is a modification of a now-classical ungraded example, which appeared in various guises in [76, Example 2.5], [81, Example 3.20], and [58, Section A.1.1] (see [61, Section 2.5] and [65, Example 2.7(1)] for a further discussion).

Let $B = k[x]$ be the polynomial algebra in one variable x of the degree $\deg x = 1$ over a field k. Let E be the free graded B-contramodule with a countable graded set of generators e_1, e_2, e_3, ..., all of them of the degrees $\deg e_i = 0$, and let F be the free graded B-contramodule with a countable set of generators f_1, f_2, f_3, ... of degrees $\deg f_i = -i$. Let $g: E \longrightarrow F$ be the unique graded B-contramodule morphism taking e_i to $x^i f_i$ for every $i \geq 1$. Consider the quotient contramodule $P = F/g(E)$. Then the formal expression $f = \sum_{i=1}^{\infty} x^i f_i$ defines an element of degree 0 in F whose image p in P belongs to $\bigcap_{m \geq 0} G^m P$. Still $f \notin g(E)$, and hence $p \neq 0$ in P.

Theorem 5.34 *Assume that the augmentation ideal $B_{\geq 1} = \bigoplus_{n=1}^{\infty} B_n$ of a nonnegatively graded ring B is finitely generated as a right ideal in B. Then the forgetful functor $B-\mathrm{contra}_{\mathrm{gr}} \longrightarrow B-\mathrm{mod}_{\mathrm{gr}}$ from the category of graded left B-contramodules to the category of graded left B-modules is fully faithful.*

Proof Let P and Q be graded left B-contramodules, and let $f: P \longrightarrow Q$ be a graded left B-module morphism. Consider the ungraded left B-contramodules ΠP and ΠQ and the morphism of ungraded left B-modules $\Pi f: \Pi P \longrightarrow \Pi Q$. By Theorem 5.17, Πf is a morphism of ungraded left B-contramodules. It follows easily that f is a morphism of graded left B-contramodules. □

5.7 The Graded Ext Comparison

Let K be a ring. Given a graded right K-module M and a graded left K-module L, we will denote by $M \otimes_K^{\Pi} L$ the graded abelian group with the grading components

$$(M \otimes_K^{\Pi} L)_n = \prod_{p+q=n} M_p \otimes_K L_q.$$

Similarly, given two graded right K-modules L and M, we denote by $\operatorname{Hom}_{K^{\mathrm{op}}}^{\Sigma}(L, M)$ the graded abelian group with the grading components

$$\operatorname{Hom}_{K^{\mathrm{op}}}^{\Sigma}(L, M)_n = \bigoplus_{p-q=n} \operatorname{Hom}_{K^{\mathrm{op}}}(L_q, M_p).$$

When one of the graded modules involved is concentrated is a finite set of degrees (so there is no difference between the direct sum and product), we will drop the superindices Π and Σ from the above notation. The superindices Π and Σ will be also absent when the usual totalization rule is applied (i.e., the bigraded group of tensor products is totalized using the direct sums, and the bigraded Hom group is totalized using the direct products).

Lemma 5.35

(a) *Let K and R be associative rings, V be a graded right R-module, D be a graded K-R-bimodule, and H be a graded left K-module. Then there is a natural map of graded abelian groups*

$$V \otimes_R (D \otimes_K^{\Pi} H) \longrightarrow (V \otimes_R D) \otimes_K^{\Pi} H.$$

This map is an isomorphism whenever V is a finitely generated projective graded right R-module.

(a) *Let K and R be associative rings, V be a graded right R-module, D be a graded R-K-bimodule, and J be a graded right K-module. Then there is a natural map of graded abelian groups*

$$\operatorname{Hom}_{K^{\mathrm{op}}}^{\Sigma}(V \otimes_R D, \ J) \longrightarrow \operatorname{Hom}_{R^{\mathrm{op}}}(V, \operatorname{Hom}_{K^{\mathrm{op}}}^{\Sigma}(D, J)).$$

This map is an isomorphism whenever V is a finitely generated projective graded right R-module. □

Let $B = \bigoplus_{n=0}^{\infty} B_n$ be a nonnegatively graded ring with the degree-zero component $R = B_0$. In the proofs below, we denote by Tor^B_* and Ext^*_B the (bi)graded Tor and Ext of graded B-modules (see Sect. 2.1 for a discussion). So for any graded left B-module L and graded right B-module M, we have $\operatorname{Tor}_i^B(M, L) = (\operatorname{Tor}_i^B(M, L)_n)_{n \in \mathbb{Z}}$, and similarly, for any graded left B-modules L and M, we have $\operatorname{Ext}_B^i(L, M) = (\operatorname{Ext}_B^i(L, M)_n)_{n \in \mathbb{Z}}$. In particular, the degree-zero components

$$\operatorname{Ext}_B^i(L, M)_0 = \operatorname{Ext}_{B-\mathrm{mod}_{\mathrm{gr}}}^i(L, M)$$

are the Ext groups in the abelian category of graded left B-modules $B-\mathsf{mod}_{\mathsf{gr}}$. More generally, for every $n \in \mathbb{Z}$,

$$\mathrm{Ext}^i_B(L, M)_n = \mathrm{Ext}^i_{B-\mathsf{mod}_{\mathsf{gr}}}(L, M(-n)),$$

where $M(-n)$ denotes the left B-module M with the shifted grading, $M(-n)_j = M_{j+n}$.

Similarly, as above in this chapter, given a graded module/abelian group F, the notation $F^+ = \mathrm{Hom}_{\mathbb{Z}}(F, \mathbb{Q}/\mathbb{Z})$ stands for the graded module/abelian group with the components $F^+_n = \mathrm{Hom}_{\mathbb{Z}}(F_{-n}, \mathbb{Q}/\mathbb{Z})$. In particular, the natural isomorphism of graded abelian groups

$$\mathrm{Ext}^i_B(L, M^+) \simeq \mathrm{Tor}^B_i(M, L)^+$$

holds for any graded right B-module M and graded left B-module L.

Theorem 5.36 *Assume that B is a weakly right finitely projective Koszul graded ring. Then the exact, fully faithful forgetful functor of abelian categories $B-\mathsf{contra}_{\mathsf{gr}} \longrightarrow B-\mathsf{mod}_{\mathsf{gr}}$ induces isomorphisms on all the Ext groups.*

Proof This is a graded version of Theorem 5.20. The forgetful functor $B-\mathsf{contra}_{\mathsf{gr}} \longrightarrow B-\mathsf{mod}_{\mathsf{gr}}$ is fully faithful by Theorem 5.34. We have to show that it induces isomorphisms of the groups Ext^i for $i > 0$.

There are enough projective objects in the abelian category $B-\mathsf{contra}_{\mathsf{gr}}$. Hence it suffices to prove the graded Ext group vanishing $\mathrm{Ext}^i_B(P, Q) = 0$ for all projective objects $P \in B-\mathsf{contra}_{\mathsf{gr}}$, all objects $Q \in B-\mathsf{contra}_{\mathsf{gr}}$, and all $i > 0$.

By Proposition 5.32, any graded left B-contramodule is the cokernel of an injective morphism of separated graded left B-contramodules. Furthermore, any separated graded left B-contramodule Q is an infinitely iterated extension, in the sense of the projective limit, of trivial graded left B-contramodules $G^m Q/G^{m+1} Q$. Using a graded version of the dual Eklof lemma [24, Proposition 18] (see [69, Lemma 4.5] for a far-reaching generalization), we can reduce the question to the case when Q is a trivial graded left B-contramodule. In other words, Q is just a graded left R-module in which the augmentation ideal $B_{\geq 1} \subset B$ acts by zero.

The graded left R-module Q has a resolution by injective graded left R-modules of the form $F^+ = \mathrm{Hom}_{\mathbb{Z}}(F, \mathbb{Q}/\mathbb{Z})$, where F ranges over free graded right R-modules. Arguing further as in the proof of Theorem 5.20, we see that it suffices to prove the graded Tor group vanishing $\mathrm{Tor}^B_i(R, P) = 0$ for all projective objects $P \in B-\mathsf{contra}_{\mathsf{gr}}$ and all integers $i > 0$ (where R is viewed as a graded right B-module with $B_{\geq 1}$ acting by zero).

A description of projective graded left B-contramodules was given in the proof of Proposition 5.27. In the notation above, we can rephrase it as follows. The projective graded left B-contramodules are precisely the direct summands of graded left B-contramodules of the form $\mathbb{M}^{\mathsf{gr}}_K(H) = B \otimes^{\Pi}_K K[X]$, where $K[X]$ is the free graded left K-module spanned by a graded set X. We put $H = K[X]$ for brevity.

Now we compute the graded groups $\operatorname{Tor}_i^B(R, B\otimes_K^{\Pi}H)$ using the graded projective resolution (5.5) of the graded right B-module R. Taking the tensor product of (5.5) with $B\otimes_K^{\Pi}H$ over B, we obtain the complex of graded abelian groups

$$\cdots \longrightarrow V_2\otimes_R(B\otimes_K^{\Pi}H) \longrightarrow V_1\otimes_R(B\otimes_K^{\Pi}H) \longrightarrow B\otimes_K^{\Pi}H \longrightarrow 0. \qquad (5.12)$$

According to Lemma 5.35(a), the complex (5.12) is isomorphic to

$$\cdots \longrightarrow (V_2\otimes_R B)\otimes_K^{\Pi}H \longrightarrow (V_1\otimes_R B)\otimes_K^{\Pi}H \longrightarrow B\otimes_K^{\Pi}H \longrightarrow 0. \qquad (5.13)$$

The complex (5.13) can be obtained by applying the exact functor $-\otimes_K^{\Pi}H$ to the resolution (5.5). Thus $\operatorname{Tor}_i^B(R, B\otimes_K^{\Pi}H) = 0$ for $i > 0$, as desired. $\square$

Theorem 5.37 *Assume that B is a weakly right finitely projective Koszul graded ring. Then the exact, fully faithful inclusion functor of abelian categories* $\mathsf{comod}_{\mathsf{gr}}-B \longrightarrow \mathsf{mod}_{\mathsf{gr}}-B$ *induces isomorphisms on all the Ext groups.*

Proof This is a graded version of Theorem 5.21. There are enough injective objects in a Grothendieck abelian category $\mathsf{comod}_{\mathsf{gr}}-B$. Hence it suffices to prove the graded Ext group vanishing $\operatorname{Ext}^i_{B^{\mathrm{op}}}(M, L) = 0$ for all objects $M \in \mathsf{comod}_{\mathsf{gr}}-B$, all injective objects $L \in \mathsf{comod}_{\mathsf{gr}}-B$, and all $i > 0$.

Every graded right B-comodule M has a natural increasing filtration $0 = G_{-1}M \subset G_0 M \subset G_1 M \subset G_2 M \subset \cdots$ by graded right B-submodules $G_m M \subset M$ with the grading component $G_m M_j \subset M_j$ consisting of all the elements $x \in M_j$ such that $x B_{\geq m+1} = 0$ in M. By the definition, we have $M = \bigcup_{m\geq 0} G_m M$. The successive quotient modules $G_m M/G_{m-1}M$ are *trivial* graded right B-comodules, that is, graded right B-modules in which $B_{\geq 1}$ acts by zero. The category of trivial graded right B-comodules is equivalent to the category of graded right R-modules $\mathsf{mod}_{\mathsf{gr}}-R$.

Using a graded version of the Eklof lemma [24, Lemma 1], we can assume that M is a trivial graded right B-comodule. Then the graded right R-module M has a resolution by free graded right R-modules. Similarly to the proof of Theorem 5.21, we see that it suffices to prove the graded Ext group vanishing $\operatorname{Ext}^i_{B^{\mathrm{op}}}(R, L) = 0$ for all injective objects $L \in \mathsf{comod}_{\mathsf{gr}}-B$ and all integers $i > 0$ (where R is viewed as a graded right R-module with $B_{\geq 1}$ acting by zero).

Injective objects of the category $\mathsf{comod}_{\mathsf{gr}}-B$ can be described as follows. The inclusion functor $\mathsf{comod}_{\mathsf{gr}}-B \longrightarrow \mathsf{mod}_{\mathsf{gr}}-B$ has a right adjoint functor $\Gamma_B \colon \mathsf{mod}_{\mathsf{gr}}-B \longrightarrow \mathsf{comod}_{\mathsf{gr}}-B$ assigning to a graded right B-module N its graded submodule $\Gamma_B(N) \subset N$ whose component $\Gamma_B(N)_j$ consists of all elements $x \in N_j$ for which there exists $m \geq 1$ such that $x B_{\geq m} = 0$. The injective objects of $\mathsf{comod}_{\mathsf{gr}}-B$ are precisely the direct summands of the objects $\Gamma_B(N)$, where N ranges over the injective objects of $\mathsf{mod}_{\mathsf{gr}}-B$.

The following description of injective graded right B-modules is convenient for our purposes. Let $H = K[X]$ be the free graded left K-module spanned by a graded set X, and let $J = K[X]^+$ be the graded character module of H. Then J is an injective graded right K-module, and all injective graded right K-modules are direct summands of modules of this form. The graded Hom module $\mathrm{Hom}_{K^{\mathrm{op}}}(B, J)$ is an injective graded right B-module, and all injective graded right B-modules are direct summands of modules of this form. One can easily check that $\Gamma_B(\mathrm{Hom}_{K^{\mathrm{op}}}(B, J)) = \mathrm{Hom}_{K^{\mathrm{op}}}^{\Sigma}(B, J)$ (see Lemma 6.13 below).

Now we compute the graded groups $\mathrm{Ext}_{B^{\mathrm{op}}}^i(R, \mathrm{Hom}_{K^{\mathrm{op}}}^{\Sigma}(B, J))$ using the graded projective resolution (5.5) of the graded right B-module R. Taking the graded right B-module Hom from (5.5) to $\mathrm{Hom}_{K^{\mathrm{op}}}^{\Sigma}(B, J)$, we obtain the complex of graded abelian groups

$$0 \longrightarrow \mathrm{Hom}_{K^{\mathrm{op}}}^{\Sigma}(B, J) \longrightarrow \mathrm{Hom}_{R^{\mathrm{op}}}(V_1, \mathrm{Hom}_{K^{\mathrm{op}}}^{\Sigma}(B, J))$$
$$\longrightarrow \mathrm{Hom}_{R^{\mathrm{op}}}(V_2, \mathrm{Hom}_{K^{\mathrm{op}}}^{\Sigma}(B, J)) \longrightarrow \cdots \qquad (5.14)$$

According to Lemma 5.35(b), the complex (5.14) is isomorphic to

$$0 \longrightarrow \mathrm{Hom}_{K^{\mathrm{op}}}^{\Sigma}(B, J) \longrightarrow \mathrm{Hom}_{K^{\mathrm{op}}}^{\Sigma}(V_1 \otimes_R B, J)$$
$$\longrightarrow \mathrm{Hom}_{K^{\mathrm{op}}}^{\Sigma}(V_2 \otimes_R B, J) \longrightarrow \cdots \qquad (5.15)$$

The complex (5.15) can be obtained by applying the exact functor $\mathrm{Hom}_{K^{\mathrm{op}}}^{\Sigma}(-, J)$ to the resolution (5.5). Thus $\mathrm{Ext}_{B^{\mathrm{op}}}^i(R, \mathrm{Hom}_{K^{\mathrm{op}}}^{\Sigma}(B, J)) = 0$ for $i > 0$, as desired. $\square$

Relative Nonhomogeneous Derived Koszul Duality: The Comodule Side

6.1 CDG-Modules

Let (B, d, h) be a curved DG-ring, as defined in Sect. 3.2, and let $(\widehat{B}, \partial)$ be the related quasi-differential graded ring constructed in Theorem 4.7.

Definition 6.1 A *left CDG-module over* (B, d, h) can be simply defined as a graded left $\widehat{B}$-module.

Equivalently, a left CDG-module (M, d_M) over (B, d, h) is a graded left B-module $M = \bigoplus_{n \in \mathbb{Z}} M^n$ endowed with a sequence of additive maps $d_{M,n} \colon M^n \longrightarrow M^{n+1}$, $n \in \mathbb{Z}$, satisfying following equations:

(i) d_M is an odd derivation of the graded left module M compatible with the odd derivation d on the graded ring B, that is $d_M(bx) = d(b)x + (-1)^{|b|}bd_M(x)$ for all $b \in B^{|b|}$ and $x \in M^{|x|}$, $|b|, |x| \in \mathbb{Z}$;

(ii) $d_M^2(x) = hx$ for all $x \in M$.

A *right CDG-module over* (B, d, h) can be simply defined as a graded right $\widehat{B}$-module. Equivalently, a right CDG-module (N, d_N) over (B, d, h) is a graded right B-module $N = \bigoplus_{n \in \mathbb{Z}} N^n$ endowed with a sequence of additive maps $d_{N,n} \colon N^n \longrightarrow N^{n+1}$, $n \in \mathbb{Z}$, satisfying the following equations:

(i) d_N is an odd derivation of the graded right module N compatible with the odd derivation d on the graded ring B, that is $d_N(yb) = d_N(y)b + (-1)^{|y|}yd(b)$ for all $b \in B^{|b|}$ and $y \in N^{|y|}$, $|b|, |y| \in \mathbb{Z}$;

(ii) $d_N^2(y) = -yh$ for all $y \in N$.

L. Positselski, *Relative Nonhomogeneous Koszul Duality*, Frontiers in Mathematics, https://doi.org/10.1007/978-3-030-89540-2_6

Notice that the graded B-B-bimodule B with its odd derivation d is *neither* a left *nor* a right CDG-module over (B, d, h), because the formula (ii) for $d^2(b)$, $b \in B$ in Sect. 3.2 does not agree with the formulas (ii) in the two definitions above. However, the diagonal graded B-B-bimodule B is naturally a *CDG-bimodule* over the CDG-ring (B, d, h), in the sense of the following definition.

Definition 6.2 Let $B = (B, d_B, h_B)$ and $C = (C, d_C, h_C)$ be two CDG-rings. A *CDG-bimodule K over B and C* is a graded B-C-bimodule $K = \bigoplus_{n \in \mathbb{Z}} K^n$ endowed with a sequence of additive maps $d_{K,n} \colon K^n \longrightarrow K^{n+1}$, $n \in \mathbb{Z}$, satisfying the following conditions:

(i) d_K is *both* an odd derivation of the graded left B-module K compatible with the odd derivation d_B on the graded ring B *and* an odd derivation of the graded right C-module K compatible with the odd derivation d_C on the graded ring C;

(ii) $d_K^2(z) = h_B z - z h_C$ for all $z \in K$.

Notice that a CDG-bimodule over B and C is *not* the same thing as a graded $\widehat{B}$-$\widehat{C}$-bimodule. We leave it to the reader to construct a CDG-ring structure on the graded ring $B \otimes_{\mathbb{Z}} C^{\mathrm{op}}$ for which a CDG-bimodule over B and C would be the same thing as a left CDG-module over $B \otimes_{\mathbb{Z}} C^{\mathrm{op}}$.

Given a graded left B-module M and an integer $n \in \mathbb{Z}$, we denote by $M[n]$ the graded left B-module with the grading components $M[n]^i = M^{n+i}$, $i \in \mathbb{Z}$, and the left action of B transformed with the sign rule $b(x[n]) = (-1)^{n|b|}(bx)[n]$ for any element $x \in M^{n+i}$ and the corresponding element $x[n] \in M[n]^i$. Given a graded right B-module N and an integer $n \in \mathbb{Z}$, we let $N[n]$ denote the graded right B-module N with the components $N[n]^i = N^{n+i}$ and the right action of B unchanged, $(y[n])b = (yb)[n]$ for $y \in N^{n+i}$ and $y[n] \in N[n]^i$. If a differential d_M or d_N is given, a grading shift transforms it with the usual sign rule $d_{M[n]}(x[n]) = (-1)^n (d_M(x))[n]$ or $d_{N[n]}(y[n]) = (-1)^n (d_N(y))[n]$. These sign rules are best understood with the hint that we write the shift-of-degree symbol $[n]$ to the right of our modules and their elements, as it is traditionally done, but tacitly presume it to be a *left* operator (as all our operators, such as the differentials, homomorphisms, etc., are generally left operators). So one may want to write $[n]x \in [n]M$ instead of $x[n] \in M[n]$ for $x \in M$ and similarly $[n]y \in [n]N$ for $y \in N$, which would explain the signs.

Given two graded left B-modules L and M, we denote by $\mathrm{Hom}_B^n(L, M)$ the abelian group of all graded left B-module homomorphisms $L \longrightarrow M[n]$. For any two left CDG-modules L and M over a CDG-ring (B, d, h), there is a natural differential d on the graded abelian group $\mathrm{Hom}_B(L, M)$ defined by the usual rule $d(f)(l) = d_M(f(l)) - (-1)^{|f|} f(d_L(l))$ for all $f \in \mathrm{Hom}_B^{|f|}(L, M)$, $|f| \in \mathbb{Z}$, and $l \in L$. Similarly, given two graded right B-modules K and N, we denote by $\mathrm{Hom}_{B^{\mathrm{op}}}^n(K, N)$ the abelian group of all graded right B-module homomorphisms $K \longrightarrow N[n]$. For any two right CDG-modules K and N over a CDG-ring (B, d, h), there is a natural differential d on the graded abelian group $\text{Hom}_{B^{\text{op}}}(K, N)$ defined by the formula $d(g)(k) = d_N(g(k)) - (-1)^{|g|}g(d_K(k))$ for all $g \in \text{Hom}_{B^{\text{op}}}^{|g|}(K, N)$, $|g| \in \mathbb{Z}$, and $k \in K$.

Definition 6.3 One easily checks that $d^2(f) = 0 = d^2(g)$ for all f and g in the notation above. Hence for any two left CDG-modules L and M over B, the *complex of morphisms* $\text{Hom}_B(L, M)$ is defined; and similarly, for any two right CDG-modules K and N over B, there is the complex of morphisms $\text{Hom}_{B^{\text{op}}}(K, N)$. These constructions produce the *DG-category of left CDG-modules* $\mathsf{DG}(B{-}\mathsf{mod})$ and the similar *DG-category of right CDG-modules* $\mathsf{DG}(\mathsf{mod}{-}B)$ *over* B. We refer to [59, Sections 1.2 and 3.1] for a discussion of these DG-categories and their triangulated *homotopy categories* $\mathsf{Hot}(B{-}\mathsf{mod}) = H^0\mathsf{DG}(B{-}\mathsf{mod})$ and $\mathsf{Hot}(\mathsf{mod}{-}B) = H^0\mathsf{DG}(\mathsf{mod}{-}B)$.

Given a right graded B-module N and a left graded B-module M, the graded tensor product $N \otimes_B M$ is a graded abelian group constructed in the usual way. So we consider the bigraded abelian group $N \otimes_{\mathbb{Z}} M$ with the components $N^i \otimes_{\mathbb{Z}} M^j$, totalize it by taking infinite direct sums along the diagonals $i + j = n$, and consider the graded quotient group by the graded subgroup spanned by the elements $yb \otimes x - y \otimes bx$, where $y \in N$, $b \in B$, and $x \in M$. For any right CDG-module N and any left CDG-module M over (B, d, h), the tensor product $N \otimes_B M$ is naturally a complex of abelian groups with the differential $d(y \otimes x) = d_N(y) \otimes x + (-1)^{|y|}y \otimes d_M(x)$. One can easily check that $d^2(y \otimes x) = 0$ in $N \otimes_B M$.

More generally, let (A, d_A, h_A), (B, d_B, h_B), and (C, d_C, h_C) be three CDG-rings. Given a graded C-B-bimodule N and a graded B-A-bimodule K, the graded C-A-bimodule structure on the tensor product $N \otimes_B K$ is defined by the usual rule $c(y \otimes z)a = (cy) \otimes (za)$ for all $c \in C$, $y \in N$, $z \in K$, and $a \in A$. Now let N be a CDG-bimodule over C and B, and let K be a CDG-bimodule over B and A. Then the graded C-A-bimodule $N \otimes_B K$ with the differential $d(y \otimes z) = d_N(y) \otimes z + (-1)^{|y|}y \otimes d_K(z)$, where $y \in N^{|y|}$ and $z \in K^{|z|}$, is a CDG-bimodule over C and A.

Given a graded B-A-bimodule K and a graded B-C-bimodule L, the graded A-C-bimodule structure on the graded abelian group $\text{Hom}_B(K, L)$ is defined by the formula $(afc)(z) = (-1)^{|a||f|+|a||z|+|c||z|}f(za)c$ for all $a \in A^{|a|}$, $c \in C^{|c|}$, $z \in K^{|z|}$, and $f \in \text{Hom}_B^{|f|}(K, L)$. Now let K be a CDG-bimodule over B and A, and let L be a CDG-bimodule over B and C. Then the graded A-C-bimodule $\text{Hom}_B(K, L)$ with the differential $d(f)(z) = d_L(f(z)) - (-1)^{|f|}f(d_K(z))$ is a CDG-bimodule over A and C.

Given a graded B-A-bimodule K and a graded C-A-bimodule M, the graded C-B-bimodule structure on the graded abelian group $\text{Hom}_{A^{\text{op}}}(K, M)$ is defined by the formula $(cgb)(z) = cg(bz)$ for all $c \in C$, $b \in B$, $z \in K$, and $g \in \text{Hom}_{A^{\text{op}}}(K, M)$. Now let K be a CDG-bimodule over B and A, and let M be a CDG-bimodule over C and A. Then the graded C-B-bimodule $\text{Hom}_{A^{\text{op}}}(K, M)$ with the differential $d(g)(z) = d_M(g(z)) - (-1)^{|g|}g(d_K(z))$, where $z \in K^{|z|}$ and $g \in \text{Hom}_{A^{\text{op}}}^{|g|}(K, M)$, is a CDG-bimodule over C and B.

Let A, B, C, and D be four CDG-rings. Then for any CDG-bimodule L over A and B, any CDG-bimodule M over B and C, and any CDG-bimodule N over C and D, the natural isomorphism of graded A-D-bimodules

$$(L \otimes_B M) \otimes_C N \simeq L \otimes_B (M \otimes_C N)$$

is a closed isomorphism of CDG-bimodules over A and D.

Similarly, for any CDG-bimodule L over B and A, any CDG-bimodule M over C and B, and any CDG-bimodule N over C and D, the natural isomorphism of graded A-D-bimodules

$$f' \in \mathrm{Hom}_C(M \otimes_B L, \ N) \simeq \mathrm{Hom}_B(L, \mathrm{Hom}_C(M, N)) \ni f''$$

given by the rule $f''(z)(x) = (-1)^{|x||z|} f'(x \otimes z)$ for all $z \in L^{|z|}$ and $x \in M^{|x|}$ is a closed isomorphism of CDG-bimodules over A and D.

For any CDG-bimodule L over A and B, any CDG-bimodule M over B and C, and any CDG-bimodule N over D and C, the natural isomorphism of graded D-A-bimodules

$$g' \in \mathrm{Hom}_{C^{\mathrm{op}}}(L \otimes_B M, \ N) \simeq \mathrm{Hom}_{B^{\mathrm{op}}}(L, \mathrm{Hom}_{C^{\mathrm{op}}}(M, N)) \ni g''$$

given by the rule $g''(z)(x) = g'(z \otimes x)$ for all $z \in L$ and $x \in M$ is a closed isomorphism of CDG-bimodules over D and A.

In particular, let K be a CDG-bimodule over B and A. Then the DG-functor

$$K \otimes_A - : \mathsf{DG}(A-\mathsf{mod}) \longrightarrow \mathsf{DG}(B-\mathsf{mod})$$

is left adjoint to the DG-functor

$$\mathrm{Hom}_B(K, -) : \mathsf{DG}(B-\mathsf{mod}) \longrightarrow \mathsf{DG}(A-\mathsf{mod}).$$

Here the DG-functor $K \otimes_A -$ acts on morphisms by the rule $(K \otimes_A f)(k \otimes l) = (-1)^{|f||k|} k \otimes f(l)$ for all L, $M \in \mathsf{DG}(A-\mathsf{mod})$, $f \in \mathrm{Hom}_A^{|f|}(L, M)$, $k \in K^{|k|}$, and $l \in L^{|l|}$. The DG-functor $\mathrm{Hom}_B(K, -)$ acts on morphisms by the rule $\mathrm{Hom}_B(K, f)(g)(k) = f(g(k))$ for all L, $M \in \mathsf{DG}(B-\mathsf{mod})$, $f \in \mathrm{Hom}_B(L, M)$, $g \in \mathrm{Hom}_B(K, L)$, and $k \in K$.

Hence the induced triangulated functors $K \otimes_A -$ and $\mathrm{Hom}_B(K, -)$ between the homotopy categories $\mathsf{Hot}(A-\mathsf{mod})$ and $\mathsf{Hot}(B-\mathsf{mod})$ are also adjoint.

Similarly, the DG-functor

$$- \otimes_B K : \mathsf{DG}(\mathsf{mod}-B) \longrightarrow \mathsf{DG}(\mathsf{mod}-A)$$

is left adjoint to the DG-functor

$$\mathrm{Hom}_{A^{\mathrm{op}}}(K, -) \colon \mathsf{DG}(\mathrm{mod}{-}A) \longrightarrow \mathsf{DG}(\mathrm{mod}{-}B).$$

Here the DG-functor $-\otimes_B K$ acts on morphisms by the rule $(f\otimes_B K)(l\otimes k) = f(l)\otimes k$ for all $L, M \in \mathsf{DG}(\mathrm{mod}{-}B)$, $f \in \mathrm{Hom}_{B^{\mathrm{op}}}(L, M)$, $l \in L$, and $k \in K$. The DG-functor $\mathrm{Hom}_{A^{\mathrm{op}}}(K, -)$ acts on morphisms by the rule $\mathrm{Hom}_{A^{\mathrm{op}}}(K, f)(g)(k) = f(g(k))$ for all L, $M \in \mathsf{DG}(\mathrm{mod}{-}A)$, $f \in \mathrm{Hom}_{A^{\mathrm{op}}}(L, M)$, $g \in \mathrm{Hom}_{A^{\mathrm{op}}}(K, L)$, and $k \in K$.

Hence the induced triangulated functors $-\otimes_B K$ and $\mathrm{Hom}_{A^{\mathrm{op}}}(K, -)$ between the homotopy categories $\mathsf{Hot}(\mathrm{mod}{-}B)$ and $\mathsf{Hot}(\mathrm{mod}{-}A)$ are also adjoint.

Now let $(f, a) \colon {''}B = ({''}B, d'', h'') \longrightarrow {'}B = ({'}B, d', h')$ be a morphism of CDG-rings (as defined in Sect. 3.2). Let (M, d'_M) be a left CDG-module over $({'}B, d', h')$. The restriction of scalars via the graded ring homomorphism $f \colon {''}B \longrightarrow {'}B$ defines a structure of graded left ${''}B$-module on M. Put $d''_M(x) = d'_M(x) + ax$ for all $x \in M$. Then (M, d''_M) is a left CDG-module over $({''}B, d'', h'')$. This construction defines the DG-functor $\mathsf{DG}({'}B{-}\mathrm{mod}) \longrightarrow \mathsf{DG}({''}B{-}\mathrm{mod})$ of restriction of scalars via (f, a).

Let (N, d'_N) be a right CDG-module over $({'}B, d', h')$. The restriction of scalars via the graded ring homomorphism $f \colon {''}B \longrightarrow {'}B$ defines a structure of graded right ${''}B$-module on N. Put $d''_N(y) = d'_N(y) - (-1)^{|y|}ya$ for all $y \in N^{|y|}$, $|y| \in \mathbb{Z}$. Then (N, d''_N) is a right CDG-module over $({''}B, d'', h'')$. This construction defines the DG-functor $\mathsf{DG}({'}B{-}\mathrm{mod}) \longrightarrow \mathsf{DG}({''}B{-}\mathrm{mod})$ of restriction of scalars via (f, a).

Similarly, let $(f, a_B) \colon {''}B = ({''}B, d''_B, h''_B) \longrightarrow {'}B = ({'}B, d'_B, h'_B)$ and $(g, a_C) \colon {''}C = ({''}C, d''_C, h''_C) \longrightarrow {'}C = ({'}C, d'_C, h'_C)$ be two morphisms of CDG-rings. Let (K, d'_K) be a CDG-bimodule over $({'}B, d'_B, h'_B)$ and $({'}C, d'_C, h'_C)$. The restriction of scalars via the graded ring homomorphisms $f \colon {''}B \longrightarrow {'}B$ and $g \colon {''}C \longrightarrow {'}C$ defines a structure of graded ${''}B$-${''}C$-bimodule on K. Put $d''_K(z) = d'_K(z) + a_B z - (-1)^{|z|}z a_C$ for all $z \in K^{|z|}$, $|z| \in \mathbb{Z}$. Then (K, d''_K) is a CDG-bimodule over $({''}B, d''_B, h''_B)$ and $({''}C, d''_C, h''_C)$. All the constructions of the tensor product complex and the tensor product and Hom CDG-bimodules above in this section agree with the transformations of CDG-(bi)modules induced by isomorphisms of CDG-rings.

6.2 Nonhomogeneous Koszul Complex/CDG-Module

Let $R \subset \widetilde{V} \subset \widetilde{A}$ be a 3-left finitely projective weak nonhomogeneous quadratic ring, as defined in Sects. 3.1 and 3.3, and let $A = \mathrm{gr}^F \widetilde{A}$ be its associated graded ring with respect to the filtration F. As in Chaps. 2–3, we put $V = A_1$ and denote by B the 3-right finitely projective quadratic graded ring quadratic dual to qA. Choose a left R-submodule of strict generators $V' \subset \widetilde{V} = F_1\widetilde{A}$, and let $(B, d, h) = (B, d', h')$ denote the related CDG-ring structure on the graded ring B, as constructed in Proposition 3.16.

Denote by $e' \in \mathrm{Hom}_R(V, R) \otimes_R \widetilde{V} = B^1 \otimes_R F_1 \widetilde{A}$ the element corresponding to the injective left R-module map $V \simeq V' \hookrightarrow \widetilde{V}$ under the construction of Lemma 2.5. Our aim is to construct the *dual nonhomogeneous Koszul CDG-module* $K^\vee(B, \widetilde{A}) = K_{e'}^\vee(B, \widetilde{A})$, which has the form

$$0 \longrightarrow \widetilde{A} \longrightarrow B^1 \otimes_R \widetilde{A} \longrightarrow B^2 \otimes_R \widetilde{A} \longrightarrow B^3 \otimes_R \widetilde{A} \longrightarrow \cdots. \qquad (6.1)$$

The differential on $K_{e'}^\vee(B, \widetilde{A})$ does not square to zero when $h \neq 0$. Rather, $K_{e'}^\vee(B, \widetilde{A})$ is a left CDG-module over the CDG-ring (B, d, h), with the obvious underlying graded left B-module structure of the tensor product $B \otimes_R \widetilde{A}$. However, the differential on $K_{e'}^\vee(B, \widetilde{A})$ preserves the right $\widetilde{A}$-module structure of the tensor product $B \otimes_R \widetilde{A}$. Summarizing, one can say that $K_{e'}^\vee(B, \widetilde{A})$ is going to be a CDG-bimodule over the CDG-rings $B = (B, d, h)$ and $\widetilde{A} = (\widetilde{A}, 0, 0)$ (where $\widetilde{A}$ is viewed as a graded ring concentrated entirely in degree zero).

When $\widetilde{A} = A = qA$ is a 3-left finitely projective quadratic graded ring and $V' = A_1 \hookrightarrow F_1 \widetilde{A} = R \oplus A_1$ is the obvious splitting of the direct summand projection $R \oplus A_1 \longrightarrow A_1$ (so the related CDG-ring structure on B is trivial, $d = 0 = h$), the construction of differential $d_{e'}$ in the next proposition reduces to that of the differential d_e on the dual Koszul complex $K_e^{\vee \bullet} = B \otimes_R A$ from Sect. 2.6.

Proposition 6.4 *The formula*

$$d_{e'}(b \otimes c) = d(b) \otimes c + (-1)^{|b|} be'c \qquad (6.2)$$

for all $b \in B^{|b|}$ and $c \in \widetilde{A}$, where $e' \in B^1 \otimes_R F_1 \widetilde{A}$ is the element mentioned above, endows the graded left B-module $K^\vee(B, \widetilde{A}) = B \otimes_R \widetilde{A}$ with a well-defined structure of left CDG-module over the CDG-ring $B = (B, d, h)$.

Proof Neither one of the two summands in the formula (6.2) produces a well-defined map $B \otimes_R \widetilde{A} \longrightarrow B \otimes_R \widetilde{A}$. It is only the sum that is well-defined as an endomorphism of the tensor product over R. However, the separate summands are well-defined as maps $B \otimes_\mathbb{Z} \widetilde{A} \longrightarrow B \otimes_R \widetilde{A}$. Indeed, the first summand is obviously well-defined as a map $B \otimes_\mathbb{Z} \widetilde{A} \longrightarrow B \otimes_\mathbb{Z} \widetilde{A}$, and one can consider the composition with the natural surjection $B \otimes_\mathbb{Z} \widetilde{A} \longrightarrow B \otimes_R \widetilde{A}$. The grading/filtration components of the second summand (without the $\pm$ sign) are the compositions

$$B^n \otimes_\mathbb{Z} F_j \widetilde{A} \xrightarrow{e'} B^n \otimes_\mathbb{Z} B^1 \otimes_R F_1 \widetilde{A} \otimes_\mathbb{Z} F_j \widetilde{A} \longrightarrow B^{n+1} \otimes_\mathbb{R} F_{j+1} \widetilde{A}.$$

Here the map $B^n \otimes_{\mathbb{Z}} F_j \widetilde{A} \longrightarrow B^n \otimes_{\mathbb{Z}} B^1 \otimes_R F_1 \widetilde{A} \otimes_{\mathbb{Z}} F_j \widetilde{A}$ is induced by the abelian group element or map $e' \colon \mathbb{Z} \longrightarrow B^1 \otimes_R F_1 \widetilde{A}$, while the map $B^n \otimes_{\mathbb{Z}} B^1 \otimes_R F_1 \widetilde{A} \otimes_{\mathbb{Z}} F_j \widetilde{A} \longrightarrow B^{n+1} \otimes_R F_{j+1} \widetilde{A}$ is the tensor product of two multiplication maps $B^n \otimes_{\mathbb{Z}} B^1 \longrightarrow B^{n+1}$ and $F_1 \widetilde{A} \otimes_{\mathbb{Z}} F_j \widetilde{A} \longrightarrow F_{j+1} \widetilde{A}$.

Let us check that the map $d_{e'}$ is well-defined on $B \otimes_R \widetilde{A}$, that is, for any $b \in B$, $r \in R$, and $c \in \widetilde{A}$ one has

$$d(br) \otimes c + (-1)^{|b|} bre'c = d(b) \otimes rc + (-1)^{|b|} be'rc$$

in $B \otimes_R \widetilde{A}$. In the pairing notation from the proof of Lemma 2.5, by the definition of the element $e' \in \mathrm{Hom}_R(V, R) \otimes_R \widetilde{V}$ we have $\langle v, e' \rangle = v'$ for every $v \in V$, where $v' \in V' \subset \widetilde{V}$ denotes the element corresponding to v under the left R-module identification $V \simeq V'$. Hence, in particular,

$$\langle v, re' \rangle = \langle vr, e' \rangle = (vr)'$$

while

$$\langle v, e'r \rangle = \langle v, e' \rangle * r = v' * r = (vr)' + q(v, r) = (vr)' + \langle v, d(r) \rangle.$$

Here, as in Sect. 3.3, the symbol $*$ denotes the multiplication in $\widetilde{A}$, and, in particular, the right action of R in $\widetilde{V}$. The notation $q(v, r)$ was defined in Eq. (3.1), and the equation $q(v, r) = \langle v, d(r) \rangle$ comes from the definition of the differential d on the graded ring B in Proposition 3.16. Therefore,

$$\langle v, e'r \rangle = \langle v, re' \rangle + \langle v, d(r) \rangle \qquad \text{for all } v \in V.$$

We have deduced the equation

$$e'r = re' + d(r) \otimes 1 \tag{6.3}$$

comparing the left and right actions of the ring R on the element $e' \in \mathrm{Hom}_R(V, R) \otimes_R \widetilde{V}$ (where $1 \in R \subset \widetilde{V}$ denotes the unit element). Now we can compute that

$$d(br) \otimes c + (-1)^{|b|} bre'c = d(b)r \otimes c + (-1)^{|b|} bd(r) \otimes c + (-1)^{|b|} bre'c$$

$$= d(b) \otimes rc + (-1)^{|b|} be'rc \in B \otimes_R \widetilde{A},$$

as desired. So the map $d_{e'} \colon B \otimes_R \widetilde{A} \longrightarrow B \otimes_R \widetilde{A}$ is well-defined.

It is obvious that $d_{e'}$ is a right $\widetilde{A}$-module map. It is also easy to check that $d_{e'}$ is an odd derivation of the graded left B-module $B \otimes_R \widetilde{A}$ compatible with the odd derivation d on the graded ring B. Indeed, for all $b, b' \in B$ and $c \in \widetilde{A}$ we have

$$d_{e'}(b'b \otimes c) = d(b'b) \otimes c + (-1)^{|b'|+|b|} b'be'c$$

$$= d(b')b \otimes c + (-1)^{|b'|} b'd(b) \otimes c + (-1)^{|b'|+|b|} b'be'c$$

$$= d(b')b \otimes c + (-1)^{|b'|} b' \left(d(b) \otimes c + (-1)^{|b|} be'c \right) = d(b')b \otimes c + (-1)^{|b'|} b'd_{e'}(b \otimes c),$$

as desired.

It remains to check the equation

$$d_{e'}^2(b \otimes c) = hb \otimes c \qquad \text{for all } b \in B \text{ and } c \in \widetilde{A}.$$

For this purpose, we choose two finite collections of elements $v_\alpha \in V$ and $u^\alpha \in B^1 = \operatorname{Hom}_R(V, R)$, indexed by the same indices α, such that $e = \sum_\alpha u^\alpha \otimes v_\alpha \in \operatorname{Hom}_R(V, R) \otimes_R V$. Then the assertion of Lemma 2.5(b) can be written as the equation

$$\sum_\alpha su^\alpha \otimes v_\alpha = \sum_\alpha u^\alpha \otimes v_\alpha s \qquad \text{for all } s \in R. \tag{6.4}$$

For any element $v \in V$, we have $\sum_\alpha \langle v, u^\alpha \rangle v_\alpha = v$.

By the definition,

$$d_{e'}(b \otimes c) = d(b) \otimes c + (-1)^{|b|} \sum_\alpha bu^\alpha \otimes v'_\alpha * c,$$

hence

$$d_{e'}^2(b \otimes c) = d^2(b) \otimes c + (-1)^{|b|} \sum_\alpha d(bu^\alpha) \otimes v'_\alpha * c$$

$$+ (-1)^{|b|+1} \sum_\alpha d(b)u^\alpha \otimes v'_\alpha * c - \sum_\alpha \sum_\beta bu^\alpha u^\beta \otimes v'_\beta * v'_\alpha * c$$

$$= (hb - bh) \otimes c + \sum_\alpha bd(u^\alpha) \otimes v'_\alpha * c - \sum_\alpha \sum_\beta bu^\alpha u^\beta \otimes v'_\beta * v'_\alpha * c.$$

So we only need to check that the equation

$$- h \otimes 1 + \sum_\alpha d(u^\alpha) \otimes v'_\alpha - \sum_\alpha \sum_\beta u^\alpha u^\beta \otimes v'_\beta * v'_\alpha = 0 \tag{6.5}$$

holds in $B^2 \otimes_R F_2 \widetilde{A}$.

We have $B^2 = \operatorname{Hom}_R(I, R)$. Hence, in order to prove Eq. (6.5), it suffices to check that the pairing of the left-hand side with any element $i \in I$ vanishes as an element of $F_2 \widetilde{A}$. As in Sect. 3.3, we denote by $\widehat{I} \subset V \otimes_{\mathbb{Z}} V$ the full preimage of the R-R-subbimodule

$I \subset V \otimes_R V$ under the natural surjective map $V \otimes_{\mathbb{Z}} V \longrightarrow V \otimes_R V$. Consider an element $i \in I$ and choose its preimage $\hat{\imath} \in \widehat{I}$. As in Sect. 3.3, we write symbolically $i = i_1 \otimes i_2$ and $\hat{\imath} = \hat{\imath}_1 \otimes \hat{\imath}_2$.

Now we deal we the three summands in (6.5) one by one. Firstly, $\langle i, h \rangle = h(i)$. Secondly,

$$
\begin{aligned}
\langle i, \sum_\alpha d(u^\alpha) \otimes v'_\alpha \rangle &= \sum_\alpha \langle i, d(u^\alpha) \rangle v'_\alpha \\
&\overset{(3.5)}{=} \sum_\alpha \langle p(\hat{\imath}_1 \otimes \hat{\imath}_2), u^\alpha \rangle v'_\alpha - \sum_\alpha q(\hat{\imath}_1, \langle \hat{\imath}_2, u^\alpha \rangle) v'_\alpha \\
&= p(\hat{\imath}_1 \otimes \hat{\imath}_2)' - \sum_\alpha q(\hat{\imath}_1, \langle \hat{\imath}_2, u^\alpha \rangle) v'_\alpha
\end{aligned}
$$

according to the formula (3.5) in Proposition 3.16. Thirdly,

$$
\begin{aligned}
\langle i, \sum_\alpha \sum_\beta u^\alpha u^\beta \otimes v'_\beta * v'_\alpha \rangle &= \sum_\alpha \sum_\beta \langle i_1 \otimes i_2, u^\alpha u^\beta \rangle v'_\beta * v'_\alpha \\
&= \sum_\alpha \sum_\beta \langle i_1, \langle i_2, u^\alpha \rangle u^\beta \rangle v'_\beta * v'_\alpha \overset{(6.4)}{=} \sum_\alpha \sum_\beta \langle i_1, u^\beta \rangle (v_\beta \langle i_2, u^\alpha \rangle)' * v'_\alpha \\
&\overset{(3.1)}{=} \sum_\alpha \sum_\beta \langle \hat{\imath}_1, u^\beta \rangle v'_\beta * \langle \hat{\imath}_2, u^\alpha \rangle * v'_\alpha - \sum_\alpha \sum_\beta \langle \hat{\imath}_1, u^\beta \rangle q(v_\beta, \langle \hat{\imath}_2, u^\alpha \rangle) * v'_\alpha \\
&\overset{(3.1)}{=} \sum_\alpha \sum_\beta \langle \hat{\imath}_1, u^\beta \rangle v'_\beta * \langle \hat{\imath}_2, u^\alpha \rangle v'_\alpha - \sum_\alpha \sum_\beta \langle \hat{\imath}_1, u^\beta \rangle q(v_\beta, \langle \hat{\imath}_2, u^\alpha \rangle) v'_\alpha \\
&= \hat{\imath}'_1 * \hat{\imath}'_2 - \sum_\alpha q(\hat{\imath}_1, \langle \hat{\imath}_2, u^\alpha \rangle) v'_\alpha,
\end{aligned}
$$

where Eq. (6.4) is being applied to the expression $e = \sum_\beta u^\beta \otimes v_\beta$ and the elements $s = \langle i_2, u^\alpha \rangle \in R$.

Finally, we recall that

$$
-h(i) + p(\hat{\imath}_1 \otimes \hat{\imath}_2)' - \hat{\imath}'_1 * \hat{\imath}'_2 = 0
$$

in $F_2 \widetilde{A}$ by Eq. (3.2). This finishes the proof of the desired Eq. (6.5) and of the whole proposition.   □

Let us show that the left CDG-module $K^\vee(B, \widetilde{A})$ which we have constructed does not depend on any arbitrary choices. Let $V'' \subset \widetilde{V}$ be another left R-submodule of strict generators, related to our original choice of the left R-submodule $V' \subset \widetilde{V}$ by the rule (3.7). Let $'B = (B, d', h')$ and $''B = (B, d'', h'')$ denote the CDG-ring structures on the graded ring B corresponding to $V' \subset \widetilde{V}$ and $V'' \subset \widetilde{V}$, as in Proposition 3.20. Let $d_{e'}$ and $d_{e''}$ denote the related two differentials on the tensor product $B \otimes_R \widetilde{A}$, endowing it with the structures of a left CDG-module over $'B$ and $''B$.

Lemma 6.5 *The restriction of scalars (as defined in Sect. 6.1) with respect to the CDG-ring isomorphism* $(\mathrm{id}, a)\colon {''}B \longrightarrow {'}B$ *from Proposition 3.20 transforms the left CDG-module* $K^\vee_{e'}(B, \widetilde{A}) = (B \otimes_R \widetilde{A}, \, d_{e'})$ *over the CDG-ring* ${'}B$ *into the left CDG-module* $K^\vee_{e''}(B, \widetilde{A}) = (B \otimes_R \widetilde{A}, \, d_{e''})$ *over the CDG-ring* ${''}B$.

Proof We use the notation of Sect. 3.5: given an element $v \in V$, the corresponding elements in V' and V'' are denoted by $v' \in V' \subset \widetilde{V}$ and $v'' \in V'' \subset \widetilde{V}$. Let e' and $e'' \in \mathrm{Hom}_R(V, R) \otimes_R \widetilde{V}$ denote the elements corresponding to the injective R-module maps $V \simeq V' \hookrightarrow \widetilde{V}$ and $V \simeq V'' \hookrightarrow \widetilde{V}$. Then for every $v \in V$ we have

$$\langle v, e' \rangle = v' \quad \text{and} \quad \langle v, e'' \rangle = v'' = v' + a(v)$$

by formula (3.7), hence

$$e'' = e' + a \otimes 1 \in B^1 \otimes_R F_1 \widetilde{A}.$$

Now for all $b \in B$ and $c \in \widetilde{A}$ we can compute that

$$d_{e''}(b \otimes c) = d''(b) \otimes c + (-1)^{|b|} b e'' c$$

$$= d'(b) \otimes c + (ab - (-1)^{|b|} ba) \otimes c + (-1)^{|b|} b e' c + (-1)^{|b|} ba \otimes c$$

$$= d'(b) \otimes c + ab \otimes c + (-1)^{|b|} b e' c = d_{e'}(b \otimes c) + ab \otimes c,$$

as desired (where the middle equation takes into account the equation (iv) from the definition of a morphism of CDG-rings in Sect. 3.2). □

Remark 6.6 The dual nonhomogeneous Koszul CDG-module $K^\vee_{e'}(B, \widetilde{A})$ which we have constructed is a nonhomogeneous generalization of the dual Koszul complex $K^\vee_\bullet(B, A)$ from Sect. 2.6. The CDG-(bi)module $K^\vee_{e'}(B, \widetilde{A})$ will be very convenient to use below in Sect. 6.5 for the purpose of constructing the nonhomogeneous Koszul duality DG-functors which we are really interested in. But the CDG-module $K^\vee_{e'}(B, \widetilde{A})$ itself, even when it happens to be a complex, has no good exactness properties for a left finitely projective nonhomogeneous Koszul ring $\widetilde{A}$ in general (see Examples 2.12 and Remark 2.40). However, applying the functor $\mathrm{Hom}_{\widetilde{A}^{\mathrm{op}}}(-, \widetilde{A})$ to $K^\vee_{e'}(B, \widetilde{A})$ produces a right CDG-module $\mathrm{Hom}_{\widetilde{A}^{\mathrm{op}}}(K^\vee_{e'}(B, \widetilde{A}), \widetilde{A})$ over (B, d, h) (in fact, a CDG-bimodule over $(\widetilde{A}, 0, 0)$ and (B, d, h)), whose underlying graded $\widetilde{A}$-B-bimodule is $\widetilde{A} \otimes_R \mathrm{Hom}_{R^{\mathrm{op}}}(B, R)$. This CDG-module, which is a nonhomogeneous generalization of the first Koszul complex $K^\tau_\bullet(B, A)$ from Sect. 2.5, will appear below in the proof of Theorem 6.14 and the subsequent discussion in Sect. 6.6.

6.3   Semicoderived Category of Modules

Let R be an associative ring. The following definition of the *coderived category of right R-modules* goes back to [58, Section 2.1] and [59, Section 3.3] (see [40, Section 2] and [4, Proposition 1.3.6(2)] for an alternative approach).

Definition 6.7 As a particular case of the notation of Sect. 6.1, we denote by $\mathsf{Hot}(\mathsf{mod}{-}R)$ the homotopy category of (unbounded complexes of infinitely generated) right R-modules. Let us consider short exact sequences of complexes of right R-modules $0 \longrightarrow L^\bullet \longrightarrow M^\bullet \longrightarrow N^\bullet \longrightarrow 0$. Any such short exact sequence can be viewed as a bicomplex with three rows, and the totalization (the total complex) $\mathrm{Tot}(L^\bullet \to M^\bullet \to N^\bullet)$ of such a bicomplex can be constructed in the usual way. A complex of right R-modules is said to be *coacyclic* if it belongs to the minimal full triangulated subcategory of $\mathsf{Hot}(\mathsf{mod}{-}R)$ containing all the totalizations of short exact sequences of complexes of right R-modules and closed under infinite direct sums.

The full triangulated subcategory of coacyclic complexes of right R-modules is denoted by $\mathsf{Acycl}^{\mathsf{co}}(\mathsf{mod}{-}R) \subset \mathsf{Hot}(\mathsf{mod}{-}R)$. The *coderived category of right R-modules* is the triangulated Verdier quotient category of the homotopy category $\mathsf{Hot}(\mathsf{mod}{-}R)$ by the triangulated subcategory of coacyclic complexes,

$$\mathsf{D}^{\mathsf{co}}(\mathsf{mod}{-}R) = \mathsf{Hot}(\mathsf{mod}{-}R)/\mathsf{Acycl}^{\mathsf{co}}(\mathsf{mod}{-}R).$$

Obviously, any coacyclic complex of R-modules is acyclic, $\mathsf{Acycl}^{\mathsf{co}}(\mathsf{mod}{-}R) \subset \mathsf{Acycl}(\mathsf{mod}{-}R)$; but the converse does not hold in general [59, Examples 3.3]. So the conventional derived category $\mathsf{D}(\mathsf{mod}{-}R)$ is a quotient category of $\mathsf{D}^{\mathsf{co}}(\mathsf{mod}{-}R)$. All acyclic complexes of right R-modules are coacyclic when the right homological dimension of the ring R is finite [58, Remark 2.1].

The following lemma is useful.

Lemma 6.8 *Let* $0 = F_{-1}M^\bullet \subset F_0M^\bullet \subset F_1M^\bullet \subset \cdots \subset M^\bullet$ *be a complex of R-modules endowed with an exhaustive increasing filtration by subcomplexes of R-modules; so* $M^\bullet = \bigcup_{n\geq 0} F_n M^\bullet$. *Assume that, for every* $n \geq 0$, *the complex of R-modules* $F_n M^\bullet/F_{n-1}M^\bullet$ *is coacyclic. Then the complex of R-modules* $M^\bullet$ *is coacyclic, too.*

Proof First one proves by induction in n that the complex of R-modules $F_n M^\bullet$ is coacyclic for every $n \geq 0$, using the fact that the totalizations of short exact sequences $0 \longrightarrow F_{n-1}M^\bullet \longrightarrow F_n M^\bullet \longrightarrow F_n M^\bullet/F_{n-1}M^\bullet \longrightarrow 0$ are coacyclic. Then one concludes that the complex of R-modules $\bigoplus_{n=0}^{\infty} F_n M^\bullet$ is coacyclic. Finally, one deduces coacyclicity of the complex $M^\bullet$ from the fact that the totalization of the telescope short exact sequence $0 \longrightarrow \bigoplus_{n=0}^{\infty} F_n M^\bullet \longrightarrow \bigoplus_{n=0}^{\infty} F_n M^\bullet \longrightarrow M^\bullet \longrightarrow 0$ is a coacyclic complex of R-modules. □

Now let $R \longrightarrow A$ be a homomorphism of associative rings. The following definition of the *A/R-semicoderived category of right A-modules* can be found in [63, Section 5].

Definition 6.9 The semicoderived category

$$\mathsf{D}_R^{\mathsf{sico}}(\mathsf{mod}-A) = \mathsf{Hot}(\mathsf{mod}-A)/\mathsf{Acycl}_R^{\mathsf{sico}}(\mathsf{mod}-A)$$

is the triangulated Verdier quotient category of the homotopy category of right A-modules by its full triangulated subcategory $\mathsf{Acycl}_R^{\mathsf{sico}}(\mathsf{mod}-A)$ consisting of all the complexes of right A-modules that are *coacyclic as complexes of right R-modules*.

Obviously, any coacyclic complex of A-modules is coacyclic as a complex of R-modules; but the converse does not need to be true. So the semicoderived category is intermediate between the derived and the coderived category of A-modules: there are natural triangulated Verdier quotient functors $\mathsf{D}^{\mathsf{co}}(\mathsf{mod}-A) \longrightarrow \mathsf{D}_R^{\mathsf{sico}}(\mathsf{mod}-A) \longrightarrow \mathsf{D}(\mathsf{mod}-A)$. (We refer to the paper [66] for a more general discussion of such intermediate triangulated quotient categories, called *pseudo-coderived categories* in [66].) The semicoderived category $\mathsf{D}_R^{\mathsf{sico}}(\mathsf{mod}-A)$ can be thought of as a mixture of the coderived category "along the variables from R" and the conventional derived category "in the direction of A relative to R."

For example, taking $R = \mathbb{Z}$, one obtains the conventional derived category of A-modules

$$\mathsf{D}_{\mathbb{Z}}^{\mathsf{sico}}(\mathsf{mod}-A) = \mathsf{D}(\mathsf{mod}-A)$$

(because all the acyclic complexes of abelian groups are coacyclic), while taking $R = A$ one obtains the coderived category of A-modules

$$\mathsf{D}_A^{\mathsf{sico}}(\mathsf{mod}-A) = \mathsf{D}^{\mathsf{co}}(\mathsf{mod}-A).$$

6.4 Coderived Category of CDG-Comodules

Let $B = (B, d, h)$ be a nonnegatively graded CDG-ring (as defined in Sect. 3.2). So $B = \bigoplus_{n=0}^{\infty} B^n$ is a nonnegatively graded associative ring, $d: B \longrightarrow B$ is an odd derivation of degree 1, and $h \in B^2$ is a curvature element for d.

Definition 6.10 Let (N, d_N) be a right CDG-module over (B, d, h) (as defined in Sect. 6.1). So $N = \bigoplus_{n \in \mathbb{Z}} N^n$ is a graded right B-module, $d_N: N \longrightarrow N$ is an odd derivation of degree 1 compatible with the derivation d on B, and the equation $d_N^2(y) = -yh$ is satisfied for all $y \in N$. We will say that the CDG-module (N, d_N) is a *CDG-comodule* (or a *right CDG-comodule*) over (B, d, h) if the graded right B-module N is a graded right B-comodule (in the sense of Sect. 5.5).

In other words, a right CDG-comodule N over (B, d, h) is a graded right B-comodule endowed with an odd coderivation $d_N \colon N \longrightarrow N$ of degree 1 compatible with the derivation d on B and satisfying the above equation for d_N^2. Here by an *odd coderivation* of a graded B-comodule N (compatible with a given odd derivation d on B) we mean an odd derivation of N as a graded B-module. Equivalently, a right CDG-comodule over (B, d, h) can be simply defined as a graded right $\widehat{B}$-comodule (where $(\widehat{B}, \partial)$ is the quasi-differential graded ring corresponding to (B, d, h), as in Theorem 4.7 and Sect. 6.1).

The DG-category $\mathsf{DG}(\mathsf{mod}{-}B)$ of right CDG-modules over $B = (B, d, h)$ was defined in Sect. 6.1. The *DG-category of right CDG-comodules* $\mathsf{DG}(\mathsf{comod}{-}B)$ *over* (B,d,h) is defined as the full DG-subcategory of the DG-category $\mathsf{DG}(\mathsf{mod}{-}B)$ whose objects are the right CDG-comodules over (B, d, h). Obviously, the full DG-subcategory $\mathsf{DG}(\mathsf{comod}{-}B) \subset \mathsf{DG}(\mathsf{mod}{-}B)$ is closed under shifts, twists, and infinite direct sums; in particular, it is closed under the cones of closed morphisms (see [59, Section 1.2] for the terminology).

Let $\mathsf{Hot}(\mathsf{comod}{-}B) = H^0\mathsf{DG}(\mathsf{mod}{-}B)$ denote the homotopy category (of the DG-category) of right CDG-comodules over B. So $\mathsf{Hot}(\mathsf{comod}{-}B)$ is a full triangulated subcategory in $\mathsf{Hot}(\mathsf{mod}{-}B)$. The following definition of the *coderived category of right CDG-comodules over B* is a variation on the definitions in [58, Sections 11.7.1–2] and [59, Sections 3.3 and 4.2].

Definition 6.11 Let us consider short exact sequences $0 \longrightarrow L \overset{i}{\longrightarrow} M \overset{p}{\longrightarrow} N \longrightarrow 0$ of right CDG-comodules over (B, d, h). Here i and p are closed morphisms of degree zero in the DG-category $\mathsf{DG}(\mathsf{comod}{-}B)$ and the short sequence $0 \longrightarrow L \longrightarrow M \longrightarrow N \longrightarrow 0$ is exact in the abelian category $\mathsf{comod}_{\mathsf{gr}}{-}B$. Using the construction of the cone of a closed morphism (or the shift, finite direct sum, and twist) in the DG-category $\mathsf{DG}(\mathsf{comod}{-}B)$, one can produce the totalization (or the *total CDG-comodule*) $\mathrm{Tot}(L \to M \to N)$ of the finite complex of CDG-comodules $L \longrightarrow M \longrightarrow N$.

A right CDG-comodule over B is said to be *coacyclic* if it belongs to the minimal full triangulated subcategory of $\mathsf{Hot}(\mathsf{comod}{-}B)$ containing all the totalizations of short exact sequences of right CDG-comodules over B and closed under infinite direct sums. The full triangulated subcategory of coacyclic CDG-comodules is denoted by $\mathsf{Acycl}^{\mathsf{co}}(\mathsf{comod}{-}B) \subset \mathsf{Hot}(\mathsf{comod}{-}B)$. The *coderived category of right CDG-comodules over B* is the triangulated Verdier quotient category of the homotopy category $\mathsf{Hot}(\mathsf{comod}{-}B)$ by its triangulated subcategory of coacyclic CDG-comodules,

$$\mathsf{D}^{\mathsf{co}}(\mathsf{comod}{-}B) = \mathsf{Hot}(\mathsf{comod}{-}B)/\mathsf{Acycl}^{\mathsf{co}}(\mathsf{comod}{-}B).$$

Notice that, generally speaking, the differential d_N on a CDG-(co)module (N, d_N) has nonzero square (if B is not a DG-ring, that is $h \neq 0$ in B). So CDG-(co)modules are *not complexes* and their cohomology (co)modules are *undefined*. For this reason, one *cannot* speak about "acyclic CDG-(co)modules," nor about "quasi-isomorphisms of CDG-(co)modules" in the usual sense of the word "quasi-isomorphism." Therefore, the conventional construction of the derived category of DG-modules is *not* applicable to curved DG-modules or curved DG-comodules. There is only the coderived category and its variations (known as *derived categories of the second kind*).

Lemma 6.12 *Let $B = (B, d, h)$ be a nonnegatively graded CDG-ring and $N = (N, d_N)$ be a right CDG-comodule over B. Let $0 = F_{-1}N \subset F_0N \subset F_1N \subset \cdots \subset N$ be an exhaustive increasing filtration of N by CDG-subcomodules, that is $(F_nN)B \subset F_nN$, $d_N(F_nN) \subset F_nN$, and $N = \bigcup_{n=0}^{\infty} F_nN$. Assume that, for every $n \geq 0$, the CDG-comodule $F_nN/F_{n-1}N$ is coacyclic over B. Then the CDG-comodule N is coacyclic, too.*

Proof This is a generalization of Lemma 6.8, provable by the same argument.   □

There is an important particular case when the nonnegatively graded ring B has only finitely many grading components, that is $B^n = 0$ for $n \gg 0$. In this case, all the (ungraded or graded) B-modules are (respectively, ungraded or graded) B-comodules, so $\mathsf{comod_{gr}}-B = \mathsf{mod_{gr}}-B$. Accordingly, the full DG-subcategory of (right) CDG-comodules over B coincides with the whole ambient DG-category of (right) CDG-modules, $\mathsf{DG}(\mathsf{comod}-B) = \mathsf{DG}(\mathsf{mod}-B)$. In this case, we will write simply $\mathsf{D^{co}}(\mathsf{mod}-B)$ instead of $\mathsf{D^{co}}(\mathsf{comod}-B)$.

6.5 Koszul Duality Functors for Modules and Comodules

Let $R \subset \widetilde{V} \subset \widetilde{A}$ be a 3-left finitely projective weak nonhomogeneous quadratic ring, and let $B = (B, d, h)$ be the nonhomogeneous quadratic qual CDG-ring. The dual nonhomogeneous Koszul CDG-module $K^{\vee}(B, \widetilde{A}) = K_{e'}^{\vee}(B, \widetilde{A}) = (B \otimes_R \widetilde{A}, d_{e'})$ constructed in Sect. 6.2 plays a key role below. We recall from the discussion in Sect. 6.2 that $K^{\vee}(B, \widetilde{A})$ is a CDG-bimodule over the CDG-rings $B = (B, d, h)$ and $\widetilde{A} = (\widetilde{A}, 0, 0)$.

According to Sect. 6.1, we have the induced pair of adjoint DG-functors between the DG-category $\mathsf{DG}(\mathsf{mod}-\widetilde{A})$ of complexes of right $\widetilde{A}$-modules and the DG-category $\mathsf{DG}(\mathsf{mod}-B)$ of right CDG-modules over (B, d, h). Here the left adjoint DG-functor

$$- \otimes_B K^{\vee}(B, \widetilde{A}) : \mathsf{DG}(\mathsf{mod}-B) \longrightarrow \mathsf{DG}(\mathsf{mod}-\widetilde{A})$$

takes a right CDG-module $N = (N, d_N)$ over (B, d, h) to the complex of right $\widetilde{A}$-modules whose underlying graded $\widetilde{A}$-module is

$$N \otimes_B K^\vee(B, \widetilde{A}) = N \otimes_B (B \otimes_R \widetilde{A}) = N \otimes_R \widetilde{A}.$$

The right adjoint DG-functor

$$\mathrm{Hom}_{\widetilde{A}^{\mathrm{op}}}(K^\vee(B, \widetilde{A}), -) \colon \mathsf{DG}(\mathsf{mod}{-}\widetilde{A}) \longrightarrow \mathsf{DG}(\mathsf{mod}{-}B)$$

takes a complex of right $\widetilde{A}$-modules $M = M^\bullet$ to the right CDG-module over (B, d, h) whose underlying graded B-module is

$$\mathrm{Hom}_{\widetilde{A}^{\mathrm{op}}}(K^\vee(B, \widetilde{A}), M) = \mathrm{Hom}_{\widetilde{A}^{\mathrm{op}}}(B \otimes_R \widetilde{A}, \ M) = \mathrm{Hom}_{R^{\mathrm{op}}}(B, M).$$

Our aim is to restrict this adjoint pair to the DG-subcategory $\mathsf{DG}(\mathsf{comod}{-}B) \subset \mathsf{DG}(\mathsf{mod}{-}B)$ of CDG-comodules over (B, d, h), as defined in Sects. 5.5 and 6.4.

We recall the notation $\mathrm{Hom}_{R^{\mathrm{op}}}^\Sigma(L, M) \subset \mathrm{Hom}_{R^{\mathrm{op}}}(L, M)$ for the direct sum totalization of the bigraded Hom group of two graded right R-modules L and M, which was introduced in Sect. 5.7. Let $\Gamma_B \colon \mathsf{mod}_{\mathrm{gr}}{-}B \longrightarrow \mathsf{comod}_{\mathrm{gr}}{-}B$ denote the functor assigning to every graded right B-module its maximal (graded) submodule which is a (graded) right B-comodule (see the proof of Theorem 5.37).

Lemma 6.13 *Let* $B = \bigoplus_{n=0}^\infty B_n$ *be a nonnegatively graded ring with the degree-zero component* $R = B_0$. *Then for any graded right R-module M one has*

$$\Gamma_B(\mathrm{Hom}_{R^{\mathrm{op}}}(B, M)) = \mathrm{Hom}_{R^{\mathrm{op}}}^\Sigma(B, M).$$

Proof We have $\mathrm{Hom}_{R^{\mathrm{op}}}^\Sigma(B, M) = \bigoplus_{n \in \mathbb{Z}} \mathrm{Hom}_{R^{\mathrm{op}}}(B, M_n(n))$, where $M \longmapsto M(n)$ denotes the functor of grading shift. Since the nonpositively graded right B-module $\mathrm{Hom}_{R^{\mathrm{op}}}(B, L)$ is a graded right B-comodule for any ungraded right R-module L, and the full subcategory of graded right B-comodules $\mathsf{comod}_{\mathrm{gr}}{-}B$ is closed under direct sums in $\mathsf{mod}_{\mathrm{gr}}{-}B$, it follows that $\mathrm{Hom}_{R^{\mathrm{op}}}^\Sigma(B, M)$ is a graded right B-comodule. Hence $\mathrm{Hom}_{R^{\mathrm{op}}}^\Sigma(B, M) \subset \Gamma_B(\mathrm{Hom}_{R^{\mathrm{op}}}(B, M))$. Conversely, let $f \in \mathrm{Hom}_{R^{\mathrm{op}}}(B, M)_n$ be a homogeneous element of degree n such that $f B_{\geq m} = 0$ for some $m \geq 1$. Then for every $b \in B_{\geq m}$ we have $f(b) = f(b \cdot 1) = (fb)(1) = 0$, where $1 \in R \subset B$ is the unit element. It follows immediately that $f \in \mathrm{Hom}_{R^{\mathrm{op}}}^\Sigma(B, M)_n \subset \mathrm{Hom}_{R^{\mathrm{op}}}(B, M)_n$. $\qquad\square$

For any complex of right A-modules $M^\bullet$, the graded right B-submodule

$$\mathrm{Hom}_{\widetilde{A}^{\mathrm{op}}}^\Sigma(K^\vee(B, \widetilde{A}), M^\bullet) = \mathrm{Hom}_{R^{\mathrm{op}}}^\Sigma(B, M^\bullet) \subset \mathrm{Hom}_{R^{\mathrm{op}}}(B, M^\bullet)$$

is preserved by the differential of the CDG-module $\mathrm{Hom}_{\widetilde{A}^{\mathrm{op}}}(K^{\vee}(B, \widetilde{A}), M^{\bullet}) = \mathrm{Hom}_{R^{\mathrm{op}}}(B, M^{\bullet})$ over (B, d, h) (so it is a CDG-submodule). In view of the above discussion and Lemma 6.13, we obtain a pair of adjoint DG-functors between the DG-category $\mathsf{DG}(\mathsf{mod}{-}\widetilde{A})$ of complexes of right $\widetilde{A}$-modules and the DG-category $\mathsf{DG}(\mathsf{comod}{-}B)$ of right CDG-comodules over (B, d, h).

Here the left adjoint DG-functor

$$-\otimes_B K^{\vee}(B, \widetilde{A}) \colon \mathsf{DG}(\mathsf{comod}{-}B) \longrightarrow \mathsf{DG}(\mathsf{mod}{-}\widetilde{A})$$

is simply the restriction of the above DG-functor to the full DG-subcategory $\mathsf{DG}(\mathsf{comod}{-}B) \subset \mathsf{DG}(\mathsf{mod}{-}B)$. We denote it by

$$N \longmapsto N\otimes_B K^{\vee}(B, \widetilde{A}) = N\otimes_R^{\tau'} \widetilde{A},$$

where the placeholder τ' is understood as a notation for the injective left R-module map $\mathrm{Hom}_{R^{\mathrm{op}}}(B^1, R) = F_1\widetilde{A}/F_0\widetilde{A} = V \simeq V' \hookrightarrow \widetilde{V} = F_1\widetilde{A}$.

The right adjoint DG-functor

$$\mathrm{Hom}_{\widetilde{A}^{\mathrm{op}}}^{\Sigma}(K^{\vee}(B, \widetilde{A}), -) \colon \mathsf{DG}(\mathsf{mod}{-}\widetilde{A}) \longrightarrow \mathsf{DG}(\mathsf{comod}{-}B)$$

takes a complex of right $\widetilde{A}$-modules $M = M^{\bullet}$ to the right CDG-comodule over (B, d, h) whose underlying graded right B-comodule is

$$\mathrm{Hom}_{\widetilde{A}^{\mathrm{op}}}^{\Sigma}(B\otimes_R \widetilde{A}, \; M) = \mathrm{Hom}_{R^{\mathrm{op}}}^{\Sigma}(B, M) = M\otimes_R \mathrm{Hom}_{R^{\mathrm{op}}}(B, R) = M\otimes_R C,$$

where C is a notation for the graded R-B-bimodule (and a graded right B-comodule) $C = \mathrm{Hom}_{R^{\mathrm{op}}}(B, R)$. We denote this right CDG-comodule over (B, d, h) by

$$\mathrm{Hom}_{\widetilde{A}^{\mathrm{op}}}^{\Sigma}(K^{\vee}(B, \widetilde{A}), M^{\bullet}) = M^{\bullet}\otimes_R^{\sigma'} \mathrm{Hom}_{R^{\mathrm{op}}}(B, R) = M^{\bullet}\otimes_R^{\sigma'} C,$$

where σ' is, for the time being, just a placeholder. The purpose of this placeholder is to remind us of the nontrivial differential $d_{e'}$ on $K^{\vee}(B, \widetilde{A})$, which is incorporated into the construction of the differential on the CDG-comodule $M^{\bullet}\otimes_R^{\sigma'} C$ (cf. the twisting cochain notation in the simpler context of [59, Section 6]). See (8.9) and Sect. 8.7 below for our suggested interpretation of σ'.

The pair of adjoint DG-functors that we have constructed induces a pair of adjoint triangulated functors

$$-\otimes_B K^{\vee}(B, \widetilde{A}) = -\otimes_R^{\tau'} \widetilde{A} \colon \mathsf{Hot}(\mathsf{comod}{-}B) \longrightarrow \mathsf{Hot}(\mathsf{mod}{-}\widetilde{A})$$

and

$$\mathrm{Hom}_{\tilde{A}^{\mathrm{op}}}^{\Sigma}(K^{\vee}(B, \tilde{A}), -) = -\otimes_R^{\sigma'} C \colon \mathsf{Hot}(\mathsf{mod}{-}\tilde{A}) \longrightarrow \mathsf{Hot}(\mathsf{comod}{-}B)$$

between the homotopy category $\mathsf{Hot}(\mathsf{mod}{-}\tilde{A})$ of complexes of right $\tilde{A}$-modules and the homotopy category $\mathsf{Hot}(\mathsf{comod}{-}B)$ of right CDG-comodules over (B, d, h).

6.6 Triangulated Equivalence

Let $R \subset \tilde{V} \subset \tilde{A}$ be a left finitely projective nonhomogeneous Koszul ring (as defined in Sect. 4.6), and let (B, d, h) be the corresponding right finitely projective Koszul CDG-ring (as constructed in Proposition 3.16; see also Corollary 4.26). The following theorem is the main result of Chap. 6.

Theorem 6.14 *The pair of adjoint triangulated functors* $-\otimes_R^{\sigma'} C \colon \mathsf{Hot}(\mathsf{mod}{-}\tilde{A}) \longrightarrow \mathsf{Hot}(\mathsf{comod}{-}B)$ *and* $-\otimes_R^{\tau'} \tilde{A} \colon \mathsf{Hot}(\mathsf{comod}{-}B) \longrightarrow \mathsf{Hot}(\mathsf{mod}{-}\tilde{A})$ *defined in Sect. 6.5 induces a pair of adjoint triangulated functors between the* $\tilde{A}/R$*-semiderived category of right* $\tilde{A}$*-modules* $\mathsf{D}_R^{\mathsf{sico}}(\mathsf{mod}{-}\tilde{A})$*, as defined in Sect. 6.3, and the coderived category* $\mathsf{D}^{\mathsf{co}}(\mathsf{comod}{-}B)$ *of right CDG-comodules over* (B, d, h)*, as defined in Sect. 6.4. Under the above Koszulity assumption, the latter two functors are mutually inverse triangulated equivalences,*

$$\mathsf{D}_R^{\mathsf{sico}}(\mathsf{mod}{-}\tilde{A}) \simeq \mathsf{D}^{\mathsf{co}}(\mathsf{comod}{-}B).$$

Concrete examples of the triangulated equivalence stated in the theorem will be presented below in Sects. 10.2–10.10. The theorem is also applicable in the situations described in Examples 3.17 and 3.22.

Proof This is essentially a generalization of [59, Theorem B.2(a)] and a particular case of [58, Theorem 11.8(b)]. The proof is similar. Let us spell out some details.

Recall from the proof of Theorem 5.37 that a graded right B-comodule N is said to be *trivial* if $N B^{\geq 1} = 0$ (where the notation is $B^{\geq m} = \bigoplus_{n \geq m} B^n \subset B$ for any integer $m \geq 0$). A right CDG-comodule (N, d_N) over (B, d, h) is said to be *trivial* if its underlying graded B-comodule is trivial. The differential d_N on a trivial CDG-comodule N squares to zero, as the curvature element h acts by zero in N. The DG-category of trivial right CDG-comodules over (B, d, h) is equivalent to the DG-category of complexes of right R-modules.

The ring $\tilde{A}$ is endowed with an increasing filtration F. Let us also introduce an increasing filtration F on the graded R-B-bimodule $C = \mathrm{Hom}_{R^{\mathrm{op}}}(B, R)$. We put $F_n C = \bigoplus_{i \leq n} \mathrm{Hom}_{R^{\mathrm{op}}}(B^i, R) \subset C$ for every $n \in \mathbb{Z}$. So $F_{-1} C = 0$ and $C = \bigcup_{n \geq 0} F_n C$. Clearly, $F_n C$ is a graded R-B-subbimodule in C. Notice that both the left R-modules $F_n \tilde{A}/F_{n-1}\tilde{A}$

and the graded left R-modules $F_n C / F_{n-1} C$ are finitely generated and projective (it is important for the argument below that they are flat).

Let $M^\bullet$ be a complex of right $\widetilde{A}$-modules that is coacyclic as a complex of right R-modules. We need to show that the right CDG-comodule $M^\bullet \otimes_R^{\sigma'} C$ is coacyclic over (B, d, h). Let F be the increasing filtration on the tensor product $M \otimes_R C$ induced by the increasing filtration F on C, that is $F_n(M \otimes_R C) = M \otimes_R F_n C$. Then F is a filtration of the CDG-comodule $M^\bullet \otimes_R^{\sigma'} C$ by CDG-subcomodules over (B, d, h). The successive quotient CDG-comodules $(M^\bullet \otimes_R^{\sigma'} F_n C) / (M^\bullet \otimes_R^{\sigma'} F_{n-1} C)$ are trivial. Viewed as complexes of right R-comodules, these are the tensor products $M^\bullet \otimes_R (F_n C / F_{n-1} C)$, where $F_n C / F_{n-1} C$ is a graded R-R-bimodule concentrated in the cohomological degree $-n$ and endowed with the zero differential.

Since the graded left R-module $F_n C / F_{n-1} C$ is flat, tensoring with it preserves short exact sequences of graded right R-modules. The tensor product functor also preserves infinite direct sums. It follows that tensoring with the graded R-R-bimodule $F_n C / F_{n-1} C$ takes coacyclic complexes of right R-modules to coacyclic complexes of right R-modules. Clearly, any coacyclic complex of right R-modules is also coacyclic as a trivial CDG-comodule over (B, d, h). It remains to make use of Lemma 6.12 in order to conclude that the right CDG-comodule $M^\bullet \otimes_R^{\sigma'} C$ over (B, d, h) is coacyclic.

For any coacyclic right CDG-comodule N over (B, d, h), the complex of right $\widetilde{A}$-modules $N \otimes_R^{\tau'} \widetilde{A}$ is coacyclic not only as a complex of right R-modules, but even as a complex of right $\widetilde{A}$-modules. This follows immediately from the fact that $\widetilde{A}$ is a flat left R-module; so the functor $- \otimes_R^{\tau'} \widetilde{A}$ takes short exact sequences of right CDG-comodules over (B, d, h) to short exact sequences of complexes of right $\widetilde{A}$-modules (and this functor also preserves infinite direct sums).

Let $M^\bullet$ be arbitrary complex of right $\widetilde{A}$-modules. We need to show the cone $Y^\bullet$ of the natural closed morphism of complexes of right $\widetilde{A}$-modules (the adjunction morphism) $M^\bullet \otimes_R^{\sigma'} C \otimes_R^{\tau'} \widetilde{A} \longrightarrow M^\bullet$ is coacyclic as a complex of right R-modules. Endow the graded right R-module $M \otimes_R C \otimes_R \widetilde{A}$ with the increasing filtration F induced by the increasing filtrations F on C and $\widetilde{A}$, that is $F_n(M \otimes_R C \otimes_R \widetilde{A}) = \sum_{i+j=n} M \otimes_R F_i C \otimes_R F_j \widetilde{A} \subset M \otimes_R C \otimes_R \widetilde{A}$. Then F is an exhaustive filtration of $M^\bullet \otimes_R^{\sigma'} C \otimes_R^{\tau'} \widetilde{A}$ by subcomplexes of right R-modules (but not of right $\widetilde{A}$-modules!). The complex $M^\bullet$ is endowed with the trivial filtration $F_{-1} M^\bullet = 0$, $F_0 M^\bullet = M^\bullet$; and the complex $Y^\bullet$ is endowed with the induced increasing filtration F.

Then the associated graded complex of right R-modules

$$\mathrm{gr}^F(M^\bullet \otimes_R^{\sigma'} C \otimes_R^{\tau'} \widetilde{A}) = \bigoplus_{n=0}^{\infty} F_n(M^\bullet \otimes_R^{\sigma'} C \otimes_R^{\tau'} \widetilde{A}) / F_{n-1}(M^\bullet \otimes_R^{\sigma'} C \otimes_R^{\tau'} \widetilde{A})$$

is naturally isomorphic to the tensor product $M^\bullet \otimes_R {}^\tau K_\bullet(B, A)$ of the complex of right R-modules $M^\bullet$ and the complex of (graded) R-R-bimodules ${}^\tau K_\bullet(B, A) = C \otimes_R^\tau A$ constructed in Sect. 2.6 (where it is called "the second Koszul complex"). Here, as usually, $A = \mathrm{gr}^F \widetilde{A}$ is the left finitely projective Koszul graded ring associated with $\widetilde{A}$.

By Proposition 2.33 or Theorem 2.35(e), every internal degree $n \geq 1$ component of the complex $^\tau K_\bullet(B, A)$ is a finite acyclic complex of R-R-bimodules whose terms are finitely generated and projective left R-modules. The internal degree $n = 0$ component of the complex $^\tau K_\bullet(B, A)$ is isomorphic to the R-R-bimodule R.

It follows that the associated graded complex $\mathrm{gr}^F Y^\bullet = \bigoplus_{n \geq 0} F_n Y^\bullet / F_{n-1} Y^\bullet$ is isomorphic to the tensor product $M^\bullet \otimes_R \mathrm{cone}(^\tau K_\bullet(B, A) \to R)$ of the complex of right R-modules $M^\bullet$ and the cone of the natural morphism of complexes of (graded) R-R-bimodules $^\tau K_\bullet(B, A) \longrightarrow R$. For every $n \in \mathbb{Z}$, the internal degree n component of the complex $\mathrm{cone}(^\tau K_\bullet(B, A) \to R)$ is a finite acyclic complex of R-R-bimodules which are finitely generated and projective as left R-modules. Consequently, the complex $\mathrm{cone}(^\tau K_\bullet(B, A) \to R)$ is "coacyclic with respect to the exact category of left R-flat (or even left R-projective) R-R-bimodules" in the sense of [58, Section 2.1]. Since the functor $M^\bullet \otimes_R -$ preserves exactness of short exact sequences of left R-flat R-R-bimodules, we can conclude that $\mathrm{gr}^F Y^\bullet \simeq M^\bullet \otimes_R \mathrm{cone}(^\tau K_\bullet(B, A) \to R)$ is a coacyclic complex of right R-modules. By Lemma 6.8, it follows that $Y^\bullet$ is a coacyclic complex of right R-modules, as desired.

Let $N = (N, d_N)$ be an arbitrary right CDG-comodule over (B, d, h). We need to show that the cone Z of the natural closed morphism of right CDG-comodules (the adjunction morphism) $N \longrightarrow N \otimes_R^{\tau'} \widetilde{A} \otimes_R^{\sigma'} C$ is coacyclic as a CDG-comodule over (B, d, h). Endow the graded right B-comodule N with the canonical increasing filtration by graded subcomodules $0 = G_{-1} N \subset G_0 N \subset G_1 N \subset G_2 N \subset \cdots \subset N$ as in the proof of Theorem 5.37: the grading component $G_m N^i \subset N^i$ consists of all the elements $y \in N^i$ such that $y B^{\geq m+1} = 0$ in N. We have $d_N(G_m N) \subset G_m N$, as $d_N(y)b = d_N(yb) - (-1)^{|y|} y d(b) = 0$ for all $y \in G_m N^{|y|}$ and $b \in B^{\geq m+1}$. So G is an exhaustive increasing filtration of N by CDG-subcomodules over (B, d, h).

Consider the induced increasing filtrations G of the right CDG-comodule $N \otimes_R^{\tau'} \widetilde{A} \otimes_R^{\sigma'} C$ and the right CDG-comodule Z over (B, d, h); so $G_m(N \otimes_R^{\tau'} \widetilde{A} \otimes_R^{\sigma'} C) = (G_m N) \otimes_R^{\tau'} \widetilde{A} \otimes_R^{\sigma'} C$ and $G_m Z = \mathrm{cone}(G_m N \to G_m N \otimes_R^{\tau'} \widetilde{A} \otimes_R^{\sigma'} C)$. By Lemma 6.12, it suffices to check that the CDG-comodules $G_m Z / G_{m-1} Z$ are coacyclic over (B, d, h) for all $m \geq 0$. Then, since $\widetilde{A}$ and C are flat (graded) left R-modules, the CDG-comodule $G_m Z / G_{m-1} Z$ is naturally isomorphic to the cone of the adjunction morphism $G_m N / G_{m-1} N \longrightarrow (G_m N / G_{m-1} N) \otimes_R^{\tau'} \widetilde{A} \otimes_R^{\sigma'} C$.

The successive quotient CDG-comodules $G_m N / G_{m-1} N$ are annihilated by $B^{\geq 1}$; so they are trivial right CDG-comodules over (B, d, h) or, which is the same, just complexes of right R-modules. Thus we have reduced our problem to the case of a trivial right CDG-comodule N, and we can assume that $N = N^\bullet$ is simply a complex of right R-modules with $B^{\geq 1}$ acting by zero. Then the right CDG-comodule $N \otimes_R^{\tau'} \widetilde{A} \otimes_R^{\sigma'} C = N^\bullet \otimes_R \widetilde{A} \otimes_R^{\sigma'} C$ over (B, d, h) is simply the tensor product of the complex of right R-modules $N^\bullet$ and the right CDG-comodule $\widetilde{A} \otimes_R^{\sigma'} C = \mathrm{Hom}_{\widetilde{A}^{\mathrm{op}}}(K_e^\vee(B, \widetilde{A}), \widetilde{A})$ (which is, in fact, a CDG-bimodule over $\widetilde{A} = (\widetilde{A}, 0, 0)$ and $B = (B, d, h)$, hence, in particular, a CDG-bimodule over $R = (R, 0, 0)$ and B).

Endow the graded right R-module $N \otimes_R \widetilde{A} \otimes_R C$ with the increasing filtration F induced by the increasing filtrations F on $\widetilde{A}$ and C, that is $F_n(N \otimes_R \widetilde{A} \otimes_R C) = \sum_{i+j=n} N \otimes_R F_j \widetilde{A} \otimes_R F_i C \subset N \otimes_R \widetilde{A} \otimes_R C$. Then F is an exhaustive filtration of $N^\bullet \otimes_R \widetilde{A} \otimes_R^{\sigma'} C$ by CDG-subcomodules over (B, d, h). Let the complex $N^\bullet$ be endowed with the trivial filtration $F_{-1} N^\bullet = 0$, $F_0 N^\bullet = N^\bullet$, and let the CDG-comodule $Z = \mathrm{cone}(N^\bullet \to N^\bullet \otimes_R \widetilde{A} \otimes_R^{\sigma'} C)$ be endowed with the induced filtration.

Then the associated graded CDG-comodule

$$\mathrm{gr}^F(N^\bullet \otimes_R \widetilde{A} \otimes_R^{\sigma'} C) = \bigoplus_{n=0}^\infty F_n(N^\bullet \otimes_R \widetilde{A} \otimes_R^{\sigma'} C) / F_{n-1}(N^\bullet \otimes_R \widetilde{A} \otimes_R^{\sigma'} C)$$

is annihilated by the action of $B^{\geq 1}$; so it is a trivial right CDG-comodule over (B, d, h). As a complex of right R-modules, it is naturally isomorphic to the tensor product $N^\bullet \otimes_R K^\tau_\bullet(B, A)$ of the complex of right R-modules $N^\bullet$ and the complex of (graded) R-R-bimodules $K^\tau_\bullet(B, A) = A \otimes_R^\sigma C$ constructed in Sect. 2.5 (where it is called "the first Koszul complex"). Here, once again, the superindex σ is just a placeholder for the time being (see Sect. 8.7 below for an explanation). By Proposition 2.32 or Theorem 2.35(d), every internal degree $n \geq 1$ component of the complex $K^\tau_\bullet(B, A)$ is a finite acyclic complex of R-R-bimodules with finitely generated projective underlying left R-modules (while the internal degree $n = 0$ component is the R-R-bimodule R).

It follows that the associated graded CDG-comodule $\mathrm{gr}^F Z = \bigoplus_{n \geq 0} F_n Z / F_{n-1} Z$ is a trivial right CDG-comodule over (B, d, h) which, as a complex of right R-modules, is naturally isomorphic to the tensor product $N^\bullet \otimes_R \mathrm{cone}(R \to K^\tau_\bullet(B, A))$. For every $n \in \mathbb{Z}$, the internal degree n component of the complex $\mathrm{cone}(R \to K^\tau_\bullet(B, A))$ is a finite acyclic complex of R-R-bimodules which are finitely generated and projective (hence flat) as left R-modules. As above, we can conclude that $\mathrm{gr}^F Z \simeq N^\bullet \otimes_R \mathrm{cone}(R \to K^\tau_\bullet(B, A))$ is a coacyclic complex of right R-modules. Hence $\mathrm{gr}^F Z$ is also coacyclic as a right CDG-comodule over (B, d, h). By Lemma 6.12, it follows that Z is a coacyclic CDG-comodule over (B, d, h), too.   □

Examples 6.15

(1) The free right R-module R can be considered as a one-term complex of right R-modules, concentrated in the cohomological degree 0 and endowed with the zero differential. This complex, in turn, can be considered as a trivial right CDG-comodule over (B, d, h). The functor $- \otimes_R^{\tau'} \widetilde{A} = - \otimes_B K^\vee_{e'}(B, \widetilde{A})$ takes this trivial right CDG-comodule R over (B, d, h) to the free right $\widetilde{A}$-module $\widetilde{A}$ (viewed as a one-term complex concentrated in the cohomological degree 0).

(2) Applying the functor $- \otimes_R^{\sigma'} C = \mathrm{Hom}_{\widetilde{A}^{\mathrm{op}}}(K^\vee_{e'}(B, \widetilde{A}), -)$ to the free right $\widetilde{A}$-module $\widetilde{A}$ produces the right CDG-comodule $\widetilde{A} \otimes_R^{\sigma'} C = \mathrm{Hom}_{\widetilde{A}^{\mathrm{op}}}(K^\vee_{e'}(B, \widetilde{A}), \widetilde{A})$, which was mentioned in the above proof and in Remark 6.6. The left $\widetilde{A}$-module structure on $\widetilde{A}$ induces a left action of $\widetilde{A}$ in $\widetilde{A} \otimes_R^{\sigma'} C$, making it a CDG-bimodule over $(\widetilde{A}, 0, 0)$

and (B, d, h). The adjunction map $R \longrightarrow \widetilde{A} \otimes_R^{\sigma'} C$ is a closed morphism of right CDG-comodules over (B, d, h) (or, if one wishes, of CDG-bimodules over $(R, 0, 0)$ and (B, d, h)). According to the above argument, its cone is a coacyclic right CDG-comodule over (B, d, h).

(3) Applying the functor $- \otimes_R^{\tau'} \widetilde{A}$ to the right CDG-comodule $\widetilde{A} \otimes_R^{\sigma'} C$ over (B, d, h) produces a complex of right $\widetilde{A}$-modules $\widetilde{A} \otimes_R^{\sigma'} C \otimes_R^{\tau'} \widetilde{A}$. The left action of $\widetilde{A}$ in $\widetilde{A} \otimes_R^{\sigma'} C$ induces a left action of $\widetilde{A}$ in the complex $\widetilde{A} \otimes_R^{\sigma'} C \otimes_R^{\tau'} \widetilde{A}$, making it a complex of $\widetilde{A}$-$\widetilde{A}$-bimodules. The adjunction map $\widetilde{A} \otimes_R^{\sigma'} C \otimes_R^{\tau'} \widetilde{A} \longrightarrow \widetilde{A}$ is a morphism of complexes of $\widetilde{A}$-$\widetilde{A}$-bimodules. According to the above proof, its cone is coacyclic as a complex of right R-modules (following the details of the argument, one can see that this cone is, in fact, coacyclic as a complex of $\widetilde{A}$-R-bimodules).

   In particular, the cone is an acyclic complex; so $\widetilde{A} \otimes_R^{\sigma'} C \otimes_R^{\tau'} \widetilde{A}$ is a bimodule resolution of the diagonal $\widetilde{A}$-$\widetilde{A}$-bimodule $\widetilde{A}$. The terms of this resolution are *not* projective $\widetilde{A} \otimes_{\mathbb{Z}} \widetilde{A}$-modules, of course; and $\widetilde{A} \otimes_R \widetilde{A}$ is, generally speaking, not even a ring! Still one can say that the terms of this resolution are the $\widetilde{A}$-$\widetilde{A}$-bimodules $\widetilde{A} \otimes_R C_i \otimes_R \widetilde{A}$ induced from the R-R-bimodules $C_i = \mathrm{Hom}_{R^{\mathrm{op}}}(B^i, R)$ (which are finitely generated and projective as left R-modules in our assumptions).

(4) Notice that the bimodule resolution $\widetilde{A} \otimes_R^{\sigma'} C \otimes_R^{\tau'} \widetilde{A} \longrightarrow \widetilde{A}$ from (3) is a quasi-isomorphism of bounded above complexes of projective left $\widetilde{A}$-modules; hence it is a homotopy equivalence of complexes of left $\widetilde{A}$-modules. It is also a quasi-isomorphism of bounded above complexes of weakly $\widetilde{A}/R$-flat right $\widetilde{A}$-modules in the sense of [63, Section 5] (cf. Sect. 2.9). In view of [63, Lemma 5.3(b)], it follows that the tensor product of this resolution with any R-flat left $\widetilde{A}$-module is still an exact complex.

   We refer to Sect. 8.7 below for a further discussion of this bimodule resolution.

(5) Now assume that $\widetilde{A}$ is a left augmented left finitely projective nonhomogeneous Koszul ring over R, in the sense of the definitions in Sects. 3.8 and 4.7. Let $\widetilde{A}^+ \subset \widetilde{A}$ be the augmentation ideal; so $\widetilde{A}^+$ is a left ideal in $\widetilde{A}$ such that $\widetilde{A} = R \oplus \widetilde{A}^+$. Choose the left R-submodule of strict generators $V' \subset \widetilde{V}$ as $V' = \widetilde{A}^+ \cap \widetilde{V}$. Then, by Theorem 3.27 (see also Corollary 4.28), we have $h = 0$; so $(B, d, h) = (B, d, 0)$ is a DG-ring. Accordingly, the CDG-bimodule $\widetilde{A} \otimes_R^{\sigma'} C$ from (2) is a DG-bimodule, i.e., a complex. This complex can be called the *nonhomogeneous Koszul complex* of a left augmented left finitely projective nonhomogeneous Koszul ring $\widetilde{A}$.

   The left augmentation of $\widetilde{A}$ endows the base ring R with a structure of left $\widetilde{A}$-module provided by the identification $R \simeq \widetilde{A}/\widetilde{A}^+$. Taking the tensor product of the bimodule resolution from (3) with this left $\widetilde{A}$-module, we obtain a morphism of complexes of left $\widetilde{A}$-modules $\widetilde{A} \otimes_R^{\sigma'} C \longrightarrow R$, which is a quasi-isomorphism by the argument in (4). Here $\widetilde{A} \otimes_R^{\sigma'} C$ is the above DG-bimodule over $(\widetilde{A}, 0)$ and (B, d).

   According to (2), we also have a natural map in the opposite direction $R \longrightarrow \widetilde{A} \otimes_R^{\sigma'} C$, whose cone is coacyclic (hence acyclic) as a right DG-bimodule over (B, d). The composition $R \longrightarrow \widetilde{A} \otimes_R^{\sigma'} C \longrightarrow R$ is the identity map. This provides another way to prove that the map $\widetilde{A} \otimes_R^{\sigma'} C \longrightarrow R$ is a quasi-isomorphism.

Clearly, the terms of the complex $\widetilde{A} \otimes_R^{\sigma'} C$ are finitely generated projective left $\widetilde{A}$-modules in our assumptions. Thus the nonhomogeneous Koszul complex $\widetilde{A} \otimes_R^{\sigma'} C$ is a resolution of the left $\widetilde{A}$-module R by finitely generated projective left $\widetilde{A}$-modules.

Remarks 6.16

(1) Consider the trivial right CDG-comodule R over (B, d, h), as in Example 6.15(1). Since the triangulated equivalence $- \otimes_R^{\tau'} \widetilde{A} \colon \mathsf{D}^{\mathsf{co}}(\mathsf{comod} - B) \longrightarrow \mathsf{D}_R^{\mathsf{sico}}(\mathsf{mod} - \widetilde{A})$ takes R to the free right $\widetilde{A}$-module $\widetilde{A}$, it follows that the graded ring of endomorphisms $R \longrightarrow R[i]$ in $\mathsf{D}^{\mathsf{co}}(\mathsf{comod} - B)$, $i \in \mathbb{Z}$, is naturally isomorphic to ring $\widetilde{A}$ (concentrated in the cohomological grading $i = 0$).

Let N be a right CDG-comodule over (B, d, h). Then the complex of right $\widetilde{A}$-modules $N \otimes_R^{\tau'} \widetilde{A}$ computes the $\widetilde{A}$-modules $\mathrm{Hom}_{\mathsf{D}^{\mathsf{co}}(\mathsf{comod} - B)}(R, N[i])$. Indeed, we have

$$\mathrm{Hom}_{\mathsf{D}^{\mathsf{co}}(\mathsf{comod} - B)}(R, N[i]) \simeq \mathrm{Hom}_{\mathsf{D}_R^{\mathsf{sico}}(\mathsf{mod} - \widetilde{A})}(\widetilde{A}, \ N \otimes_R^{\tau'} \widetilde{A}[i]) = H^i(N \otimes_R^{\tau'} \widetilde{A}),$$

because

$$\mathrm{Hom}_{\mathsf{D}_R^{\mathsf{sico}}(\mathsf{mod} - \widetilde{A})}(\widetilde{A}, M^{\bullet}[i]) = \mathrm{Hom}_{\mathsf{Hot}(\mathsf{mod} - \widetilde{A})}(\widetilde{A}, M^{\bullet}[i]) = H^i(M^{\bullet})$$

for any complex of right $\widetilde{A}$-modules $M^{\bullet}$. The latter isomorphism holds since all complexes in $\mathsf{Acycl}_R^{\mathsf{sico}}(\mathsf{mod} - \widetilde{A})$ are acyclic.

(2) Now assume that $\widetilde{A}$ is left augmented over R, as in Example 6.15(5). Then, for any complex of right $\widetilde{A}$-modules $M^{\bullet}$, the CDG-(co)module $M^{\bullet} \otimes_R^{\sigma'} C$ is a DG-module over $(B, d, h) = (B, d, 0)$, i.e., a complex. We have a natural isomorphism $M^{\bullet} \otimes_R^{\sigma'} C \simeq M^{\bullet} \otimes_{\widetilde{A}} (\widetilde{A} \otimes_R^{\sigma'} C)$. Since $\widetilde{A} \otimes_R^{\sigma'} C$ is a projective resolution of the left $\widetilde{A}$-module R, it follows that the complex of abelian groups $M^{\bullet} \otimes_R^{\sigma'} C$ computes the derived tensor product $M^{\bullet} \otimes_{\widetilde{A}}^{\mathbb{L}} R$. In fact, the DG-ring (B, d) computes (the opposite ring to) the graded ring of endomorphisms $R \longrightarrow R[i]$ in $\mathsf{D}(\widetilde{A} - \mathsf{mod})$, $i \in \mathbb{Z}$, and the DG-module $M^{\bullet} \otimes_R^{\sigma'} C$ computes the homology $\mathrm{Tor}_*^{\widetilde{A}}(M^{\bullet}, R) = H_*(M^{\bullet} \otimes_{\widetilde{A}}^{\mathbb{L}} R)$ as a graded module over this graded ring of endomorphisms.

Example 6.17 Let R be a commutative ring and $\mathfrak{g}$ be a Lie algebroid over R such that $\mathfrak{g}$ is finitely generated and projective as an R-module. Then the enveloping ring $U(R, \mathfrak{g})$ with its natural filtration F is a left augmented left finitely projective Koszul ring over R (see the discussion in Sect. 10.9 below; cf. Example 3.15(1) above). The Koszul resolution of the left $U(R, \mathfrak{g})$-module R constructed in Example 6.15(5) was introduced and discussed in [73, Lemma 4.1] (in the greater generality of a projective, but not necessarily finitely generated R-module $\mathfrak{g}$). This resolution is also well-known in the D-module theory; see the discussion and references in Sect. 10.2.

6.7 Reduced Coderived Category

We keep the assumptions of Sect. 6.6. So $R \subset \widetilde{V} \subset \widetilde{A}$ is a left finitely projective nonhomogeneous Koszul ring and (B, d, h) is the corresponding right finitely projective Koszul CDG-ring. The following result is a special case of Theorem 6.14.

Corollary 6.18 *Assume additionally that the right homological dimension of the ring R (that is, the homological dimension of the abelian category $\mathsf{mod}{-}R$) is finite. Then the pair of adjoint triangulated functors $-\otimes_R^{\sigma'} C\colon \mathsf{Hot}(\mathsf{mod}{-}\widetilde{A}) \longrightarrow \mathsf{Hot}(\mathsf{comod}{-}B)$ and $-\otimes_R^{\tau'} \widetilde{A}\colon \mathsf{Hot}(\mathsf{comod}{-}B) \longrightarrow \mathsf{Hot}(\mathsf{mod}{-}\widetilde{A})$ induces mutually inverse triangulated equivalences*

$$\mathsf{D}(\mathsf{mod}{-}\widetilde{A}) \simeq \mathsf{D}^{\mathsf{co}}(\mathsf{comod}{-}B) \tag{6.6}$$

between the derived category of right $\widetilde{A}$-modules and the coderived category of right CDG-comodules over (B, d, h).

Concrete examples of the triangulated equivalence (6.6) will be given below in Sects. 10.2–10.5 and 10.8–10.10.

Proof When the right homological dimension of the ring R is finite, all acyclic complexes of right R-modules are coacyclic by [58, Remark 2.1]. So the $\widetilde{A}/R$-semicoderived category of right $\widetilde{A}$-modules coincides with their derived category, $\mathsf{D}_R^{\mathsf{sico}}(\mathsf{mod}{-}\widetilde{A}) = \mathsf{D}(\mathsf{mod}{-}\widetilde{A})$. This means, specifically, that the right CDG-comodule $M^\bullet \otimes_R^{\sigma'} C$ over (B, d, h) is coacyclic for any acyclic complex of right $\widetilde{A}$-modules $M^\bullet$. The rest is explained in Theorem 6.14 and its proof.   □

We would like to obtain a description of the derived category $\mathsf{D}(\mathsf{mod}{-}\widetilde{A})$ in terms of some kind of exotic derived category of right CDG-comodules over (B, d, h) *without* assuming finiteness of the right homological dimension of R. The following definition serves this purpose.

Definition 6.19 Consider the full subcategory of acyclic complexes of right R-modules $\mathsf{Acycl}(\mathsf{mod}{-}R) \subset \mathsf{Hot}(\mathsf{mod}{-}R)$. All complexes of right R-modules, and, in particular, acyclic complexes of right R-modules, can be viewed as trivial right CDG-comodules over (B, d, h).

By a kind of abuse of notation, let us denote simply by $\langle \mathsf{Acycl}(\mathsf{mod}{-}R)\rangle_\oplus$ the minimal full triangulated subcategory in $\mathsf{D}^{\mathsf{co}}(\mathsf{comod}{-}B)$ containing all the acyclic complexes of right R-modules (viewed as trivial CDG-comodules) and closed under infinite direct sums. Consider the triangulated Verdier quotient category

$$\mathsf{D}_{R-\mathsf{red}}^{\mathsf{co}}(\mathsf{comod}{-}B) = \mathsf{D}^{\mathsf{co}}(\mathsf{comod}{-}B)/\langle \mathsf{Acycl}(\mathsf{mod}{-}R)\rangle_\oplus.$$

We will call this quotient category the *reduced coderived category of right CDG-co-modules over* (B, d, h) *relative to* R.

Theorem 6.20 *The pair of adjoint triangulated functors* $-\otimes_R^{\sigma'} C \colon \mathsf{Hot}(\mathsf{mod}-\widetilde{A}) \longrightarrow$ $\mathsf{Hot}(\mathsf{comod}-B)$ *and* $-\otimes_R^{\tau'} \widetilde{A} \colon \mathsf{Hot}(\mathsf{comod}-B) \longrightarrow \mathsf{Hot}(\mathsf{mod}-\widetilde{A})$ *induces mutually inverse triangulated equivalences*

$$\mathsf{D}(\mathsf{mod}-\widetilde{A}) \simeq \mathsf{D}^{\mathsf{co}}_{R-\mathsf{red}}(\mathsf{comod}-B)$$

between the derived category of right $\widetilde{A}$*-modules and the reduced coderived category of right CDG-comodules over* (B, d, h) *relative to* R.

Examples of the triangulated equivalence stated in the theorem will be given below in Sects. 10.6–10.10. Some other situations in which the theorem is applicable are described in Examples 3.17 and 3.22.

Proof In view of Theorem 6.14, only two things still need to be checked. Firstly, we have to show that, for any acyclic complex of right $\widetilde{A}$-modules $M^\bullet$, the CDG-comodule $M^\bullet \otimes_R^{\sigma'} C$ over (B, d, h) belongs to the triangulated subcategory $\langle \mathsf{Acycl}(\mathsf{mod}-R) \rangle_\oplus \subset$ $\mathsf{D}^{\mathsf{co}}(\mathsf{comod}-B)$. Secondly, it needs to be demonstrated that, for any CDG-comodule $N \in$ $\langle \mathsf{Acycl}(\mathsf{mod}-R) \rangle_\oplus$, the complex of right $\widetilde{A}$-modules $N \otimes_R^{\tau'} \widetilde{A}$ is acyclic.

Firstly, let $M = M^\bullet$ be an acyclic complex of right $\widetilde{A}$-modules. Consider the increasing filtration F on the graded right R-module $M \otimes_R C$ induced by the increasing filtration F on the R-R-bimodule C, as in the proof of Theorem 6.14. Then F is a filtration of the CDG-comodule $M^\bullet \otimes_R^{\sigma'} C$ over (B, d, h) by CDG-subcomodules. The successive quotient CDG-comodules are trivial CDG-comodules which, viewed as complexes of R-modules, can be computed as the tensor products $M^\bullet \otimes_R (F_n C / F_{n-1} C)$. Here $F_n C / F_{n-1} C$ is a graded R-R-bimodule, finitely generated and projective as a left R-module, and concentrated in the single cohomological degree $-n$.

It follows that the complexes of right R-modules $M^\bullet \otimes_R (F_n C / F_{n-1} C)$ are acyclic. Now the iterated extension and telescope sequence argument from the proofs of Lemmas 6.8 and 6.12 shows that the right CDG-comodule $M^\bullet \otimes_R^{\sigma'} C$ over (B, d, h) belongs to the triangulated subcategory $\langle \mathsf{Acycl}(\mathsf{mod}-R) \rangle_\oplus$ in $\mathsf{D}^{\mathsf{co}}(\mathsf{comod}-B)$.

Secondly, since the functor $-\otimes_R^{\tau'} \widetilde{A}$ is triangulated and preserves infinite direct sums, it suffices to check that for any acyclic complex of right R-modules $N^\bullet$, viewed as a trivial right CDG-comodule over (B, d, h), the complex of right $\widetilde{A}$-modules $N^\bullet \otimes_R^{\tau'} \widetilde{A}$ is acyclic. As $N^\bullet$ is a trivial CDG-comodule, the complex of right $\widetilde{A}$-modules $N^\bullet \otimes_R^{\tau'} \widetilde{A} = N^\bullet \otimes_R \widetilde{A}$ is simply obtained from the complex of right R-modules $N^\bullet$ by applying the tensor product functor $-\otimes_R \widetilde{A}$ termwise. Since $\widetilde{A}$ is a projective (hence flat) left R-module, it follows that the complex $N^\bullet \otimes_R \widetilde{A}$ is acyclic. $\qquad \square$

Relative Nonhomogeneous Derived Koszul Duality: The Contramodule Side

7

7.1 Semicontraderived Category of Modules

Let R be an associative ring. The following definition of the *contraderived category of left R-modules* is dual to the definition of the coderived category in Sect. 6.3. It goes back to [58, Section 4.1] and [59, Section 3.3] (see [35] and [4, Proposition 1.3.6(1)] for an alternative approach).

Definition 7.1 Similarly to Sect. 6.3, we consider short exact sequences of left R-modules $0 \longrightarrow L^\bullet \longrightarrow M^\bullet \longrightarrow N^\bullet \longrightarrow 0$ and their totalizations $\mathrm{Tot}(L^\bullet \to M^\bullet \to N^\bullet)$. A complex of left R-modules is said to be *contraacyclic* if it belongs to the minimal full triangulated subcategory of the homotopy category of complexes of left R-modules $\mathsf{Hot}(R\text{–}\mathsf{mod})$ containing all the totalizations of short exact sequences of left R-modules and closed under infinite products.

The full subcategory of contraacyclic complexes of left R-modules is denoted by $\mathsf{Acycl}^{\mathrm{ctr}}(R\text{–}\mathsf{mod}) \subset \mathsf{Hot}(R\text{–}\mathsf{mod})$. The *contraderived category of left R-modules* is the triangulated Verdier quotient category of the homotopy category $\mathsf{Hot}(R\text{–}\mathsf{mod})$ by the triangulated subcategory of contraacyclic complexes,

$$\mathsf{D}^{\mathrm{ctr}}(R\text{–}\mathsf{mod}) = \mathsf{Hot}(R\text{–}\mathsf{mod})/\mathsf{Acycl}^{\mathrm{ctr}}(R\text{–}\mathsf{mod}).$$

Clearly, any contraacyclic complex of R-modules is acyclic, $\mathsf{Acycl}^{\mathrm{ctr}}(R\text{–}\mathsf{mod}) \subset \mathsf{Acycl}(R\text{–}\mathsf{mod})$; but the converse does not hold in general [59, Examples 3.3]. So the derived category $\mathsf{D}(R\text{–}\mathsf{mod})$ is a quotient category of $\mathsf{D}^{\mathrm{ctr}}(R\text{–}\mathsf{mod})$. All acyclic complexes of left R-modules are contraacyclic when the left homological dimension of the ring R is finite [58, Remark 2.1].

L. Positselski, *Relative Nonhomogeneous Koszul Duality*, Frontiers in Mathematics, https://doi.org/10.1007/978-3-030-89540-2_7

The following lemma is dual-analogous to and very slightly more complicated than Lemma 6.8.

Lemma 7.2 *Let* $P^\bullet = F^0 P^\bullet \supset F^1 P^\bullet \supset F^2 P^\bullet \supset \cdots$ *be a complex of R-modules that is separated and complete with respect to a decreasing filtration by subcomplexes of R-modules, that is* $P^\bullet = \varprojlim_{n \geq 0} P^\bullet / F^{n+1} P^\bullet$. *Assume that, for every* $n \geq 0$, *the complex of R-modules* $F^n P^\bullet / F^{n+1} P^\bullet$ *is contraacyclic. Then the complex of R-modules* $P^\bullet$ *is contraacyclic.*

Proof First one proves by induction in n that the complex of R-modules $P^\bullet / F^{n+1} P^\bullet$ is contraacyclic for every $n \geq 0$, using the fact that the totalizations of short exact sequences $0 \longrightarrow F^n P^\bullet / F^{n+1} P^\bullet \longrightarrow P^\bullet / F^{n+1} P^\bullet \longrightarrow P^\bullet / F^n P^\bullet \longrightarrow 0$ are contraacyclic. Then one concludes that the complex of R-modules $\prod_{n=0}^\infty P^\bullet / F^{n+1} P^\bullet$ is contraacyclic. Finally, one observes that the telescope sequence $0 \longrightarrow P^\bullet \longrightarrow \prod_{n=0}^\infty P^\bullet / F^{n+1} P^\bullet \longrightarrow \prod_{n=0}^\infty P^\bullet / F^{n+1} P^\bullet \longrightarrow 0$ of the projective system $(P^\bullet / F^{n+1} P^\bullet)_{n \geq 0}$ is exact, since the transition maps $P^\bullet / F^{n+1} P^\bullet \longrightarrow P^\bullet / F^n P^\bullet$ are surjective (so the derived projective limit of this projective system vanishes, $\varprojlim_{n \geq 0}^1 P^\bullet / F^{n+1} P^\bullet = 0$). Hence the totalization of the telescope sequence is a contraacyclic complex of R-modules, and contraacyclicity of the complex $P^\bullet$ follows. $\square$

Definition 7.3 Now let $R \longrightarrow A$ be a homomorphism of associative rings. The following definition of the *A/R-semicontraderived category of left A-modules* can be found in [63, Section 5]. The semicontraderived category

$$\mathsf{D}_R^{\mathrm{sictr}}(A\text{–mod}) = \mathsf{Hot}(A\text{–mod})/\mathsf{Acycl}_R^{\mathrm{sictr}}(A\text{–mod})$$

is the triangulated Verdier quotient category of the homotopy category of left A-modules by its full triangulated subcategory $\mathsf{Acycl}_R^{\mathrm{sictr}}(A\text{–mod})$ consisting of all the complexes of left A-modules that are *contraacyclic as complexes of left R-modules*.

Clearly, any contraacyclic complex of A-modules is contraacyclic as a complex of R-modules; but the converse does not need to be true. So the semicontraderived category is intermediate between the derived and the contraderived category of A-modules: there are natural triangulated Verdier quotient functors $\mathsf{D}^{\mathrm{ctr}}(A\text{–mod}) \longrightarrow \mathsf{D}_R^{\mathrm{sictr}}(A\text{–mod}) \longrightarrow \mathsf{D}(A\text{–mod})$. (See the paper [66] for a more general discussion of such intermediate triangulated quotient categories, called *pseudo-contraderived categories* in [66].) The semicontraderived category $\mathsf{D}_R^{\mathrm{sictr}}(A\text{–mod})$ can be thought of as a mixture of the contraderived category "along the variables from R" and the conventional derived category "in the direction of A relative to R."

For example, taking $R = \mathbb{Z}$, one obtains the conventional derived category of A-modules

$$\mathsf{D}_{\mathbb{Z}}^{\mathsf{sictr}}(A\text{–mod}) = \mathsf{D}(A\text{–mod})$$

(because all the acyclic complexes of abelian groups are contraacyclic), while taking $R = A$ one obtains the contraderived category of A-modules

$$\mathsf{D}_A^{\mathsf{sictr}}(A\text{–mod}) = \mathsf{D}^{\mathsf{ctr}}(A\text{–mod}).$$

7.2 Contraderivations

Let $B = \bigoplus_{n=0}^{\infty} B^n$ be a nonnegatively graded ring and K be a ring endowed with a ring homomorphism $K \longrightarrow B^0$.

Let $k \in \mathbb{Z}$ be a fixed integer and $d: B \longrightarrow B$ be a fixed derivation of degree k (even or odd, depending on the parity of k). This means that we have additive maps $d_n: B^n \longrightarrow B^{n+k}$ given for all $n \geq 0$ and satisfying the equation $d(bc) = d(b)c + (-1)^{k|b|}bd(c)$ for all $b \in B^{|b|}$ and $c \in B^{|c|}$. Assume that composition $K \longrightarrow B^0 \longrightarrow B^k$ of the maps $K \longrightarrow B^0$ and $d_0: B^0 \longrightarrow B^k$ vanishes (e. g., one can always take $K = \mathbb{Z}$). So $d: B \longrightarrow B$ is a left and right K-linear map.

Let $M = \bigoplus_{n \in \mathbb{Z}} M^n$ be a graded left B-module. A homogeneous additive map $d_M: M \longrightarrow M$ of degree k (that is, $d_M(M^n) \subset M^{n+k}$ for all $n \in \mathbb{Z}$) is said to be an (even or odd) derivation of M compatible with the derivation d on B if the equation $d_M(bx) = d(b)x + (-1)^{k|b|}bd_M(x)$ holds for all $b \in B^{|b|}$ and $x \in M^{|x|}$.

Let C and D be graded right K-modules, P and Q be graded left K-modules, $g: C \longrightarrow D$ be a homogeneous K-linear map of degree l, and $f: P \longrightarrow Q$ be a homogeneous K-linear map of degree m. Consider the graded abelian groups $C \otimes_K^{\Pi} P$ and $D \otimes_K^{\Pi} Q$ (in the notation of Sect. 5.7), and construct the homogeneous map

$$g \otimes^{\Pi} f: C \otimes_K^{\Pi} P \longrightarrow D \otimes_K^{\Pi} Q$$

of degree $l + m$ by taking the infinite products along the diagonals of the bigrading for the bihomogeneous map $g \otimes f: C \otimes_K P \longrightarrow D \otimes_K Q$ with the components $g \otimes f: C^i \otimes_K P^j \longrightarrow D^{i+l} \otimes_K Q^{j+m}$ given by the rule $(g \otimes f)(c \otimes p) = (-1)^{mi} g(c) \otimes f(p)$ for all $c \in C^i$ and $p \in P^j$.

Definition 7.4 Let $P = \bigoplus_{n \in \mathbb{Z}} P^n$ be a graded left B-contramodule (in the sense of the definition in Sect. 5.6), and let $d_P: P \longrightarrow P$ be a homogeneous K-linear map of degree k. We will say that d_P is an (even or odd) contraderivation of P compatible with the derivation d on B if the contraaction map $\pi_P: \mathbb{M}_K^{\mathrm{gr}}(P) = B \otimes_K^{\Pi} P \longrightarrow P$ forms a commutative square diagram with the maps

$$d\otimes^{\Pi}\mathrm{id}_P + \mathrm{id}_B\otimes^{\Pi}d_P : B\otimes_K^{\Pi} P \longrightarrow B\otimes_K^{\Pi} P$$

and $d_P : P \longrightarrow P$, that is

$$d_P \circ \pi_P = \pi_P \circ (d\otimes^{\Pi}\mathrm{id}_P + \mathrm{id}_B\otimes^{\Pi}d_P).$$

Obviously, any contraderivation of a graded left B-contramodule P compatible with the given derivation d on B is a derivation of the underlying graded left B-module of P compatible with the same derivation d on B.

Proposition 7.5 *Assume that the forgetful functor* B–$\mathsf{contra}_{\mathrm{gr}} \longrightarrow B$–$\mathsf{mod}_{\mathrm{gr}}$ *from the category of graded left B-contramodules to the category of graded left B-modules is fully faithful. Let Q be a graded left B-contramodule. Then any derivation of the underlying graded left B-module of Q (compatible with the given derivation d on B) is a contraderivation of Q (compatible with the same derivation d on B).*

Proof The argument is based on a buildup of auxiliary definitions. Given a graded left B-module M and an integer n, the graded left B-module $M[n]$ has the grading components $M[n]^i = M^{n+i}$; the action of B in $M[n]$ is defined in terms of the left action of B in M with the sign rule written down in Sect. 6.1. Given a graded left B-contramodule P, the graded left B-contramodule $P[n]$ has the grading components $P[n]^i = P^{n+i}$. We leave it to the reader to spell out the construction of the left contraaction of B in $P[n]$ involving the same sign rule.

Let $f : M \longrightarrow N$ be a morphism of graded left B-modules, and let $d_{N,M} : M \longrightarrow N$ be a homogeneous additive map of degree k. We will say that $d_{N,M}$ is a *relative derivation compatible with the derivation d on B and the morphism f* if the equation $d_{N,M}(bx) = d(b)f(x) + (-1)^{k|b|}bd_{N,M}(x)$ holds in $N^{|b|+|x|+k}$ for all $b \in B^{|b|}$ and $x \in M^{|x|}$. A derivation $d_M : M \longrightarrow M$ compatible with the given derivation d on B is the same thing as a relative derivation $d_{M,M} : M \longrightarrow M$ compatible with d and the identity morphism $\mathrm{id}_M : M \longrightarrow M$.

For any two relative derivations $d'_{N,M}$ and $d''_{N,M} : M \longrightarrow N$ compatible with the same derivation d on B and the same morphism $f : M \longrightarrow N$, the difference $d''_{N,M} - d'_{N,M}$ is a morphism of graded left B-modules $M \longrightarrow N[k]$. Conversely, the sum $d'_{N,M} = d_{N,M} + g$ of a relative derivation $d_{N,M} : M \longrightarrow N$ compatible with d and f and a morphism of graded left B-modules $g : M \longrightarrow N[k]$ is a relative derivation compatible with d and f.

Let $f : M \longrightarrow N$ and $g : L \longrightarrow M$ be two morphisms of graded left B-modules, and let $d_{N,M} : M \longrightarrow N$ be a relative derivation compatible with the derivation d on B and the morphism f. Then $d_{N,M} \circ g : L \longrightarrow N$ is a relative derivation compatible with the derivation d on B and the morphism $f \circ g : L \longrightarrow N$.

Let $f: P \longrightarrow Q$ be a morphism of graded left B-contramodules, and let $d_{Q,P}: P \longrightarrow Q$ be a homogeneous K-linear map of degree k. Consider the square diagram formed by the map

$$(d \otimes^{\Pi} f + \mathrm{id}_B \otimes^{\Pi} d_{Q,P}): B \otimes^{\Pi} P \longrightarrow B \otimes^{\Pi} Q$$

together with the map $d_{Q,P}: P \longrightarrow Q$ and the contraaction maps $\pi_P: B \otimes^{\Pi} P \longrightarrow P$ and $\pi_Q: B \otimes^{\Pi} Q \longrightarrow Q$. We will say that $d_{Q,P}$ is a *relative contraderivation compatible with the derivation d on B and the morphism f* if this square diagram is commutative, that is

$$d_{Q,P} \circ \pi_P = \pi_Q \circ (d \otimes^{\Pi} f + \mathrm{id}_B \otimes^{\Pi} d_{Q,P}).$$

A contraderivation $d_P: P \longrightarrow P$ compatible with the given derivation d on B is the same thing as a relative contraderivation $d_{P,P}: P \longrightarrow P$ compatible with d and the identity morphism $\mathrm{id}_P: P \longrightarrow P$.

For any two relative contraderivations $d'_{Q,P}$ and $d''_{Q,P}: P \longrightarrow Q$ compatible with the same derivation d on B and the same morphism $f: P \longrightarrow Q$, the difference $d''_{Q,P} - d'_{Q,P}$ is a morphism of graded left B-contramodules $P \longrightarrow Q[k]$. Conversely, the sum $d'_{Q,P} = d_{Q,P} + g$ of a relative contraderivation $d_{Q,P}: P \longrightarrow Q$ compatible with d and f and a morphism of graded left B-contramodules $g: P \longrightarrow Q[k]$ is a relative contraderivation compatible with d and f.

Let $f: P \longrightarrow Q$ and $g: S \longrightarrow P$ be two morphisms of graded left B-contramodules, and let $d_{Q,P}: P \longrightarrow Q$ be a relative contraderivation compatible with the derivation d on B and the morphism f. Then $d_{Q,P} \circ g: S \longrightarrow Q$ is a relative contraderivation compatible with the derivation d on B and the morphism $f \circ g: S \longrightarrow Q$.

Let L be a graded left K-module and $M = B \otimes_K L$ be the induced graded left B-module (so there is a natural graded left K-module morphism $L \longrightarrow M$). Let $f: M \longrightarrow N$ a morphism of graded left B-modules. Then a relative derivation $d_{N,M}: M \longrightarrow N$ compatible with the given derivation d on B and the morphism f is uniquely determined by its composition with the map $L \longrightarrow M$.

The latter composition $L \longrightarrow M \xrightarrow{d_{N,M}} N$ can be an arbitrary homogeneous K-linear map $t: L \longrightarrow N$ of degree k. Given such a map t, and denoting by $f': L \longrightarrow N$ the composition $L \longrightarrow M \xrightarrow{f} N$ (which is a homogeneous K-linear map of degree 0), the related relative derivation $d_{N,M}: M \longrightarrow N$ can be recovered by the formula $d_{N,M}(b \otimes l) = d(b) f'(l) + (-1)^{k|b|} b t(l)$ for all $b \in B^{|b|}$ and $l \in K^{|l|}$.

In particular, let $X = \coprod_{i \in \mathbb{Z}} X_i$ be a graded set and $K[X]$ be the free graded left K-module with the components $K[X]^i = K^{(X_i)}$ as in the proof of Proposition 5.27. Let $M = B[X] = B \otimes_K K[X]$ be the free graded left B-module spanned by X, and let $f: M \longrightarrow N$ be a morphism of graded left B-modules. Then a relative derivation $d_{N,M}: M \longrightarrow N$ compatible with the given derivation d on B and the morphism f is uniquely determined by its restriction to the subset of free generators $X \subset M$, and this restriction can be an arbitrary homogeneous map $X \longrightarrow N$ of degree k (i.e., a collection of maps $X_i \longrightarrow N^{i+k}$, $i \in \mathbb{Z}$).

Consider the graded left B-contramodule $P = \mathbb{M}_K^{\mathrm{gr}}(L) = B \otimes_K^\Pi L$ freely generated by L. Then, once again, there is a natural graded left K-module morphism $L \longrightarrow P$. Let $f: P \longrightarrow Q$ be a morphism of graded left B-contramodules. Then a relative contraderivation $d_{Q,P}: P \longrightarrow Q$ compatible with the given derivation d on B and the morphism f is uniquely determined by its composition with the map $L \longrightarrow P$.

The latter composition $L \longrightarrow P \xrightarrow{d_{Q,P}} Q$ can be an arbitrary homogeneous K-linear map $t: L \longrightarrow Q$ of degree k. Given such a map t, and denoting by $f': L \longrightarrow Q$ the composition $L \longrightarrow P \xrightarrow{f} Q$ (which is a homogeneous K-linear map of degree 0), the related relative contraderivation $d_{Q,P}: P \longrightarrow Q$ can be recovered by the formula $d_{Q,P} = \pi_Q \circ (d \otimes^\Pi f' + \mathrm{id}_B \otimes^\Pi t)$,

$$ P \longrightarrow B \otimes_K^\Pi Q \longrightarrow Q. $$

In particular, let $P = \mathbb{M}_K^{\mathrm{gr}}(K[X]) = B \otimes_K^\Pi K[X]$ be the free graded left B-contramodule spanned by X, and let $f: P \longrightarrow Q$ be a morphism of graded left B-contramodules. Then a relative contraderivation $d_{Q,P}: P \longrightarrow Q$ compatible with the given derivation d on B and the morphism f is uniquely determined by its restriction to the subset of free generators $X \subset P$, and this restriction can be an arbitrary homogeneous map $X \longrightarrow Q$ of degree k.

Now that we are done with the auxiliary material, we can prove the proposition. Let $d_Q: Q \longrightarrow Q$ be a derivation of the underlying graded left B-module of Q compatible with the given derivation d on B. Choose a free graded left B-contramodule P (spanned by some graded set X) together with a surjective graded left B-contramodule morphism $f: P \longrightarrow Q$. Then the composition $d_Q \circ f: P \longrightarrow Q$ is a relative derivation of the underlying graded B-modules of P and Q compatible with the derivation d on B and the morphism $f: P \longrightarrow Q$.

Consider the restriction of the map $d_Q \circ f: P \longrightarrow Q$ to the subset of free generators $X \subset P$, and extend the resulting map $X \longrightarrow Q$ to a relative contraderivation $d_{Q,P}: P \longrightarrow Q$ of the graded B-contramodules P and Q compatible with the derivation d on B and the contramodule morphism $f: P \longrightarrow Q$. Then the difference $d_Q \circ f - d_{Q,P}$ is a morphism of underlying graded B-modules $P \longrightarrow Q[k]$ of the graded left B-contramodules P and $Q[k]$.

By the assumption of the proposition, any morphism between the underlying graded left B-modules of two graded left B-contramodules is a graded left B-contramodule morphism. Hence $d_Q \circ f - d_{Q,P}$ must be, in fact, a graded left B-contramodule morphism. Since $d_{Q,P}$ is a relative contraderivation compatible with the derivation d on B and the morphism f, it follows that $d_Q \circ f$ is a relative contraderivation compatible with d and f, too.

Finally, since the map f is surjective, we can conclude that d_Q is a contraderivation of Q compatible with the derivation d on B.   □

7.3   **Contraderived Category of CDG-Contramodules**

Let (B, d, h) be a nonnegatively graded CDG-ring, as defined in Sect. 3.2, and let $(\widehat{B}, \partial)$ be the related quasi-differential graded ring constructed in Theorem 4.7. So $h \in B^2$ is a curvature element for an odd derivation $d \colon B \longrightarrow B$ of degree 1.

Definition 7.6 A *left CDG-contramodule over (B,d,h)* can be simply defined as a graded left $\widehat{B}$-contramodule. Equivalently, a left CDG-contramodule (P, d_P) over (B, d, h) is a graded left B-contramodule $P = \bigoplus_{n \in \mathbb{Z}} P^n$ endowed with a sequence of additive maps $d_{P,n} \colon P^n \longrightarrow P^{n+1}$ satisfying the following conditions:

(i) d_P is an odd contraderivation of the graded left B-contramodule P compatible with the odd derivation d on B (in the sense of Sect. 7.2);

(ii) $d_P^2(x) = hx$ for all $x \in P$.

According to Proposition 7.5, when the forgetful functor $B\text{–contra}_{\mathsf{gr}} \longrightarrow B\text{–mod}_{\mathsf{gr}}$ is fully faithful, condition (i) is equivalent to the seemingly weaker condition that d_P is an odd derivation of the underlying graded left B-module of P compatible with the odd derivation d on B. In particular, by Theorem 5.34, this holds whenever the augmentation ideal $B^{\geq 1} = \bigoplus_{n=1}^{\infty} B^n$ is finitely generated as a right ideal in B. This includes all 2-right finitely projective quadratic graded rings B (see Sect. 5.1).

Given two graded left B-contramodules P and Q, we denote by $\operatorname{Hom}^{B,n}(P, Q)$ the abelian group of all graded left B-contramodule homomorphisms $P \longrightarrow Q[n]$ (where a sign rule is needed in the definition of the graded left B-contramodule structure on the grading shift $Q[n]$ of a graded left B-contramodule Q; see the proof of Proposition 7.5 for a brief discussion). For any two left CDG-contramodules P and Q over (B, d, h), there is a natural differential d on the graded abelian group $\operatorname{Hom}^B(P, Q)$. In fact, $\operatorname{Hom}^B(P, Q)$ is a subcomplex in the complex $\operatorname{Hom}_B(P, Q)$ of morphisms between the underlying graded left B-modules of P and Q (which was constructed in Sect. 6.1).

Of course, one has $\operatorname{Hom}^B(P, Q) = \operatorname{Hom}_B(P, Q)$ for all left CDG-contramodules P and Q over (B, d, h) whenever the forgetful functor $B\text{–contra}_{\mathsf{gr}} \longrightarrow B\text{–mod}_{\mathsf{gr}}$ is fully faithful. But generally speaking, the conditions that d_P and d_Q are contraderivations (rather than just derivations of the underlying graded left B-modules of P and Q) is needed in order to check that $d(f) \in \operatorname{Hom}^{B,n+1}(P, Q) \subset \operatorname{Hom}_B^{n+1}(P, Q)$ for every $f \in \operatorname{Hom}^{B,n}(P, Q) \subset \operatorname{Hom}_B^n(P, Q)$.

The above construction produces the *DG-category of left CDG-contramodules* $\mathsf{DG}(B\text{–contra})$ *over* $B = (B, d, h)$. The DG-category of left CDG-contramodules comes endowed with the forgetful DG-functor $\mathsf{DG}(B\text{–contra}) \longrightarrow \mathsf{DG}(B\text{–mod})$.

All the shifts, twists, and infinite products (in the sense of the discussion in [59, Section 1.2]) exist in the DG-category $\mathsf{DG}(B\text{–contra})$; in particular, all the cones of closed morphisms exist. Passing to the degree-zero cohomology groups of the complexes of morphisms, we obtain the triangulated *homotopy category of CDG-contramodules* $\mathsf{Hot}(B\text{–contra}) = H^0\mathsf{DG}(B\text{–contra})$ *over* (B, d, h) and the triangulated forgetful functor $\mathsf{Hot}(B\text{–contra}) \longrightarrow \mathsf{Hot}(B\text{–mod})$.

Definition 7.7 Similarly to Sect. 6.4, we consider short exact sequences $0 \longrightarrow P \xrightarrow{i} Q \xrightarrow{p} S \longrightarrow 0$ of left CDG-contramodules over (B, d, h). Here i and p are closed morphisms of degree zero in the DG-category $\mathsf{DG}(B\text{–contra})$ and the short sequence $0 \longrightarrow P \longrightarrow Q \longrightarrow S \longrightarrow 0$ is exact in the abelian category $B\text{–contra}_{\mathsf{gr}}$. Using the construction of the cone of a closed morphism (or the shift, finite direct sum, and twist) in the DG-category $\mathsf{DG}(B\text{–contra})$, one can produce the totalization (or the *total CDG-contramodule*) $\mathrm{Tot}(P \to Q \to S)$ of the finite complex of CDG-contramodules $P \longrightarrow Q \longrightarrow S$.

A left CDG-contramodule over B is said to be *contraacyclic* if it belongs to the minimal full triangulated subcategory of $\mathsf{Hot}(B\text{–contra})$ containing all the totalizations of short exact sequences of left CDG-contramodules over B and closed under infinite products. The full triangulated subcategory of contraacyclic CDG-contramodules is denoted by $\mathsf{Acycl}^{\mathsf{ctr}}(B\text{–contra}) \subset \mathsf{Hot}(B\text{–contra})$. The *contraderived category of left CDG-contramodules over B* is the triangulated Verdier quotient category of the homotopy category $\mathsf{Hot}(B\text{–contra})$ by its triangulated subcategory of contraacyclic CDG-contramodules,

$$\mathsf{D}^{\mathsf{ctr}}(B\text{–contra}) = \mathsf{Hot}(B\text{–contra})/\mathsf{Acycl}^{\mathsf{ctr}}(B\text{–contra}).$$

Similarly to the discussion in Sect. 6.4, CDG-contramodules, generally speaking, are *not complexes* (if $h \neq 0$ in B) and their cohomology (contra)modules are *undefined*. So the conventional construction of the derived category is *not* applicable to curved DG-contramodules. There is only the contraderived category and its variations (known as derived categories of the second kind).

Lemma 7.8 *Let $B = (B, d, h)$ be a nonnegatively graded CDG-ring and $Q = (Q, d_Q)$ be a left CDG-contramodule over B. Let $Q = F^0Q \supset F^1Q \supset F^2Q \supset \cdots$ be a decreasing filtration of Q by CDG-subcontramodules $F^nQ \subset Q$ such that Q is separated and complete with respect to this filtration, that is $Q = \varprojlim_{n \geq 0} Q/F^{n+1}Q$. Assume that, for every $n \geq 0$, the CDG-contramodule $F^nQ/F^{n+1}Q$ is contraacyclic over B. Then the CDG-contramodule Q is contraacyclic, too.*

Proof This is a generalization of Lemma 7.2, provable by the same argument. Let us only point out that the forgetful functor $\mathsf{DG}(B\text{–contra}) \longrightarrow \mathsf{DG}(B\text{–mod})$ preserves infinite products (cf. Proposition 5.27); so the projective limits of CDG-contramodules over B agree with those of CDG-modules over B (or just of graded abelian groups). Hence the derived countable projective limit vanishes on sequences of surjective morphisms of CDG-contramodules, which makes our argument work. $\square$

There is an important particular case when the nonnegatively graded ring B has only finitely many grading components, that is $B^n = 0$ for $n \gg 0$. In this case, the forgetful functors from the categories of (ungraded or graded) B-contramodules to the categories of (ungraded or graded) B-modules are equivalences of categories, so, in particular, B-contra$_{\mathsf{gr}} \simeq B$-mod$_{\mathsf{gr}}$. Consequently, the forgetful DG-functor $\mathsf{DG}(B\text{-contra}) \longrightarrow \mathsf{DG}(B\text{-mod})$ is an equivalence of DG-categories. In this case, we will write simply $\mathsf{D}^{\mathsf{ctr}}(B\text{-mod})$ instead of $\mathsf{D}^{\mathsf{ctr}}(B\text{-contra})$.

7.4   Koszul Duality Functors for Modules and Contramodules

Let $R \subset \widetilde{V} \subset \widetilde{A}$ be a 3-left finitely projective weak nonhomogeneous quadratic ring, and let $B = (B, d, h)$ be the nonhomogeneous quadratic dual CDG-ring, as constructed in Proposition 3.16. As in Sect. 6.5, the dual nonhomogeneous Koszul CDG-module $K^{\vee}(B, \widetilde{A}) = K_{e'}^{\vee}(B, \widetilde{A}) = (B \otimes_R \widetilde{A}, d_{e'})$ constructed in Sect. 6.2 plays a key role.

According to Sect. 6.1, the CDG-bimodule $K^{\vee}(B, \widetilde{A})$ over the DG-rings $B = (B, d, h)$ and $\widetilde{A} = (\widetilde{A}, 0, 0)$ induces a pair of adjoint DG-functors between the DG-category $\mathsf{DG}(\widetilde{A}\text{-mod})$ of complexes of left $\widetilde{A}$-modules and the DG-category $\mathsf{DG}(B\text{-mod})$ of left CDG-modules over (B, d, h). Here the right adjoint DG-functor

$$\mathrm{Hom}_B(K^{\vee}(B, \widetilde{A}), -) \colon \mathsf{DG}(B\text{-mod}) \longrightarrow \mathsf{DG}(\widetilde{A}\text{-mod})$$

takes a left CDG-module $Q = (Q, d_Q)$ over (B, d, h) to the complex of left $\widetilde{A}$-modules whose underlying graded $\widetilde{A}$-module is

$$\mathrm{Hom}_B(K^{\vee}(B, \widetilde{A}), Q) = \mathrm{Hom}_B(B \otimes_R \widetilde{A}, Q) = \mathrm{Hom}_R(\widetilde{A}, Q).$$

The left adjoint DG-functor

$$K^{\vee}(B, \widetilde{A}) \otimes_{\widetilde{A}} - \colon \mathsf{DG}(\widetilde{A}\text{-mod}) \longrightarrow \mathsf{DG}(B\text{-mod})$$

takes a complex of left $\widetilde{A}$-modules $P = P^{\bullet}$ to the left CDG-module over (B, d, h) whose underlying graded B-module is

$$K^{\vee}(B, \widetilde{A}) \otimes_{\widetilde{A}} P = (B \otimes_R \widetilde{A}) \otimes_{\widetilde{A}} P = B \otimes_R P.$$

Our aim is to replace the DG-category of CDG-modules $\mathsf{DG}(B\text{–mod})$ with the DG-category of CDG-contramodules $\mathsf{DG}(B\text{–contra})$ in this adjoint pair. In fact, the augmentation ideal $B^{\geq 1}$ is finitely generated as a right ideal in B for any 2-right finitely projective quadratic graded ring B (according to the discussion in Sect. 5.1). Therefore, the forgetful DG-functor $\mathsf{DG}(B\text{–contra}) \longrightarrow \mathsf{DG}(B\text{–mod})$ is fully faithful in our assumptions (see the discussion in Sect. 7.3). So the situation is formally somewhat similar to that in Sect. 6.5.

Lemma 7.9 *Let* $B = \bigoplus_{n=0}^{\infty} B_n$ *be a nonnegatively graded ring with the degree-zero component* $R = B_0$. *Then the forgetful functor (between the abelian categories)* $B\text{–contra}_{\mathsf{gr}} \longrightarrow B\text{–mod}_{\mathsf{gr}}$ *has a left adjoint functor* $\Delta_B \colon B\text{–mod}_{\mathsf{gr}} \longrightarrow B\text{–contra}_{\mathsf{gr}}$. *Furthermore, for any graded left* R*-module* L *one has*

$$\Delta_B(B\otimes_R L) = B\otimes_R^{\Pi} L.$$

Proof For any graded left B-contramodule Q, the group of morphisms $B\otimes_R L \longrightarrow Q$ in $B\text{–mod}_{\mathsf{gr}}$ is isomorphic to the group of morphisms $L \longrightarrow Q$ in $R\text{–mod}_{\mathsf{gr}}$. The latter group is isomorphic to the group of morphisms $B\otimes_R^{\Pi} L = \mathbb{M}_R^{\mathsf{gr}}(L) \longrightarrow Q$ in $B\text{–contra}_{\mathsf{gr}}$ (see the discussion in Sect. 5.2 and Proposition 5.27). This shows that the functor Δ_B is defined on the full subcategory in $B\text{–mod}_{\mathsf{gr}}$ formed by the graded left B-modules $B\otimes_R L$ induced from graded left R-modules L, and proves the second assertion of the lemma. To prove the first assertion, present an arbitrary graded left B-module N as the cokernel of a morphism $f \colon B\otimes_R L \longrightarrow B\otimes_R M$ of graded left B-modules induced from graded left R-modules L and M (e.g., one can always take L and M to be free graded left R-modules). Then the graded left B-contramodule $\Delta_B(N)$ can be computed as the cokernel of the related morphism of graded left B-contramodules $\Delta_B(f) \colon B\otimes_R^{\Pi} L \longrightarrow B\otimes_R^{\Pi} M$. (It is helpful to keep in mind that the functor Δ_B, being a left adjoint, has to preserve cokernels.) $\square$

Lemma 7.10 *Let* (B, d, h) *be a nonnegatively graded CDG-ring. Then, for any left CDG-module* (M, d_M) *over* (B, d, h), *the graded left* B*-contramodule* $\Delta_B(M)$ *admits a unique odd contraderivation* $d_{\Delta_B(M)}$ *of degree* 1 *compatible with the derivation* d *on* B *and forming a commutative square diagram with the odd derivation* d_M *and the adjunction morphism* $M \longrightarrow \Delta_B(M)$. *The pair* $(\Delta_B(M), d_{\Delta_B(M)})$ *is a CDG-contramodule over* (B, d, h). *The assignment* $(M, d_M) \longmapsto (\Delta_B(M), d_{\Delta_B(M)})$ *is a DG-functor* $\mathsf{DG}(B\text{–mod}) \longrightarrow \mathsf{DG}(B\text{–contra})$ *left adjoint to the forgetful DG-functor* $\mathsf{DG}(B\text{–contra}) \longrightarrow \mathsf{DG}(B\text{–mod})$.

Proof To check uniqueness, suppose that $d'_{\Delta_B(M)}$ and $d''_{\Delta_B(M)}$ are two odd contraderivations of degree 1 on $\Delta_B(M)$ compatible with the derivation d on B. Then the difference $d''_{\Delta_B(M)} - d'_{\Delta_B(M)}$ is a morphism of graded left B-contramodules $\Delta_B(M) \longrightarrow \Delta_B(M)[1]$. If both $d'_{\Delta_B(M)}$ and $d''_{\Delta_B(M)}$ form commutative square diagrams with the odd derivation d_M

and the adjunction morphism $M \longrightarrow \Delta_B(M)$, then the difference $d''_{\Delta_B(M)} - d'_{\Delta_B(M)}$ is annihilated by the composition with the adjunction morphism. In view of the adjunction, it follows that $d''_{\Delta_B(M)} - d'_{\Delta_B(M)} = 0$.

Now let (Q, d_Q) be a left CDG-contramodule over (B, d, h). Then the group of all graded left B-contramodule morphisms $\Delta_B(M) \longrightarrow Q[n]$ is naturally isomorphic to the group of all graded left B-module morphisms $M \longrightarrow Q[n]$ for every $n \in \mathbb{Z}$. Hence we obtain an isomorphism between the underlying graded abelian groups of the complexes $\mathrm{Hom}^B(\Delta_B(M), Q)$ and $\mathrm{Hom}_B(M, Q)$. Since the adjunction map $M \longrightarrow \Delta_B(M)$ commutes with the differentials, it follows that the $\mathrm{Hom}^B(\Delta_B(M), Q) \simeq \mathrm{Hom}_B(M, Q)$ is an isomorphism of complexes of abelian groups. This establishes the DG-adjunction, which implies the DG-functoriality.

It remains to prove existence of the desired CDG-contramodule structure on $\Delta_B(M)$. For this purpose, consider the quasi-differential ring $(\widehat{B}, \partial)$ assigned to the CDG-ring (B, d, h) by the construction of Theorem 4.7. The key observation is that the functors $\Delta_{\widehat{B}}$ and Δ_B agree with each other, that is, they form a commutative square diagram with the functors of restriction of scalars $\widehat{B}\text{–mod}_{\mathrm{gr}} \longrightarrow B\text{–mod}_{\mathrm{gr}}$ and $\widehat{B}\text{–contra}_{\mathrm{gr}} \longrightarrow B\text{–contra}_{\mathrm{gr}}$. The latter two functors, essentially, assign to a CDG-(contra)module over (B, d, h) its underlying graded B-(contra)module.

Since both the functors $\Delta_{\widehat{B}}$ and Δ_B are right exact (while the functors of restriction of scalars are exact), it suffices to check that the two functors agree on free graded $\widehat{B}$-modules. It is helpful to observe that the restriction of scalars $\widehat{B}\text{–mod}_{\mathrm{gr}} \longrightarrow B\text{–mod}_{\mathrm{gr}}$ takes free graded $\widehat{B}$-modules to free graded B-modules. The rest is a straightforward computation based on Lemma 7.9.

Finally, given a CDG-module (M, d_M) over (B, d, h), in order to produce the induced differential $d_{\Delta_B(M)}$ on the graded B-contramodule $\Delta_B(M)$, one simply applies the functor $\Delta_{\widehat{B}}$ to the graded $\widehat{B}$-module M. $\qquad\square$

Now we consider the composition of the pair of adjoint DG-functors $\mathrm{Hom}_B(K^\vee(B, \widetilde{A}), -)\colon \mathsf{DG}(B\text{–mod}) \longrightarrow \mathsf{DG}(\widetilde{A}\text{–mod})$ and $K^\vee(B, \widetilde{A}) \otimes_{\widetilde{A}} -\colon \mathsf{DG}(\widetilde{A}\text{–mod}) \longrightarrow \mathsf{DG}(B\text{–mod})$ from the above discussion with the pair of adjoint DG-functors $\mathsf{DG}(B\text{–contra}) \longrightarrow \mathsf{DG}(B\text{–mod})$ and $\mathsf{DG}(B\text{–mod}) \longrightarrow \mathsf{DG}(B\text{–contra})$ provided by Lemma 7.10. This produces a pair of adjoint DG-functors between the DG-category $\mathsf{DG}(\widetilde{A}\text{–mod})$ of complexes of left $\widetilde{A}$-modules and the DG-category $\mathsf{DG}(B\text{–contra})$ of left CDG-contramodules over (B, d, h).

Here the right adjoint DG-functor

$$\mathrm{Hom}_B(K^\vee(B, \widetilde{A}), -)\colon \mathsf{DG}(B\text{–contra}) \longrightarrow \mathsf{DG}(\widetilde{A}\text{–mod})$$

is the composition $\mathsf{DG}(B\text{–contra}) \longrightarrow \mathsf{DG}(B\text{–mod}) \longrightarrow \mathsf{DG}(\widetilde{A}\text{–mod})$ of the forgetful DG-functor $\mathsf{DG}(B\text{–contra}) \longrightarrow \mathsf{DG}(B\text{–mod})$ with the functor $\mathrm{Hom}_B(K^\vee(B, \widetilde{A}), -)$: $\mathsf{DG}(B\text{–mod}) \longrightarrow \mathsf{DG}(\widetilde{A}\text{–mod})$. We denote it by

$$Q \longmapsto \mathrm{Hom}_B(K^\vee(B, \widetilde{A}), Q) = \mathrm{Hom}_R^{\tau'}(\widetilde{A}, Q),$$

where τ' is the same placeholder as in Sect. 6.5.

The left adjoint DG-functor

$$K^\vee(B, \widetilde{A}) \otimes_{\widetilde{A}}^{\Pi} - : \mathsf{DG}(\widetilde{A}\text{–mod}) \longrightarrow \mathsf{DG}(B\text{–contra})$$

is the composition $\mathsf{DG}(\widetilde{A}\text{–mod}) \longrightarrow \mathsf{DG}(B\text{–mod}) \longrightarrow \mathsf{DG}(B\text{–contra})$ of the DG-functor $K^\vee(B, \widetilde{A}) \otimes_{\widetilde{A}} - : \mathsf{DG}(\widetilde{A}\text{–mod}) \longrightarrow \mathsf{DG}(B\text{–mod})$ with the DG-functor $\Delta_B : \mathsf{DG}(B\text{–mod}) \longrightarrow \mathsf{DG}(B\text{–contra})$. This DG-functor takes a complex of left $\widetilde{A}$-modules $P = P^\bullet$ to the left CDG-contramodule over (B, d, h) whose underlying graded left B-contramodule is

$$K^\vee(B, \widetilde{A}) \otimes_{\widetilde{A}}^{\Pi} P = B \otimes_R^{\Pi} P = \mathrm{Hom}_R(\mathrm{Hom}_{R^{\mathrm{op}}}(B, R), P) = \mathrm{Hom}_R(C, P),$$

where C is a notation for the graded R-B-bimodule $C = \mathrm{Hom}_{R^{\mathrm{op}}}(B, R)$, as in Sect. 6.5. We denote this CDG-contramodule over (B, d, h) by

$$K^\vee(B, \widetilde{A}) \otimes_{\widetilde{A}}^{\Pi} P^\bullet = \mathrm{Hom}_R^{\sigma'}(\mathrm{Hom}_{R^{\mathrm{op}}}(B, R), P^\bullet) = \mathrm{Hom}_R^{\sigma'}(C, P^\bullet),$$

where σ' is also the same placeholder as in Sect. 6.5 (see Sect. 8.7 below for a further discussion of these placeholders).

Here the computation of $\Delta_B(B \otimes_R P)$ as $B \otimes_R^{\Pi} P$ is provided by Lemma 7.9. It is only the differential on $K^\vee(B, \widetilde{A}) \otimes_{\widetilde{A}}^{\Pi} P^\bullet = \Delta_B(K^\vee(B, \widetilde{A}) \otimes_{\widetilde{A}} P^\bullet)$ that remains to be explained. Lemma 7.10 establishes existence of a unique odd contraderivation of degree 1 on $B \otimes_R^{\Pi} P = K^\vee(B, \widetilde{A}) \otimes_{\widetilde{A}}^{\Pi} P$ compatible with the derivation d on B and agreeing with the differential on $K^\vee(B, \widetilde{A}) \otimes_{\widetilde{A}} P^\bullet$. Now it is straightforward to check that the differential $d_{e'} \otimes^{\Pi} \mathrm{id}_P + \mathrm{id}_{K^\vee(B,\widetilde{A})} \otimes^{\Pi} d_{P^\bullet}$ on $K^\vee(B, \widetilde{A}) \otimes_{\widetilde{A}}^{\Pi} P^\bullet$ satisfies these conditions. Hence we have

$$d_{\Delta_B(K^\vee(B,\widetilde{A}) \otimes_{\widetilde{A}} P^\bullet)} = d_{e'} \otimes^{\Pi} \mathrm{id}_P + \mathrm{id}_{K^\vee(B,\widetilde{A})} \otimes^{\Pi} d_{P^\bullet},$$

as one would expect (where $d_{e'}$ is the differential on the dual nonhomogeneous Koszul CDG-module $K^\vee(B, \widetilde{A})$ and $d_{P^\bullet}$ denotes the differential on the complex $P^\bullet$). The placeholder σ' is supposed to remind us of the first summand in this formula.

The pair of adjoint DG-functors that we have constructed induces a pair of adjoint triangulated functors

$$\mathrm{Hom}_B(K^\vee(B, \widetilde{A}), -) = \mathrm{Hom}_R^{\tau'}(\widetilde{A}, -)\colon \mathsf{Hot}(B\text{–}\mathsf{contra}) \longrightarrow \mathsf{Hot}(\widetilde{A}\text{–}\mathsf{mod})$$

and

$$K^\vee(B, \widetilde{A}) \otimes_{\widetilde{A}}^{\Pi} - = \mathrm{Hom}_R^{\sigma'}(C, -)\colon \mathsf{Hot}(\widetilde{A}\text{–}\mathsf{mod}) \longrightarrow \mathsf{Hot}(B\text{–}\mathsf{contra})$$

between the homotopy category $\mathsf{Hot}(\widetilde{A}\text{–}\mathsf{mod})$ of complexes of left $\widetilde{A}$-modules and the homotopy category $\mathsf{Hot}(B\text{–}\mathsf{contra})$ of left CDG-contramodules over (B, d, h).

7.5 Triangulated Equivalence

Let $R \subset \widetilde{V} \subset \widetilde{A}$ be a left finitely projective nonhomogeneous Koszul ring, and let (B, d, h) be the corresponding right finitely projective Koszul CDG-ring (see Sects. 3.4 and 4.6–4.7). The following theorem is the main result of Chap. 7.

Theorem 7.11 *The pair of adjoint triangulated functors* $\mathrm{Hom}_R^{\sigma'}(C, -)\colon \mathsf{Hot}(\widetilde{A}\text{–}\mathsf{mod})$ $\longrightarrow$ $\mathsf{Hot}(B\text{–}\mathsf{contra})$ *and* $\mathrm{Hom}_R^{\tau'}(\widetilde{A}, -)\colon \mathsf{Hot}(B\text{–}\mathsf{contra})$ $\longrightarrow$ $\mathsf{Hot}(\widetilde{A}\text{–}\mathsf{mod})$ *defined in Sect. 7.4 induces a pair of adjoint triangulated functors between the* $\widetilde{A}/R$-*semicontraderived category of left* $\widetilde{A}$-*modules* $\mathsf{D}_R^{\mathsf{sictr}}(\widetilde{A}\text{–}\mathsf{mod})$, *as defined in Sect. 7.1, and the contraderived category* $\mathsf{D}^{\mathsf{ctr}}(B\text{–}\mathsf{contra})$ *of left CDG-contramodules over* (B, d, h), *as defined in Sect. 7.3. Under the above Koszulity assumption, the latter two functors are mutually inverse triangulated equivalences,*

$$\mathsf{D}_R^{\mathsf{sictr}}(\widetilde{A}\text{–}\mathsf{mod}) \simeq \mathsf{D}^{\mathsf{ctr}}(B\text{–}\mathsf{contra}).$$

Concrete examples of the triangulated equivalence stated in the theorem will be presented below in Sects. 10.2–10.10. The theorem is also applicable in the situations described in Examples 3.17 and 3.22.

Proof This is a generalization of [59, Theorem B.2(b)] and a particular case of [58, Theorem 11.8(c)]. The proof is similar to the proofs of these results, and dual-analogous to the proof of Theorem 6.14.

Recall from the discussion in Section 5.6 that any graded left B-contramodule Q is endowed with a canonical decreasing filtration by graded B-subcontramodules $Q = G^0 Q \supset G^1 Q \supset G^2 Q \supset \cdots$, where $G^m Q \subset Q$ is the image of the restriction

$$B^{\geq m} \otimes_R^{\Pi} Q \xrightarrow{\pi_Q} Q$$

of the contraaction map $\pi_Q: B \otimes_R^{\Pi} Q \longrightarrow Q$ to the graded submodule $B^{\geq m} \otimes_R^{\Pi} Q \subset B \otimes_R^{\Pi} Q$. Here, as usually, the notation $B^{\geq m}$ stands for the homogeneous two-sided ideal $\bigoplus_{n \geq m} B^n \subset B$. The canonical decreasing filtration G on a graded B-contramodule Q is always complete, but it does not have to be separated; see Proposition 5.32 and Example 5.33. The successive quotient contramodules $G^m Q / G^{m+1} Q$ have trivial graded B-contramodule structures, in the sense explained in Sect. 5.6.

A left CDG-contramodule (Q, d_Q) over (B, d, h) is said to be *trivial* if its underlying graded B-contramodule is trivial. The square of the differential d_Q on a trivial CDG-contramodule Q is zero, as the curvature element $h \in B^2$ acts by zero in Q. The DG-category of trivial left CDG-contramodules over (B, d, h) is equivalent to the DG-category of complexes of left R-modules. For any left CDG-contramodule (Q, d_Q) over (B, d, h), the canonical decreasing filtration G of the underlying graded B-contramodule of Q is a filtration by CDG-subcontramodules with trivial CDG-contramodule successive quotients.

The ring $\widetilde{A}$ is endowed with an increasing filtration F, which was defined in Sect. 3.1. We also endow the graded R-B-bimodule $C = \mathrm{Hom}_{R^{\mathrm{op}}}(B, R)$ with the increasing filtration F induced by the grading, as in the proof of Theorem 6.14. It is important for the argument below that both $F_n \widetilde{A}/F_{n-1}\widetilde{A}$ and $F_n C/F_{n-1}C$ are projective (graded) left R-modules for all $n \geq 0$ (in fact, they are finitely generated projective left R-modules in our assumptions). In particular, $\widetilde{A}$ is a projective left R-module and C is a projective graded left R-module. Hence both the functors $\mathrm{Hom}_R^{\tau'}(\widetilde{A}, -)$ and $\mathrm{Hom}_R^{\sigma'}(C, -)$ preserve short exact sequences (of left CDG-contramodules over (B, d, h) and of complexes of left $\widetilde{A}$-modules, respectively). Both the functors also obviously preserve infinite products.

Let $P^\bullet$ be a complex of left $\widetilde{A}$-modules that is contraacyclic as a complex of left R-modules. We need to show that the left CDG-contramodule $\mathrm{Hom}_R^{\sigma'}(C, P^\bullet)$ over (B, d, h) is contraacyclic. Let F be the decreasing filtration on the Hom module $\mathrm{Hom}_R(C, P)$ induced by the increasing filtration F on C, that is $F^n \mathrm{Hom}_R(C, P) = \mathrm{Hom}_R(C/F_{n-1}C, P) \subset \mathrm{Hom}_R(C, P)$. Then F is a complete, separated filtration of the CDG-contramodule $\mathrm{Hom}_R^{\sigma'}(C, P^\bullet)$ by CDG-subcontramodules over (B, d, h). The successive quotient CDG-contramodules $F^n \mathrm{Hom}_R^{\sigma'}(C, P^\bullet)/F^{n+1} \mathrm{Hom}_R^{\sigma'}(C, P^\bullet)$ are trivial. Viewed as complexes of left R-modules, these are the Hom complexes $\mathrm{Hom}_R(F_n C/F_{n-1}C, P^\bullet)$, where $F_n C/F_{n-1}C$ is a graded R-R-bimodule concentrated in the cohomological degree $-n$ and endowed with the zero differential.

Since the graded left R-module $F_n C/F_{n-1}C$ is projective, the functor $\mathrm{Hom}_R(F_n C/F_{n-1}C, -)$ takes contraacyclic complexes of left R-modules to contraacyclic complexes of left R-modules. Clearly, any contraacyclic complex of left R-modules is also contraacyclic as a CDG-contramodule over (B, d, h) with the trivial contramodule structure. It remains to use Lemma 7.8 in order to conclude that $\mathrm{Hom}_R^{\sigma'}(C, P^\bullet)$ is a contraacyclic left CDG-contramodule over (B, d, h).

For any contraacyclic left CDG-contramodule Q over (B, d, h), the complex of left $\widetilde{A}$-modules $\mathrm{Hom}_R^{\tau'}(\widetilde{A}, Q)$ is contraacyclic not only as a complex of left R-modules, but even as a complex of left $\widetilde{A}$-modules. This follows immediately from the fact that $\widetilde{A}$ is a projective left R-module; so the functor $\mathrm{Hom}_R^{\tau'}(\widetilde{A}, -)$ takes short exact sequences of left CDG-contramodules over (B, d, h) to short exact sequences of complexes of left $\widetilde{A}$-modules (and this functor also preserves infinite products).

Let $P^\bullet$ be an arbitrary complex of left $\widetilde{A}$-modules. We need to show that the cone $Y^\bullet$ of the adjunction morphism of complexes of left $\widetilde{A}$-modules $P^\bullet \longrightarrow \mathrm{Hom}_R^{\tau'}(\widetilde{A}, \mathrm{Hom}_R^{\sigma'}(C, P^\bullet))$ is contraacyclic as a complex of left R-modules. Endow the graded left R-module $\mathrm{Hom}_R(\widetilde{A}, \mathrm{Hom}_R(C, P))$ with the decreasing filtration F induced by the increasing filtrations F on C and $\widetilde{A}$, that is $F^n \mathrm{Hom}_R(\widetilde{A}, \mathrm{Hom}_R(C, P)) = \sum_{i+j=n-1} \mathrm{Hom}_R(\widetilde{A}/F_j\widetilde{A}, \mathrm{Hom}_R(C/F_iC, P)) \subset \mathrm{Hom}_R(\widetilde{A}, \mathrm{Hom}_R(C, P))$. Then F is a complete, separated filtration of the complex $\mathrm{Hom}_R^{\tau'}(\widetilde{A}, \mathrm{Hom}_R^{\sigma'}(C, P^\bullet))$ by subcomplexes of left R-modules (but not of left $\widetilde{A}$-modules). The complex $P^\bullet$ is endowed with the trivial filtration $F^0 P^\bullet = P^\bullet$, $F^1 P^\bullet = 0$; and the complex $Y^\bullet$ is endowed with the induced decreasing filtration F.

Then the associated graded complex of left R-modules

$$\mathrm{gr}_F \mathrm{Hom}_R^{\tau'}(\widetilde{A}, \mathrm{Hom}_R^{\sigma'}(C, P^\bullet))$$
$$= \prod_{n=0}^\infty F^n \mathrm{Hom}_R^{\tau'}(\widetilde{A}, \mathrm{Hom}_R^{\sigma'}(C, P^\bullet)) / F^{n+1} \mathrm{Hom}_R^{\tau'}(\widetilde{A}, \mathrm{Hom}_R^{\sigma'}(C, P^\bullet))$$

is naturally isomorphic to the Hom complex $\mathrm{Hom}_R({}^\tau K_\bullet(B, A), P^\bullet)$ from the complex of (graded) R-R-bimodules ${}^\tau K_\bullet(B, A) = C \otimes_R^\tau A$ constructed in Sect. 2.6 into the complex of left R-modules $P^\bullet$. Here $A = \mathrm{gr}^F \widetilde{A}$ is the left finitely projective Koszul graded ring associated with $\widetilde{A}$. Arguing further as in the proof of Theorem 6.14, we conclude that $\mathrm{gr}_F Y^\bullet = \mathrm{Hom}_R(\mathrm{cone}({}^\tau K_\bullet(B, A) \to R)[-1], P^\bullet)$ is a contraacyclic complex of left R-modules. By Lemma 7.2, it follows that $Y^\bullet$ is a contraacyclic complex of left R-modules, too.

Let $Q = (Q, d_Q)$ be an arbitrary left CDG-contramodule over (B, d, h). We need to show that the cone Z of the adjunction morphism of left CDG-contramodules $\mathrm{Hom}_R^{\sigma'}(C, \mathrm{Hom}_R^{\tau'}(\widetilde{A}, Q)) \longrightarrow Q$ is contraacyclic as a CDG-contramodule over (B, d, h).

First of all we observe that the class of all left CDG-contramodules Q having the desired property is preserved by the passages to the cokernels of injective closed morphisms of CDG-contramodules (as well as to the kernels of surjective closed morphisms; but we need the cokernels). This holds because the class of all contraacyclic left CDG-contramodules over (B, d, h) is closed under the cokernels of injective closed morphisms (as well as the kernels of surjective closed morphisms).

Consider our CDG-contramodule Q as a graded left contramodule over the graded ring $\widehat{B}$ from Theorem 4.7, and present it as a quotient of a free graded $\widehat{B}$-contramodule. All free graded contramodules are separated, all subcontramodules of separated graded contramodules are separated, and the restriction of scalars takes separated graded $\widehat{B}$-contramodules to separated graded B-contramodules (besides, it takes free graded $\widehat{B}$-contramodules to free graded B-contramodules). So any subcontramodule of a free graded $\widehat{B}$-contramodule is separated as a graded B-contramodule. This allows to present an arbitrary CDG-contramodule Q over (B, d, h) as the cokernel of an injective morphism of CDG-contramodules whose underlying graded B-contramodules are separated. Thus we can assume that Q is a separated graded B-contramodule.

Now we endow Q with the canonical decreasing filtration $Q = G^0 Q \supset G^1 Q \supset G^2 Q \supset \cdots$; so, as mentioned above, $G^m Q$ are CDG-subcontramodules in Q and $G^m Q / G^{m+1} Q$ are trivial CDG-contramodules over (B, d, h). The filtration G is separated and complete. Consider the induced decreasing filtrations G on the left CDG-contramodule $\mathrm{Hom}_R^{\sigma'}(C, \mathrm{Hom}_R^{\tau'}(\widetilde{A}, Q))$ and the left CDG-contramodule Z over (B, d, h). In particular, $G^m \mathrm{Hom}_R^{\sigma'}(C, \mathrm{Hom}_R^{\tau'}(\widetilde{A}, Q)) = \mathrm{Hom}_R^{\sigma'}(C, \mathrm{Hom}_R^{\tau'}(\widetilde{A}, G^m Q))$. These decreasing filtrations are also separated and complete. The successive quotient CDG-contramodules $\mathrm{Hom}_R^{\sigma'}(C, \mathrm{Hom}_R^{\tau'}(\widetilde{A}, G^m Q)) / \mathrm{Hom}_R^{\sigma'}(C, \mathrm{Hom}_R^{\tau'}(\widetilde{A}, G^{m+1} Q))$ are computable as the CDG-contramodules $\mathrm{Hom}_R^{\sigma'}(C, \mathrm{Hom}_R^{\tau'}(\widetilde{A}, G^m Q / G^{m+1} Q))$, and similarly for the CDG-contramodules $G^m Z / G^{m+1} Z$. Arguing as in the proof of Theorem 6.14 and using Lemma 7.8, we reduce the question to the case of a trivial left CDG-contramodule Q over (B, d, h). So we can assume that $Q = Q^\bullet$ is simply a complex of left R-modules endowed with the trivial graded B-contramodule structure.

Next we endow the graded left R-module $\mathrm{Hom}_R(C, \mathrm{Hom}_R(\widetilde{A}, Q))$ with the decreasing filtration F induced by the increasing filtrations F on $\widetilde{A}$ and C, that is $F^n \mathrm{Hom}_R(C, \mathrm{Hom}_R(\widetilde{A}, Q)) = \sum_{i+j=n-1} \mathrm{Hom}_R(C/F_i C, \mathrm{Hom}_R(\widetilde{A}/F_j \widetilde{A}, Q)) \subset \mathrm{Hom}_R(C, \mathrm{Hom}_R(\widetilde{A}, Q))$. Then F is a complete, separated filtration of the CDG-contramodule $\mathrm{Hom}_R^{\sigma'}(C, \mathrm{Hom}_R^{\tau'}(\widetilde{A}, Q^\bullet))$ by CDG-subcontramodules with trivial CDG-contramodule successive quotients. As a complex of left R-modules, the trivial CDG-contramodule $\mathrm{gr}_F \mathrm{Hom}_R^{\sigma'}(C, \mathrm{Hom}_R^{\tau'}(\widetilde{A}, Q^\bullet))$ is naturally isomorphic to the Hom complex $\mathrm{Hom}_R(K_\bullet^\tau(B, A), Q^\bullet)$ from the complex of (graded) R-R-bimodules $K_\bullet^\tau(B, A) = A \otimes_R^\sigma C$ constructed in Sect. 2.5 into the complex of left R-modules $Q^\bullet$.

Continuing to argue as in the proof of Theorem 6.14, we conclude that $\mathrm{gr}_F Z = \prod_{n=0}^\infty F^n Z / F^{n+1} Z \simeq \mathrm{Hom}_R(\mathrm{cone}(R \to K_\bullet^\tau(B, A))[-1], Q^\bullet)$ is a contraacyclic complex of left R-modules. Here it is important that the complex $\mathrm{cone}(R \to K_\bullet^\tau(B, A))$ is "coacyclic with respect to the exact category of left R-projective R-R-bimodules" in the sense of [58, Section 2.1]. Hence $\mathrm{gr}_F Z$ is also contraacyclic as a left CDG-contramodule over (B, d, h). By Lemma 7.8, it follows that Z is a contraacyclic CDG-contramodule over (B, d, h) as well. $\square$

Examples 7.12

(1) The free left R-module R can be considered as a one-term complex of left R-modules, concentrated in the cohomological degree 0 and endowed with the zero differential. This one-term complex of left R-modules can be then considered as a trivial left CDG-contramodule over (B, d, h). The functor $\mathrm{Hom}_R^{\tau'}(\widetilde{A}, -) = \mathrm{Hom}_B(K_{e'}^\vee(B, \widetilde{A}), -)$ takes this trivial left CDG-contramodule R over (B, d, h) to

the left $\widetilde{A}$-module $\mathrm{Hom}_R(\widetilde{A}, R)$ (viewed as a one-term complex concentrated in the cohomological degree 0).

(2) Assume that $\widetilde{A}$ is a left augmented left finitely projective nonhomogeneous Koszul ring, in the sense of the definitions in Sects. 3.8 and 4.7. Then the left augmentation of $\widetilde{A}$ endows the base ring R with a structure of left $\widetilde{A}$-module.

Furthermore, by Theorem 3.27 we have $h = 0$, so $(B, d, h) = (B, d)$ is a DG-ring. As such, (B, d) is naturally a left DG-module over itself. Being nonnegatively cohomologically graded, this DG-module has a unique structure of DG-contramodule over (B, d) compatible with its DG-module structure (see Example 5.30(1)). (Here, by definition, a *DG-contramodule* over a nonnegatively graded DG-ring (B, d) is the same thing as a CDG-contramodule over $(B, d, 0)$.)

In this setting, the functor $\mathrm{Hom}_R^{\sigma'}(C, -) = K_{e'}^\vee(B, \widetilde{A}) \otimes_{\widetilde{A}}^{\Pi} -$ takes the left $\widetilde{A}$-module R to the left DG-contramodule (B, d) over the DG-ring (B, d).

Remark 7.13 Assume that $\widetilde{A}$ is left augmented over R, as in Remark 6.16(2) and Example 7.12(2). Then, for any complex of left $\widetilde{A}$-modules $P^\bullet$, the CDG-(contra)module $\mathrm{Hom}_R^{\sigma'}(C, P^\bullet)$ is a DG-module over $(B, d, h) = (B, h)$, i.e., a complex. We have a natural isomorphism of complexes of abelian groups $\mathrm{Hom}_{\mathscr{B}}^{\sigma'}(C, P^\bullet) \simeq \mathrm{Hom}_{\widetilde{A}}(\widetilde{A} \otimes_R^{\sigma'} C, P^\bullet)$. Since $\widetilde{A} \otimes_R^{\sigma'} C$ is a projective resolution of the left $\widetilde{A}$-module R by Example 6.15(5), it follows that the complex of abelian groups $\mathrm{Hom}_R^{\sigma'}(C, P^\bullet)$ computes $\mathbb{R}\,\mathrm{Hom}_{\widetilde{A}}(R, P^\bullet)$. In fact, the DG-module $\mathrm{Hom}_R^{\sigma'}(C, P^\bullet)$ computes the cohomology $\mathrm{Ext}_{\widetilde{A}}^i(R, P^\bullet) = \mathrm{Hom}_{\mathsf{D}(\widetilde{A}-\mathsf{mod})}(R, P^\bullet[i])$, $i \in \mathbb{Z}$, as a graded module over the graded ring of endomorphisms of the derived category object $R \in \mathsf{D}(\widetilde{A}-\mathsf{mod})$.

7.6 Reduced Contraderived Category

We keep the assumptions of Sect. 7.5 (which coincide with the assumptions of Sections 6.6–6.7). The following corollary is a special case of Theorem 7.11.

Corollary 7.14 *Assume additionally that the left homological dimension of the ring R (that is, the homological dimension of the abelian category $R-\mathsf{mod}$) is finite. Then the pair of adjoint triangulated functors $\mathrm{Hom}_R^{\sigma'}(C, -)\colon \mathsf{Hot}(\widetilde{A}-\mathsf{mod}) \longrightarrow \mathsf{Hot}(B-\mathsf{contra})$ and $\mathrm{Hom}_R^{\tau'}(\widetilde{A}, -)\colon \mathsf{Hot}(B-\mathsf{contra}) \longrightarrow \mathsf{Hot}(\widetilde{A}-\mathsf{mod})$ induces mutually inverse triangulated equivalences*

$$\mathsf{D}(\widetilde{A}-\mathsf{mod}) \simeq \mathsf{D}^{\mathrm{ctr}}(B-\mathsf{contra}) \tag{7.1}$$

between the derived category of left $\widetilde{A}$-modules and the contraderived category of left CDG-contramodules over (B, d, h).

Concrete examples of the triangulated equivalence (7.1) will be given below in Sects. 10.2–10.5 and 10.8–10.10.

Proof When the left homological dimension of the ring R is finite, all acyclic complexes of left R-modules are contraacyclic by [58, Remark 2.1]. So the $\widetilde{A}/R$-semicontraderived category of left $\widetilde{A}$-modules coincides with their derived category, $\mathsf{D}_R^{\mathsf{sictr}}(\widetilde{A}\text{–mod}) = \mathsf{D}(\widetilde{A}\text{–mod})$, and the assertion follows from Theorem 7.11.   □

We would like to obtain a description of the derived category $\mathsf{D}(\widetilde{A}\text{–mod})$ in terms of some kind of exotic derived category of left CDG-contramodules over (B, d, h) *without* assuming finiteness of the left homological dimension of R. The next definition serves this purpose.

Definition 7.15 Consider the full subcategory of acyclic complexes of left R-modules $\mathsf{Acycl}(R\text{–mod}) \subset \mathsf{Hot}(R\text{–mod})$. All complexes of left R-modules, and, in particular, acyclic complexes of left R-modules, can be viewed as trivial left CDG-contramodules over (B, d, h).

By an abuse of notation, let us denote simply by $\langle \mathsf{Acycl}(R\text{–mod})\rangle_\sqcap$ the minimal full triangulated subcategory in $\mathsf{D}^{\mathsf{ctr}}(B\text{–contra})$ containing all the acyclic complexes of left R-modules (viewed as trivial CDG-contramodules) and closed under infinite products. Consider the triangulated Verdier quotient category

$$\mathsf{D}_{R\text{-red}}^{\mathsf{ctr}}(B\text{–contra}) = \mathsf{D}^{\mathsf{ctr}}(B\text{–contra})/\langle \mathsf{Acycl}(R\text{–mod})\rangle_\sqcap.$$

We will call it the *reduced contraderived category of left CDG-contramodules over* (B, d, h) *relative to* R.

Theorem 7.16 *The pair of adjoint triangulated functors* $\mathsf{Hom}_R^{\sigma'}(C, -)\colon \mathsf{Hot}(\widetilde{A}\text{–mod})$ $\longrightarrow \mathsf{Hot}(B\text{–contra})$ *and* $\mathsf{Hom}_R^{\tau'}(\widetilde{A}, -)\colon \mathsf{Hot}(B\text{–contra}) \longrightarrow \mathsf{Hot}(\widetilde{A}\text{–mod})$ *induces mutually inverse triangulated equivalences*

$$\mathsf{D}(\widetilde{A}\text{–mod}) \simeq \mathsf{D}_{R\text{-red}}^{\mathsf{ctr}}(B\text{–contra})$$

between the derived category of left $\widetilde{A}$-modules *and the reduced contraderived category of left CDG-contramodules over* (B, d, h) *relative to* R.

Examples of the triangulated equivalence stated in the theorem will be given below in Sects. 10.6–10.10. Some other situations in which the theorem is applicable are described in Examples 3.17 and 3.22.

Proof This is the dual-analogous assertion to Theorem 6.20. In view of Theorem 7.11, only two things still need to be checked. Firstly, one has to show that, for any acyclic complex of left $\widetilde{A}$-modules $P^\bullet$, the CDG-contramodule $\operatorname{Hom}_R^{\sigma'}(C, P^\bullet)$ over (B, d, h) belongs to the triangulated subcategory $\langle\mathsf{Acycl}(R\text{–mod})\rangle_\sqcap \subset \mathsf{D}^{\mathrm{ctr}}(B\text{–contra})$. Secondly, it needs to be established that, for any CDG-contramodule $Q \in \langle\mathsf{Acycl}(R\text{–mod})\rangle_\sqcap$, the complex of left $\widetilde{A}$-modules $\operatorname{Hom}_R^{\tau'}(\widetilde{A}, Q)$ is acyclic.

Firstly, let $P = P^\bullet$ be an acyclic complex of left $\widetilde{A}$-modules. Consider the decreasing filtration F on the graded left R-module $\operatorname{Hom}_R(C, P)$ induced by the increasing filtration F on the R-R-bimodule C, that is $F^n \operatorname{Hom}_R(C, P) = \operatorname{Hom}_R(C/F_{n-1}C, P)$ for all $n \geq 0$. Then F is a complete, separated filtration of the CDG-contramodule $\operatorname{Hom}_R^{\sigma'}(C, P^\bullet)$ over (B, d, h) by its CDG-subcontramodules. The successive quotient CDG-contramodules $F^n \operatorname{Hom}_R^{\sigma'}(C, P)/F^{n+1} \operatorname{Hom}_R^{\sigma'}(C, P)$ are trivial CDG-contramodules which, viewed as complexes of left R-modules, can be computed as the Hom complexes $\operatorname{Hom}_R(F_n C/F_{n-1}C, P^\bullet)$.

As in the proof of Theorem 6.20, one can see that the complexes of left R-modules $\operatorname{Hom}_R(F_n C/F_{n-1}C, P^\bullet)$ are acyclic. Now the iterated extension and telescope sequence argument from the proofs of Lemmas 7.2 and 7.8 shows that the left CDG-contramodule $\operatorname{Hom}_R^{\sigma'}(C, P^\bullet)$ over (B, d, h) belongs to the triangulated subcategory $\langle\mathsf{Acycl}(R\text{–mod})\rangle_\sqcap \subset \mathsf{D}^{\mathrm{ctr}}(R\text{–mod})$.

Secondly, since the functor $\operatorname{Hom}_R^{\tau'}(\widetilde{A}, -)$ is triangulated and preserves infinite products, it suffices to check that for any acyclic complex of left R-modules $Q^\bullet$, viewed as a trivial left CDG-contramodule over (B, d, h), the complex of left $\widetilde{A}$-modules $\operatorname{Hom}_R^{\tau'}(\widetilde{A}, Q^\bullet)$ is acyclic. As $Q^\bullet$ is a trivial CDG-contramodule, the complex of left $\widetilde{A}$-modules $\operatorname{Hom}_R^{\tau'}(\widetilde{A}, Q^\bullet) = \operatorname{Hom}_R(\widetilde{A}, Q^\bullet)$ is simply obtained from the complex of left R-modules $Q^\bullet$ by applying the Hom functor $\operatorname{Hom}_R(\widetilde{A}, -)$ termwise. Since $\widetilde{A}$ is a projective left R-module, it follows that the complex $\operatorname{Hom}_R(\widetilde{A}, Q^\bullet)$ is acyclic.   □

The Co-Contra Correspondence

8

8.1   Contratensor Product

Let $B = \bigoplus_{n=0}^{\infty} B_n$ be a nonnegatively graded ring, and let K be a ring endowed with a ring homomorphism $K \longrightarrow B_0$. We start from the case of ungraded comodules and contramodules before passing to the graded ones.

Definition 8.1 Let N be an ungraded right B-comodule (in the sense of Sect. 5.1) and P be an ungraded left B-contramodule (in the sense of Sect. 5.3). The *contratensor product* $N \odot_B P$ is an abelian group constructed as the cokernel of (the difference of) a natural pair of abelian group homomorphisms

$$N \otimes_K \prod_{n=0}^{\infty} (B_n \otimes_K P) \rightrightarrows N \otimes_K P.$$

Here the first map is simply $N \otimes_K \pi_P : N \otimes_K \prod_{n=0}^{\infty} (B_n \otimes_K P) \longrightarrow N \otimes_K P$, where $\pi_P : \prod_{n=0}^{\infty} B_n \otimes_K P \longrightarrow P$ is the contraaction map. The second map is constructed using the right (co)action of B in N.

Specifically, consider a pair of elements $y \in N$ and $w \in \prod_{n=0}^{\infty} B_n \otimes_K P$. Choose an integer $m \geq 0$ such that $y B_{\geq m+1} = 0$ in N. Then the image of $y \otimes w$ under the second map $N \otimes_K \left(\prod_{n=0}^{\infty} B_n \otimes_K P \right) \longrightarrow N \otimes_K P$ is, by definition, equal to the image of $y \otimes w$ under the composition

$$N \otimes_K \prod_{n=0}^{\infty} (B_n \otimes_K P) \longrightarrow N \otimes_K \bigoplus_{n=0}^{m} (B_n \otimes_K P)$$

$$= \bigoplus_{n=0}^{m} N \otimes_K B_n \otimes_K P \longrightarrow N \otimes_K P.$$

© The Author(s), under exclusive license to Springer Nature Switzerland AG 2021   183
L. Positselski, *Relative Nonhomogeneous Koszul Duality*, Frontiers in Mathematics,
https://doi.org/10.1007/978-3-030-89540-2_8

Here the map $N \otimes_K \prod_{n=0}^{\infty} (B_n \otimes_K P) \longrightarrow N \otimes_K \bigoplus_{n=0}^{m} (B_n \otimes_K P)$ is the direct summand projection, while the map $\bigoplus_{n=0}^{m} N \otimes_K B_n \otimes_K P \longrightarrow N \otimes_K P$ is induced by the right action maps $N \otimes_K B_n \longrightarrow N$.

Notice that, by construction, the contratensor product $N \odot_B P$ is a quotient group of the tensor product $N \otimes_B P$. In other words, there is a natural surjective map of abelian groups

$$N \otimes_B P \longrightarrow N \odot_B P. \tag{8.1}$$

One can also easily see that the contratensor product $N \odot_B P$ does not depend on the choice of a ring K. (So our notation is unambiguous, and one can take, at one's convenience, $K = \mathbb{Z}$, or $K = B_0$, etc.)

Lemma 8.2 *Let E be an associative ring and N be an E-B-bimodule such that the underlying right B-module of N is an (ungraded) right B-comodule. Then the functor*

$$\operatorname{Hom}_E(N, -)\colon E\text{--}\mathsf{mod} \longrightarrow B\text{--}\mathsf{contra}$$

constructed in Example 5.14 is right adjoint to the contratensor product functor

$$N \odot_B -\colon B\text{--}\mathsf{contra} \longrightarrow E\text{--}\mathsf{mod}.$$

In other words, for any ungraded left B-contramodule P and any left E-module U there is a natural isomorphism of abelian groups

$$\operatorname{Hom}_E(N \odot_B P, \, U) \simeq \operatorname{Hom}^B(P, \operatorname{Hom}_E(N, U)).$$

Proof The claim is that, under the adjunction isomorphism $\operatorname{Hom}_E(N \otimes_B P, \, U) \simeq \operatorname{Hom}_B(P, \operatorname{Hom}_E(N, U))$, the subgroup $\operatorname{Hom}_E(N \odot_B P, \, U) \subset \operatorname{Hom}_E(N \otimes_B P, \, U)$ provided by the surjective left E-module map (8.1) corresponds to the subgroup $\operatorname{Hom}^B(P, \operatorname{Hom}_E(N, U)) \subset \operatorname{Hom}_B(P, \operatorname{Hom}_E(N, U))$ provided by the faithful forgetful functor $B\text{--}\mathsf{contra} \longrightarrow B\text{--}\mathsf{mod}$. One can see this by comparing the above construction of the contratensor product $N \odot_B P$ with the construction of the B-contramodule structure on the Hom group $\operatorname{Hom}_E(N, U)$ in Example 5.14. $\square$

Definition 8.3 Now let N be a graded right B-comodule (in the sense of Sect. 5.5) and P be a graded left B-contramodule (in the sense of Sect. 5.6). Then the *contratensor product* $N \odot_B P$ is a graded abelian group constructed as the cokernel of (the difference of) a natural pair of homomorphisms of graded abelian groups

$$N \otimes_K (B \otimes_K^{\Pi} P) \rightrightarrows N \otimes_K P.$$

Here the first map is $N \otimes_K \pi_P \colon N \otimes_K (B \otimes_K^{\Pi} P) \longrightarrow N \otimes_K P$, where $\pi_P \colon B \otimes_K^{\Pi} P = \mathbb{M}_K^{\mathrm{gr}}(P) \longrightarrow P$ is the contraaction map. The second map is constructed using the right (co)action of B in N.

Specifically, consider a pair of elements $y \in N_i$ and $w \in (B \otimes_K^{\Pi} P)_j = \prod_{n \in \mathbb{Z}} (B_n \otimes_K P_{j-n})$, $i, j \in \mathbb{Z}$. Choose an integer $m \geq 0$ such that $y B_{\geq m+1} = 0$. Then the image of $y \otimes w$ under the second map $N \otimes_K (B \otimes_K^{\Pi} P) \longrightarrow N \otimes_K P$ is, by definition, equal to the image of $y \otimes w$ under the composition

$$N_i \otimes_K \prod_{n=0}^{\infty} (B_n \otimes_K P_{j-n}) \longrightarrow N_i \otimes_K \bigoplus_{n=0}^{m} (B_n \otimes_K P_{j-n})$$

$$= \bigoplus_{n=0}^{m} N_i \otimes_K B_n \otimes_K P_{j-n} \longrightarrow (N \otimes_K P)_{i+j}.$$

Here the map $N_i \otimes_K \prod_{n=0}^{\infty} (B_n \otimes_K P_{j-n}) \longrightarrow N_i \otimes_K \bigoplus_{n=0}^{m} (B_n \otimes_K P_{j-n})$ is the direct summand projection, while the map $\bigoplus_{n=0}^{m} N_i \otimes_K B_n \otimes_K P_{j-n} \longrightarrow (N \otimes_K P)_{i+j}$ is induced by the right action maps $N_i \otimes_K B_n \longrightarrow N_{i+n}$.

By construction, the contratensor product $N \odot_B P$ is a homogeneous quotient group of the tensor product $N \otimes_B P$. In other words, there is a natural surjective map of graded abelian groups $N \otimes_B P \longrightarrow N \odot_B P$. One can also easily see that the contratensor product $N \odot_B P$ does not depend on the choice of a ring K (so our notation is unambiguous).

For any graded left B-contramodules P and Q, we denote by $\mathrm{Hom}^B(P, Q)$ the graded B-contramodule Hom group from P to Q, that is, the graded abelian group with the components

$$\mathrm{Hom}^B(P, Q)_n = \mathrm{Hom}_{B\text{-contragr}}(P, Q(-n)),$$

where $Q(-n)$ denotes the left B-contramodule Q with the shifted grading, $Q(-n)_j = Q_{j+n}$ (cf. the notation in Sect. 5.7).

Lemma 8.4 *Let $E = \bigoplus_{n \in \mathbb{Z}} E_n$ be a graded associative ring and $N = \bigoplus_{n \in \mathbb{Z}} N_n$ be a graded E-B-bimodule such that the underlying graded right B-module of N is a (graded) right B-comodule. Then the functor*

$$\mathrm{Hom}_E(N, -) \colon E\text{-mod}_{\mathrm{gr}} \longrightarrow B\text{-contra}_{\mathrm{gr}}$$

constructed in Example 5.31 is right adjoint to the contratensor product functor

$$N \odot_B - \colon B\text{-contra}_{\mathrm{gr}} \longrightarrow E\text{-mod}_{\mathrm{gr}}.$$

In other words, for any graded left B-contramodule P and any graded left E-module U there is a natural isomorphism of graded abelian groups

$$\operatorname{Hom}_E(N \odot_B P, \ U) \simeq \operatorname{Hom}^B(P, \operatorname{Hom}_E(N, U)).$$

Proof Similar to the proof of Lemma 8.2.   □

Proposition 8.5

(a) *Assume that the forgetful functor B–contra $\longrightarrow$ B–mod is fully faithful. Then the natural map of abelian groups $N \otimes_B P \longrightarrow N \odot_B P$ (8.1) is an isomorphism for any ungraded right B-comodule N and any ungraded left B-contramodule P.*
(b) *Assume that the forgetful functor B–contra$_{\mathrm{gr}}$ $\longrightarrow$ B–mod$_{\mathrm{gr}}$ is fully faithful. Then the natural map of graded abelian groups $N \otimes_B P \longrightarrow N \odot_B P$ is an isomorphism for any graded right B-comodule N and any graded left B-contramodule P.*

Proof We will prove part (b); part (a) is similar. Let U be a graded abelian group. Then we have

$$\operatorname{Hom}_{\mathbb{Z}}(N \otimes_B P, \ U) \simeq \operatorname{Hom}_B(P, \operatorname{Hom}_{\mathbb{Z}}(N, U)),$$

$$\operatorname{Hom}_{\mathbb{Z}}(N \odot_B P, \ U) \simeq \operatorname{Hom}^B(P, \operatorname{Hom}_{\mathbb{Z}}(N, U)).$$

Applying the functor $\operatorname{Hom}_{\mathbb{Z}}(-, U)$ to the natural map $N \otimes_B P \longrightarrow N \odot_B P$ produces the map $\operatorname{Hom}^B(P, \operatorname{Hom}_{\mathbb{Z}}(N, U)) \longrightarrow \operatorname{Hom}_B(P, \operatorname{Hom}_{\mathbb{Z}}(N, U))$ induced by the forgetful functor B–contra$_{\mathrm{gr}}$ $\longrightarrow$ B–mod$_{\mathrm{gr}}$. Since the latter map is an isomorphism for all graded abelian groups U by assumption, it follows that the former map is an isomorphism, too.

□

8.2   CDG-Contramodules of the Induced Type

Let $B = \bigoplus_{n=0}^{\infty} B^n$ be a nonnegatively graded ring, and let K be a fixed associative ring endowed with a ring homomorphism $K \longrightarrow B^0$.

Definition 8.6 Let $L = \bigoplus_{n \in \mathbb{Z}} L^n$ be a graded left K-module. By the graded left B-contramodule *induced from* (or *freely generated by*) a graded left K-module L we mean the graded left B-contramodule $\mathbb{M}^{\mathrm{gr}}_K(L) = B \otimes^{\Pi}_K L$. These were called the "free modules over the monad $\mathbb{M} = \mathbb{M}^{\mathrm{gr}}_K$" in Sect. 5.2. Subsequently we described such graded B-contramodules in Lemma 7.9.

Assume that B is a flat graded right K-module. Then, given a short exact sequence of graded left K-modules $0 \longrightarrow L' \longrightarrow L \longrightarrow L'' \longrightarrow 0$, one can apply the functor $\mathbb{M}_K^{\mathrm{gr}}$, producing a short exact sequence of graded left B-contramodules $0 \longrightarrow$ $\mathbb{M}_K^{\mathrm{gr}}(L') \longrightarrow \mathbb{M}_K^{\mathrm{gr}}(L) \longrightarrow \mathbb{M}_K^{\mathrm{gr}}(L'') \longrightarrow 0$. The resulting short exact sequence of graded left B-contramodules is said to be *induced from* the original short exact sequence of graded left K-modules.

Now let $B = (B, d, h)$ be a nonnegatively graded CDG-ring.

Definition 8.7 We will say that a left CDG-contramodule (Q, d_Q) over (B, d, h) is *of the induced type* if its underlying graded left B-contramodule Q is (isomorphic to a graded left B-contramodule) induced from a graded left K-module. (Of course, the meaning of this definition depends on the choice of a ring K and a ring homomorphism $K \longrightarrow B^0$, which we presume to be fixed.)

Similarly, assume that B is a flat graded right K-module. Then we will say that a short exact sequence $0 \longrightarrow (Q', d_{Q'}) \longrightarrow (Q, d_Q) \longrightarrow (Q'', d_{Q''}) \longrightarrow 0$ of left CDG-contramodules over (B, d, h) (with closed morphisms between them) is a *left CDG-contramodule short exact sequence of the induced type* if the underlying short exact sequence of graded left B-contramodules $0 \longrightarrow Q' \longrightarrow Q \longrightarrow Q'' \longrightarrow 0$ is isomorphic to a short exact sequence of graded left B-contramodules induced from a short exact sequence of graded left K-modules.

As in the proof of Proposition 5.27, by the *free* graded left B-contramodule spanned by a graded set X we mean the graded left B-contramodule $\mathbb{M}_K^{\mathrm{gr}}(K[X])$ induced from the free graded left K-module $K[X]$. The *projective* graded left B-contramodules are the projective objects of the abelian category B–contra$_{\mathrm{gr}}$; these are the direct summands of the free graded left B-contramodules. A CDG-contramodule (P, d_P) over (B, d, h) is said to be *of the projective* (resp., *free*) *type* if its underlying graded left B-contramodule is projective (resp., free).

Denote the full DG-subcategories in the DG-category $\mathsf{DG}(B$–contra$)$ formed by CDG-contramodules of the induced, free, and projective type by $\mathsf{DG}(B$–contra$_{K\text{-ind}})$, $\mathsf{DG}(B$–contra$_{\mathrm{free}})$, and $\mathsf{DG}(B$–contra$_{\mathrm{proj}})$, respectively. Clearly, all the three DG-subcategories are closed under shifts, twists, and finite direct sums; in particular, they are closed under the cones of closed morphisms. Therefore, their homotopy categories $\mathsf{Hot}(B$–contra$_{K\text{-ind}}) = H^0\mathsf{DG}(B$–contra$_{K\text{-ind}})$, $\mathsf{Hot}(B$–contra$_{\mathrm{free}}) = H^0\mathsf{DG}(B$–contra$_{\mathrm{free}})$, and $\mathsf{Hot}(B$–contra$_{\mathrm{proj}}) = H^0\mathsf{DG}(B$–contra$_{\mathrm{proj}})$ are full triangulated subcategories in $\mathsf{Hot}(B$–contra$)$.

The main results of this Sect. 8.2, following below, constitute our version of [58, Theorem 5.5(b)].

Lemma 8.8 *The two full triangulated subcategories* $\mathsf{Hot}(B$–contra$_{\mathrm{free}})$ *and* Hot $(B$–contra$_{\mathrm{proj}}) \subset \mathsf{Hot}(B$–contra$)$ *coincide.*

Proof The argument is based on a version of Eilenberg's cancellation trick. Let (Q, d_Q) be a left CDG-contramodule over (B, d, h) whose underlying graded left B-contramodule Q is a direct summand of the free graded left B-contramodule $B \otimes_K^{\Pi} K[X]$ spanned by a graded set X. Let $Y = X \times \mathbb{Z}$ be the graded set with the components $Y_n = X_n \times \mathbb{Z}$. Consider the quasi-differential graded ring $(\widehat{B}, \partial)$ corresponding to the CDG-ring (B, d, h), and let $P = \widehat{B} \otimes_K^{\Pi} K[Y]$ be the free graded contramodule over $\widehat{B}$ spanned by the graded set Y. Then P can be viewed as a CDG-contramodule (P, d_P) over (B, d, h). Moreover, (P, d_P) is a contractible CDG-contramodule (i. e., it represents a zero object in the homotopy category $\mathsf{Hot}(B\text{–contra})$); essentially, the homogeneous map $\partial \colon P \longrightarrow P$ of degree -1 induced by the differential ∂ on $\widehat{B}$ provides a contracting homotopy. As a graded B-contramodule, P is isomorphic to $B \otimes_K^{\Pi} K[Y] \oplus (B \otimes_K^{\Pi} K[Y])[-1]$. Hence $P \oplus Q$ is a free graded B-contramodule, and the object $(Q, d_Q) \in \mathsf{Hot}(B\text{–contra}_{\mathsf{proj}})$ is isomorphic to the object $(Q, d_Q) \oplus (P, d_P) \in \mathsf{Hot}(B\text{–contra}_{\mathsf{free}})$. $\square$

Proposition 8.9 *Assume that B is a flat graded right K-module and the left homological dimension of the ring K is finite. Then the inclusion $\mathsf{Hot}(B\text{–contra}_{\mathsf{free}}) \longrightarrow \mathsf{Hot}(B\text{–contra}_{K\text{-ind}})$ induces a triangulated equivalence between $\mathsf{Hot}(B\text{–contra}_{\mathsf{free}})$ and the triangulated quotient category of $\mathsf{Hot}(B\text{–contra}_{K\text{-ind}})$ by its minimal full triangulated subcategory containing all the totalizations of left CDG-contramodule short exact sequences of the induced type over (B, d, h).*

Proof Essentially, the claim is that the full subcategory of CDG-contramodules of the free type and the minimal full triangulated subcategory containing the totalizations of CDG-contramodule short exact sequences of the induced type form a semiorthogonal decomposition of $\mathsf{Hot}(B\text{–contra}_{K\text{-ind}})$. Firstly, this means that the complex of abelian groups $\mathrm{Hom}^B(P, Z)$ is acyclic for any $P \in \mathsf{Hot}(B\text{–contra}_{\mathsf{free}})$ and the totalization Z of any CDG-contramodule short exact sequence $0 \longrightarrow Q' \longrightarrow Q \longrightarrow Q'' \longrightarrow 0$ of the induced type. This is a particular case of a standard observation (cf. [59, Theorem 3.5(b)]), and the assumption that the short exact sequence is of the induced type is not needed for it; see the proof of Theorem 8.11 below.

Secondly, it is claimed that for any object $Q \in \mathsf{Hot}(B\text{–contra}_{K\text{-ind}})$ there exists a distinguished triangle $P \longrightarrow Q \longrightarrow Z \longrightarrow P[1]$ in $\mathsf{Hot}(B\text{–contra}_{K\text{-ind}})$ with $P \in \mathsf{Hot}(B\text{–contra}_{\mathsf{free}})$ and Z belonging to the minimal triangulated subcategory of $\mathsf{Hot}(B\text{–contra}_{K\text{-ind}})$ containing the totalizations of CDG-contramodule short exact sequences of the induced type. This is provable by a variation of the standard argument from [59, proof of Theorem 3.6].

Let (Q, d_Q) be a CDG-contramodule structure on the graded left B-contramodule $Q = B \otimes_K^{\Pi} L$ induced from a graded left K-module L. Choose a free graded left K-module $G = K[X]$ together with a surjective morphism of graded left K-modules $g \colon G \longrightarrow L$. Consider the free graded left $\widehat{B}$-contramodule $F = \widehat{B} \otimes_K^{\Pi} G$ induced from G. The CDG-contramodule (Q, d_Q) over (B, d, h) can be also viewed as a graded $\widehat{B}$-contramodule. Let $f \colon F \longrightarrow Q$ be the graded left $\widehat{B}$-contramodule morphism corresponding to the composition of graded left K-module maps $G \longrightarrow L \longrightarrow Q$. Then the restriction of f to the graded B-subcontramodule $B \otimes_K^\Pi G \subset \widehat{B} \otimes_K^\Pi G$ is the graded B-contramodule map $B \otimes_K^\Pi G \longrightarrow B \otimes_K^\Pi L = Q$ obtained by applying the functor $B \otimes_K^\Pi -$ to the graded left K-module map $g \colon G \longrightarrow L$.

The key observation is that the whole map $f \colon F \longrightarrow Q$, viewed as a morphism of graded left B-contramodules, is induced from a morphism of graded left K-modules. Put $F' = B \otimes_K^\Pi G \subset \widehat{B} \otimes_K^\Pi G = F$. Then the quotient graded left B-contramodule $F'' = F/F'$ is isomorphic to $F'[-1]$; so it is also a free graded left B-contramodule. Hence the short exact sequence of graded left B-contramodules $0 \longrightarrow F' \longrightarrow F \longrightarrow F'' \longrightarrow 0$ splits; we can choose a splitting and consider F'' as a graded left B-subcontramodule in F. Put $f' = f|_{F'} \colon F' \longrightarrow Q$ and $f'' = f|_{F''} \colon F'' \longrightarrow Q$; so $f = (f', f'') \colon F' \oplus F'' \longrightarrow Q$. Then $f' = B \otimes_K^\Pi g$ is a surjective morphism of graded left B-contramodules (as $g \colon G \longrightarrow L$ is a surjective morphism of graded left K-modules). Since F'' is a projective graded left B-contramodule, there exists a graded left B-contramodule morphism $t \colon F'' \longrightarrow F'$ such that $f'' = f' \circ t$. Using the map t, one can construct an automorphism of the graded left B-contramodule $F = F' \oplus F''$ such that the composition $F' \oplus F'' \longrightarrow F' \oplus F'' \overset{f}{\longrightarrow} Q$ is equal to the map $(f', 0) \colon F' \oplus F'' \longrightarrow Q$.

Put $F_0 = F$ and $G_0 = G \oplus G[-1]$; and denote by Q_1 the kernel of the graded left $\widehat{B}$-contramodule morphism $f \colon F_0 \longrightarrow Q$. Then $0 \longrightarrow Q_1 \longrightarrow F_0 \longrightarrow Q \longrightarrow 0$ is a short exact sequence of graded left $\widehat{B}$-contramodules. We have shown that, as a short exact sequence of graded left B-contramodules, it is induced from a short exact sequence of graded left K-modules $0 \longrightarrow L_1 \longrightarrow G_0 \longrightarrow L \longrightarrow 0$ (where L_1 is the kernel of the morphism $(g, 0) \colon G_0 = G \oplus G[-1] \longrightarrow L$). Now we can consider $0 \longrightarrow Q_1 \longrightarrow F_0 \longrightarrow Q \longrightarrow 0$ as a CDG-contramodule short exact sequence of the induced type over (B, d, h).

It remains to iterate this construction, applying it to the CDG-contramodule Q_1 in place of Q, etc. Proceeding in this way, we produce an exact sequence of CDG-contramodules $\cdots \longrightarrow F_2 \longrightarrow F_1 \longrightarrow F_0 \longrightarrow Q \longrightarrow 0$ such that its underlying exact sequence of graded left B-contramodules is induced from an exact sequence of graded left K-modules $\cdots \longrightarrow G_2 \longrightarrow G_1 \longrightarrow G_0 \longrightarrow L \longrightarrow 0$. Here G_i are free graded left K-modules by construction. Since the left homological dimension of the ring K is finite by assumption, there exists $k \geq 0$ such that the image L_k of the morphism $G_k \longrightarrow G_{k-1}$ is a projective graded left K-module.

Consider the finite exact sequence of CDG-contramodules $0 \longrightarrow Q_k \longrightarrow F_{k-1} \longrightarrow \cdots \longrightarrow F_0 \longrightarrow Q \longrightarrow 0$ over (B, d, h), where Q_k is the image of the CDG-contramodule morphism $F_k \longrightarrow F_{k-1}$. Denote by (Z, d_Z) the total CDG-contramodule of this finite exact sequence of CDG-contramodules; and denote by (P, d_P) the total CDG-contramodule of the complex of CDG-contramodules $Q_k \longrightarrow F_{k-1} \longrightarrow \cdots \longrightarrow F_0$. Then we have a distinguished triangle $P \longrightarrow Q \longrightarrow Z \longrightarrow P[1]$ in $\mathsf{Hot}(B\text{--}\mathsf{contra}_{K\text{-ind}})$.

The underlying exact sequence of graded left B-contramodules $0 \longrightarrow Q_k \longrightarrow$ $F_{k-1} \longrightarrow \cdots \longrightarrow F_0 \longrightarrow Q \longrightarrow 0$ is induced from the exact sequence of graded left K-modules $0 \longrightarrow L_k \longrightarrow G_{k-1} \longrightarrow \cdots \longrightarrow G_0 \longrightarrow L \longrightarrow 0$. Hence it follows that the CDG-contramodule (Z, d_Z) over (B, d, h) belongs to the minimal triangulated subcategory in $\mathsf{Hot}(B\text{--contra}_{K\text{-ind}})$ containing the totalizations of CDG-contramodule short exact sequences of the induced type. Finally, the underlying graded left B-contramodule P of the CDG-contramodule (P, d_P) is projective, since the graded left B-contramodule Q_k is induced from the projective graded left K-module L_k. By Lemma 8.8, the CDG-contramodule (P, d_P) is homotopy equivalent to a CDG-contramodule of the free type over (B, d, h). $\square$

Proposition 8.10 *Assume that the right K-module B^n is finitely generated and projective for every $n \geq 0$. Then the inclusion $\mathsf{Hot}(B\text{--contra}_{K\text{-ind}}) \longrightarrow \mathsf{Hot}(B\text{--contra})$ induces a triangulated equivalence between the contraderived category $\mathsf{D}^{\mathrm{ctr}}(B\text{--contra})$ (as defined in Sect. 7.3) and the triangulated quotient category of the homotopy category $\mathsf{Hot}(B\text{--contra}_{K\text{-ind}})$ by its intersection with the full triangulated subcategory $\mathsf{Acycl}^{\mathrm{ctr}}(B\text{--contra})$ of contraacyclic CDG-contramodules in $\mathsf{Hot}(B\text{--contra})$,*

$$\mathsf{Hot}(B\text{--contra}_{K\text{-ind}})/(\mathsf{Acycl}^{\mathrm{ctr}}(B\text{--contra}) \cap \mathsf{Hot}(B\text{--contra}_{K\text{-ind}})) \simeq \mathsf{D}^{\mathrm{ctr}}(B\text{--contra}).$$

Proof In view of a well-known lemma ([58, Lemma 2.6] or [59, Lemma 1.6(a)]), it suffices to show that for any left CDG-contramodule (S, d_S) over (B, d, h) there exists a left CDG-contramodule (Q, d_Q) of the induced type together with a closed morphism of CDG-contramodules $(Q, d_Q) \longrightarrow (S, d_S)$ whose cone is a contraacyclic CDG-contramodule over (B, d, h). A standard argument (cf. [59, Theorem 3.8]) reduces the question to showing that countable products of free graded left B-contramodules are induced graded left B-contramodules.

Let us spell out some details. Consider the CDG-contramodule (S, d_S) as a graded left $\widehat{B}$-contramodule. Let $\cdots \longrightarrow P_2 \longrightarrow P_1 \longrightarrow P_0 \longrightarrow S \longrightarrow 0$ be a resolution of the object $S \in \widehat{B}\text{--contra}_{\mathrm{gr}}$ by free graded left $\widehat{B}$-contramodules P_i. Consider this resolution as a complex of left CDG-contramodules $\cdots \longrightarrow P_2 \longrightarrow P_1 \longrightarrow P_0 \longrightarrow 0$ over (B, d, h) and totalize it by taking infinite products along the diagonals. We refer to [59, Section 1.2] for a discussion of totalizations of complexes in DG-categories. Notice that the infinite products in $B\text{--contra}_{\mathrm{gr}}$ agree with those in $K\text{--mod}_{\mathrm{gr}}$, so there is no ambiguity in how the totalization construction needs to be applied. Let (Q, d_Q) be the resulting total CDG-contramodule over (B, d, h). Then the underlying graded left B-contramodule Q of (Q, d_Q) is simply the countable product $Q = \prod_{i=0}^{\infty} P_i[i]$. Once again we recall that the forgetful functor $\widehat{B}\text{--contra}_{\mathrm{gr}} \longrightarrow B\text{--contra}_{\mathrm{gr}}$ takes free graded left $\widehat{B}$-contramodules to free graded left B-contramodules.

The cone (Y, d_Y) of the natural closed morphism of CDG-contramodules $(Q, d_Q) \longrightarrow (S, d_S)$ is the infinite product totalization of the exact complex $\cdots \longrightarrow P_2 \longrightarrow P_1 \longrightarrow P_0 \longrightarrow S \longrightarrow 0$. As such, it admits a complete, separated decreasing filtration F by the totalizations of the subcomplexes of canonical filtration of the complex $\cdots \longrightarrow P_2 \longrightarrow P_1 \longrightarrow P_0 \longrightarrow S \longrightarrow 0$. The successive quotient CDG-contramodules $F^n Y / F^{n+1} Y$ are contractible. In fact, the CDG-contramodules $F^n Y / F^{n+1} Y$ are the cones of identity endomorphisms of the CDG-contramodules of cocycles of our acyclic complex of CDG-contramodules (up to some shifts). By Lemma 7.8, it follows that the CDG-contramodule (Y, d_Y) is contraacyclic.

It remains to check that the class of all induced graded left B-contramodules is closed under infinite products in B–$\mathsf{contra}_{\mathsf{gr}}$. This is where we use the assumption that B^n is a finitely generated projective right K-module for every $n \geq 0$. Under this assumption, one can easily see that the functor $B \otimes_K^\Pi -$ (viewed as a functor K–$\mathsf{mod}_{\mathsf{gr}} \longrightarrow K$–$\mathsf{mod}_{\mathsf{gr}}$ or K–$\mathsf{mod}_{\mathsf{gr}} \longrightarrow B$–$\mathsf{contra}_{\mathsf{gr}}$) preserves infinite products. In fact, it suffices that B^n be a finitely presentable right K-module for every $n \geq 0$. $\square$

Theorem 8.11 *Assume that B^n is a finitely generated projective right K-module for every $n \geq 0$ and the left homological dimension of the ring K is finite. Then the composition of the fully faithful inclusion* $\mathsf{Hot}(B$–$\mathsf{contra}_{\mathsf{proj}}) \longrightarrow \mathsf{Hot}(B$–$\mathsf{contra})$ *with the triangulated quotient functor* $\mathsf{Hot}(B$–$\mathsf{contra}) \longrightarrow \mathsf{D}^{\mathsf{ctr}}(B$–$\mathsf{contra})$ *is a triangulated equivalence*

$$\mathsf{Hot}(B\text{–}\mathsf{contra}_{\mathsf{proj}}) \simeq \mathsf{D}^{\mathsf{ctr}}(B\text{–}\mathsf{contra})$$

between the homotopy category of left CDG-contramodules of the projective type over (B, d, h) and the contraderived category of left CDG-contramodules over (B, d, h). Furthermore, the intersection of two full subcategories

$$\mathsf{Acycl}^{\mathsf{ctr}}(B\text{–}\mathsf{contra}) \cap \mathsf{Hot}(B\text{–}\mathsf{contra}_{K\text{-ind}}) \subset \mathsf{Hot}(B\text{–}\mathsf{contra})$$

coincides with the minimal full triangulated subcategory in $\mathsf{Hot}(B$–$\mathsf{contra}_{K\text{-ind}})$ containing all the totalizations of CDG-contramodule short exact sequences of the induced type over (B, d, h).

Proof In the first assertion of the theorem, the claim is that the full subcategory of CDG-contramodules of the projective type $\mathsf{Hot}(B$–$\mathsf{contra}_{\mathsf{proj}})$ and the full subcategory of contraacyclic CDG-contramodules $\mathsf{Acycl}^{\mathsf{ctr}}(B$–$\mathsf{contra})$ form a semiorthogonal decomposition of the homotopy category $\mathsf{Hot}(B$–$\mathsf{contra})$. First of all, this means that the complex of abelian groups $\mathsf{Hom}^B(P, Z)$ is acyclic for any $P \in \mathsf{Hot}(B$–$\mathsf{contra}_{\mathsf{proj}})$ and $Z \in \mathsf{Acycl}^{\mathsf{ctr}}(B$–$\mathsf{contra})$. A particular case of this observation was already mentioned in the proof of Proposition 8.9.

Indeed, let us fix a CDG-contramodule $(P, d_P) \in \mathsf{Hot}(B$–$\mathsf{contra}_{\mathsf{proj}})$. Then the class of all CDG-contramodules $(Z, d_Z) \in \mathsf{Hot}(B$–$\mathsf{contra})$ for which the complex $\mathsf{Hom}^B(P, Z)$ is acyclic is closed under shifts, cones, and infinite products in $\mathsf{Hot}(B$–$\mathsf{contra})$. Thus it remains to check that the totalization $Z = \mathsf{Tot}(Q' \to Q \to Q'')$ of any short exact sequence of CDG-contramodules $0 \longrightarrow (Q', d_{Q'}) \longrightarrow (Q, d_Q) \longrightarrow (Q'', d_{Q''}) \longrightarrow 0$

over (B, d, h) has the desired property. Here it suffices to notice that the complex of abelian groups $\mathrm{Hom}^B(P, Z)$ is the totalization of the bicomplex with three rows $\mathrm{Hom}^B(P, Q') \longrightarrow \mathrm{Hom}^B(P, Q) \longrightarrow \mathrm{Hom}^B(P, Q'')$. The short sequence of complexes of abelian groups $0 \longrightarrow \mathrm{Hom}^B(P, Q') \longrightarrow \mathrm{Hom}^B(P, Q) \longrightarrow \mathrm{Hom}^B(P, Q'') \longrightarrow 0$ is exact, since the graded left B-contramodule P is projective. The totalization of any short exact sequence of complexes of abelian groups is an exact complex.

Now we can prove the second assertion of the theorem. Let (Z, d_Z) be a left CDG-contramodule of the induced type over (B, d, h) which is contraacyclic as a left CDG-contramodule over (B, d, h). Then, according to the above argument, the complex $\mathrm{Hom}^B(P, Z)$ is acyclic for any left CDG-contramodule (P, d_P) of the free type over (B, d, h). In view of the semiorthogonal decomposition of the category $\mathsf{Hot}(B\mathrm{-contra}_{K\text{-ind}})$ constructed in the proof of Proposition 8.9, it follows that (Z, d_Z) belongs to the minimal triangulated subcategory of $\mathsf{Hot}(B\mathrm{-contra}_{K\text{-ind}})$ containing the totalizations of CDG-contramodule short exact sequences of the induced type.

Finally, the assertion that the functor $\mathsf{Hot}(B\mathrm{-contra}_{\mathsf{proj}}) \longrightarrow \mathsf{D}^{\mathsf{ctr}}(B\mathrm{-contra})$ is a triangulated equivalence can be now obtained by comparing the results of Lemma 8.8, Propositions 8.9, and 8.10. This also implies the semiorthogonal decomposition promised in the beginning of this proof (given that the semiorthogonality of the two subcategories in question has been established already). $\qquad\square$

8.3   CDG-Comodules of the Coinduced Type

Let $B = \bigoplus_{n=0}^{\infty} B^n$ be a nonnegatively graded ring, and let K be a fixed ring endowed with a ring homomorphism $K \longrightarrow B^0$. Similarly to the definition of graded right B-comodules in Sect. 5.5, one can define graded left B-comodules. So a graded left B-module $M = \bigoplus_{n\in\mathbb{Z}} M^n$ is said to be a B-comodule (or a graded left B-comodule) if for every element $x \in M^j$, $j \in \mathbb{Z}$, there exists an integer $m \geq 0$ such that $B^{\geq m+1}x = 0$. We denote the full subcategory of graded left B-comodules by $B\mathrm{-comod}_{\mathsf{gr}} \subset B\mathrm{-mod}_{\mathsf{gr}}$.

Definition 8.12 Let $L = \bigoplus_{n\in\mathbb{Z}} L^n$ be a graded left K-module. By the graded left B-comodule *coinduced from* (or *cofreely cogenerated by*) a graded left K-module L we mean the graded left B-comodule $\mathrm{Hom}_K^{\Sigma}(B, L)$, in the notation of Sect. 5.7 (cf. Lemma 6.13).

Assume that B is a projective graded left K-module. Then, given a short exact sequence of graded left K-modules $0 \longrightarrow L' \longrightarrow L \longrightarrow L'' \longrightarrow 0$, one can apply the functor $\mathrm{Hom}_K^{\Sigma}(B, -)$, producing a short exact sequence of graded left B-comodules $0 \longrightarrow \mathrm{Hom}_K^{\Sigma}(B, L') \longrightarrow \mathrm{Hom}_K^{\Sigma}(B, L) \longrightarrow \mathrm{Hom}_K^{\Sigma}(B, L'') \longrightarrow 0$. The resulting short exact sequence of graded left B-comodules is said to be *coinduced from* the original short exact sequence of graded left K-modules.

Now let $B = (B, d, h)$ be a nonnegatively graded CDG-ring. Similarly to the definition of right CDG-comodules over (B, d, h) in Sect. 6.4, one can define left CDG-comodules over (B, d, h). So a left CDG-module (M, d_M) over (B, d, h) is said to be a *CDG-comodule* (or a *left CDG-comodule*) over (B, d, h) if the graded left B-module M is a graded left B-comodule. The full DG-subcategory in $\mathsf{DG}(B\text{–mod})$ whose objects are the left CDG-comodules over (B, d, h) is called the *DG-category of left CDG-comodules over* (B, d, h) and denoted by $\mathsf{DG}(B\text{–comod})$. Its triangulated homotopy category is denoted by $\mathsf{Hot}(B\text{–comod}) = H^0\mathsf{DG}(B\text{–comod})$.

Definition 8.13 We will say that a left CDG-comodule (M, d_M) over (B, d, h) is *of the coinduced type* if its underlying graded left B-comodule M is (isomorphic to a graded left B-comodule) coinduced from a graded left K-module. (The meaning of this definition depends on the choice of a ring K and a ring homomorphism $K \longrightarrow B^0$, which we presume to be fixed.)

Assume that B is a projective graded left K-module. Then we will say that a short exact sequence $0 \longrightarrow (M', d_{M'}) \longrightarrow (M, d_M) \longrightarrow (M'', d_{M''}) \longrightarrow 0$ of left CDG-comodules over (B, d, h) (with closed morphisms between them) is a *left CDG-comodule short exact sequence of the coinduced type* if the underlying short exact sequence of graded left B-comodules $0 \longrightarrow M' \longrightarrow M \longrightarrow M'' \longrightarrow 0$ is isomorphic to a short exact sequence of graded left B-comodules coinduced from a short exact sequence of graded left K-modules.

By the *cofree* graded left B-comodule spanned by a graded set X we mean the graded left B-comodule $\mathrm{Hom}_K^\Sigma(B, K[X]^+)$ coinduced from the cofree graded left K-module $K[X]^+$. Here F^+ is the notation for the graded character group $\mathrm{Hom}_{\mathbb{Z}}(F, \mathbb{Q}/\mathbb{Z})$ of an abelian group F (see Sect. 5.7). The *injective* graded left B-comodules are the injective objects of the abelian category $B\text{–comod}_{\mathrm{gr}}$; these are the direct summands of the cofree graded left B-comodules (see the discussion in the proof of Theorem 5.37).

A CDG-comodule (J, d_J) over (B, d, h) is said to be *of the injective* (resp., *cofree*) *type* if its underlying graded left B-comodule is injective (resp., cofree). Denote the full DG-subcategories in the DG-category $\mathsf{DG}(B\text{–comod})$ formed by CDG-comodules of the coinduced, cofree, and injective type by $\mathsf{DG}(B\text{–comod}_{K\text{-coind}})$, $DG(B\text{–comod}_{\mathrm{cofr}})$, and $\mathsf{DG}(B\text{–comod}_{\mathrm{inj}})$, respectively. All the three DG-subcategories are closed under shifts, twists, and finite direct sums; in particular, they are closed under the cones of closed morphisms. Therefore, their homotopy categories $\mathsf{Hot}(B\text{–comod}_{K\text{-coind}}) = H^0\mathsf{DG}(B\text{–comod}_{K\text{-coind}})$, $\mathsf{Hot}(B\text{–comod}_{\mathrm{cofr}}) = H^0\mathsf{DG}(B\text{–comod}_{\mathrm{cofr}})$, and $\mathsf{Hot}(B\text{–comod}_{\mathrm{inj}}) = H^0\mathsf{DG}(B\text{–comod}_{\mathrm{inj}})$ are full triangulated subcategories in $\mathsf{Hot}(B\text{–comod})$.

The main results of this Sect. 8.3, following below, form our version of [58, Theorem 5.5(a)].

Lemma 8.14 *The two full triangulated subcategories* $\mathsf{Hot}(B\text{–comod}_{\mathrm{cofr}})$ *and* $\mathsf{Hot}(B\text{–comod}_{\mathrm{inj}}) \subset \mathsf{Hot}(B\text{–comod})$ *coincide.*

Proof This is a dual-analogous version of Lemma 8.8, provable by a similar argument using the cancellation trick.   □

Proposition 8.15 *Assume that B is a projective graded left K-module and the left homological dimension of the ring K is finite. Then the inclusion* $\mathsf{Hot}(B\text{–comod}_{\mathrm{cofr}}) \longrightarrow \mathsf{Hot}(B\text{–comod}_{K\text{-coind}})$ *induces a triangulated equivalence between* $\mathsf{Hot}(B\text{–comod}_{\mathrm{cofr}})$ *and the triangulated quotient category of* $\mathsf{Hot}(B\text{–comod}_{K\text{-coind}})$ *by its minimal full triangulated subcategory containing all the totalizations of left CDG-comodule short exact sequences of the coinduced type over* (B, d, h).

Proof Essentially, the claim is that the full subcategory of CDG-comodules of the cofree type and the minimal full triangulated subcategory containing the totalizations of CDG-comodule short exact sequences of the coinduced type form a semiorthogonal decomposition of $\mathsf{Hot}(B\text{–comod}_{K\text{-coind}})$. Firstly, this means that the complex of abelian groups $\mathsf{Hom}_B(Z, J)$ is acyclic for any $J \in \mathsf{Hot}(B\text{–comod}_{\mathrm{cofr}})$ and the totalization Z of any CDG-comodule short exact sequence $0 \longrightarrow M' \longrightarrow M \longrightarrow M'' \longrightarrow 0$ of the coinduced type. This is a particular case of a standard observation (cf. [59, Theorem 3.6(a)]), and the assumption that the short exact sequence is of the coinduced type is not needed for it; see the proof of Theorem 8.17 below.

Secondly, it is claimed that for any object $N \in \mathsf{Hot}(B\text{–comod}_{K\text{-coind}})$ there exists a distinguished triangle $Z \longrightarrow N \longrightarrow J \longrightarrow Z[1]$ in $\mathsf{Hot}(B\text{–comod}_{K\text{-coind}})$ with $J \in \mathsf{Hot}(B\text{–comod}_{\mathrm{cofr}})$ and Z belonging to the minimal triangulated subcategory of $\mathsf{Hot}(B\text{–comod}_{K\text{-coind}})$ containing the totalizations of CDG-comodule short exact sequences of the coinduced type. This is provable by a variation of the standard argument form [59, proof of Theorem 3.6]. We skip the further details, which are very similar to the proof of the dual-analogous result in Proposition 8.9.   □

Proposition 8.16 *Assume that the left K-module B^n is finitely generated for every $n \geq 0$. Then the inclusion* $\mathsf{Hot}(B\text{–comod}_{K\text{-coind}}) \longrightarrow \mathsf{Hot}(B\text{–comod})$ *induces a triangulated equivalence between the coderived category* $\mathsf{D}^{\mathrm{co}}(B\text{–comod})$ *(as defined in Sect. 6.4) and the triangulated quotient category of the homotopy category* $\mathsf{Hot}(B\text{–comod}_{K\text{-coind}})$ *by its intersection with the full triangulated subcategory* $\mathsf{Acycl}^{\mathrm{co}}(B\text{–comod})$ *of coacyclic CDG-comodules in* $\mathsf{Hot}(B\text{–comod})$,

$$\mathsf{Hot}(B\text{–comod}_{K\text{-coind}})/(\mathsf{Acycl}^{\mathrm{co}}(B\text{–comod}) \cap \mathsf{Hot}(B\text{–comod}_{K\text{-coind}}) \simeq \mathsf{D}^{\mathrm{co}}(B\text{–comod}).$$

Proof In view of [58, Lemma 2.6] or [59, Lemma 1.6(b)], it suffices to show that for any left CDG-comodule (M, d_M) over (B, d, h) there exists a left CDG-comodule (N, d_N) of the coinduced type together with a closed morphism of CDG-comodules $(M, d_M) \longrightarrow$

(N, d_N) whose cone is a coacyclic CDG-comodule over (B, d, h). A standard argument using Lemma 6.12 (cf. [59, proof of Theorem 3.7]) reduces the question to showing that countable direct sums of cofree graded left B-comodules are coinduced graded left B-comodules. We skip the details, which are very similar to the ones in the proof of the dual-analogous result in Proposition 8.10.

Hence it remains to check that the class of all coinduced graded left B-comodules is closed under infinite direct sums in B–$\mathsf{comod}_{\mathsf{gr}}$. This is where we use the assumption that B^n is a finitely generated left K-module for every $n \geq 0$. Under this assumption, one can easily see that the functor $\mathrm{Hom}_K^{\Sigma}(B, -)$ (viewed as a functor K–$\mathsf{mod}_{\mathsf{gr}} \longrightarrow K$–$\mathsf{mod}_{\mathsf{gr}}$ or K–$\mathsf{mod}_{\mathsf{gr}} \longrightarrow B$–$\mathsf{comod}_{\mathsf{gr}}$) preserves infinite direct sums. $\qquad \square$

Theorem 8.17 *Assume that B^n is a finitely generated projective left K-module for every $n \geq 0$ and the left homological dimension of the ring K is finite. Then the composition of the fully faithful inclusion $\mathsf{Hot}(B$–$\mathsf{comod}_{\mathsf{inj}}) \longrightarrow \mathsf{Hot}(B$–$\mathsf{comod})$ with the triangulated quotient functor $\mathsf{Hot}(B$–$\mathsf{comod}) \longrightarrow \mathsf{D}^{\mathsf{co}}(B$–$\mathsf{comod})$ is a triangulated equivalence*

$$\mathsf{Hot}(B\text{–}\mathsf{comod}_{\mathsf{inj}}) \simeq \mathsf{D}^{\mathsf{co}}(B\text{–}\mathsf{comod})$$

between the homotopy category of left CDG-comodules of the injective type over (B, d, h) and the coderived category of left CDG-comodules over (B, d, h). Furthermore, the intersection of two full subcategories

$$\mathsf{Acycl}^{\mathsf{co}}(B\text{–}\mathsf{comod}) \cap \mathsf{Hot}(B\text{–}\mathsf{comod}_{K\text{-coind}}) \subset \mathsf{Hot}(B\text{–}\mathsf{comod})$$

coincides with the minimal full triangulated subcategory in $\mathsf{Hot}(B$–$\mathsf{comod}_{K\text{-coind}})$ containing all the totalizations of CDG-comodule short exact sequences of the coinduced type over (B, d, h).

Proof In the first assertion of the theorem, the claim is that the full subcategory of CDG-comodules of the injective type $\mathsf{Hot}(B$–$\mathsf{comod}_{\mathsf{inj}})$ and the full subcategory of coacyclic CDG-comodules $\mathsf{Acycl}^{\mathsf{co}}(B$–$\mathsf{comod})$ form a semiorthogonal decomposition of the homotopy category $\mathsf{Hot}(B$–$\mathsf{comod})$. First of all, this means that the complex of abelian groups $\mathrm{Hom}_B(Z, J)$ is acyclic for any $Z \in \mathsf{Acycl}^{\mathsf{co}}(B$–$\mathsf{comod})$ and $J \in \mathsf{Hot}(B$–$\mathsf{comod}_{\mathsf{inj}})$. A particular case of this observation was already mentioned in the proof of Proposition 8.15.

The proof of the theorem is based on Lemma 8.14, Proposition 8.15, and Proposition 8.16. We skip the details, which are very similar to the proof of the dual-analogous Theorem 8.11. $\qquad \square$

8.4   The Diagonal CDG-Bicomodule

Let $B = \bigoplus_{n=0}^{\infty} B^n$ be a nonnegatively graded ring, and let K be a ring endowed with a ring homomorphism $K \longrightarrow B^0$. Put $C_n = C^{-n} = \mathrm{Hom}_{K^{op}}(B^n, K)$. We will use the pairing notation

$$\langle c, b \rangle = c(b) \in K \quad \text{for all } c \in C_n, \ b \in B^n, \ n \geq 0,$$

$$\langle c, b \rangle = 0 \quad \text{for all } c \in C_j, \ b \in B^i, \ i \neq j.$$

The graded abelian group $C = \bigoplus_{n \leq 0} C^n$ has a natural structure of graded K-B-bimodule, with the left action of K given by the rule

$$\langle kc, b \rangle = k\langle c, b \rangle \quad \text{for all } k \in K, \ c \in C^{-n}, \ b \in B^n$$

and the right action of B given by the rule

$$\langle cb, b' \rangle = \langle c, bb' \rangle \quad \text{for all } c \in C, \ b, b' \in B.$$

Furthermore, assume that B^n is a finitely generated projective right K-module for every $n \geq 0$. Then a straightforward generalization of the construction from Remark 2.13 endows C with the structure of a graded coring over K. (The discussion in Remark 2.13 presumes the base ring $K = R = B^0$.) The grading components $\mu_{i,j} \colon C_{i+j} \longrightarrow C_i \otimes_K C_j$, $i, j \geq 0$, of the comultiplication map $\mu \colon C \longrightarrow C \otimes_K C$ are obtained by applying the functor $\mathrm{Hom}_{K^{op}}(-, K)$ to the multiplication maps $B^j \otimes_K B^i \longrightarrow B^{i+j}$ and taking into account the natural isomorphisms of Lemma 1.1(b). In the pairing notation,

$$\langle \mu_{i,j}(c), b'' \otimes b' \rangle = \langle c, b''b' \rangle \quad \text{for all } c \in C_{i+j}, \ b' \in B^i, \ b'' \in B^j,$$

where the pairing $\langle \, , \, \rangle \colon C_i \otimes_K C_j \otimes_K B^j \otimes_K B^i \longrightarrow K$ is defined by the rule

$$\langle c' \otimes c'', \ b'' \otimes b' \rangle = \langle c' \langle c'', b'' \rangle, b' \rangle = \langle c', \langle c'', b'' \rangle b' \rangle.$$

The counit map $\varepsilon \colon C_0 \longrightarrow K$ is obtained by applying the functor $\mathrm{Hom}_{K^{op}}(-, K)$ to the ring homomorphism $K \longrightarrow B^0$,

$$\varepsilon(c) = \langle c, 1 \rangle \quad \text{for all } c \in C,$$

where $1 \in B^0$ is the unit element.

Generalizing Remark 2.13 again, we put ${}^{\#}B^n = \mathrm{Hom}_{K^{\mathrm{op}}}(C_n, K)$ for every $n \geq 0$. The related pairing notation is

$$\langle {}^{\#}b, c \rangle = {}^{\#}b(c) \quad \text{for all } {}^{\#}b \in {}^{\#}B^n, \ c \in C_n, \ n \geq 0,$$

$$\langle {}^{\#}b, c \rangle = 0 \quad \text{for all } {}^{\#}b \in {}^{\#}B^i, \ c \in C_j, \ i \neq j.$$

Then ${}^{\#}B^n$ is a K-K-bimodule with the left and right actions of K given by the usual rules

$$\langle k{}^{\#}b, c \rangle = k \langle {}^{\#}b, c \rangle, \quad \langle {}^{\#}bk, c \rangle = \langle {}^{\#}b, kc \rangle \quad \text{for all } k \in K, \ {}^{\#}b \in {}^{\#}B^n, \ c \in C_n.$$

Moreover, ${}^{\#}B = \bigoplus_{n=0}^{\infty} {}^{\#}B^n$ is a nonnegatively graded ring with the multiplication given by the rule

$$\langle {}^{\#}b'{}^{\#}b'', c \rangle = \langle {}^{\#}b' \otimes {}^{\#}b'', \mu(c) \rangle \quad \text{for all } {}^{\#}b', {}^{\#}b'' \in {}^{\#}B, \ c \in C,$$

where the pairing $\langle \, , \, \rangle : {}^{\#}B^j \otimes_K {}^{\#}B^i \otimes_K C_i \otimes_K C_j \longrightarrow K$ is defined by

$$\langle {}^{\#}b'' \otimes {}^{\#}b', c' \otimes c'' \rangle = \langle {}^{\#}b'' \langle {}^{\#}b', c' \rangle, c'' \rangle = \langle {}^{\#}b'', \langle {}^{\#}b', c' \rangle c'' \rangle.$$

Applying the functor $\mathrm{Hom}_{K^{\mathrm{op}}}(-, K)$ to the counit map $C_0 \longrightarrow K$, we obtain a ring homomorphism $K \longrightarrow {}^{\#}B^0$ (which, together with the graded ring structure on ${}^{\#}B$, induces the K-K-bimodule structures on ${}^{\#}B^n$ mentioned above). In the pairing notation, the formula

$$\langle 1, c \rangle = \varepsilon(c) \quad \text{for all } c \in C$$

defines the unit element $1 \in {}^{\#}B^0$.

Furthermore, the graded left K-module structure on $C = \bigoplus_{n \geq 0} C^n$ extends naturally to a graded left ${}^{\#}B$-module structure, making C a graded ${}^{\#}B$-B-bimodule (as it was mentioned already in Remark 2.13). In the pairing notation, the right action of B in C (defined above) can be expressed in terms of the comultiplication in C by the formula

$$cb = \langle \mu_{i,j}(c), b \rangle \quad \text{for all } c \in C_{i+j} \text{ and } b \in B^j,$$

where $\langle \, , \, \rangle$ denotes the map $C_i \otimes_K C_j \otimes_K B^j \longrightarrow C_i$ induced by the pairing $C_j \otimes_K B^j \longrightarrow K$. Similarly, the left action of ${}^{\#}B$ in C is defined by the formula

$$\,{}^{\#}bc = \langle {}^{\#}b, \mu_{j,i}(c) \rangle \quad \text{for all } c \in C_{i+j} \text{ and } {}^{\#}b \in {}^{\#}B^j,$$

where $\langle \, , \, \rangle$ denotes the map ${}^{\#}B^j \otimes_K C_j \otimes_K C_i \longrightarrow C_i$ induced by the pairing ${}^{\#}B^j \otimes_K C_j \longrightarrow K$.

By construction, we know that C_n is a finitely generated projective left K-module for every $n \geq 0$. Now we make an additional assumption.

Proposition 8.18 *Assume additionally that C_n is a finitely generated projective right K-module for every $n \geq 0$. Then there is a natural bijective correspondence between CDG-ring structures (B, d, h) on the graded ring B, viewed up to change-of-connection isomorphisms $(id, a): (B, d', h') \longrightarrow (B, d, h)$, and CDG-ring structures $({}^\#B, {}^\#d, {}^\#h)$ on the graded ring ${}^\#B$, viewed up to change-of-connection isomorphisms $(id, {}^\#a): ({}^\#B, {}^\#d', {}^\#h') \longrightarrow ({}^\#B, {}^\#d, {}^\#h)$.*

Proof We use the result of Theorem 4.7 and identify CDG-ring structures (B, d, h) on B with quasi-differential structures $(\widehat{B}, \partial)$ (that is, quasi-differential graded rings $(\widehat{B}, \partial)$ endowed with an injective graded ring homomorphism $B \longrightarrow \widehat{B}$ with the image equal to $\ker \partial \subset \widehat{B}$). Similarly, we identify CDG-ring structures $({}^\#B, {}^\#d, {}^\#h)$ on the graded ring ${}^\#B$ with quasi-differential structures $({}^\#\widehat{B}, \partial)$ on ${}^\#B$, and construct a bijection between the isomorphism classes of $(\widehat{B}, \partial)$ and $({}^\#\widehat{B}, \partial)$.

Let $(\widehat{B}, \partial)$ be a quasi-differential structure on B. Recall that the differential $\partial : \widehat{B}^n \longrightarrow \widehat{B}^{n-1}$ is a morphism of B^0-B^0-bimodules (since $\partial(B^0) \subset B^{-1} = 0$), hence also a morphism of K-K-bimodules. Therefore, in view of the short exact sequences $0 \longrightarrow B^n \longrightarrow \widehat{B}^n \stackrel{\partial}{\longrightarrow} B^{n-1} \longrightarrow 0$, the graded right K-module $\widehat{B}^n$ is also finitely generated and projective for every $n \geq 0$.

Applying the above construction to the graded ring $\widehat{B}$, we produce a graded coring $\widehat{C}$ with the components $\widehat{C}_n = \widehat{C}^{-n} = \operatorname{Hom}_{K^{\mathrm{op}}}(\widehat{B}^n, K)$. Moreover, the differential $\partial : \widehat{B}^n \longrightarrow \widehat{B}^{n-1}$ induces a differential $\partial : \widehat{C}_{n-1} \longrightarrow \widehat{C}_n$, $n \geq 0$, which we define with the sign rule so that

$$\langle \partial(c), b \rangle + (-1)^{|c|} \langle c, \partial(b) \rangle = 0 \quad \text{for all } c \in \widehat{C}^{|c|} \text{ and } b \in \widehat{B}^{|b|}.$$

By construction, $\partial : \widehat{C} \longrightarrow \widehat{C}$ is a homogeneous K-K-bimodule map. As explained in the second proof of Theorem 4.25 in Sect. 4.6, it is also an odd coderivation of the coring $\widehat{C}$ (of degree -1 in the upper indices). Since $\partial^2 = 0$ on $\widehat{B}$, we also have $\partial^2 = 0$ on $\widehat{C}$. The inclusion of graded rings $B \longrightarrow \widehat{B}$ induces a graded coring homomorphism $\widehat{C} \longrightarrow C$, whose composition with the odd coderivation ∂ on $\widehat{C}$ obviously vanishes.

Furthermore, since the short exact sequences of K-K-bimodules $0 \longrightarrow B^n \longrightarrow \widehat{B}^n \stackrel{\partial}{\longrightarrow} B^{n-1} \longrightarrow 0$ are split exact as short sequences of right K-modules, we have short exact sequences of K-K-bimodules $0 \longrightarrow C_{n-1} \longrightarrow \widehat{C}_n \longrightarrow C_n \longrightarrow 0$. So the graded coring homomorphism $\widehat{C} \longrightarrow C$ is surjective and its kernel is equal to the image of the odd coderivation ∂ on C, as well as to the kernel of the odd coderivation ∂.

Continuing to apply the above construction, we produce a graded ring $^\#\widehat{B}$ with the components $^\#\widehat{B}^n = \operatorname{Hom}_{K^{\mathrm{op}}}(\widehat{C}_n, K)$. The differential $\partial \colon \widehat{C}_{n-1} \longrightarrow \widehat{C}_n$ induces a differential $\partial \colon {}^\#\widehat{B}^n \longrightarrow {}^\#\widehat{B}^{n-1}$, which we define using the sign rule

$$\langle \partial({}^\#b), c \rangle + (-1)^{|{}^\#b|} \langle {}^\#b, \partial(c) \rangle = 0 \quad \text{for all } {}^\#b \in {}^\#\widehat{B}^{|{}^\#b|} \text{ and } c \in \widehat{C}^{|c|}.$$

The map $\partial \colon {}^\#\widehat{B} \longrightarrow {}^\#\widehat{B}$ is an odd derivation of degree -1 on the graded ring $^\#\widehat{B}$. Since $\partial^2 = 0$ on $\widehat{C}$, we also have $\partial^2 = 0$ on $^\#\widehat{B}$. The surjective homomorphism of graded corings $\widehat{C} \longrightarrow C$ induces an injective homomorphism of graded rings $^\#B \longrightarrow {}^\#\widehat{B}$, whose composition with the coderivation ∂ on $^\#\widehat{B}$ vanishes.

By the additional assumption, all the K-K-bimodules in the short exact sequence $0 \longrightarrow C_{n-1} \longrightarrow \widehat{C}_n \longrightarrow C_n \longrightarrow 0$ are finitely generated and projective as right K-modules, so the sequence splits as a short exact sequence of right K-modules. Hence we have short exact sequences of K-K-bimodules $0 \longrightarrow {}^\#B^n \longrightarrow {}^\#\widehat{B}^n \overset{\partial}{\longrightarrow} {}^\#B^{n-1} \longrightarrow 0$. Thus the kernel of the odd derivation ∂ on $^\#\widehat{B}$ is equal to the image of the injective ring homomorphism $^\#B \longrightarrow {}^\#\widehat{B}$, as well as the to the image of the odd derivation $\partial \colon {}^\#\widehat{B} \longrightarrow {}^\#\widehat{B}$.

We have constructed the quasi-differential graded ring $({}^\#\widehat{B}, \partial)$ corresponding to a quasi-differential graded ring $(\widehat{B}, \partial)$. To check that this is a bijective correspondence, it suffices to reverse the construction, recovering the graded coring $\widehat{C}$ with its odd coderivation ∂ as the graded Hom bimodule $\widehat{C} = \operatorname{Hom}_K({}^\#\widehat{B}, K)$, and the graded ring $\widehat{B}$ with its odd derivation ∂ as the graded Hom group/bimodule $\widehat{B} = \operatorname{Hom}_K(\widehat{C}, K)$. $\qquad \square$

Lemma 8.19 *In the context of Proposition 8.18, for any pair of CDG-ring structures $B = (B, d, h)$ and $^\#B = ({}^\#B, {}^\#d, {}^\#h)$ corresponding to each other under the construction of the proposition, there is a natural structure of CDG-bimodule over $^\#B$ and B on the graded $^\#B$-B-bimodule C.*

Proof Saying that $B = (B, d, h)$ and $^\#B = ({}^\#B, {}^\#d, {}^\#h)$ correspond to each other under the construction of Proposition 8.18 means the following. Let $(\widehat{B}, \partial)$ and $({}^\#\widehat{B}, \partial)$ be a pair of quasi-differential graded rings corresponding to each other under the construction from the proof of the proposition. (This presumes that $\widehat{B}$ and $^\#\widehat{B}$ are nonnegatively graded, and that we are given ring homomorphisms $K \longrightarrow \widehat{B}^0$ and $K \longrightarrow {}^\#\widehat{B}^0$ such that $\widehat{B}^n$ is a finitely generated projective right K-module and $^\#\widehat{B}^n$ is a finitely generated projective left K-module for every $n \geq 0$.)

Choose arbitrary (unrelated) elements $\delta \in \widehat{B}^1$ such that $\partial(\delta) = 1$ in $\widehat{B}^0$ and $^\#\delta \in {}^\#\widehat{B}^1$ such that $\partial({}^\#\delta) = 1$ in $^\#\widehat{B}^0$. Define the differential d on $B = \ker \partial \subset \widehat{B}$ and the curvature element $h \in B^2$ as in the proof of Theorem 4.7,

$$d(b) = [\delta, b] \quad \text{for all } b \in B, \quad \text{and} \quad h = \delta^2,$$

where the bracket [,] denotes the graded commutator. Similarly, consider the graded ring $^{\#}B = \ker \partial \subset {}^{\#}\widehat{B}$, and put

$$^{\#}d(^{\#}b) = [^{\#}\delta, {}^{\#}b] \quad \text{for all } {}^{\#}b \in {}^{\#}B, \quad \text{and} \quad {}^{\#}h = {}^{\#}\delta^2.$$

In this context, we define the differential $d_{\widehat{C}}$ on the graded $^{\#}\widehat{B}\text{-}\widehat{B}$-bimodule $\widehat{C}$ by the rule

$$d_{\widehat{C}}(c) = {}^{\#}\delta c - (-1)^{|c|} c\delta \quad \text{for all } c \in \widehat{C}.$$

The claim is that the map $d_{\widehat{C}} \colon \widehat{C} \longrightarrow \widehat{C}$ induces a well-defined map $d_C \colon C \longrightarrow C$. In other words, there exists an (obviously unique) homogeneous map $d_C \colon C \longrightarrow C$ (of degree 1 in the upper indices) forming a commutative square diagram with the map $d_{\widehat{C}}$ and the surjective coring homomorphism $\widehat{C} \longrightarrow C$.

In order to prove this assertion, we will check that $d_{\widehat{C}}$ preserves the kernel $\partial(C)$ of the surjective map $\widehat{C} \longrightarrow C$, that is $d_{\widehat{C}}(\partial(C)) \subset \partial(C)$. Moreover, we will show that the two differentials ∂ and $d_{\widehat{C}}$ on $\widehat{C}$ anti-commute, that is

$$\partial \circ d_{\widehat{C}} + d_{\widehat{C}} \circ \partial = 0 \quad \text{on } \widehat{C}. \tag{8.2}$$

First of all, let us check that the differential ∂ on the graded right $\widehat{B}$-module $\widehat{C}$ is an odd derivation compatible with the odd derivation ∂ on $\widehat{B}$, that is

$$\partial(cb) = \partial(c)b + (-1)^{|c|} c\partial(b) \in \widehat{C} \quad \text{for all } c \in \widehat{C}^{|c|} \text{ and } b \in \widehat{B}^{|b|}.$$

Indeed, for any $b' \in \widehat{B}$ we have

$$\langle \partial(cb), b' \rangle = -(-1)^{|c|+|b|} \langle cb, \partial(b') \rangle = -(-1)^{|c|+|b|} \langle c, b\partial(b') \rangle$$

$$= -(-1)^{|c|} \langle c, \partial(bb') \rangle + (-1)^{|c|} \langle c, \partial(b)b' \rangle$$

$$= \langle \partial(c), bb' \rangle + (-1)^{|c|} \langle c, \partial(b)b' \rangle = \langle \partial(c)b, b' \rangle + (-1)^{|c|} \langle c\partial(b), b' \rangle.$$

Similarly one can check that the differential ∂ on the graded left $^{\#}\widehat{B}$-module $\widehat{C}$ is also an odd derivation compatible with the odd derivation ∂ on $^{\#}\widehat{B}$, that is

$$\partial(^{\#}bc) = \partial(^{\#}b)c + (-1)^{|{}^{\#}b|} {}^{\#}b\partial(c) \in \widehat{C} \quad \text{for all } {}^{\#}b \in {}^{\#}\widehat{B}^{|{}^{\#}b|} \text{ and } c \in \widehat{C}^{|c|}.$$

Now for any $c \in \widehat{C}$ we have

$$\partial(^{\#}\delta c) = \partial(^{\#}\delta)c - {}^{\#}\delta\partial(c) = c - {}^{\#}\delta\partial(c)$$

and

$$\partial(c\delta) = \partial(c)\delta + (-1)^{|c|}c\partial(\delta) = (-1)^{|c|}c + \partial(c)\delta,$$

hence

$$\partial(d_{\widehat{C}}(c)) = \partial(^{\#}\delta c - (-1)^{|c|}c\delta) = -^{\#}\delta\partial(c) - (-1)^{|c|}\partial(c)\delta = -d_{\widehat{C}}(\partial(c)),$$

and Eq. (8.2) is deduced.

Thus the desired differential $d_C \colon C \longrightarrow C$ is well-defined. Checking that d_C is an odd derivation on the graded left $^{\#}B$-module C and the graded right B-module C compatible with the odd derivations $^{\#}d = [^{\#}\delta, -]$ on $^{\#}B$ and $d = [\delta, -]$ on B is easy. The same applies to checking the equation $d_C^2(c) = {}^{\#}hc - ch$ for all $c \in C$. $\qquad\square$

8.5 Comodule-Contramodule Correspondence

We formulate two lemmas of rather general character before specializing to the situation we are interested in.

Lemma 8.20 *Let* $B = \bigoplus_{n=0}^{\infty} B^n$ *be a nonnegatively graded ring, and let* K *be an associative ring endowed with a ring homomorphism* $K \longrightarrow B^0$. *Then*

(a) *For any graded left* K-*module* L *and any graded left* B-*comodule* M, *the graded abelian group of graded left* B-*comodule (or graded left* B-*module) Hom from* M *to the coinduced graded left* B-*comodule* $\mathrm{Hom}_K^{\Sigma}(B, L)$ *can be computed as*

$$\mathrm{Hom}_B(M, \mathrm{Hom}_K^{\Sigma}(B, L)) \simeq \mathrm{Hom}_K(L, M).$$

(b) *For any graded left* K-*module* L *and any graded right* B-*comodule* N, *the contratensor product of* N *with the induced graded left* B-*contramodule* $\mathbb{M}_K^{\mathrm{gr}}(L) = B\otimes_K^{\Pi} L$ *can be computed as*

$$N \odot_B (B\otimes_K^{\Pi} L) \simeq N \otimes_K L.$$

Proof The natural isomorphism of graded abelian groups in part (a) follows from the similar natural isomorphism for the graded Hom group of graded B-modules, $\mathrm{Hom}_B(M, \mathrm{Hom}_K(B, L)) \simeq \mathrm{Hom}_K(L, M)$, together with Lemma 6.13.

To construct the natural isomorphism of graded abelian groups in part (b), let U be a graded abelian group. Then we have natural isomorphisms

$$\operatorname{Hom}_{\mathbb{Z}}(N \odot_B (B \otimes_K^{\Pi} L),\ U) \overset{8.4}{\simeq} \operatorname{Hom}^B(B \otimes_K^{\Pi} L,\ \operatorname{Hom}(N, U))$$

$$\overset{7.9}{\simeq} \operatorname{Hom}_K(L, \operatorname{Hom}(N, U)) \simeq \operatorname{Hom}_K(N \otimes_K L,\ U),$$

where the first isomorphism is provided by Lemma 8.4 (applied to the graded ring $E = \mathbb{Z}$ concentrated in degree 0), while the second one holds in view of the discussion in Sect. 5.2 or by Lemma 7.9. It remains to say that any isomorphism of representable functors comes from a unique isomorphism of the representing objects. (In fact, the isomorphism we have constructed is provided by the map $N \otimes_K L \longrightarrow N \odot_B (B \otimes_K^{\Pi} L)$ induced by the unit element of B.) □

Lemma 8.21 *Let* $E = (E, d_E, h_E)$ *be a CDG-ring and* $B = (B, d, h)$ *be a nonnegatively graded CDG-ring. Let* $N = (N, d_N)$ *be a CDG-bimodule over* E *and* B *whose underlying graded right* B-module is a graded right B-comodule. Then*

(a) *for any left CDG-module* $M = (M, d_M)$ *over* E, *the graded left* B-contramodule* $\operatorname{Hom}_E(N, M)$ *constructed in Example 5.31 endowed with the differential of the left CDG-module* $\operatorname{Hom}_E(N, M)$ *over* B *constructed in Sect. 6.1 is a left CDG-contra-module over* B *in the sense of the definition in Sect. 7.3.*
(b) *for any left CDG-contramodule* (Q, d_Q) *over* B, *the differential of the left CDG-module* $N \otimes_B Q$ *over* E *constructed in Sect. 6.1 induces a well-defined differential on the graded quotient* E-module* $N \odot_B Q$ *of* $N \otimes_B Q$ *constructed in Sect. 8.1, making* $N \odot_B Q$ *a left CDG-module over* E.*

Proof The proofs of both (a) and (b) are straightforward. □

Now we assume that $B = (B, d, h)$ and $^{\#}B = (^{\#}B, {}^{\#}d, {}^{\#}h)$ are two CDG-rings corresponding to each other under the construction of Proposition 8.18. This means, in particular, that $B = \bigoplus_{n=0}^{\infty} B^n$ and $^{\#}B = \bigoplus_{n=0}^{\infty} {}^{\#}B^n$ are nonnegatively graded, that we are given a ring K together with ring homomorphisms $K \longrightarrow B^0$ and $K \longrightarrow {}^{\#}B^0$, that the right K-module B^n is finitely generated and projective for every $n \geq 0$, and that the left K-module $^{\#}B^n$ is finitely generated and projective for every $n \geq 0$.

The following theorem is the first main result of Chap. 8. It is a generalization of [59, Theorem B.3]. It also solves a particular case of the problem posed in [58, Question at the end of Chapter 11].

Theorem 8.22 *Assume additionally that the left homological dimension of the ring* K *is finite. Then there is a natural triangulated equivalence between the coderived category of left CDG-comodules over* $^{\#}B = (^{\#}B, {}^{\#}d, {}^{\#}h)$ *and the contraderived category of left*

CDG-contramodules over (B, d, h),

$$\mathbb{R}\operatorname{Hom}_{\#B}(C, -)\colon \mathsf{D}^{\mathrm{co}}(^{\#}B\text{–comod}) \simeq \mathsf{D}^{\mathrm{ctr}}(B\text{–contra})\colon C\odot_B^{\mathbb{L}}-, \tag{8.3}$$

provided by derived functors of comodule Hom and contratensor product with the CDG-bi(co)module (C, d_C) *over* $^{\#}B$ *and* B *constructed in Lemma 8.19.*

Proof First of all, the CDG-bimodule (C, d_C) is nonpositively cohomologically graded. According to the discussion in Sect. 5.5, it follows that C is a graded left $^{\#}B$-comodule and a graded right B-comodule.

The DG-functor

$$\operatorname{Hom}_{\#B}(C, -)\colon \mathsf{DG}(^{\#}B\text{–comod}) \longrightarrow \mathsf{DG}(B\text{–contra}) \tag{8.4}$$

assigns to a left CDG-comodule M over $(^{\#}B, {^{\#}d}, {^{\#}h})$ the graded left B-contramodule $\operatorname{Hom}_{\#B}(C, M)$, as constructed in Example 5.31, endowed with the structure of a left CDG-contramodule over (B, d, h) described in Lemma 8.21(a). Since $\operatorname{Hom}_{\#B}(C, -)$ is a DG-functor $\mathsf{DG}(^{\#}B\text{–mod}) \longrightarrow \mathsf{DG}(B\text{–mod})$ according to the discussion in Sect. 6.1, it follows that it is also a DG-functor $\mathsf{DG}(^{\#}B\text{–comod}) \longrightarrow \mathsf{DG}(B\text{–contra})$.

The DG-functor

$$C\odot_B-\colon \mathsf{DG}(B\text{–contra}) \longrightarrow \mathsf{DG}(^{\#}B\text{–comod}) \tag{8.5}$$

assigns to a left CDG-contramodule Q over (B, d, h) the graded abelian group of the contratensor product $C\odot_B Q$, as constructed in Section 8.1, endowed with the obvious graded left $^{\#}B$-comodule structure induced by the graded left $^{\#}B$-comodule structure on C and with the structure of a left CDG-comodule over $(^{\#}B, {^{\#}d}, {^{\#}h})$ described in Lemma 8.21(b). Since $C\otimes_B-$ is a DG-functor $\mathsf{DG}(B\text{–mod}) \longrightarrow \mathsf{DG}(^{\#}B\text{–mod})$ according to the discussion in Sect. 6.1, it follows that $C\odot_B-$ is a DG-functor $\mathsf{DG}(B\text{–contra}) \longrightarrow \mathsf{DG}(^{\#}B\text{–comod})$.

The adjunction of Lemma 8.4 makes the DG-functors (8.4) and (8.5) adjoint to each other. Hence the induced triangulated functors between the homotopy categories

$$\operatorname{Hom}_{\#B}(C, -)\colon \mathsf{Hot}(^{\#}B\text{–comod}) \longrightarrow \mathsf{Hot}(B\text{–contra}) \tag{8.6}$$

and

$$C\odot_B-\colon \mathsf{Hot}(B\text{–contra}) \longrightarrow \mathsf{Hot}(^{\#}B\text{–comod}) \tag{8.7}$$

are also adjoint.

In order to construct the derived functor $\mathbb{R}\,\mathrm{Hom}_{\#B}(C, -)\colon \mathsf{D}^{\mathsf{co}}(^\#B\text{–comod}) \longrightarrow$ $\mathsf{D}^{\mathsf{ctr}}(B\text{–contra})$, we restrict the triangulated functor $\mathrm{Hom}_{\#B}(C, -)$ (8.6) to the full triangulated subcategory of CDG-comodules of the coinduced type

$$\mathsf{Hot}(^\#B\text{–comod}_{K\text{-coind}}) \subset \mathsf{Hot}(^\#B\text{–comod}),$$

which was defined in Sect. 8.3. Similarly, in order to construct the derived functor $C\odot_B^{\mathbb{L}}-\colon \mathsf{D}^{\mathsf{ctr}}(B\text{–contra}) \longrightarrow \mathsf{D}^{\mathsf{co}}(B\text{–comod})$, we restrict the triangulated functor $C\odot_B -$ (8.7) to the full triangulated subcategory of CDG-contramodules of the induced type

$$\mathsf{Hot}(B\text{–contra}_{K\text{-ind}}) \subset \mathsf{Hot}(B\text{–contra}),$$

which was defined in Sect. 8.2.

Now it is claimed that the functors $\mathrm{Hom}_{\#B}(C, -)$ and $C\odot_B -$ take the full subcategories $\mathsf{Hot}(^\#B\text{–comod}_{K\text{-coind}}) \subset \mathsf{Hot}(^\#B\text{–comod})$ and $\mathsf{Hot}(B\text{–contra}_{K\text{-ind}}) \subset \mathsf{Hot}(B\text{–contra})$ into each other and induce mutually inverse triangulated equivalences between them

$$\mathrm{Hom}_{\#B}(C, -)\colon \mathsf{Hot}(^\#B\text{–comod}_{K\text{-coind}}) \simeq \mathsf{Hot}(B\text{–contra}_{K\text{-ind}}) \colon C\odot_B -. \qquad (8.8)$$

Moreover, we have mutually inverse DG-equivalences

$$\mathrm{Hom}_{\#B}(C, -)\colon \mathsf{DG}(^\#B\text{–comod}_{K\text{-coind}}) \simeq \mathsf{DG}(B\text{–contra}_{K\text{-ind}}) \colon C\odot_B -.$$

It suffices to check the latter assertion on the level of the full subcategory of coinduced graded comodules $^\#B\text{–comod}_{K\text{-coind}} \subset {}^\#B\text{–comod}$ in the abelian category of graded left $^\#B$-comodules $^\#B\text{–comod}$ and the full subcategory of induced graded contramodules $B\text{–contra}_{K\text{-ind}} \subset B\text{–contra}$ in the abelian category of graded left contramodules $B\text{–contra}$. So we need to check that the adjunction of Lemma 8.4 restricts to a pair of mutually inverse equivalences of additive categories

$$\mathrm{Hom}_{\#B}(C, -)\colon {}^\#B\text{–comod}_{K\text{-coind}} \simeq B\text{–contra}_{K\text{-ind}} \colon C\odot_B -.$$

This amounts to a simple computation based on Lemma 8.20. To show that two adjoint functors are mutually inverse equivalences, it suffices to check that the adjunction morphisms (i.e., the adjunction unit and counit) are isomorphisms.

Let L be a graded left K-module, $\mathrm{Hom}_K^\Sigma(^\#B, L)$ be the graded left $^\#B$-comodule coinduced from L, and $B\otimes_K^\Pi L$ be the graded left B-contramodule induced from L. Then we have

$$\mathrm{Hom}_{\#B}(C, \mathrm{Hom}_K^\Sigma(^\#B, L)) \overset{8.20(a)}{\simeq} \mathrm{Hom}_K(C, L) \simeq \mathrm{Hom}_K(C, K)\otimes_K^\Pi L = B\otimes_K^\Pi L,$$

since $C^{-n} = \mathrm{Hom}_{K^{\mathrm{op}}}(B^n, K)$ is a finitely generated projective left K-module for every $n \geq 0$. Similarly,

$$C \odot_B (B \otimes_K^{\Pi} L) \overset{8.20(b)}{\simeq} C \otimes_K L \simeq \mathrm{Hom}_K^{\Sigma}(\mathrm{Hom}_{K^{\mathrm{op}}}(C, K), L) = \mathrm{Hom}_K^{\Sigma}({}^{\#}B, L),$$

since $C^{-n} = \mathrm{Hom}_K({}^{\#}B^n, K)$ is a finitely generated projective right K-module for every $n \geq 0$. This finishes the proof of the triangulated equivalence (8.8).

Finally, in order to show that the triangulated equivalence (8.8) descends to the desired triangulated equivalence (8.3), we use the descriptions of the coderived category $\mathsf{D}^{\mathrm{co}}({}^{\#}B\text{-}\mathsf{comod})$ and the contraderived category $\mathsf{D}^{\mathrm{ctr}}(B\text{-}\mathsf{contra})$ provided by Theorems 8.17 and 8.11 together with Propositions 8.16 and 8.10.

Specifically, by Proposition 8.16 and the second assertion of Theorem 8.17, the coderived category $\mathsf{D}^{\mathrm{co}}({}^{\#}B\text{-}\mathsf{comod})$ is equivalent to the triangulated quotient category of $\mathsf{Hot}({}^{\#}B\text{-}\mathsf{comod}_{K\text{-coind}})$ by its minimal triangulated subcategory containing all the totalizations of left CDG-comodule short exact sequences of the coinduced type over $({}^{\#}B, {}^{\#}d, {}^{\#}h)$. Similarly, by Proposition 8.10 and the second assertion of Theorem 8.11, the contraderived category $\mathsf{D}^{\mathrm{ctr}}(B\text{-}\mathsf{contra})$ is equivalent to the triangulated quotient category of $\mathsf{Hot}(B\text{-}\mathsf{contra}_{K\text{-ind}})$ by its minimal triangulated subcategory containing all the totalizations of left CDG-contramodule short exact sequences of the induced type over (B, d, h).

In view of these results, it remains to observe that the DG-functors $\mathrm{Hom}_{{}^{\#}B}(C, -)$ and $C \odot_B -$ take CDG-comodule short exact sequences of the coinduced type over $({}^{\#}B, {}^{\#}d, {}^{\#}h)$ to CDG-contramodule short exact sequences of the induced type over (B, d, h) and back. This is also an assertion on the level of the additive categories ${}^{\#}B\text{-}\mathsf{comod}_{K\text{-coind}}$ and $B\text{-}\mathsf{contra}_{K\text{-ind}}$, and it follows from the above computation based on Lemma 8.20. $\square$

Let us recall that, whenever the forgetful functor $B\text{-}\mathsf{contra}_{\mathrm{gr}} \longrightarrow B\text{-}\mathsf{mod}_{\mathrm{gr}}$ is fully faithful, the contratensor product functor $C \odot_B -$ appearing in the above theorem is isomorphic to the tensor product functor $C \otimes_B -$ by Proposition 8.5(b). In particular, by Theorem 5.34, this holds whenever the augmentation ideal $B^{\geq 1} = \bigoplus_{n=1}^{\infty} B^n$ is finitely generated as a right ideal in B.

8.6 Two-Sided Finitely Projective Koszul Rings

Let $A = \bigoplus_{n=0}^{\infty} A_n$ be a nonnegatively graded ring with the degree-zero component $R = A_0$.

Definition 8.23 We will say that A is *two-sided finitely projective Koszul* if it is *both* left finitely projective Koszul and right finitely projective Koszul (in the sense of Sect. 2.10).

Let A be a two-sided finitely projective Koszul graded ring. Denote by B the right finitely projective Koszul graded ring quadratic dual to the left finitely projective Koszul ring A, and denote by $^{\#}B$ the left finitely projective Koszul graded ring quadratic dual to the right finitely projective Koszul ring A (as in Chap. 1 and Proposition 2.37). So we have

$$B = \mathrm{Ext}_A(R, R)^{\mathrm{op}} \quad \text{and} \quad {}^{\#}B = \mathrm{Ext}_{A^{\mathrm{op}}}(R, R),$$

where Ext_A is taken in the category of graded left A-modules and $\mathrm{Ext}_{A^{\mathrm{op}}}$ in the category of graded right A-modules, as usually.

Lemma 8.24 *For any two-sided finitely projective Koszul graded ring A, the nonnegatively graded rings B and $^{\#}B$ are related to each other by the construction of Remark 2.13 and Sect. 8.4 (with $K = R$). In other words, the grading components of B and $^{\#}B$ can be obtained from each other as the R-R-bimodules*

$$^{\#}B^n = \mathrm{Hom}_{R^{\mathrm{op}}}(\mathrm{Hom}_{R^{\mathrm{op}}}(B^n, R), R) \quad \text{and} \quad B^n = \mathrm{Hom}_R(\mathrm{Hom}_R({}^{\#}B^n, R), R)$$

for every $n \geq 0$, and the multiplicative structures are related accordingly.

Proof Put $V = A_1$, and denote by $I \subset V \otimes_R V$ the kernel of the multiplication map $A_1 \otimes_R A_1 \longrightarrow A_2$. The graded R-R-bimodule C with the components

$$C_n = \mathrm{Tor}_n^A(R, R) = \mathrm{Tor}_{n,n}^A(R, R)$$

plays a key role. By Proposition 2.1(b), we have $C_0 = R$, $C_1 = V$, $C_2 = I$, and

$$C_n = \bigcap_{k=1}^{n-1} V^{\otimes_R k-1} \otimes_R I \otimes_R V^{\otimes_R n-k-1} \subset V^{\otimes_R n}, \qquad n \geq 2.$$

Since A is a left finitely projective Koszul graded ring, Theorem 2.35(c) tells, in particular, that C_n is a finitely generated projective left R-module for every $n \geq 0$. As A is also a right finitely projective Koszul graded ring, C_n is also a finitely generated projective right R-module.

According to Remark 2.3, C has a natural graded coring structure. The grading components of the comultiplication map $\mu_{i,j} \colon C_{i+j} \longrightarrow C_i \otimes_R C_j$, $i, j \geq 0$, can be described explicitly as the R-R-subbimodule inclusions

$$C_{i+j} = \bigcap_{k=1}^{i+j-1} V^{\otimes_R k-1} \otimes_R I \otimes_R V^{\otimes_R i+j-k-1}$$

$$\longrightarrow \bigcap_{\substack{1 \leq k \leq i+j-1 \\ k \neq i}} V^{\otimes_R k-1} \otimes_R I \otimes_R V^{\otimes_R i+j-k-1} \simeq C_i \otimes_R C_j,$$

where the latter isomorphism holds by Lemma 2.24(b,d,e).

Computing the Ext in terms of the relative cobar complex as in Proposition 2.2, one obtains the isomorphisms

$$B^n = \text{Ext}_A^n(R, R) = \text{Ext}_A^{n,n}(R, R) \simeq \text{Hom}_R(C_n, R)$$

and

$$ {}^\#B^n = \text{Ext}_{A^{\text{op}}}^n(R, R) = \text{Ext}_{A^{\text{op}}}^{n,n}(R, R) \simeq \text{Hom}_{R^{\text{op}}}(C_n, R), $$

and these isomorphisms are compatible with the multiplicative structures.   □

Definition 8.25 Let $(\widetilde{A}, F)$ be an associative ring with an increasing filtration $0 = F_{-1}\widetilde{A} \subset R = F_0\widetilde{A} \subset \widetilde{V} = F_1\widetilde{A} \subset F_2\widetilde{A} \subset \cdots$. We recall that the filtered ring $(\widetilde{A}, F)$ is said to be *left finitely projective nonhomogeneous Koszul* if the graded ring $A = \text{gr}^F \widetilde{A} = \bigoplus_{n=0}^{\infty} F_n\widetilde{A}/F_{n-1}\widetilde{A}$ is left finitely projective Koszul (see Sect. 4.6).

Dually, let us say that $(\widetilde{A}, F)$ is *right finitely projective nonhomogeneous Koszul* if $(\widetilde{A}^{\text{op}}, F^{\text{op}})$ is left finitely projective nonhomogeneous Koszul, that is, in other words, the graded ring A is right finitely projective Koszul. We will say that $(\widetilde{A}, F)$ is *two-sided finitely projective nonhomogeneous Koszul* if it is both left and right finitely projective nonhomogeneous Koszul; in other words, this means that the graded ring A is two-sided finitely projective Koszul.

Let B and ${}^\#B$ be the Koszul dual rings to A on the two sides, as in Lemma 8.24. Recall the notation τ' introduced in Sect. 6.5 for a left R-linear splitting

$$\text{Hom}_{R^{\text{op}}}(B^1, R) \simeq F_1\widetilde{A}/F_0\widetilde{A} = V \hookrightarrow \widetilde{V} = F_1\widetilde{A}$$

of the surjective R-R-bimodule map $\widetilde{V} = F_1\widetilde{A} \longrightarrow F_1\widetilde{A}/F_0\widetilde{A} = V$. Such a splitting was first considered (under the assumption of projectivity of the left R-module V, guaranteeing its existence) in Sect. 3.3.

For a two-sided finitely projective nonhomogeneous Koszul ring $\widetilde{A}$, one can also choose a right R-linear splitting $V \longrightarrow \widetilde{V}$ of the surjective R-R-bimodule map $\widetilde{V} \longrightarrow \widetilde{V}/R = V$. Denote by σ' the resulting right R-module map

$$\text{Hom}_R({}^\#B^1, R) \simeq F_1\widetilde{A}/F_0\widetilde{A} = V \hookrightarrow \widetilde{V} = F_1\widetilde{A}. \tag{8.9}$$

Having chosen a one-sided splitting τ', one can apply the construction of Sects. 3.3–3.4 to the left finitely projective nonhomogeneous Koszul ring $\widetilde{A}$ and produce a CDG-ring structure (B, d, h) on the right finitely projective Koszul graded ring B. Having chosen a splitting σ' on the other side, one can apply the opposite version of the same construction (with the left and right sides switched) to the right finitely projective nonhomogeneous

Koszul ring $\widetilde{A}$ and produce a CDG-ring structure $(^{\#}B, {}^{\#}d, {}^{\#}h)$ on the left finitely projective Koszul graded ring $^{\#}B$.

Lemma 8.26 *For any two-sided finitely projective nonhomogeneous Koszul ring $\widetilde{A}$, the nonnegatively graded CDG-rings (B, d, h) and $(^{\#}B, {}^{\#}d, {}^{\#}h)$ are related by the construction of Proposition 8.18 (with $K = R$).*

Proof We use the interpretation of nonhomogeneous quadratic duality in terms of quasi-differential corings, as formulated in Theorem 4.22. The graded ring $\widehat{A} = \bigoplus_{n=0}^{\infty} \widetilde{A}_n$ is left finitely projective Koszul by Lemma 4.3 and Theorem 4.4. By the opposite assertion, the graded ring $\widehat{A}$ is also right finitely projective Koszul.

Let $\widehat{B}$ be the right finitely projective Koszul graded ring quadratic dual to the left finitely projective Koszul graded ring $\widehat{A}$, and let $^{\#}\widehat{B}$ be the left finitely projective Koszul graded ring quadratic dual to the right finitely projective Koszul graded ring $\widehat{A}$. By Lemma 8.24 applied to the two-sided finitely projective Koszul graded ring $\widehat{A}$, the nonnegatively graded rings $\widehat{B}$ and $^{\#}\widehat{B}$ are related to each other by the construction from the beginning of Sect. 8.4.

It remains to observe that the acyclic odd derivations ∂ of degree -1 on the graded rings $\widehat{B}$ and $^{\#}\widehat{B}$, both originating from the canonical central element $t \in \widehat{A}_1$ via the construction of Lemma 4.14(b) and its opposite version, correspond to each other under the construction from the proof of Proposition 8.18.   □

8.7   Bimodule Resolution Revisited

Now, under the more restrictive assumption of two-sided (rather than just left) finite projective Koszulity, we can offer an explanation of the "twisting cochain σ' notation" of Sects. 6.5 and 7.4. Let us start with the "twisting cochain σ," which is the homogeneous Koszul version (it was briefly mentioned in the proofs of Theorems 6.14 and 7.11).

Let A be a left finitely projective Koszul graded ring and B be the quadratic dual right finitely projective Koszul graded ring. In this context, in the discussion of homogeneous Koszul complexes in Sects. 2.5–2.6, the symbol τ was used as a notation for the structure isomorphism of quadratic duality

$$\tau : \operatorname{Hom}_{R^{\mathrm{op}}}(B^1, R) \xrightarrow{\;\simeq\;} A_1.$$

Assume that A is also a right finitely projective Koszul graded ring, and let $^{\#}B$ be the quadratic dual left finitely projective Koszul graded ring. Denote by σ the structure isomorphism

$$\sigma : \operatorname{Hom}_R(^{\#}B^1, R) \xrightarrow{\;\simeq\;} A_1.$$

Moreover, in the context of the discussion of the dual Koszul complex $K_e^{\vee\bullet}(B, A) = (B \otimes_R A, \ d_e)$ in Sect. 2.6, we denoted by $e \in B^1 \otimes_R A_1$ the element corresponding to the map τ under the construction of Lemma 2.5. Let us now denote by $u \in A_1 \otimes_R {}^\#B^1$ the element corresponding to the map σ under the opposite version of the same construction. Denote the related dual Koszul complex by $K_u^{\vee\bullet}(A, {}^\#B) = (A \otimes_R {}^\#B, \ d_u)$.

For any left A-module M and any graded right B-module N (which we would prefer to assume additionally to be a B-comodule) we define the complex $N \otimes_R^\tau M$ as the tensor product

$$N \otimes_R^\tau M = N \otimes_B K_e^{\vee\bullet}(B, A) \otimes_A M$$

with the differential induced by the differential d_e on $K_e^{\vee\bullet}$. So $N \otimes_R^\tau M$ is a complex of abelian groups whose underlying graded abelian group is $N \otimes_R M$. Additional structures on a left A-module M and a graded right B-comodule N may induce additional structures on the complex $N \otimes_R^\tau M$, as usually.

Similarly, for any right A-module M and any graded left ${}^\#B$-module N (which we would rather assume to be a graded left ${}^\#B$-comodule) we define the complex $M \otimes_R^\sigma N$ as the tensor product

$$M \otimes_R^\sigma N = M \otimes_A K_u^{\vee\bullet}(A, {}^\#B) \otimes_{{}^\#B} N$$

with the differential induced by the differential d_u on $K_u^{\vee\bullet}$. So $M \otimes_R^\sigma N$ is a complex of abelian groups whose underlying graded abelian group is $M \otimes_R N$.

Now we recall the notation C for the graded R-R-bimodule (in fact, a graded coring) $\mathrm{Tor}^A(R, R)$ which was used and discussed in the proof of Lemma 8.24 and the references therein. In particular, according to Remark 2.13 and Sect. 8.4, C is a graded ${}^\#B$-B-bimodule (in fact, even a comodule on both sides, as explained in the beginning of the proof of Theorem 8.22).

With these preparations, we can interpret the second Koszul complex ${}^\tau K_\bullet(B, A) = (\mathrm{Hom}_{R^{op}}(B, R) \otimes_R A, \ {}^\tau\partial)$ (2.18) as

$$ {}^\tau K_\bullet(B, A) = C \otimes_R^\tau A.$$

The first Koszul complex $K_\bullet^\tau(B, A) = (\mathrm{Hom}_{R^{op}}(B, A), \partial^\tau) = (A \otimes_R \mathrm{Hom}_{R^{op}}(B, R), \partial^\tau)$ (2.16) gets interpreted as

$$K_\bullet^\tau(B, A) = A \otimes_R^\sigma C,$$

as mentioned in the proofs of Theorems 6.14 and 7.11. Looking on A as primarily a right finitely projective Koszul graded ring rather than a left one, the first and second Koszul complexes switch their roles, of course.

Let us turn to the nonhomogeneous situation. Let $(\widetilde{A}, F)$ be a two-sided finitely projective nonhomogeneous Koszul ring, and let (B, d, h) and $({}^{\#}B, {}^{\#}d, {}^{\#}h)$ be its two nonhomogeneous quadratic dual CDG-rings on the right and left sides, as in Sect. 8.6. The notation σ' for the right R-linear map between R-R-bimodules $\mathrm{Hom}_R({}^{\#}B^1, R) \longrightarrow F_1\widetilde{A}$ was already introduced in (8.9).

Denote by $u' \in F_1\widetilde{A}\otimes_R{}^{\#}B^1$ the element corresponding to σ' under the opposite version of the construction of Lemma 2.5. The opposite version of the construction of Sect. 6.2 produces the dual nonhomogeneous Koszul CDG-module $K^{\vee}(\widetilde{A}, {}^{\#}B) = K^{\vee}_{u'}(\widetilde{A}, {}^{\#}B)$, which has the form

$$0 \longrightarrow \widetilde{A} \longrightarrow \widetilde{A}\otimes_R{}^{\#}B^1 \longrightarrow \widetilde{A}\otimes_R{}^{\#}B^2 \longrightarrow \cdots$$

The differential on $K^{\vee}_{u'}(\widetilde{A}, {}^{\#}B)$ does not square to zero when ${}^{\#}h \neq 0$. Rather, $K^{\vee}_{u'}(\widetilde{A}, {}^{\#}B)$ is a right CDG-module over $({}^{\#}B, {}^{\#}d, {}^{\#}h)$, and in fact a CDG-bimodule over $(\widetilde{A}, 0, 0)$ and $({}^{\#}B, {}^{\#}d, {}^{\#}h)$. The underlying graded $\widetilde{A}$-${}^{\#}B$-bimodule of $K^{\vee}_{u'}(\widetilde{A}, {}^{\#}B)$ is the bimodule tensor product $\widetilde{A}\otimes_R{}^{\#}B$.

Now let $D = (D, d_D, h_D)$ and $E = (E, d_E, h_E)$ be two CDG-rings. Let M be a CDG-bimodule over $(\widetilde{A}, 0, 0)$ and (E, d_E, h_E), and let N be a CDG-bimodule over (D, d_D, h_D) and (B, d, h). We would prefer to assume N to be a graded right B-comodule rather than an arbitrary graded right B-module. In this setting, we define the CDG-bimodule $N\otimes_R^{\tau'}M$ over D and E as the tensor product of CDG-bimodules

$$N\otimes_R^{\tau'}M = N\otimes_B K^{\vee}_{e'}(B, \widetilde{A})\otimes_{\widetilde{A}}M.$$

The underlying graded D-E-bimodule of $N\otimes_R^{\tau'}M$ is the bimodule tensor product $N\otimes_R M$. If D is nonnegatively graded and N is a D-comodule, or if E is nonnegatively graded and M is an E-comodule, then the tensor product $N\otimes_R M$ is obviously a comodule on the respective side.

Similarly, let M be a CDG-bimodule over (E, d_E, h_E) and $(\widetilde{A}, 0, 0)$, and let N be a CDG-bimodule over $({}^{\#}B, {}^{\#}d, {}^{\#}h)$ and (D, d_D, h_D). We would rather assume the graded left ${}^{\#}B$-module N to be a ${}^{\#}B$-comodule. Then we define the CDG-bimodule $M\otimes_R^{\sigma'}N$ over E and D as the tensor product of CDG-bimodules

$$M\otimes_R^{\sigma'}N = M\otimes_{\widetilde{A}}K^{\vee}_{u'}(\widetilde{A}, {}^{\#}B)\otimes_{{}^{\#}B}N.$$

The underlying graded E-D-bimodule of $M\otimes_R^{\sigma'}N$ is the bimodule tensor product $M\otimes_R N$. If D is nonnegatively graded and N is a D-comodule, or if E is nonnegatively graded and M is an E-comodule, then the tensor product $M\otimes_R N$ is obviously a comodule on the respective side.

Now we recall that, by Lemma 8.19 (which is applicable to the situation at hand in view of Lemma 8.26), there is a natural differential d_C on the graded ${}^{\#}B$-B-bimodule C, making

(C, d_C) a CDG-bimodule over $(^\#B, {}^\#d, {}^\#h)$ and (B, d, h). With these preparations, we can interpret the nonhomogeneous Koszul duality functors

$$M^\bullet \longmapsto M^\bullet \otimes^{\sigma'}_R C \quad \text{and} \quad N \longmapsto N \otimes^{\tau'}_R \widetilde{A}$$

of Sect. 6.5 as particular cases of the above constructions of twisted tensor products. This explains the meaning of the "placeholder σ'" in Sect. 6.5.

Let M be a CDG-bimodule over $(\widetilde{A}, 0, 0)$ and (D, d_D, h_D), and let Q be a left CDG-module over (B, d, h). We would prefer to assume Q to be a left CDG-contramodule over (B, d, h). In this setting, we define the left CDG-module $\operatorname{Hom}^{\tau'}_R(M, Q)$ over (D, d_D, h_D) as the Hom CDG-module

$$\operatorname{Hom}^{\tau'}_R(M, Q) = \operatorname{Hom}_B(K^\vee_{e'}(B, \widetilde{A}) \otimes_{\widetilde{A}} M, \; Q) = \operatorname{Hom}_{\widetilde{A}}(M, \operatorname{Hom}_B(K^\vee_{e'}(B, \widetilde{A}), Q)).$$

The underlying graded left D-module of $\operatorname{Hom}^{\tau'}_R(M, Q)$ is the Hom module $\operatorname{Hom}_R(M, Q)$ with the left D-module structure induced by the right D-module structure on M. If D is nonnegatively graded and M is a D-comodule, then $\operatorname{Hom}^{\tau'}_R(M, Q)$ is a left CDG-contramodule over (D, d_D, h_D) by Lemma 8.21(a).

Let N be a CDG-bimodule over $(^\#B, {}^\#d, {}^\#h)$ and (D, d_D, h_D), and let $P = P^\bullet$ be a complex of left $\widetilde{A}$-modules. We would rather assume the graded left $^\#B$-module N to be a $^\#B$-comodule. Then we define the left CDG-module $\operatorname{Hom}^{\sigma'}_R(N, P^\bullet)$ over (D, d_D, h_D) as the Hom CDG-module

$$\operatorname{Hom}^{\sigma'}_R(N, P^\bullet) = \operatorname{Hom}_{\widetilde{A}}(K^\vee_{u'}(\widetilde{A}, {}^\#B) \otimes_{\#B} N, \; P^\bullet) = \operatorname{Hom}_{\#B}(N, \operatorname{Hom}_{\widetilde{A}}(K^\vee_{u'}(\widetilde{A}, {}^\#B), P^\bullet)).$$

The underlying graded left D-module of $\operatorname{Hom}^{\sigma'}_R(N, P^\bullet)$ is the Hom module $\operatorname{Hom}_R(N, P)$ with the left D-module structure induced by the right D-module structure on N. If D is nonnegatively graded and N is a D-comodule, then $\operatorname{Hom}^{\sigma'}_R(N, P^\bullet)$ is a left CDG-contramodule over (D, d_D, h_D) by Lemma 8.21(a).

With these preparations, we can interpret the nonhomogeneous Koszul duality functors

$$P^\bullet \longmapsto \operatorname{Hom}^{\sigma'}_R(C, P^\bullet) \quad \text{and} \quad Q \longmapsto \operatorname{Hom}^{\tau'}_R(\widetilde{A}, Q)$$

of Sect. 7.4 as particular cases of the above constructions of twisted Hom CDG-modules. This explains the meaning of the "placeholder σ'" in Sect. 7.4.

To conclude, let us say a few words about the bimodule resolution $\widetilde{A} \otimes^{\sigma'}_R C \otimes^{\tau'}_R \widetilde{A}$ of the diagonal $\widetilde{A}$-$\widetilde{A}$-bimodule $\widetilde{A}$, which was constructed in Examples 6.15(3–4). It was explained in Examples 6.15 why this is indeed a bimodule resolution, but the construction itself was decidedly not left-right symmetric, and the meaning of the placeholder σ' remained mysterious.

Now this is explained, and the symmetry is restored. (Under the additional assumption that $\widetilde{A}$ is two-sided finitely projective nonhomogeneous Koszul.) As a corollary of the above discussion, we have

$$\widetilde{A} \otimes_R^{\sigma'} C \otimes_R^{\tau'} \widetilde{A} \;=\; K_{u'}^{\vee}(\widetilde{A}, {}^{\#}B) \otimes_{{}^{\#}B} C \otimes_B K_{e'}^{\vee}(B, \widetilde{A}),$$

where C is a CDG-bimodule over $({}^{\#}B, {}^{\#}d, {}^{\#}h)$ and (B, d, h), as explained above, and the tensor products in the right-hand side are the CDG-bimodule tensor products of Sect. 6.1.

8.8 Koszul Triality

A "Koszul triality" is a commutative triangle diagram of triangulated equivalences between a derived category of modules, a coderived category of comodules, and a contraderived category of contramodules. This phenomenon was first observed in [59] (see [59, Sections 6.3–6.5]; cf. [59, Sections 6.7–6.8]). It is *not* supposed to manifest itself in the more general setting of [58, Chapter 11], which has coalgebra variables in the base. (It is expected to be a "quadrality" there; cf. Sect. 9.7.) But it still exists in the context of the present book, under the assumptions listed in Theorem 8.29 in this section.

Let $R \subset \widetilde{V} \subset \widetilde{A}$ be a right finitely projective nonhomogeneous Koszul ring, and let ${}^{\#}B = ({}^{\#}B, {}^{\#}d, {}^{\#}h)$ be the corresponding left finitely projective Koszul CDG-ring under the opposite version of the construction of Proposition 3.16 and Corollary 4.26. Then, as a particular case of the discussion in Sect. 8.7, there are DG-functors

$$\widetilde{A} \otimes_R^{\sigma'} - \;:\; \mathsf{DG}({}^{\#}B\text{–comod}) \longrightarrow \mathsf{DG}(\widetilde{A}\text{–mod})$$

and

$$C \otimes_R^{\tau'} - \;:\; \mathsf{DG}(\widetilde{A}\text{–mod}) \longrightarrow \mathsf{DG}({}^{\#}B\text{–comod}).$$

By the opposite version of the discussion in Sect. 6.5, the DG-functor $\widetilde{A} \otimes_R^{\sigma'} -$ is left adjoint to the DG-functor $C \otimes_R^{\tau'} -$. Hence the induced triangulated functors between the homotopy categories,

$$\widetilde{A} \otimes_R^{\sigma'} - \;:\; \mathsf{Hot}({}^{\#}B\text{–comod}) \longrightarrow \mathsf{Hot}(\widetilde{A}\text{–mod})$$

and

$$C \otimes_R^{\tau'} - \;:\; \mathsf{Hot}(\widetilde{A}\text{–mod}) \longrightarrow \mathsf{Hot}({}^{\#}B\text{–comod}),$$

are also adjoint.

The following assertions are the opposite versions of Theorem 6.14 and Corollary 6.18.

Theorem 8.27 *The above pair of adjoint triangulated functors* $C\otimes_R^{\tau'}-$ *and* $\widetilde{A}\otimes_R^{\sigma'}-$ *induces a pair of adjoint triangulated functors between the* $\widetilde{A}/R$-*semicoderived category of left* $\widetilde{A}$-*modules* $D_R^{\text{sico}}(\widetilde{A}\text{-mod})$ *and the coderived category* $D^{\text{co}}(^{\#}B\text{-comod})$ *of left CDG-comodules over* $(^{\#}B, {}^{\#}d, {}^{\#}h)$, *which are mutually inverse triangulated equivalences*

$$D_R^{\text{sico}}(\widetilde{A}\text{-mod}) \simeq D^{\text{co}}(^{\#}B\text{-comod}).$$

□

Corollary 8.28 *Assume additionally that the left homological dimension of the ring* R *is finite. Then the pair of adjoint triangulated functors* $\widetilde{A}\otimes_R^{\sigma'}- : \text{Hot}(^{\#}B\text{-comod}) \longrightarrow \text{Hot}(\widetilde{A}\text{-mod})$ *and* $C\otimes_R^{\tau'}- : \text{Hot}(\widetilde{A}\text{-mod}) \longrightarrow \text{Hot}(^{\#}B\text{-comod})$ *induces mutually inverse triangulated equivalences*

$$D(\widetilde{A}\text{-mod}) \simeq D^{\text{co}}(^{\#}B\text{-comod}) \tag{8.10}$$

between the derived category of left $\widetilde{A}$-*modules and the coderived category of left CDG-comodules over* $(^{\#}B, {}^{\#}d, {}^{\#}h)$. □

The next theorem is the second main result of Chap. 8.

Theorem 8.29 *Let* $R \subset \widetilde{V} \subset \widetilde{A}$ *be a two-sided finitely projective nonhomogeneous Koszul ring, let* $B = (B, d, h)$ *be the right finitely projective Koszul CDG-ring nonhomogeneous quadratic dual to* $\widetilde{A}$, *and let* $^{\#}B = (^{\#}B, {}^{\#}d, {}^{\#}h)$ *be the left finitely projective Koszul CDG-ring nonhomogeneous quadratic dual to* $\widetilde{A}$, *as in Lemma 8.26. Assume that the left homological dimension of the ring* R *is finite. Then the triangulated equivalence of derived nonhomogeneous Koszul duality on the comodule side (8.10) of Corollary 8.28 (cf. the right (co)module side version in (6.6) of Corollary 6.18),*

$$C\otimes_R^{\tau'}- : D(\widetilde{A}\text{-mod}) \simeq D^{\text{co}}(^{\#}B\text{-comod}) : \widetilde{A}\otimes_R^{\sigma'}-,$$

the triangulated equivalence of derived nonhomogeneous Koszul duality on the contramodule side (7.1) of Corollary 7.14,

$$\text{Hom}_R^{\sigma'}(C, -) : D(\widetilde{A}\text{-mod}) \simeq D^{\text{ctr}}(B\text{-contra}) : \text{Hom}_R^{\tau'}(\widetilde{A}, -),$$

and the triangulated equivalence of derived comodule-contramodule correspondence (8.3) of Theorem 8.22,

$$\mathbb{R}\,\text{Hom}_{{}^{\#}B}(C, -) : D^{\text{co}}(^{\#}B\text{-comod}) \simeq D^{\text{ctr}}(B\text{-contra}) : C\odot_B^{\mathbb{L}}-$$

form a commutative triangle diagram of triangulated category equivalences,

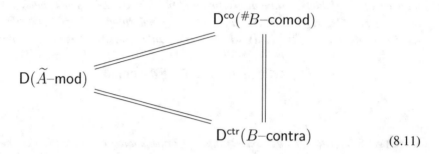

$$D^{co}(^{\#}B\text{–comod})$$

$$D(\widetilde{A}\text{–mod})$$

$$D^{ctr}(B\text{–contra}) \tag{8.11}$$

Proof As a preliminary observation, it is worth pointing out that the contratensor product functor $C \odot_B -$ appearing in the third displayed formula is isomorphic to the tensor product functor $C \otimes_B -$ in our present assumptions. This is the result of Proposition 8.5(b), which is applicable in view of Theorem 5.34.

To prove the theorem, we notice first of all that, for any complex of left $\widetilde{A}$-modules $M = M^\bullet$, one has

$$\mathbb{R}\operatorname{Hom}_{^{\#}B}(C, \ C\otimes_R^{\tau'} M^\bullet) = \operatorname{Hom}_{^{\#}B}(C, \ C\otimes_R^{\tau'} M^\bullet)$$

and similarly,

$$C\odot_B^{\mathbb{L}} \operatorname{Hom}_R^{\sigma'}(C, M^\bullet) = C\odot_B \operatorname{Hom}_R^{\sigma'}(C, M^\bullet),$$

that is, the derived functors $\mathbb{R}\operatorname{Hom}_{^{\#}B}(C, -)$ and $C\odot_B^{\mathbb{L}}-$ coincide with the respective underived functors on these objects. Indeed, the graded left $^{\#}B$-module $C\otimes_R M = \operatorname{Hom}_R^\Sigma(^{\#}B, M)$ is coinduced (from the graded left R-module M), and the graded left B-contramodule $\operatorname{Hom}_R(C, M) = B\otimes_R^\Pi M$ is induced (from the graded left R-module M). So the objects in question are adjusted to the respective functors in view of the construction of the derived functors in the proof of Theorem 8.22.

Finally, the main claim is that, for any complex of left $\widetilde{A}$-modules $M = M^\bullet$, there is a natural isomorphism of left CDG-contramodules over (B, d, h)

$$\operatorname{Hom}_{^{\#}B}(C, \ C\otimes_R^{\tau'} M^\bullet) \simeq \operatorname{Hom}_R^{\sigma'}(C, M^\bullet),$$

as well as a natural isomorphism of left CDG-comodules over $(^{\#}B, {}^{\#}d, {}^{\#}h)$

$$C\odot_B \operatorname{Hom}_R^{\sigma'}(C, M^\bullet) \simeq C\otimes_R^{\tau'} M^\bullet.$$

Indeed, on the level of the underlying graded co/contramodules we have

$$\operatorname{Hom}_{^{\#}B}(C, \ C\otimes_R M) \simeq \operatorname{Hom}_{^{\#}B}(C, \operatorname{Hom}_R^\Sigma(^{\#}B, M)) \simeq \operatorname{Hom}_R(C, M)$$

and

$$C \odot_B \mathrm{Hom}_R(C, M) \simeq C \odot_B (B \otimes_R^{\Pi} M) \simeq C \otimes_R M$$

(cf. Lemma 8.20 and the computation in the proof of Theorem 8.22). We leave it to the reader to check that the differentials agree. $\square$

Koszul Duality and Conversion Functor

9

9.1 Relatively Frobenius Rings

Let K be an associative ring and B be a K-K-bimodule. Assume that B is a finitely generated projective left K-module. Then for any left K-module L there is a natural isomorphism of left K-modules $\operatorname{Hom}_K(B, L) \simeq \operatorname{Hom}_K(B, K) \otimes_K L$.

Let T be another K-K-bimodule which is finitely generated and projective as a left K-module. Then there is a natural morphism of K-K-bimodules $K \longrightarrow \operatorname{Hom}_K(T, T) \simeq \operatorname{Hom}_K(T, K) \otimes_K T$ induced by the right action of K in T. There is also a natural evaluation morphism of K-K-bimodules $T \otimes_K \operatorname{Hom}_K(T, K) \longrightarrow K$ given by the rule $t \otimes f \longmapsto f(t)$ for all $t \in T$ and $f \in \operatorname{Hom}_K(T, K)$. The pairing notation $\langle t, f \rangle = f(t)$ is convenient.

Definition 9.1 We will say that a K-K-bimodule T is *invertible* if T is projective and finitely generated as a left K-module, and both the above morphisms $K \longrightarrow \operatorname{Hom}_K(T, K) \otimes_K T$ and $T \otimes_K \operatorname{Hom}_K(T, K) \longrightarrow K$ are isomorphisms.

In this case, the functors $T \otimes_K - : K\text{–mod} \longrightarrow K\text{–mod}$ and $\operatorname{Hom}_K(T, -) = \operatorname{Hom}_K(T, K) \otimes_K - : K\text{–mod} \longrightarrow K\text{–mod}$ are mutually inverse auto-equivalences of the category of left K-modules. Similarly, the functors $- \otimes_K T : \text{mod–}K \longrightarrow \text{mod–}K$ and $- \otimes_K \operatorname{Hom}_K(T, K) : \text{mod–}K \longrightarrow \text{mod–}K$ are mutually inverse auto-equivalences of the category of right K-modules.

It follows that T is also a finitely generated projective right K-module, and the notion of an invertible K-K-bimodule is left-right symmetric. Moreover, we obtain a natural isomorphism of K-K-bimodules $\operatorname{Hom}_K(T, K) \simeq \operatorname{Hom}_{K^{\mathrm{op}}}(T, K)$.

Furthermore, for any invertible K-K-bimodule T and any finitely generated projective left K-module L the natural evaluation map $L \longrightarrow \operatorname{Hom}_{K^{\mathrm{op}}}(\operatorname{Hom}_K(L, T), T)$ is an isomorphism of left K-modules (as one can see, e.g., by reducing the question to the case of a free module with one generator $L = K$).

© The Author(s), under exclusive license to Springer Nature Switzerland AG 2021   217
L. Positselski, *Relative Nonhomogeneous Koszul Duality*, Frontiers in Mathematics,
https://doi.org/10.1007/978-3-030-89540-2_9

Definition 9.2 Let B' and B'' be K-K-bimodules, T be an invertible K-K-bimodule, and $\phi\colon B'\otimes_K B'' \longrightarrow T$ be a K-K-bimodule morphism. We will say that ϕ is a *perfect pairing* if B' is a finitely generated projective left K-module and the K-K-bimodule map $\check{\phi}\colon B'' \longrightarrow \operatorname{Hom}_K(B',T) \simeq \operatorname{Hom}_K(B',K)\otimes_K T$ induced by ϕ is an isomorphism.

In this case, $\operatorname{Hom}_K(B',K)$ is a finitely generated projective right K-module, and it follows that B'' is a finitely generated projective right K-module. Furthermore, the K-K-bimodule map $\hat{\phi}\colon B' \longrightarrow \operatorname{Hom}_{K^{\mathrm{op}}}(B'',T) \simeq T\otimes_K \operatorname{Hom}_{K^{\mathrm{op}}}(B'',K)$ induced by ϕ is an isomorphism, because $\hat{\phi} \simeq \operatorname{Hom}_{K^{\mathrm{op}}}(\check{\phi},T)$ (while $\check{\phi} \simeq \operatorname{Hom}_K(\hat{\phi},T)$). Therefore, the notion of a perfect pairing ϕ is left-right symmetric as well.

Definition 9.3 Let $K \longrightarrow B$ be a ring homomorphism, T be an invertible K-K-bimodule, and $t\colon B \longrightarrow T$ be a morphism of K-K-bimodules. Take $\phi\colon B\otimes_K B \longrightarrow K$ to be equal to the composition of the multiplication map $B\otimes_K B \longrightarrow B$ with the map t.

Then the map $\check{\phi}\colon B \longrightarrow \operatorname{Hom}_K(B,T)$ is a morphism of left B-modules, while the map $\hat{\phi}\colon B \longrightarrow \operatorname{Hom}_{K^{\mathrm{op}}}(B,T)$ is a morphism of right B-modules. We will say that the ring B is *relatively Frobenius over K* (with respect to the map t) if the pairing $\phi\colon B\otimes_K B \longrightarrow T$ induced by the map $t\colon B \longrightarrow T$ is perfect. Following the discussion above, any relatively Frobenius ring B over K is a finitely generated and projective left and right K-module, and both the maps $\check{\phi}$ and $\hat{\phi}$ are isomorphisms in this case.

Let B be a relatively Frobenius ring over K with respect to a K-K-bimodule morphism $t\colon B \longrightarrow T$. Then for any left K-module L there is a natural isomorphism of left B-modules $B\otimes_K L \simeq \operatorname{Hom}_K(B,T)\otimes_K L \simeq \operatorname{Hom}_K(B,K)\otimes_K T\otimes_K L \simeq \operatorname{Hom}_K(B,T\otimes_K L)$. Consequently, for any left K-module M there is a natural isomorphism of left B-modules $\operatorname{Hom}_K(B,M) \simeq B\otimes_K \operatorname{Hom}_K(T,M)$. So the class of left B-modules induced from left K-modules coincides with that of left B-modules coinduced from left K-modules.

9.2   Relatively Frobenius Graded Rings

Let $B = \bigoplus_{n=0}^{\infty} B^n$ be a nonnegatively graded ring, and let $K \longrightarrow B^0$ be a ring homomorphism. Let T be an invertible K-K-bimodule.

Definition 9.4 Assume that there is an integer $m \geq 0$ such that $B^n = 0$ for all $n > m$, and a morphism of K-K-bimodules $t\colon B^m \longrightarrow T$ is given. Then we will say that B is a *relatively Frobenius graded ring over K* (with respect to the map t) if the underlying ungraded ring of B is relatively Frobenius, in the sense of the definition in Sect. 9.1, with respect to the composition of the direct summand projection $B = \bigoplus_{n=0}^{m} B^n \longrightarrow B^m$ with the map t.

Equivalently, this means that the composition $\phi_n \colon B^n \otimes_K B^{m-n} \longrightarrow T$ of the multiplication map $B^n \otimes_K B^{m-n} \longrightarrow B^m$ with the map $t \colon B^m \longrightarrow T$ is a perfect pairing (in the sense of the definition in Sect. 9.1) for every $0 \leq n \leq m$. In this case, B^n is a finitely generated projective left and right K-module for every $0 \leq n \leq m$.

Example 9.5 Let R be a commutative ring and V be a free R-module with m generators. Consider the exterior algebra $B = \Lambda_R(V)$, as defined in Example 1.3(5). Then the R-module $T = B^m = \Lambda_R^m(V)$, viewed as an R-R-bimodule in the usual way, is invertible in the sense of Sect. 9.1. The exterior algebra $B = \Lambda_R(V)$ is a Frobenius graded ring over $K = R$ (with respect to the identity map $t \colon B^m \longrightarrow B^m = T$). Notice that $\Lambda_R(V)$ is also a left and right finitely projective Koszul graded ring over R (see Example 2.39(3)). We refer to Sect. 10.1 for a further discussion.

Since $B^n = 0$ for all $n > m$, all (graded or ungraded, right or left) B-modules are B-comodules, as mentioned at the end of Sect. 6.4. So, in particular, we have $B\text{-comod}_{\mathsf{gr}} = B\text{-mod}_{\mathsf{gr}}$, in the notation of Sect. 8.3, and therefore we will use $\mathsf{D}^{\mathsf{co}}(B\text{-mod})$ instead of $\mathsf{D}^{\mathsf{co}}(B\text{-comod})$ as the notation for the coderived category of left CDG-(co)modules over B when B is endowed with a CDG-ring structure $B = (B, d, h)$. Similarly, the forgetful functors from the categories of (graded or ungraded) B-contramodules to the similar categories of B-modules are category equivalences, as mentioned at the end of Sect. 7.3. So we will use $\mathsf{D}^{\mathsf{ctr}}(B\text{-mod})$ instead of $\mathsf{D}^{\mathsf{ctr}}(B\text{-contra})$ as the notation for the contraderived category of left CDG-(contra)modules over B when B is endowed with a CDG-ring structure.

The definitions of the induced graded contramodules and coinduced graded comodules given in Sects. 8.2 and 8.3 simplify similarly: for any graded left K-module L one has $B\otimes_K^\Pi L = B\otimes_K L$ and $\mathrm{Hom}_K^\Sigma(B, L) = \mathrm{Hom}_K(B, L)$. So we will speak of *induced* and *coinduced graded left B-modules* instead of induced graded left B-contramodules and coinduced graded left B-comodules.

Finally, since B is a relatively Frobenius graded ring over K, the class of left graded B-modules induced from graded left K-modules coincides with that of graded left B-modules coinduced from graded left K-modules, essentially as explained in Sect. 9.1. More precisely, for any graded left K-modules L and M there are natural isomorphisms of graded left B-modules $B\otimes_K L \simeq \mathrm{Hom}_K(B, T\otimes_K L[-m])$ and $\mathrm{Hom}_K(B, M) \simeq B\otimes_K \mathrm{Hom}_K(T, M[m])$, where $[i]$, $i \in \mathbb{Z}$, denotes the usual (cohomological) grading shift.

9.3 Relatively Frobenius Co-contra Correspondence

Let $B = (B, d, h)$ be a nonnegatively graded CDG-ring, so $B = \bigoplus_{n=0}^\infty B^n$, and let $K \longrightarrow B^0$ be a ring homomorphism. We will say that the CDG-ring (B, d, h) is *relatively Frobenius over K* if the graded ring B is relatively Frobenius (with respect to some morphism of K-K-bimodules $t \colon B^m \longrightarrow T$).

Let (B, d, h) be a relatively Frobenius CDG-ring over K. Similarly to Sects. 8.2 and 8.3, one can speak of *left CDG-modules of the induced type* and *left CDG-modules of the coinduced type* over (B, d, h). Moreover, following the discussion in Sect. 9.2, these two classes of left CDG-modules coincide.

Furthermore, similarly to Sects. 8.2 and 8.3, one can speak of *left CDG-module short exact sequences of the induced type* and *of the coinduced type* over (B, d, h). Once again, following the discussion in Sect. 9.2, these two classes of short exact sequences of left CDG-modules over (B, d, h) coincide.

Hence we arrive to the following theorem, which is the first main result of Chap. 9. It is to be compared with [59, Theorems 3.9–3.10], and it is a generalization of [59, Theorem B.3]. It is also a simplified version of Theorem 8.22 above.

Theorem 9.6 *Let* $B = (B, d, h)$ *be a relatively Frobenius CDG-ring over a ring* K. *Assume that the left homological dimension of* K *is finite. Then there is a natural triangulated equivalence between the coderived and contraderived categories of left CDG-modules over* (B, d, h),

$$\mathsf{D}^{\mathrm{co}}(B\text{–}\mathsf{mod}) \simeq \mathsf{D}^{\mathrm{ctr}}(B\text{–}\mathsf{mod}). \tag{9.1}$$

Proof By Proposition 8.16 and the second assertion of Theorem 8.17, the coderived category $\mathsf{D}^{\mathrm{co}}(B\text{–}\mathsf{mod})$ is equivalent to the triangulated quotient category of the full triangulated subcategory in $\mathsf{Hot}(B\text{–}\mathsf{mod})$ formed by left CDG-modules of the coinduced type over (B, d, h) by its minimal triangulated subcategory containing all the totalizations of left CDG-module short exact sequences of the coinduced type. Similarly, by Proposition 8.10 and the second assertion of Theorem 8.11, the contraderived category $\mathsf{D}^{\mathrm{ctr}}(B\text{–}\mathsf{mod})$ is equivalent to the triangulated quotient category of the full subcategory in $\mathsf{Hot}(B\text{–}\mathsf{mod})$ formed by left CDG-modules of the induced type over (B, d, h) by its minimal triangulated subcategory containing all the totalizations of left CDG-module short exact sequences of the induced type. Following the above discussion, it is one and the same triangulated quotient category. $\quad\square$

9.4 Relatively Frobenius Koszul Graded Rings

Let $A = \bigoplus_{n=0}^{\infty} A_n$ be a left finitely projective Koszul graded ring with the degree-zero component $R = A_0$, and let $B = \bigoplus_{n=0}^{\infty} B^n$ be the quadratic dual right finitely projective Koszul graded ring, as per the construction of Chap. 1 and Proposition 2.37.

Assume that there is an integer $m \geq 0$ such that $B^n = 0$ for $n > m$, the R-R-bimodule $T = B^m$ is invertible, and the graded ring B is relatively Frobenius over the ring $K = R$ with respect to the identity morphism of R-R-bimodules $B^m \longrightarrow T$. In other words, this means that the multiplication map $B^n \otimes_R B^{m-n} \longrightarrow B^m$ is a perfect pairing for every $0 \leq n \leq m$. Then we will say that B is a *relatively Frobenius Koszul graded ring*.

Consider the dual Koszul complex $K_e^{\vee\bullet}(B, A)$ (2.17) from Section 2.6:

$$0 \longrightarrow A \longrightarrow B^1 \otimes_R A \longrightarrow \cdots \longrightarrow B^{m-1} \otimes_R A \longrightarrow B^m \otimes_R A \longrightarrow 0, \qquad (9.2)$$

and consider also the second Koszul complex ${}^{\tau}K_\bullet(B, A) = C \otimes_B K_e^{\vee\bullet}(B, A) = C \otimes_R^{\tau} A$ from formula (2.18) and Remark 2.13 (where $C = \mathrm{Hom}_{R^{\mathrm{op}}}(B, R)$).

According to Theorem 2.35(e), the complex ${}^{\tau}K_\bullet(B, A)$ is a graded right A-module resolution of the graded right A-module R. In view of the isomorphism of graded right B-modules $B \simeq \mathrm{Hom}_{R^{\mathrm{op}}}(B, T) \simeq T \otimes_R \mathrm{Hom}_{R^{\mathrm{op}}}(B, R)[-m] = T \otimes_R C[-m]$ (see Sects. 9.1–9.2), the dual Koszul complex (9.2) is a graded right A-module resolution of the graded right A-module $T = B^m$ (placed in the cohomological degree m). In other words, the complex (9.2) is acyclic at every term except the rightmost one, and at the rightmost term $B^m \otimes_R A$ its cohomology module is the internal grading component $B^m \otimes_R A_0 = T$. (See Example 2.12(1) for a classical example.)

9.5 Two-Sided Koszul CDG-Rings

Let $B = \bigoplus_{n=0}^{\infty} B^n$ be a two-sided finitely projective Koszul graded ring (in the sense of Sect. 8.6) with the degree-zero component $R = B^0$. Denote by A the left finitely projective Koszul graded ring quadratic dual to the right finitely projective Koszul ring B, and denote by $A^{\#}$ the right finitely projective Koszul graded ring quadratic dual to the left finitely projective Koszul ring B. So, according to Proposition 2.37, we have

$$A = \mathrm{Ext}_{B^{\mathrm{op}}}(R, R) \quad \text{and} \quad A^{\#} = \mathrm{Ext}_B(R, R)^{\mathrm{op}},$$

where $\mathrm{Ext}_{B^{\mathrm{op}}}$ is taken in the category of graded right B-modules and Ext_B is taken in the category of graded left B-modules. Following Lemma 8.24, there are natural isomorphisms of R-R-bimodules

$$A_n^{\#} = \mathrm{Hom}_R(\mathrm{Hom}_R(A_n, R), R) \quad \text{and} \quad A_n = \mathrm{Hom}_{R^{\mathrm{op}}}(\mathrm{Hom}_{R^{\mathrm{op}}}(A_n^{\#}, R), R)$$

for all $n \geq 0$, and the multiplicative structures on A and $A^{\#}$ are related by the construction of Remark 2.13 and Sect. 8.4.

Now let (B, d, h) be a nonnegatively graded CDG-ring. We recall that (B, d, h) is said to be right finitely projective Koszul if the graded ring $B = \bigoplus_{n=0}^{\infty} B^n$ is right finitely projective Koszul. Dually, let us say that (B, d, h) is *left finitely projective Koszul* if the graded ring B is left finitely projective Koszul. We will say that a CDG-ring (B, d, h) is *two-sided finitely projective Koszul* if the graded ring B is two-sided finitely projective Koszul.

Let (B, d, h) be a two-sided finitely projective Koszul CDG-ring. Applying the Poincaré–Birkhoff–Witt Theorem 4.25 (see also Corollary 4.26) to the right finitely projective Koszul CDG-ring (B, d, h), we obtain a left finitely projective nonhomogeneous Koszul ring $\widetilde{A}$ with an increasing filtration $0 = F_{-1}\widetilde{A} \subset R = F_0\widetilde{A} \subset \widetilde{V} = F_1\widetilde{A} \subset F_2\widetilde{A} \subset \cdots$ and a natural isomorphism of graded rings $A \simeq \mathrm{gr}^F \widetilde{A} = \bigoplus_{n=0}^{\infty} F_n\widetilde{A}/F_{n-1}\widetilde{A}$. The filtered ring $\widetilde{A}$ comes together with a left R-linear splitting

$$\mathrm{Hom}_{R^{\mathrm{op}}}(B^1, R) \simeq F_1\widetilde{A}/F_0\widetilde{A} \simeq A_1 = V \hookrightarrow \widetilde{V} = F_1\widetilde{A},$$

which we denote by τ', as in Sects. 6.5 and 7.4.

Applying the opposite version of Theorem 4.25 and Corollary 4.26 to the left finitely projective Koszul CDG-ring (B, d, h), we obtain a right finitely projective nonhomogeneous Koszul ring $\widetilde{A}^{\#}$ with an increasing filtration $0 = F_{-1}\widetilde{A}^{\#} \subset R = F_0\widetilde{A}^{\#} \subset \widetilde{V}^{\#} = F_1\widetilde{A}^{\#} \subset F_2\widetilde{A}^{\#} \subset \cdots$ and a natural isomorphism of graded rings $A^{\#} \simeq \mathrm{gr}^F \widetilde{A}^{\#} = \bigoplus_{n=0}^{\infty} F_n\widetilde{A}^{\#}/F_{n-1}\widetilde{A}^{\#}$. The filtered ring $\widetilde{A}^{\#}$ comes together with a right R-linear splitting

$$\mathrm{Hom}_R(B^1, R) \simeq F_1\widetilde{A}^{\#}/F_0\widetilde{A}^{\#} \simeq A_1^{\#} = V^{\#} \hookrightarrow \widetilde{V}^{\#} = F_1\widetilde{A}^{\#},$$

which we will denote by ρ'.

The dual nonhomogeneous Koszul CDG-module $K^{\vee}(B, \widetilde{A})$ was constructed in Sect. 6.2. We recall that $K^{\vee}(B, \widetilde{A})$ is a CDG-bimodule over the CDG-rings $B = (B, d, h)$ and $\widetilde{A} = (\widetilde{A}, 0, 0)$ whose underlying graded B-$\widetilde{A}$-bimodule is $B \otimes_R \widetilde{A}$. The opposite construction produces a CDG-bimodule $K^{\vee}(\widetilde{A}^{\#}, B)$ over the CDG-rings $\widetilde{A}^{\#} = (\widetilde{A}^{\#}, 0, 0)$ and $B = (B, d, h)$. So the underlying graded $\widetilde{A}^{\#}$-B-bimodule of $K^{\vee}(\widetilde{A}^{\#}, B)$ is $\widetilde{A}^{\#} \otimes_R B$.

From now on we will assume for simplicity that $B^n = 0$ for $n \gg 0$; so there is no difference between B-modules and B-comodules, and B is a finitely generated projective graded left and right R-module. Let $D = (D, d_D, h_D)$ and $E = (E, d_E, h_E)$ be two CDG-rings. Let M be a CDG-bimodule over $(\widetilde{A}, 0, 0)$ and (E, d_E, h_E), and let N be a CDG-bimodule over (D, d_D, h_D) and (B, d, h). Following the constructions of Sects. 6.5 and 8.7, the CDG-bimodule $N \otimes_R^{\tau'} M$ over D and E is defined as the tensor product of CDG-bimodules

$$N \otimes_R^{\tau'} M = N \otimes_B K^{\vee}(B, \widetilde{A}) \otimes_{\widetilde{A}} M.$$

Furthermore, let M be a CDG-bimodule over (E, d_E, h_E) and $(\widetilde{A}, 0, 0)$. Then the CDG-bimodule $M \otimes_R^{\sigma'} C$ over (E, d_E, h_E) and (B, d, h) is defined as the Hom CDG-bimodule

$$M \otimes_R^{\sigma'} C = \mathrm{Hom}_{\widetilde{A}^{\mathrm{op}}}(K^{\vee}(B, \widetilde{A}), M),$$

where σ' is the placeholder introduced in Sect. 6.5 and explained in Sect. 8.7, while C is the graded coring $C = \mathrm{Hom}_{R^{\mathrm{op}}}(B, R)$.

Let us introduce notation for the opposite constructions. Let M be a CDG-bimodule over (E, d_E, h_E) and $(\widetilde{A}^{\#}, 0, 0)$, and let N be a CDG-bimodule over (B, d, h) and (D, d_D, h_D). We define the CDG-bimodule $M \otimes_R^{\rho'} N$ over E and D as the tensor product of CDG-bimodules

$$M \otimes_R^{\rho'} N = M \otimes_{\widetilde{A}^{\#}} K^{\vee}(\widetilde{A}^{\#}, B) \otimes_B N.$$

So the underlying graded E-D-bimodule of $M \otimes_R^{\rho'} N$ is $M \otimes_R N$.

Furthermore, let M be a CDG-bimodule over $(\widetilde{A}^{\#}, 0, 0)$ and (E, d_E, h_E). Then the CDG-bimodule $C^{\#} \otimes_R^{\pi'} M$ over (B, d, h) and (E, d_E, h_E) is defined as the Hom CDG-bimodule

$$C^{\#} \otimes_R^{\pi'} M = \operatorname{Hom}_{\widetilde{A}^{\#}}(K^{\vee}(\widetilde{A}^{\#}, B), M),$$

where $C^{\#}$ is the graded coring $C^{\#} = \operatorname{Hom}_R(B, R)$ over R, while π' is another placeholder. The underlying graded B-E-bimodule of $C^{\#} \otimes_R^{\pi'} M$ is $\operatorname{Hom}_{\widetilde{A}^{\#}}(\widetilde{A}^{\#} \otimes_R B, M) \simeq \operatorname{Hom}_R(B, M) \simeq C^{\#} \otimes_R M$.

9.6 Conversion Bimodule

Let $B = (B, d, h)$ be a two-sided finitely projective Koszul nonnegatively graded CDG-ring (in the sense of Sect. 9.5) with the degree-zero component $R = B^0$. We will say that (B, d, h) is a *relatively Frobenius Koszul CDG-ring* if the graded ring B is relatively Frobenius over R in the sense of Sect. 9.4. This means that there is an integer $m \geq 0$ such that $B^n = 0$ for $n > m$, the R-R-bimodule $T = B^m$ is invertible, and the multiplication map $B^n \otimes_R B^{m-n} \longrightarrow B^m$ is a perfect pairing for every $0 \leq n \leq m$ (in the sense of Sect. 9.1).

Let (B, d, h) be a relatively Frobenius Koszul CDG-ring. We will use the notation of Sect. 9.5. In particular, $\widetilde{A}$ is the left finitely projective nonhomogeneous Koszul ring corresponding to the right finitely projective Koszul CDG-ring (B, d, h), while $\widetilde{A}^{\#}$ is the right finitely projective nonhomogeneous Koszul ring corresponding to the left finitely projective Koszul CDG-ring (B, d, h).

Lemma 9.7

(a) *There is a natural morphism $T[-m] \longrightarrow B \otimes_R^{\tau'} \widetilde{A} = K^{\vee}(B, \widetilde{A})$ of left CDG-modules over (B, d, h), where the one-term complex of left R-modules $T[-m]$ is endowed with the trivial structure of left CDG-(co)module over (B, d, h). The cone of this morphism is a coacyclic left CDG-(co)module over (B, d, h).*

(b) *There is a natural morphism* $T[-m] \longrightarrow \widetilde{A}^{\#} \otimes_R^{\rho'} B = K^{\vee}(\widetilde{A}^{\#}, B)$ *of right CDG-modules over* (B, d, h), *where the one-term complex of right R-modules* $T[-m]$ *is endowed with the trivial structure of right CDG-(co)module over* (B, d, h). *The cone of this morphism is a coacyclic right CDG-(co)module over* (B, d, h).

Proof Let us prove part (a); part (b) is the opposite version. The left finitely projective nonhomogeneous Koszul ring $\widetilde{A}$ is endowed with an increasing filtration F. Define a decreasing filtration F on the graded ring B by the rule $F^n B = \bigoplus_{i \geq n} B^i[-i] \subset B$; so F is the decreasing filtration on B induced by the grading. Define an increasing filtration F on B by the rule $F_{-n} B = F^n B$; so $0 = F_{-m-1} B \subset F_{-m} B \subset \cdots \subset F_0 B = B$. Consider the increasing filtration F on $K^{\vee}(B, \widetilde{A}) = B \otimes_R \widetilde{A}$ induced by the increasing filtrations F on B and $\widetilde{A}$, i.e., $F_n(B \otimes_R \widetilde{A}) = \sum_{i+j=n} F_i B \otimes_R F_j \widetilde{A}$.

It is clear from the definition of the differential $d_{e'}$ on $K^{\vee}(B, \widetilde{A})$ that $F_n K^{\vee}(B, \widetilde{A})$ is a CDG-submodule of the left CDG-module $K^{\vee}(B, \widetilde{A})$ over (B, d, h) for every $-m \leq n < \infty$. Furthermore, the CDG-submodule $F_{-m} K^{\vee}(B, \widetilde{A}) = B^m[-m] \otimes_R F_0 A = B^m[-m] \subset K^{\vee}(B, \widetilde{A})$ is isomorphic to $T[-m]$; hence the desired (injective) morphism of CDG-modules $T[-m] \longrightarrow K^{\vee}(B, \widetilde{A})$. In view of Lemma 6.12, in order to show that the cone (equivalently, the cokernel) of this morphism is coacyclic, it suffices to check that the CDG-(co)module $F_n K^{\vee}(B, \widetilde{A})/F_{n-1} K^{\vee}(B, \widetilde{A})$ over (B, d, h) is coacyclic for every $n \geq -m + 1$.

In fact, $F_n K^{\vee}(B, \widetilde{A})/F_{n-1} K^{\vee}(B, \widetilde{A})$ is a trivial CDG-comodule over (B, d, h), that is, the ideal $B^{\geq 1} \subset B$ acts by zero in this quotient CDG-module. The whole associated graded CDG-module $\mathrm{gr}^F K^{\vee}(B, \widetilde{A}) = \bigoplus_{n=-m}^{\infty} F_n K^{\vee}(B, \widetilde{A})/F_{n-1} K^{\vee}(B, \widetilde{A})$ is the complex of left R-modules $K_e^{\vee \bullet}(B, A) = B \otimes_R A$ (2.17) from Sect. 2.6, viewed as a left CDG-module over (B, d, h) with the trivial CDG-(co)module structure. In view of the discussion in Sect. 9.4, this means that, for every $n \geq -m + 1$, the complex of left R-modules $F_n K^{\vee}(B, \widetilde{A})/F_{n-1} K^{\vee}(B, \widetilde{A})$ is acyclic. This is a finite complex of finitely generated projective left R-modules; so it is contractible as a complex of left R-modules, hence it is also contractible (and consequently, coacyclic) as a trivial left CDG-(co)module over (B, d, h).   □

Lemma 9.8

(a) *For any complex of right* $\widetilde{A}^{\#}$-*modules* $M^{\bullet}$, *there is a natural quasi-isomorphism of complexes of right R-modules*

$$M^{\bullet} \otimes_R T[-m] \longrightarrow M^{\bullet} \otimes_R^{\rho'} B \otimes_R^{\tau'} \widetilde{A},$$

where $M^{\bullet} \otimes_R^{\rho'} B \otimes_R^{\tau'} \widetilde{A} = M^{\bullet} \otimes_{\widetilde{A}^{\#}} K^{\vee}(\widetilde{A}^{\#}, B) \otimes_B K^{\vee}(B, \widetilde{A})$ *is the complex of right* $\widetilde{A}$-*modules provided by the constructions of Sect. 9.5.*

(b) *For any complex of left $\widetilde{A}$-modules $L^\bullet$, there is a natural quasi-isomorphism of complexes of left R-modules*

$$T\otimes_R L^\bullet[-m] \longrightarrow \widetilde{A}^\#\otimes_R^{\rho'} B\otimes_R^{\tau'} L^\bullet,$$

where $\widetilde{A}^\#\otimes_R^{\rho'} B\otimes_R^{\tau'} L^\bullet = K^\vee(\widetilde{A}^\#, B)\otimes_B K^\vee(B, \widetilde{A})\otimes_{\widetilde{A}} L^\bullet$ *is the complex of left* $\widetilde{A}^\#$*-modules provided by the constructions of Sect. 9.5.*

Proof Let us explain part (a); part (b) is opposite. The desired morphism is obtained by applying the functor $M^\bullet\otimes_R^{\rho'} -$ to the morphism of CDG-modules in Lemma 9.7(a). It is important here that the latter morphism is right R-linear; so it is, in fact, a morphism of CDG-bimodules over (B, d, h) and $(R, 0, 0)$. Furthermore, following the proof of Lemma 9.7(a), the cone of the morphism $T[-m] \longrightarrow B\otimes_R^{\tau'}\widetilde{A}$ is coacyclic as a CDG-bimodule over (B, d, h) and $(R, 0, 0)$, and in fact even as a CDG-bimodule over (B, d, h) and $(R, 0, 0)$ with a projective underlying graded left R-module (because any finite acyclic complex of R-R-bimodules is coacyclic). Hence the tensor product functor $M^\bullet\otimes_R^{\rho'} -$ preserves coacyclicity of the cone, so the cone of the resulting morphism $M^\bullet\otimes_R T[-m] \longrightarrow M^\bullet\otimes_R^{\rho'} B\otimes_R^{\tau'}\widetilde{A}$ is not only acyclic but even coacyclic as a complex of right R-modules.   □

We are interested in the finite complex of $\widetilde{A}^\#$-$\widetilde{A}$-bimodules $\widetilde{A}^\#\otimes_R^{\rho'} B\otimes_R^{\tau'}\widetilde{A}$, or more specifically, in its top cohomology bimodule $E = H^m(\widetilde{A}^\#\otimes_R^{\rho'} B\otimes_R^{\tau'}\widetilde{A})$. According to Lemma 9.8 applied to the one-term complexes $M^\bullet = \widetilde{A}^\#$ and $L^\bullet = \widetilde{A}$, we have $H^n(\widetilde{A}^\#\otimes_R^{\rho'} B\otimes_R^{\tau'}\widetilde{A}) = 0$ for all $n \neq m$, while the $\widetilde{A}^\#$-$\widetilde{A}$-bimodule E is naturally isomorphic to $\widetilde{A}^\#\otimes_R T$ as an $\widetilde{A}^\#$-R-bimodule and naturally isomorphic to $T\otimes_R\widetilde{A}$ as an R-$\widetilde{A}$-bimodule. The following theorem tells that E is a Morita equivalence bimodule for the rings A and $A^\#$.

Theorem 9.9 *Let (B, d, h) be a relatively Frobenius Koszul CDG-ring, and let $\widetilde{A}$ and $\widetilde{A}^\#$ be the two nonhomogeneous Koszul rings quadratic dual to B on the two sides, as above. Then*

(a) *the tensor product and Hom functors*

$$E\otimes_{\widetilde{A}}- : \widetilde{A}\text{--mod} \simeq \widetilde{A}^\#\text{--mod} : \mathrm{Hom}_{\widetilde{A}^\#}(E, -)$$

are mutually inverse equivalences of abelian categories;

(b) *the tensor product and Hom functors*

$$-\otimes_{\widetilde{A}^{\#}} E \colon \operatorname{\mathsf{mod}-}\widetilde{A}^{\#} \simeq \operatorname{\mathsf{mod}-}\widetilde{A} \colon \operatorname{Hom}_{\widetilde{A}^{\mathrm{op}}}(E, -)$$

are mutually inverse equivalences of abelian categories.

Proof The left $\widetilde{A}^{\#}$-module $E \simeq \widetilde{A}^{\#}\otimes_R T$ is a finitely generated projective generator of $\widetilde{A}^{\#}$–mod, since the left R-module T is a finitely generated projective generator of R–mod. Similarly, the right $\widetilde{A}$-module $E \simeq T\otimes_R \widetilde{A}$ is a finitely generated projective generator of mod–$\widetilde{A}$, since the right R-module T is a finitely generated projective generator of mod–R. Finally, one computes that $\operatorname{Hom}_{\widetilde{A}^{\#}}(E, E) \simeq \operatorname{Hom}_{\widetilde{A}^{\#}}(\widetilde{A}^{\#}\otimes_R T, E) \simeq \operatorname{Hom}_R(T, E) \simeq \operatorname{Hom}_R(T, T\otimes_R \widetilde{A}) \simeq \operatorname{Hom}_R(T, T)\otimes_R \widetilde{A} \simeq \widetilde{A}^{\mathrm{op}}$, and similarly $\operatorname{Hom}_{\widetilde{A}^{\mathrm{op}}}(E, E) \simeq \widetilde{A}^{\#}$. These observations imply the assertions of the theorem. $\square$

The result of Theorem 9.9 can be rephrased as follows. For any left $\widetilde{A}$-module L, the tensor product $T\otimes_R L$ has a natural left $\widetilde{A}^{\#}$-module structure. For any right $\widetilde{A}^{\#}$-module M, the tensor product $M\otimes_R T$ has a natural right $\widetilde{A}$-module structure. For any left $\widetilde{A}^{\#}$-module N, the left R-module $\operatorname{Hom}_R(T, N)$ has a natural left $\widetilde{A}$-module structure. For any right $\widetilde{A}$-module N, the right R-module $\operatorname{Hom}_{R^{\mathrm{op}}}(T, N)$ has a natural right $\widetilde{A}^{\#}$-module structure.

Remark 9.10 When (B, d, h) is a graded commutative DG-ring (so the graded ring B is graded commutative and $h = 0$), the two filtered rings $\widetilde{A}$ and $\widetilde{A}^{\#}$ are naturally opposite to each other, that is $\widetilde{A}^{\#} \simeq \widetilde{A}^{\mathrm{op}}$. This isomorphism is induced by the natural identity isomorphism between the CDG-ring $B = (B, d, h)$ and its opposite CDG-ring $B^{\mathrm{op}} = (B^{\mathrm{op}}, d^{\mathrm{op}}, -h^{\mathrm{op}})$. So the Morita equivalence bimodule E can be viewed as having two right $\widetilde{A}$-module structures which commute with each other, and Theorem 9.9 provides an equivalence between the categories of left and right $\widetilde{A}$-modules, $\widetilde{A}$–mod $\simeq$ mod–$\widetilde{A}$, in this case.

Here, for an arbitrary CDG-ring (B, d, h), the CDG-ring $(B^{\mathrm{op}}, d^{\mathrm{op}}, -h^{\mathrm{op}})$ is defined as follows. As a graded abelian group, B^{op} is identified with B by the map denoted by $b \longmapsto b^{\mathrm{op}} \colon B^n \longrightarrow B^{\mathrm{op},n}$, $n \in \mathbb{Z}$. The multiplication in B^{op} is given by the formula $b^{\mathrm{op}}c^{\mathrm{op}} = (-1)^{|b||c|}(cb)^{\mathrm{op}}$ for all $b, c \in B$, the differential is $d^{\mathrm{op}}(b^{\mathrm{op}}) = d(b)^{\mathrm{op}}$, and the curvature element is $-h^{\mathrm{op}} \in B^{\mathrm{op},2}$. The construction of the opposite CDG-ring $(B^{\mathrm{op}}, d^{\mathrm{op}}, -h^{\mathrm{op}})$ has the expected property that the DG-category of right CDG-modules over (B, d, h) is equivalent to the DG-category of left CDG-modules over $(B^{\mathrm{op}}, d^{\mathrm{op}}, -h^{\mathrm{op}})$ and vice versa. Notice that the formulas above imply that the map $b \longmapsto b^{\mathrm{op}} \colon B \longrightarrow B^{\mathrm{op}}$ is *not* an isomorphism of CDG-rings when the ring B is graded commutative, but $2h \neq 0$ in B^2. Therefore, there is *no* natural isomorphism between the rings $\widetilde{A}^{\#}$ and $\widetilde{A}^{\mathrm{op}}$ in the situation with a nonzero curvature in characteristic different from 2 (generally speaking).

9.7 Relatively Frobenius Koszul Quadrality

We keep the assumptions and notation of Sects. 9.5–9.6. So $B = (B, d, h)$ is a relatively Frobenius Koszul CDG-ring, $\widetilde{A}$ is the left finitely projective nonhomogeneous Koszul ring nonhomogeneous quadratic dual to the right finitely projective Koszul CDG-ring (B, d, h), and $\widetilde{A}^{\#}$ is the right finitely projective nonhomogeneous Koszul ring nonhomogeneous quadratic dual to the left finitely projective Koszul CDG-ring (B, d, h). The $\widetilde{A}^{\#}$-$\widetilde{A}$-bimodule E was constructed in the paragraph before Theorem 9.9.

Following the discussion in Sect. 9.5, there are DG-functors

$$\widetilde{A}^{\#} \otimes_R^{\rho'} - : \mathsf{DG}(B\text{–mod}) \longrightarrow \mathsf{DG}(\widetilde{A}^{\#}\text{–mod})$$

and

$$C^{\#} \otimes_R^{\pi'} - : \mathsf{DG}(\widetilde{A}^{\#}\text{–mod}) \longrightarrow \mathsf{DG}(B\text{–mod})$$

between the DG-category of complexes of left $\widetilde{A}^{\#}$-modules and the DG-category of left CDG-modules over (B, d, h). Following the discussion in the beginning of Sect. 8.8, the DG-functor $\widetilde{A}^{\#} \otimes_R^{\rho'} -$ is left adjoint to the DG-functor $C^{\#} \otimes_R^{\pi'} -$. Hence the induced triangulated functors between the homotopy categories,

$$\widetilde{A}^{\#} \otimes_R^{\rho'} - : \mathsf{Hot}(B\text{–mod}) \longrightarrow \mathsf{Hot}(\widetilde{A}^{\#}\text{–mod})$$

and

$$C^{\#} \otimes_R^{\pi'} - : \mathsf{Hot}(\widetilde{A}^{\#}\text{–mod}) \longrightarrow \mathsf{Hot}(B\text{–mod}),$$

are also adjoint on the respective sides.

Lemma 9.11

(a) *For any complex of right $\widetilde{A}^{\#}$-modules $M^{\bullet}$, there is a natural closed isomorphism of right CDG-modules over (B, d, h)*

$$M^{\bullet} \otimes_{\widetilde{A}^{\#}} E[-m] \otimes_R^{\sigma'} C \simeq M^{\bullet} \otimes_R^{\rho'} B.$$

(b) *For any complex of left $\widetilde{A}$-modules $L^{\bullet}$, there is a natural closed isomorphism of left CDG-modules over (B, d, h)*

$$C^{\#} \otimes_R^{\pi'} E[-m] \otimes_{\widetilde{A}} L^{\bullet} \simeq B \otimes_R^{\tau'} L^{\bullet}.$$

Proof Let us prove part (b); part (a) is opposite. We will construct a closed isomorphism $B \otimes_R^{\tau'} \widetilde{A} \longrightarrow C^{\#} \otimes_R^{\pi'} E[-m]$ of CDG-bimodules over (B, d, h) and $(\widetilde{A}, 0, 0)$. Then the desired closed isomorphism of left CDG-modules in (b) will be produced by applying the DG-functor $- \otimes_{\widetilde{A}} L^{\bullet}$.

Recall that, by the definition, the $\widetilde{A}^{\#}$-$\widetilde{A}$-bimodule E is constructed as $E = H^m(\widetilde{A}^{\#} \otimes_R^{\rho'} B \otimes_R^{\tau'} \widetilde{A})$. Moreover, there is a natural quasi-isomorphism of finite complexes of $\widetilde{A}^{\#}$-$\widetilde{A}$-bimodules $\widetilde{A}^{\#} \otimes_R^{\rho'} B \otimes_R^{\tau'} \widetilde{A} \longrightarrow E[-m]$. Applying the DG-functor $C^{\#} \otimes_R^{\pi'} -$, we obtain a closed morphism

$$C^{\#} \otimes_R^{\pi'} \widetilde{A}^{\#} \otimes_R^{\rho'} B \otimes_R^{\tau'} \widetilde{A} \longrightarrow C^{\#} \otimes_R^{\pi'} E[-m] \tag{9.3}$$

of CDG-bimodules over (B, d, h) and $(\widetilde{A}, 0, 0)$. In fact, the cone of the closed morphism (9.3) is a coacyclic CDG-bimodule (since $C^{\#}$ is a projective graded right R-module), but we will not need to use this observation.

For any left CDG-module N over (B, d, h), there is a natural adjunction morphism $N \longrightarrow C^{\#} \otimes_R^{\pi'} \widetilde{A}^{\#} \otimes_R^{\rho'} N$, which is a closed morphism of left CDG-modules over (B, d, h). In particular, we are interested in the adjunction morphism

$$B \otimes_R^{\tau'} \widetilde{A} \longrightarrow C^{\#} \otimes_R^{\pi'} \widetilde{A}^{\#} \otimes_R^{\rho'} B \otimes_R^{\tau'} \widetilde{A}, \tag{9.4}$$

which is a closed morphism of CDG-bimodules over (B, d, h) and $(\widetilde{A}, 0, 0)$. The composition of (9.4) and (9.3) is the desired closed morphism of CDG-bimodules

$$B \otimes_R^{\tau'} \widetilde{A} \longrightarrow C^{\#} \otimes_R^{\pi'} E[-m]. \tag{9.5}$$

In order to show that (9.5) is an isomorphism, one can define filtrations on the left-hand and the right-hand side and check that the associated graded map is an isomorphism. The nonhomogeneous Koszul rings $\widetilde{A}$ and $\widetilde{A}^{\#}$ are endowed with increasing filtrations F. A finite increasing filtration F on the graded ring B was defined in the proof of Lemma 9.7. Hence one obtains an induced increasing filtration on the complex of $\widetilde{A}^{\#}$-$\widetilde{A}$-bimodules $\widetilde{A}^{\#} \otimes_R^{\rho'} B \otimes_R^{\tau'} \widetilde{A}$, and consequently on its cohomology $\widetilde{A}^{\#}$-$\widetilde{A}$-bimodule E. Finally, a finite increasing filtration F on the graded coring C was defined in the proof of Theorem 6.14 (as the filtration induced by the grading); a filtration F on the graded coring $C^{\#}$ is constructed similarly.

Both the left-hand and the right-hand sides of the maps (9.3) and (9.4) acquire induced filtrations, which are preserved by both the maps. It is straightforward to check that the composition (9.5) becomes an isomorphism after the passage to the associated graded bimodules with respect to F.   □

The following assertion is a restatement of Corollary 8.28 with $\widetilde{A}$ replaced by $\widetilde{A}^{\#}$ and $^{\#}B$ replaced by B, and with the notation taking into account the assumption that $B^n = 0$ for $n > m$. (Recall that Corollary 8.28, in turn, is the opposite version of Corollary 6.18.)

Corollary 9.12 *Assume that the left homological dimension of the ring R is finite. Then the pair of adjoint triangulated functors $\widetilde{A}^{\#}\otimes_R^{\rho'} -: \mathsf{Hot}(B\text{–mod}) \longrightarrow \mathsf{Hot}(\widetilde{A}^{\#}\text{–mod})$ and $C^{\#}\otimes_R^{\pi'} -: \mathsf{Hot}(\widetilde{A}^{\#}\text{–mod}) \longrightarrow \mathsf{Hot}(B\text{–mod})$ induces mutually inverse triangulated equivalences*

$$\mathsf{D}(\widetilde{A}^{\#}\text{–mod}) \simeq \mathsf{D}^{\mathsf{co}}(B\text{–mod}) \tag{9.6}$$

between the derived category of left $\widetilde{A}^{\#}$-modules and the coderived category of left CDG-modules over (B, d, h). □

The next theorem is the second main result of Chap. 9.

Theorem 9.13 *Let $B = (B, d, h)$ be a relatively Frobenius Koszul CDG-ring whose degree-zero component $R = B^0$ is a ring of finite left homological dimension. Let $\widetilde{A}$ and $\widetilde{A}^{\#}$ be two nonhomogeneous Koszul rings quadratic dual to B on the two sides, as above. Then there is a commutative square diagram of triangulated equivalences*

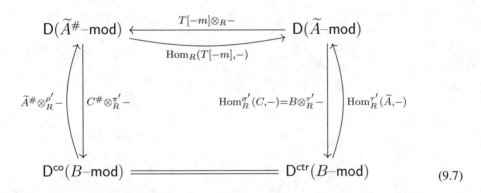

Here the triangulated equivalence in the upper line is induced by the equivalence of abelian categories from Theorem 9.9(a) (up to the cohomological shift by $[-m]$). The triangulated equivalence in the leftmost column is the assertion of Corollary 9.12. The triangulated equivalence in the rightmost column is a particular case of Corollary 7.14. The triangulated equivalence in the lower line is provided by Theorem 9.6.

Proof Recall that, by the definition, for any complex of left $\widetilde{A}$-modules $L^\bullet$ we have $\mathrm{Hom}_R^{\sigma'}(C, L^\bullet) = K_{e'}^\vee(B, \widetilde{A}) \otimes_{\widetilde{A}} L^\bullet = B \otimes_R^{\tau'} L^\bullet$, since $B^n = 0$ for $n > m$ (see Sects. 7.4 and 8.7). Hence the equality in the label at the rightmost straight arrow in the diagram. We also have $B\text{–contra} = B\text{–mod} = B\text{–comod}$ because $B^n = 0$ for $n > m$.

It remains to explain why the diagram is commutative. It is convenient to check commutativity of the diagram of triangulated functors between the homotopy categories formed by the three straight arrows and the equality in the lower horizontal line. This follows from the natural closed isomorphism of left CDG-modules in Lemma 9.11(b). Then one observes that both sides of the latter closed isomorphism are CDG-modules of the induced and coinduced type; so they are adjusted to the derived functors in the construction of the triangulated equivalence in Theorem 9.6.   □

For examples of a relatively Frobenius Koszul quadrality picture as in (9.7), we refer to Sects. 10.2, 10.3, 10.5, 10.8, and 10.9 below.

Examples

10.1   Symmetric and Exterior Algebras

In this section we recall the main points of the discussion of the tensor rings and, more importantly, the symmetric and exterior algebras, which was presented in the examples given throughout Chaps. 1–2; see, in particular, Examples 1.7, 2.31, and 2.39.

Let us start with the tensor ring. Let R be an associative ring and V be an R-R-bimodule. Consider the tensor ring $T_R(V) = \bigoplus_{n=0}^{\infty} V^{\otimes_R n}$, as defined in Chap. 1 (see the notation in Sect. 2.2). By the definition, the graded ring $A = T_R(V)$ is quadratic.

It is clear from, e.g., Theorem 2.30(b) that the graded ring A is left flat Koszul whenever V is a flat left R-module. By Theorem 2.35(a), the graded ring A is left finitely projective Koszul whenever V is a finitely generated projective left R-module. In the latter case, the quadratic dual right finitely projective Koszul graded ring B has the components $B_0 = R$, $B_1 = \mathrm{Hom}_R(V, R)$, and $B_n = 0$ for all $n \geq 2$.

Now let R be a commutative ring and V be an R-module, viewed as an R-R-bimodule in which the left and right actions of R coincide. Then $T_R(V)$ is a graded R-algebra (or in other words, the degree-zero component $R = T_{R,0}(V)$ lies in the center of $T_R(V)$). By the definition, the *symmetric algebra* $\mathrm{Sym}_R(V) = \bigoplus_{n=0}^{\infty} \mathrm{Sym}_R^n(V)$ is the largest commutative quotient algebra of $T_R(V)$. Equivalently, $\mathrm{Sym}_R(V) = T_R(V)/(I)$ is the quadratic algebra over R with the submodule of quadratic relations $I \subset V \otimes_R V$ spanned by all the tensors $v \otimes w - w \otimes v \in V \otimes_R V$ with $v, w \in V$.

The *exterior algebra* $\Lambda_R(V) = \bigoplus_{n=0}^{\infty} \Lambda_R^n(V)$ is the largest *strictly* graded commutative quotient algebra of $T_R(V)$, that is, the largest quotient algebra of $T_R(V)$ in which the identities $ab = (-1)^{|a||b|}ba$ and $c^2 = 0$ hold for all elements a of degree $|a|$, b of degree $|b|$, and c of odd degree. Equivalently, $\Lambda_R(V) = T_R(V)/(I)$ is the quadratic algebra over R with the submodule of quadratic relations $I \subset V \otimes_R V$ spanned by all the

© The Author(s), under exclusive license to Springer Nature Switzerland AG 2021   231
L. Positselski, *Relative Nonhomogeneous Koszul Duality*, Frontiers in Mathematics,
https://doi.org/10.1007/978-3-030-89540-2_10

tensors $v \otimes v \in V \otimes_R V$ with $v \in V$. When the element $2 \in R$ is invertible, the submodule $I \subset V \otimes_R V$ is also spanned by the tensors $v \otimes w + w \otimes v$ with $v, w \in V$.

When V is a (finitely generated) free R-module, the grading components of the quadratic algebras $\mathrm{Sym}_R(V)$ and $\Lambda_R(V)$ are also (finitely generated) free R-modules with explicit bases which are easy to construct. It follows by passing to retracts that the R-modules $\mathrm{Sym}_R^n(V)$ and $\Lambda_R^n(V)$ are (finitely generated) projective whenever the R-module V is (finitely generated) projective. Passing to the filtered direct limits, one can see that the grading components of $\mathrm{Sym}_R(V)$ and $\Lambda_R(V)$ are flat R-modules whenever V is a flat R-module.

It is worth noticing that the naïve definition of an exterior algebra as the quotient algebra of the tensor algebra by the relations $v \otimes w + w \otimes v$ with $v, w \in V$ (equivalently, the maximal quotient algebra of $T_R(V)$ in which the identity $ab = (-1)^{|a||b|}ba$ holds) does *not* have these flatness/projectivity properties. For a counterexample, it suffices to take R to be the ring of integers $\mathbb{Z}$ or the ring of 2-adic integers $\mathbb{Z}_2$, and V to be the free R-module with one generator.

Let V be a finitely generated projective R-module, and let $V^\vee = \mathrm{Hom}_R(V, R)$ denote the dual finitely generated projective R-module. Then the quadratic graded rings $A = \mathrm{Sym}_R(V)$ and $B = \Lambda_R(V^\vee)$ are quadratic dual to each other (in the sense of Propositions 1.6 and 1.8). Moreover, both the quadratic algebras $\mathrm{Sym}_R(V)$ and $\Lambda_R(V^\vee)$ are (left and right) finitely projective Koszul, as one can see, e. g., from exactness of the Koszul complex associated with any chosen basis in the module of generators V of the polynomial algebra $\mathrm{Sym}_R(V)$ (use Theorem 2.35(d) or (e)).

As the class of (left or right) flat Koszul graded rings is closed under filtered direct limits, it follows that the quadratic graded rings $\mathrm{Sym}_R(V)$ and $\Lambda_R(V)$ are (left and right) flat Koszul for any flat R-module V.

Let us say that a finitely generated projective R-module V is *everywhere of rank m* if, for every prime ideal $\mathfrak{p}$ in R, the localization $V_\mathfrak{p}$ is a free $R_\mathfrak{p}$-module with m generators. An R-module T is said to be *invertible* if, viewed as an R-R-bimodule in which the left and right actions of R coincide, T is invertible in the sense of Sect. 9.1.

Lemma 10.1 *Let R be a commutative ring and V be a finitely generated projective R-module everywhere of rank m. Then $\Lambda_R^n(V) = 0$ for $n > m$, the R-module $\Lambda_R^m(V)$ is invertible, and the multiplication map $\Lambda_R^n(V) \otimes_R \Lambda_R^{m-n}(V) \longrightarrow \Lambda_R^m(V)$ is a perfect pairing for every $0 \le n \le m$ (in the sense of Sect. 9.1). In other words, $\Lambda_R(V)$ is a relatively Frobenius Koszul graded ring over R in the sense of Sect. 9.4.*

Proof First one observes that the assertions of the lemma hold for a free R-module V with m generators. Secondly, for any finitely generated projective R-module V, the graded ring $\Lambda_R(V)$ is finitely projective Koszul as explained above. Now all the modules we are dealing with are finitely generated and projective, and all the constructions involved commute with localizations at prime ideals $\mathfrak{p} \subset R$. One has $\Lambda_R^n(V) = 0$ for $n > m$, since $\Lambda_R^n(V)_\mathfrak{p} = \Lambda_{R_\mathfrak{p}}^n(V_\mathfrak{p}) = 0$. The R-module $T = \Lambda_R^m(V)$ being invertible means that the natural morphisms $R \longrightarrow T^\vee \otimes_R T$ and $T \otimes_R T^\vee \longrightarrow R$ are isomorphisms; this holds because these maps become isomorphisms after localization at every prime ideal. Similarly one shows that the multiplication maps taking values in $\Lambda_R^m(V)$ are perfect pairings. $\square$

10.2   Algebraic Differential Operators

Let X be a smooth affine algebraic variety over a field k (which we will eventually assume to have characteristic 0 in this section). Denote by $O(X)$ the finitely generated commutative k-algebra of regular functions on X.

Construction 10.2 *Differential operators on X* (in the sense of Grothendieck) form a subring in the ring $\mathrm{End}_k(O(X))$ of k-linear endomorphisms of the k-vector space $O(X)$ (with respect to the composition multiplication of linear operators).

The differential operators of order 0 are the $O(X)$-linear endomorphisms of $O(X)$, that is the operators of multiplication with regular functions $f \in O(X)$. A k-linear map $D \colon O(X) \longrightarrow O(X)$ is said to be a *differential operator of order $\leq n$* if, for any $f \in O(X)$, the operator $[D, f] = D \circ f - f \circ D \colon O(X) \longrightarrow O(X)$ is a differential operator of order $\leq n - 1$. A *differential operator* $D \colon O(X) \longrightarrow O(X)$ is a map which is a differential operator of some finite order $n \geq 0$.

We denote the subspace of differential operators by $\mathrm{Diff}(X) \subset \mathrm{End}_k(O(X))$ and the subspace of differential operators of order $\leq n$ by $F_n \mathrm{Diff}(X) \subset \mathrm{Diff}(X)$ for every integer $n \geq 0$. It is straightforward to check that $\mathrm{Diff}(X)$ is a subring in $\mathrm{End}_k(O(X))$ and F is a multiplicative increasing filtration on $\mathrm{Diff}(X) = \bigcup_n F_n \mathrm{Diff}(X)$.

Let T denote the tangent bundle to X and T^* denote the cotangent bundle. The global sections of the tangent bundle are called the *vector fields* on X and the global sections of the cotangent bundle are called the *differential 1-forms*.

Construction 10.3 The k-vector space of vector fields $T(X)$ can be constructed as the subspace $T(X) \subset F_1 \mathrm{Diff}(X)$ of all differential operators v of order ≤ 1 on X which annihilate the constant functions, that is $v(1) = 0$. Equivalently, $T(X)$ is the space of all *derivations* of the k-algebra $O(X)$, that is k-linear maps $v \colon O(X) \longrightarrow O(X)$ such that $v(fg) = v(f)g + fv(g)$ for all $f, g \in O(X)$. The subspace $T(X) \subset F_1 \mathrm{Diff}(X)$ is preserved by the left (but not right) multiplications of differential operators by the functions; so the rule $(fv)(g) = f(v(g))$ defines an $O(X)$-module structure on $T(X)$.

Construction 10.4 The $O(X)$-module of differential 1-forms $T^*(X)$ is produced by the construction of *Kähler differentials*. Consider the free $O(X)$-module spanned by the symbols $d(f)$ with $f \in O(X)$ and take its quotient module by the submodule spanned by all elements of the form $d(a)$ with $a \in k$ and $d(fg) - fd(g) - gd(f)$ with $f, g \in O(X)$; this quotient module is $T^*(X)$.

Since we are assuming that X is smooth, both $T(X)$ and $T^*(X)$ are finitely generated projective $O(X)$-modules (of the rank equal to the dimension of X over k). They are also naturally dual to each other: it follows immediately from the definitions one has $T(X) = \mathrm{Hom}_{O(X)}(T^*(X), O(X))$.

We are interested in the symmetric powers of the $O(X)$-module $T(X)$ and the exterior powers of the $O(X)$-module $T^*(X)$. Following the notation in Sect. 10.1, the former are denoted by $\mathrm{Sym}^n_{O(X)}(T(X))$ and the latter by $\Lambda^n_{O(X)}(T^*(X)) = \Omega^n(X)$, where $n \geq 0$; so, in particular, $\mathrm{Sym}^0_{O(X)}(T(X)) = O(X) = \Omega^0(X)$, $\mathrm{Sym}^1_{O(X)}(T(X)) = T(X)$, and $\Omega^1(X) = T^*(X)$. These are the global sections of the symmetric/exterior powers of the vector bundles T and T^* on X. The elements of $\Omega^n(X)$ are called the *differential n-forms* on X.

Construction 10.5 There exists a unique odd derivation d of degree 1 with $d^2 = 0$ on the graded algebra $\Omega(X) = \bigoplus_{n \geq 0} \Omega^n(X)$ whose restriction to $\Omega^0(X) = O(X)$ is the map $d: O(X) \longrightarrow T^*(X) = \Omega^1(X)$ appearing in Construction 10.4. The differential $d: \Omega^n(X) \longrightarrow \Omega^{n+1}(X)$, $n \geq 0$, is called the *de Rham differential*. So $(\Omega(X), d)$ is a DG-ring. More precisely, it is a DG-algebra over k; it is called the *de Rham DG-algebra*.

When the characteristic of k is equal to 0, the associated graded ring $\mathrm{gr}^F \mathrm{Diff}(X) = \bigoplus_{n=0}^{\infty} F_n \mathrm{Diff}(X)/F_{n-1} \mathrm{Diff}(X)$ is naturally isomorphic to the graded $O(X)$-algebra $\mathrm{Sym}_{O(X)}(T(X))$. Since the latter is left (and right) finitely projective Koszul, the filtered ring $(\mathrm{Diff}(X), F)$ is left finitely projective nonhomogeneous Koszul in the sense of Sect. 4.6. Furthermore, the ring $\mathrm{Diff}(X)$ is left augmented over its subring $O(X)$ in the sense of Sect. 3.8. The natural left action of $\mathrm{Diff}(X)$ in $O(X)$ (by the differential operators) provides the augmentation.

One can check that the left augmented left finitely projective nonhomogeneous Koszul ring $\mathrm{Diff}(X)$ (with the above filtration and augmentation) corresponds to the right finitely projective Koszul DG-ring $\Omega(X)$ (with the de Rham differential) under the anti-equivalence of categories from Corollary 4.28.

Notice that the commutative ring $R = O(X)$ has finite homological dimension (since X is a smooth algebraic variety by assumption). Furthermore, the graded ring $\Omega(X)$ has only finitely many grading components (indeed, $\Omega^n(X) = 0$ for $n > \dim_k X$); so there is no difference between graded *comodules*, graded *contramodules*, and the conventional graded *modules* over $\Omega(X)$. Hence the results of Chaps. 6 and 7 lead to the following theorem.

Theorem 10.6 *For any smooth affine algebraic variety X over a field k of characteristic 0, the construction of Corollary 6.18 provides a triangulated equivalence between the derived category of right modules over the ring of algebraic differential operators $\mathrm{Diff}(X)$ and the* *coderived category of right DG-modules over the de Rham DG-algebra* $(\Omega(X), d)$,

$$\mathsf{D}(\mathsf{mod}\text{–}\mathrm{Diff}(X)) \simeq \mathsf{D}^{\mathsf{co}}(\mathsf{mod}\text{–}(\Omega(X), d)). \tag{10.1}$$

In the same context, the construction of Corollary 7.14 provides an equivalence between the derived category of left $\mathrm{Diff}(X)$-*modules and the contraderived category of left DG-modules over* $(\Omega(X), d)$,

$$\mathsf{D}(\mathrm{Diff}(X)\text{–}\mathsf{mod}) \simeq \mathsf{D}^{\mathsf{ctr}}((\Omega(X), d)\text{–}\mathsf{mod}). \tag{10.2}$$

Proof The assertions are particular cases of the respective corollaries. Let us just describe very briefly the functors providing the triangulated equivalences. The mutually inverse functors in (10.1) assign to a complex of right $\mathrm{Diff}(X)$-modules $M^\bullet$ the right DG-module $M^\bullet \otimes_{O(X)} \Lambda_{O(X)}(T(X))$ over the de Rham DG-ring $\Omega(X) = \Lambda_{O(X)}(T^*(X))$, and conversely, to a right DG-module $N^\bullet$ over $\Omega(X)$ the complex of right $\mathrm{Diff}(X)$-modules $N^\bullet \otimes_{O(X)} \mathrm{Diff}(X)$ is assigned. The mutually inverse functors in (10.2) assign to a complex of left $\mathrm{Diff}(X)$-modules $P^\bullet$ the left DG-module $\Omega(X) \otimes_{O(X)} P^\bullet$ over $\Omega(X)$, and conversely, to a left DG-module $Q^\bullet$ over $\Omega(X)$ the complex of left $\mathrm{Diff}(X)$-modules $\mathrm{Hom}_{O(X)}(\mathrm{Diff}(X), Q^\bullet)$ is assigned. This formulation only describes the underlying (graded) module structures; the constructions of the differentials are spelled out in Chaps. 6 and 7. $\square$

The abelian categories of left and right modules over the ring $\mathrm{Diff}(X)$ are naturally equivalent (see the discussion in Sect. 0.11). The results of Chap. 9 provide a square diagram of triangulated equivalences connecting (10.1) with (10.2).

Theorem 10.7 *For any smooth affine algebraic variety X over a field k of characteristic 0, the constructions of Theorem 9.13 (with Remark 9.10 taken into account) provide a commutative square diagram of triangulated equivalences*

$$
\begin{array}{ccc}
\mathsf{D}(\mathsf{mod}\text{–}\mathrm{Diff}(X)) & \longleftrightarrow & \mathsf{D}(\mathrm{Diff}(X)\text{–}\mathsf{mod}) \\
\Big\Updownarrow & & \Big\Updownarrow \\
\mathsf{D}^{\mathsf{co}}((\Omega(X), d)\text{–}\mathsf{mod}) & = & \mathsf{D}^{\mathsf{ctr}}((\Omega(X), d)\text{–}\mathsf{mod})
\end{array}
\tag{10.3}
$$

where the equivalence of derived categories in the upper line is induced by the conversion equivalence of abelian categories $\mathsf{mod}\text{–}\mathrm{Diff}(X) \simeq \mathrm{Diff}(X)\text{–}\mathsf{mod}$ *(up to a cohomological shift by* $[-\dim_k X]$*), while the co-contra correspondence in the lower line is the result of Theorem 9.6. The vertical equivalences are (10.1) and (10.2).* $\square$

Modules over the ring of differential operators Diff(X) (or more generally, sheaves of modules over the sheaf of rings of differential operators over a nonaffine smooth algebraic variety X) are known colloquially as "D-modules" [5, 8, 11, 29]. The complex $M \otimes_{O(X)} \Lambda_{O(X)}(T(X))$ (for a right D-module M) or $\Omega(X) \otimes_{O(X)} P$ (for a left D-module P) appearing in the proof of Theorem 10.6, known as "the de Rham complex of a D-module," is a familiar construction in the D-module theory; see, e.g., [11, Section VI.1.8]. An approach to the theory of D-modules based on DG-modules over the de Rham DG-algebra was developed in the paper [75].

10.3   Crystalline Differential Operators

Let X be a smooth affine algebraic variety over a field k of arbitrary characteristic.

Construction 10.8 The ring of *crystalline differential operators* $\text{Diff}^{cr}(X)$ is defined by generators and relations as follows [10, Section 1.2].

The generators are the elements of the ring of functions $O(X)$ and the $O(X)$-module of vector fields $T(X)$ (as defined in Sect. 10.2). The sum of any two elements of $O(X)$ in $\text{Diff}^{cr}(X)$ equals their sum in $O(X)$; and the sum of any two elements of $T(X)$ in $\text{Diff}^{cr}(X)$ equals their sum in $T(X)$. Concerning the product, let us use the notation of Sect. 3.3 and denote by $*$ the product of any two elements in $\text{Diff}^{cr}(X)$, to be distinguished from their product in $O(X)$ or $T(X)$. Then the relations

$$f * g = fg \qquad \text{for all } f, g \in O(X),$$

$$f * v = fv \qquad \text{for all } f \in O(X) \text{ and } v \in T(X), \tag{10.4}$$

$$v * f = fv + v(f) \qquad \text{for all } f \in O(X) \text{ and } v \in T(X)$$

are imposed. Here $fg \in O(X)$ denotes the product in $O(X)$ and $fv \in T(X)$ denotes the action of elements of $O(X)$ in the $O(X)$-module $T(X)$, while $v(f) \in O(X)$ denotes the action of vector fields by differential operators (or more precisely derivations) on the functions; so $v(f)$ is "the derivative of f along v."

Finally, we need to recall that derivations form a Lie algebra: for any vector fields v and $w \in T(X)$, there exists a unique vector field $[v, w] \in T(X)$, called the *commutator* of v and w, such that $[v, w](f) = v(w(f)) - w(v(f))$ for all $f \in O(X)$. The relation

$$v * w - w * v = [v, w] \qquad \text{for all } v, w \in T(X) \tag{10.5}$$

is also imposed in $\text{Diff}^{cr}(X)$.

One considers the increasing filtration F on the ring $\text{Diff}^{cr}(X)$ generated by $F_1 \text{Diff}^{cr}(X) = O(X) \oplus T(X)$ over $F_0 \text{Diff}^{cr}(X) = O(X)$ (cf. the discussion of generated filtrations in Sect. 3.1). Then, irrespectively of the characteristic of k, the associated graded ring $\mathrm{gr}^F \mathrm{Diff}^{\mathrm{cr}}(X) = \bigoplus_{n=0}^{\infty} F_n \mathrm{Diff}^{\mathrm{cr}}(X)/F_{n-1} \mathrm{Diff}^{\mathrm{cr}}(X)$ is naturally isomorphic to the symmetric algebra $\mathrm{Sym}_{O(X)}(T(X))$. This is provable as a particular case of Theorem 4.25 or Corollary 4.28 (see the discussion below).

There is a natural homomorphism of filtered rings $\mathrm{Diff}^{\mathrm{cr}}(X) \longrightarrow \mathrm{Diff}(X)$ uniquely defined by the condition that it acts by the identity maps on the subring $O(X) \subset \mathrm{Diff}^{\mathrm{cr}}(X)$ (taking it to the subring $O(X) \subset \mathrm{Diff}(X)$) and on the subspace $T(X) \subset \mathrm{Diff}^{\mathrm{cr}}(X)$ (taking it to the subspace $T(X) \subset \mathrm{Diff}(X)$). Over a field k of characteristic 0, this is a ring isomorphism, and in fact an isomorphism of filtered rings. But it is *neither* surjective *nor* injective in prime characteristic.

Examples 10.9

(1) To give an example of noninjectivity, let Y be the affine line over a field k of prime characteristic p; so $O(Y) = k[y]$ is the ring of polynomials in one variable. Then $d/dy: k[y] \longrightarrow k[y]$ is a vector field on Y. One would expect $(d/dy)^p: k[y] \longrightarrow k[y]$ to be a differential operator of order p, but in fact it is a zero map. So $(d/dy)^p \in \mathrm{Diff}^{\mathrm{cr}}(Y)$ is an element of $F_p \mathrm{Diff}^{\mathrm{cr}}(Y)$ not belonging to $F_{p-1} \mathrm{Diff}^{\mathrm{cr}}(Y)$, but belonging to the kernel of the ring homomorphism $\mathrm{Diff}^{\mathrm{cr}}(Y) \longrightarrow \mathrm{Diff}(Y)$.

(2) To give an example of nonsurjectivity, consider the ring of polynomials with integer coefficients $R = \mathbb{Z}[y]$. Then, for any element $f \in \mathbb{Z}[y]$, the element $f^{(p)}(y) = d^p f/dy^p \in \mathbb{Z}[y]$ is divisible by p; so $\frac{1}{p}(d/dy)^p$ is a well-defined map $R \longrightarrow R$. Taking the tensor product $k \otimes_{\mathbb{Z}} -$, one obtains a differential operator of order p on $k[y]$ which can be denoted by "$\frac{1}{p}(d/dy)^p$": $k[y] \longrightarrow k[y]$. This is an element of $F_p \mathrm{Diff}(Y)$ not belonging to the sum of $F_{p-1} \mathrm{Diff}(Y)$ with the image of the ring homomorphism $\mathrm{Diff}^{\mathrm{cr}}(Y) \longrightarrow \mathrm{Diff}(Y)$.

(3) Here is another example of nonsurjectivity. Let Z be the punctured affine line over k; so $O(Z) = k[z, z^{-1}]$. Then $z\frac{d}{dz}$ is a vector field on Z. One would expect $\left(z\frac{d}{dz}\right)^p: k[z, z^{-1}] \longrightarrow k[z, z^{-1}]$ to be a differential operator of order p, but in fact it is the same map as $z\frac{d}{dz}$. So $\left(z\frac{d}{dz}\right)^p - z\frac{d}{dz} \in \mathrm{Diff}^{\mathrm{cr}}(Z)$ is an element of $F_p \mathrm{Diff}^{\mathrm{cr}}(Z)$ not belonging to $F_{p-1} \mathrm{Diff}^{\mathrm{cr}}(Z)$, but belonging to the kernel of the ring homomorphism $\mathrm{Diff}^{\mathrm{cr}}(Z) \longrightarrow \mathrm{Diff}(Z)$. Similarly to the construction above, one can define a differential operator "$\frac{1}{p}\left(\left(z\frac{d}{dz}\right)^p - z\frac{d}{dz}\right)$": $k[z, z^{-1}] \longrightarrow k[z, z^{-1}]$. This is an element of $F_p \mathrm{Diff}(Z)$ not belonging to the sum of $F_{p-1} \mathrm{Diff}(Z)$ with the image of the ring homomorphism $\mathrm{Diff}^{\mathrm{cr}}(Z) \longrightarrow \mathrm{Diff}(Z)$.

Returning to the general case, let us point out that the map $F_1 \mathrm{Diff}^{\mathrm{cr}}(X) \longrightarrow F_1 \mathrm{Diff}(X)$ is still an isomorphism, for any smooth affine variety X over a field k of any characteristic. In fact, the map $F_{p-1} \mathrm{Diff}^{\mathrm{cr}}(X) \longrightarrow F_{p-1} \mathrm{Diff}(X)$ is an isomorphism when k has characteristic p. But the ring $\mathrm{Diff}(X)$ is not generated by $F_1 \mathrm{Diff}(X)$ in the latter case: the subring in $\mathrm{Diff}(X)$ generated by $F_1 \mathrm{Diff}(X)$ does not contain $F_p \mathrm{Diff}(X)$ (whenever the dimension of X is more than zero); moreover, the ring $\mathrm{Diff}(X)$ is not finitely generated [77, Section 3]. The ring $\mathrm{Diff}^{\mathrm{cr}}(X)$ has very different properties: over a field k of prime characteristic, $\mathrm{Diff}^{\mathrm{cr}}(X)$ is finitely generated as a module over its center [10, Sections 1.3 and 2].

On the other hand, $(\mathrm{Diff}^{\mathrm{cr}}(X), F)$ is a left finitely projective nonhomogeneous Koszul ring in the sense of Sect. 4.6, irrespectively of the characteristic of k. Composing the action of $\mathrm{Diff}(X)$ in $O(X)$ with the ring homomorphism $\mathrm{Diff}^{\mathrm{cr}}(X) \longrightarrow \mathrm{Diff}(X)$, one defines a left action of $\mathrm{Diff}^{\mathrm{cr}}(X)$ in $O(X)$ making $\mathrm{Diff}^{\mathrm{cr}}(X)$ a left augmented ring over its subring $O(X)$. The left augmented left finitely projective nonhomogeneous Koszul ring of crystalline differential operators $\mathrm{Diff}^{\mathrm{cr}}(X)$ (with the above filtration and augmentation) corresponds to the right finitely projective Koszul DG-ring of differential forms $\Omega(X)$ (with the above differential) under the anti-equivalence of categories from Corollary 4.28.

Hence we have the following generalization of Theorem 10.6.

Theorem 10.10 *For any smooth affine algebraic variety X over a field k, the construction of Corollary 6.18 provides a triangulated equivalence between the derived category of right modules over the ring of crystalline differential operators $\mathrm{Diff}^{\mathrm{cr}}(X)$ and the coderived category of right DG-modules over the de Rham DG-algebra $(\Omega(X), d)$,*

$$\mathsf{D}(\mathrm{mod}\text{--}\mathrm{Diff}^{\mathrm{cr}}(X)) \simeq \mathsf{D}^{\mathrm{co}}(\mathrm{mod}\text{--}(\Omega(X), d)). \tag{10.6}$$

In the same context, the construction of Corollary 7.14 provides an equivalence between the derived category of left $\mathrm{Diff}^{\mathrm{cr}}(X)$-modules and the contraderived category of left DG-modules over $(\Omega(X), d)$,

$$\mathsf{D}(\mathrm{Diff}^{\mathrm{cr}}(X)\text{--}\mathrm{mod}) \simeq \mathsf{D}^{\mathrm{ctr}}((\Omega(X), d)\text{--}\mathrm{mod}). \tag{10.7}$$

Proof The proof is similar to that of Theorem 10.6 (cf. Theorem 10.55 below). In particular, similarly to the characteristic 0 case of Sect. 10.2, the ring of functions $R = O(X)$ on any smooth affine algebraic variety over a field k of any characteristic has finite homological dimension, and the graded ring $\Omega(X)$ has only finitely many grading components. So there is no difference between graded modules, graded comodules, and graded contramodules over $\Omega(X)$. □

Similarly to Sect. 10.2, there is a natural equivalence between the abelian categories of left and right modules over $\mathrm{Diff}^{\mathrm{cr}}(X)$, provided by the mutually inverse functors of tensor product with $\Omega^m(X)$ and $\Lambda^m_{O(X)}(T(X))$ over $O(X)$ (where $m = \dim_k X$). The results of Chap. 9 provide a comparison between (10.6) and (10.7).

Theorem 10.11 *For any smooth affine algebraic variety X over a field k, the constructions of Theorem 9.13 (with Remark 9.10 taken into account) provide a commutative square diagram of triangulated equivalences*

$$
\begin{array}{ccc}
\mathsf{D}(\mathrm{mod\text{--}Diff}^{\mathrm{cr}}(X)) & \longleftarrow & \mathsf{D}(\mathrm{Diff}^{\mathrm{cr}}(X)\text{--mod}) \\
\Big\updownarrow & & \Big\updownarrow \\
\mathsf{D}^{\mathrm{co}}((\Omega(X),d)\text{--mod}) & =\!\!\!= & \mathsf{D}^{\mathrm{ctr}}((\Omega(X),d)\text{--mod})
\end{array}
\qquad (10.8)
$$

where the equivalence of derived categories in the upper line is induced by the conversion equivalence of abelian categories $\mathrm{mod\text{--}Diff}^{\mathrm{cr}}(X) \simeq \mathrm{Diff}^{\mathrm{cr}}(X)\text{--mod}$ *(up to a cohomological shift by* $[-\dim_k X]$*), while the co-contra correspondence in the lower line is the result of Theorem 9.6. The vertical equivalences are* (10.6) *and* (10.7).

Proof Notice that $\Omega(X)$ is a relatively Frobenius Koszul graded ring (as per the discussion in Lemma 10.1); so Theorem 9.13 is indeed applicable. (See Theorem 10.56 below for a generalization.) $\qquad\square$

10.4 Differential Operators in a Vector Bundle

Let E be a vector bundle over a smooth affine algebraic variety X (over a field k). Then the space of global sections $E(X)$ of the vector bundle E is a finitely generated projective $O(X)$-module. Conversely, any finitely generated projective $O(X)$-module arises from a (uniquely defined) vector bundle on X.

Construction 10.12 Given two vector bundles E' and E'' over X, one can consider differential operators acting from $E'(X)$ to $E''(X)$. The space of such differential operators $\mathrm{Diff}(X, E', E'')$ is a filtered k-vector subspace in $\mathrm{Hom}_k(E'(X), E''(X))$.

Specifically, the differential operators of order 0 are the $O(X)$-linear maps $E'(X) \longrightarrow E''(X)$. A k-linear map $D\colon E'(X) \longrightarrow E''(X)$ is said to be a *differential operator of order* $\leq n$ if, for every regular function $f \in O(X)$, the map $[D, f] = D \circ f - f \circ D\colon E'(X) \longrightarrow E''(X)$ is a differential operator of order $\leq n - 1$. Here f acts in $E'(X)$ and $E''(X)$ as in $O(X)$-modules. A *differential operator* $E'(X) \longrightarrow E''(X)$ is a map which is a differential operator of some finite order $n \geq 0$.

We denote the subspace of differential operators of order $\leq n$ by $F_n \mathrm{Diff}(X, E', E'') \subset \mathrm{Diff}(X, E', E'')$. When the two vector bundles are the same, $E' = E = E''$, we write simply $\mathrm{Diff}(X, E)$ instead of $\mathrm{Diff}(X, E', E'')$. The k-vector space $\mathrm{Diff}(X, E)$ with its increasing filtration F is a filtered ring and a subring in $\mathrm{End}_k(E(X))$. The subring of differential operators of order 0 in $\mathrm{Diff}(X, E)$ is $F_0\,\mathrm{Diff}(X, E) = \mathrm{Hom}_{O(X)}(E(X), E(X))$; it can be described as the ring of global sections $\mathrm{End}(E)(X)$ of the vector bundle $\mathrm{End}(E)$ of endomorphisms of the vector bundle E over X.

Proposition 10.13 *Assuming that the characteristic of k is equal to 0, the associated graded ring $\mathrm{gr}^F\,\mathrm{Diff}(X, E)$ is naturally isomorphic to the tensor product of the $O(X)$-algebra $\mathrm{End}(E)(X)$ and the graded $O(X)$-algebra $\mathrm{Sym}_{O(X)}(T(X))$,*

$$\mathrm{gr}^F\,\mathrm{Diff}(X, E) \simeq \mathrm{End}(E)(X)\otimes_{O(X)} \mathrm{Sym}_{O(X)}(T(X)).$$

Proof The construction of the principal symbol of the differential operator provides the isomorphism. We omit the (fairly well-known) details. □

Both $\mathrm{End}(E)(X)$ and the grading components of $\mathrm{Sym}_{O(X)}(T(X))$ are finitely generated projective $O(X)$-modules, while the filtration components of the ring $\mathrm{Diff}(X, E)$ have the left and the right $O(X)$-module structures, induced by the inclusion $O(X) \longrightarrow \mathrm{End}(E)(X) = F_0\,\mathrm{Diff}(X, E)$. In particular, we have a short exact sequence of $O(X)$-$O(X)$-bimodules

$$0 \longrightarrow \mathrm{End}(E)(X) \longrightarrow F_1\,\mathrm{Diff}(X, E) \longrightarrow \mathrm{End}(E)(X)\otimes_{O(X)}T(X) \longrightarrow 0.$$

$$(10.9)$$

In fact, this is even a short exact sequence of bimodules over the (noncommutative) $O(X)$-algebra $\mathrm{End}(E)(X)$, which exists irrespectively of the characteristic of k (cf. the discussion in Sect. 10.3).

The notion of a *connection* on a vector bundle is of key importance. It can be defined in many ways; we present four (equivalent) definitions below and briefly explain how they are equivalent.

Definition 10.14.A A *connection* ∇ in a vector bundle E is a splitting of (10.9) as a short exact sequence of *left* $\mathrm{End}(E)(X)$-modules.

All the three terms of (10.9) are projective as left (as well as right) $\mathrm{End}(E)(X)$-modules; so a connection in a vector bundle E over a smooth affine algebraic variety X always exists. In fact, the connections in E form an affine space (in a different language, a principal homogeneous space) over the vector space $T^*(X)\otimes_{O(X)} \mathrm{End}(E)(X)$. In other words, the difference $\nabla'' - \nabla'$ of any two connections in E is an element of $T^*(X)\otimes_{O(X)} \mathrm{End}(E)(X)$, and conversely, to any connection ∇ one can add any element of $T^*(X)\otimes_{O(X)} \mathrm{End}(E)(X)$ and obtain a new connection in E.

The inclusion of rings $O(X) \longrightarrow \mathrm{End}(E)(X)$ induces an inclusion of $O(X)$-modules $T(X) \longrightarrow \mathrm{End}(E)(X) \otimes_{O(X)} T(X)$. Taking the pull-back of the short exact sequence (10.9) with respect to the latter map, we obtain a short exact sequence of $O(X)$-$O(X)$-bimodules

$$0 \longrightarrow \mathrm{End}(E)(X) \longrightarrow \overline{F}_1 \mathrm{Diff}(X, E) \longrightarrow T(X) \longrightarrow 0. \tag{10.10}$$

Here $\overline{F}_1 \mathrm{Diff}(X, E)$ is a certain $O(X)$-$O(X)$-subbimodule in $F_1 \mathrm{Diff}(X, E)$.

A splitting of (10.9) as a short exact sequence of left $\mathrm{End}(E)(X)$-modules is equivalent to a left $O(X)$-linear splitting of (10.10). So the following definition is equivalent to the previous one.

Definition 10.14.B A connection ∇ in a vector bundle E is a splitting of (10.10) as a short exact sequence of $O(X)$-modules in the *left $O(X)$-module* structure.

Unwinding this definition in explicit terms, we arrive to the next one.

Definition 10.14.C A connection ∇ in E is a map assigning to every vector field $v \in T(X)$ a k-linear operator $\nabla_v \colon E(X) \longrightarrow E(X)$ in such a way that the following two equations are satisfied:

(i) $\nabla_{fv}(e) = f\nabla_v(e)$ for all $f \in O(X)$, $v \in T(X)$, and $e \in E(X)$;
(ii) $\nabla_v(fe) = v(f)e + f\nabla_v(e)$ for all $f \in O(X)$, $v \in T(X)$, and $e \in E(X)$.

Here $v(f) \in O(X)$ is the derivative of f along v.

A connection ∇ in E can be further interpreted as a k-linear map assigning to every section $e \in E(X)$ an $O(X)$-linear map from $T(X)$ to $E(X)$, or which is the same, an element of the tensor product $\Omega^1(X) \otimes_{O(X)} E(X)$. Hence, to a connection ∇ in the sense of Definition 10.14.C one assigns a k-linear map $E(X) \longrightarrow \Omega^1(X) \otimes_{O(X)} E(X)$ defined by the rule $\langle v, \nabla(e) \rangle = \nabla_v(e)$, where $\langle \, , \, \rangle$ denotes the $O(X)$-linear map $T(V) \otimes_{O(X)} T^*(X) \otimes_{O(X)} E(X) \longrightarrow E(X)$ induced by the natural pairing $T(X) \otimes_{O(X)} T^*(X) \longrightarrow O(X)$. Then the identity (ii) takes the form that is spelled out in our last equivalent definition immediately below.

Definition 10.14.D A connection ∇ in E is a k-linear map between the two $O(X)$-modules $E(X) \longrightarrow \Omega^1(X) \otimes_{O(X)} E(X)$ satisfying the equation

$$\nabla(fe) = d(f) \otimes e + f\nabla(e) \quad \text{for all } f \in O(X) \text{ and } e \in E(X),$$

where d denotes the de Rham differential $d \colon O(X) \longrightarrow \Omega^1(X)$.

Construction 10.15 If vector bundles E' and E'' over X are endowed with connections ∇' and ∇'', then any vector bundle produced naturally from E' and E'', such as $E' \oplus E''$ and $E' \otimes E''$, acquires an induced connection. In particular, the vector bundle $E' \otimes E''$ is defined by the rule $(E' \otimes E'')(X) = E'(X) \otimes_{O(X)} E''(X)$, and the induced connection ∇ on $E' \otimes E''$ is given by the rule $\nabla_v(e' \otimes e'') = \nabla'_v(e') \otimes e'' + e' \otimes \nabla''_v(e'')$ for all $e' \in E'(X)$, $e'' \in E''(X)$, and $v \in T(X)$. Similarly, the vector bundle $\mathrm{Hom}(E', E'')$ is defined by the rule $\mathrm{Hom}(E', E'')(X) = \mathrm{Hom}_{O(X)}(E'(X), E''(X))$, and the induced connection ∇ on $\mathrm{Hom}(E', E'')$ is given by the rule $\nabla_v(g)(e') = \nabla''_v(g(e')) - g(\nabla'_v(e'))$ for all $g \in \mathrm{Hom}_{O(X)}(E'(X), E''(X))$, $e' \in E'(X)$ and $v \in T(X)$.

Construction 10.16 Let $\nabla = \nabla_E$ be a connection in a vector bundle E over X. Consider the graded left $\Omega(X)$-module $\Omega(X) \otimes_{O(X)} E(X)$ of *differential forms on X with the coefficients in E*. Then the connection ∇ on E induces an odd derivation d_∇ on the graded module $\Omega(X) \otimes_{O(X)} E(X)$ compatible with the odd derivation d (the de Rham differential) on the graded ring $\Omega(X)$, in the sense of Sect. 6.1. The map $d_\nabla \colon \Omega^n(X) \otimes_{O(X)} E(X) \longrightarrow \Omega^{n+1}(X) \otimes_{O(X)} E(X)$ is given by the formula

$$d_\nabla(\omega \otimes e) = d(\omega) \otimes e + (-1)^n \omega \wedge \nabla(e) \tag{10.11}$$

for all $\omega \in \Omega^n(X)$ and $e \in E(X)$. Here the wedge $\wedge$ denotes the map $\Omega^n(X) \otimes_k \Omega^1(X) \otimes_{O(X)} E(X) \longrightarrow \Omega^{n+1}(X) \otimes_{O(X)} E(X)$ induced by the multiplication map $\Omega^1(X) \otimes_k \Omega^n(X) \longrightarrow \Omega^{n+1}(X)$. One needs to check that the map $d_\nabla = d_{\nabla_E}$ is well-defined, that is $d_\nabla(f\omega \otimes e) = d_\nabla(\omega \otimes fe)$ for all $f \in O(X)$.

Construction 10.17 The square d_∇^2 of the differential d_∇ on $\Omega(X) \otimes_{O(X)} E(X)$ is an $\Omega(X)$-linear map. Hence there exists an element $h_\nabla \in \Omega^2(X) \otimes_{O(X)} \mathrm{End}(E)(X)$ such that $d_\nabla^2(\phi) = h(\phi)$ for all $\phi \in \Omega(X) \otimes_{O(X)} E(X)$. Here the left action of $\Omega(X) \otimes_{O(X)} \mathrm{End}(E)(X)$ in $\Omega(X) \otimes_{O(X)} E(X)$ is induced by the multiplication in $\Omega(X)$ and the left action of $\mathrm{End}(E)(X)$ in $E(X)$. The element $h_\nabla = h_{\nabla_E} \in \Omega^2(X) \otimes_{O(X)} \mathrm{End}(E)(X)$ is called the *curvature* of the connection ∇ in a vector bundle E.

Construction 10.18 The tensor product $\Omega(X) \otimes_{O(X)} \mathrm{End}(E)(X)$ of the graded $O(X)$-algebra $\Omega(X)$ with the $O(X)$-algebra $\mathrm{End}(E)(X)$ has a natural structure of graded $O(X)$-algebra. The connection $\nabla = \nabla_E$ on the vector bundle E induces a connection $\nabla_{\mathrm{End}(E)}$ on the vector bundle $\mathrm{End}(E)$ on X, as explained in Construction 10.15. The induced differential $d_{\nabla_{\mathrm{End}(E)}}$ on the graded $\Omega(X)$-module $\Omega(X) \otimes_{O(X)} \mathrm{End}(E)(X)$ is, in fact, an odd derivation of the graded ring $\Omega(X) \otimes_{O(X)} \mathrm{End}(E)(X)$ (since the connection $\nabla_{\mathrm{End}(E)}$ is compatible with the composition multiplication in $\mathrm{End}(E)$, in the appropriate sense). The triple

$$(\Omega(X) \otimes_{O(X)} \mathrm{End}(E)(X), d_{\nabla_{\mathrm{End}(E)}}, h_{\nabla_E}) \tag{10.12}$$

is a curved DG-ring. The pair

$$(\Omega(X) \otimes_{O(X)} E(X), \ d_{\nabla_E}) \tag{10.13}$$

is a left CDG-module over (10.12).

The graded ring $A = \mathrm{End}(E)(X) \otimes_{O(X)} \mathrm{Sym}_{O(X)}(T(X))$ is left (and right) finitely projective Koszul (over its degree-zero component $R = \mathrm{End}(E)(X)$). The quadratic dual right (and left) finitely projective Koszul graded ring to A, as per the construction of Propositions 1.6 and 1.8, is $B = \Omega(X) \otimes_{O(X)} \mathrm{End}(E)(X)$.

Proposition 10.19 *Assume that the characteristic of k is equal to 0. Then the filtered ring* $(\mathrm{Diff}(X, E), F)$ *is a left finitely projective nonhomogeneous Koszul ring. The choice of a connection ∇ on E means the choice of a left $\mathrm{End}(E)(X)$-linear splitting of the short exact sequence (10.9); in the terminology of Sect. 3.3, this is the choice of a submodule of strict generators for* $\mathrm{Diff}(X, E)$. *The corresponding right finitely projective Koszul CDG-ring produced by the construction of Proposition 3.16 is the CDG-ring* $\Omega(X) \otimes_{O(X)} \mathrm{End}(E)(X)$ *(10.12). Replacing a connection ∇_E in E with another connection ∇'_E leads to a CDG-ring* $(\Omega(X) \otimes_{O(X)} \mathrm{End}(E)(X), \ d_{\nabla'_{\mathrm{End}(E)}}, \ h_{\nabla'_E})$ *connected with (10.12) by a natural change-of-connection isomorphism of CDG-rings, as per the discussion in Sects. 3.2 and 3.5.* $\square$

The ring of endomorphisms $R = \mathrm{End}(E)(X)$ of any vector bundle E on a smooth affine algebraic variety X has finite left and right homological dimensions; in fact, assuming that E is nonzero on all the connected components of X, both the abelian categories of left and right R-modules are equivalent to $O(X)$–mod. Furthermore, just as in Sect. 10.2, the graded ring $\Omega(X) \otimes_{O(X)} \mathrm{End}(E)(X)$ has only finitely many grading components. Hence we obtain the following theorem.

Theorem 10.20 *For any smooth affine algebraic variety X over a field k of characteristic 0 and any vector bundle E over X with a chosen connection ∇_E, the construction of Corollary 6.18 provides a triangulated equivalence between the derived category of right modules over the ring of differential operators $\mathrm{Diff}(X, E)$ acting in the sections of E and the coderived category of right CDG-modules over the CDG-algebra (10.12) of differential forms on X with the coefficients in $\mathrm{End}(E)$,*

$$\mathsf{D}(\mathrm{mod}\text{–}\mathrm{Diff}(X, E)) \simeq \mathsf{D}^{\mathsf{co}}(\mathrm{mod}\text{–}(\Omega(X) \otimes_{O(X)} \mathrm{End}(E)(X), \ d_{\nabla_{\mathrm{End}(E)}}, \ h_{\nabla_E})).$$
$$\tag{10.14}$$

In the same context, the construction of Corollary 7.14 provides an equivalence between the derived category of left $\mathrm{Diff}(X, E)$*-modules and the contraderived category of left CDG-modules over the CDG-ring* (10.12)

$$\mathsf{D}(\mathrm{Diff}(X, E)\text{--mod}) \simeq \mathsf{D}^{\mathrm{ctr}}((\Omega(X)\otimes_{O(X)}\mathrm{End}(E)(X), d_{\nabla_{\mathrm{End}(E)}}, h_{\nabla_E})\text{--mod}).$$
(10.15)

Proof This is also a generalization of Theorem 10.6, provable in a similar way. The assertions of the theorem are particular cases of the respective corollaries; and we will only describe very briefly the functors providing the triangulated equivalences. The functors in (10.14) assign to a complex of right $\mathrm{Diff}(X, E)$-modules $M^\bullet$ the right CDG-module $M^\bullet\otimes_{\mathrm{End}(E)(X)}(\mathrm{End}(E)(X)\otimes_{O(X)}\Lambda_{O(X)}(T(X))) = M^\bullet\otimes_{O(X)}\Lambda_{O(X)}(T(X))$ over the CDG-ring $\Omega(X)\otimes_{O(X)}\mathrm{End}(E)(X) = \Lambda_{O(X)}(T^*(X))\otimes_{O(X)}\mathrm{End}(E)(X)$, and conversely, to a right CDG-module N over $\Omega(X)\otimes_{O(X)}\mathrm{End}(E)(X)$ the complex of right $\mathrm{Diff}(X, E)$-modules $N\otimes_{\mathrm{End}(E)}\mathrm{Diff}(X, E)$ is assigned. The functors in (10.15) assign to a complex of left $\mathrm{Diff}(X, E)$-modules $P^\bullet$ the left CDG-module $(\Omega(X)\otimes_{O(X)}\mathrm{End}(E)(X))\otimes_{\mathrm{End}(E)(X)}P^\bullet = \Omega(X)\otimes_{O(X)}P^\bullet$ over the CDG-ring $\Omega(X)\otimes_{O(X)}\mathrm{End}(E)(X)$, and conversely, to a left CDG-module Q over $\Omega(X)\otimes_{O(X)}\mathrm{End}(E)(X)$ the complex of left $\mathrm{Diff}(X, E)$-modules $\mathrm{Hom}_{\mathrm{End}(E)(X)}(\mathrm{Diff}(X, E), Q)$ is assigned.   □

For example, the equivalence of categories (10.15) assigns the left CDG-module $(\Omega(X)\otimes_{O(X)}E(X), d_{\nabla_E})$ (10.13) over the CDG-ring (10.12) to the left $\mathrm{Diff}(X, E)$-module $E(X)$. The result of Theorem 10.20 is the affine, characteristic 0 particular case of the $\mathcal{D}$–Ω duality theorem of [59, Theorem B.2].

10.5 Twisted Differential Operators

As in the previous sections, we consider a smooth affine algebraic variety X over a field k. Let us start with specializing the discussion in Sect. 10.4 to the case of a *line bundle* $E = L$.

Examples 10.21

(0) For any line bundle L over X, the module of sections $L(X)$ in an invertible finitely generated projective $O(X)$-module: one has $L(X)\otimes_{O(X)}L^*(X) \simeq O(X)$, where L^* is the dual line bundle to L. Notice that for any vector bundle E one has $\mathrm{End}(E) = E\otimes E^*$, so $\mathrm{End}(E)(X) = E(X)\otimes_{O(X)}E^*(X)$ (where E^* is the dual vector bundle to E, so $E^*(X) = \mathrm{Hom}_{O(X)}(E(X), O(X))$). For a line bundle L, this means that $\mathrm{End}(L)$ is the trivial line bundle, $\mathrm{End}(L)(X) = O(X)$.

(1) Choose a connection $\nabla = \nabla_L$ in L. Then the induced connection $\nabla_{\mathrm{End}(L)}$ is the trivial (canonical) connection in the trivial line bundle $\mathrm{End}(L)$. Therefore, the graded ring $\Omega(X) \otimes_{O(X)} \mathrm{End}(L)(X)$ is simply the ring of differential forms $\Omega(X)$, and the differential $d_{\nabla_{\mathrm{End}(L)}}$ in the CDG-ring (10.12) is equal to the standard de Rham differential, $d_{\nabla_{\mathrm{End}(L)}} = d$. However, the curvature form $h_{\nabla_L} \in \Omega^2(X)$ of the connection ∇_L in L can well be nontrivial. It is always a *closed* differential form: $d(h_{\nabla_L}) = 0$ in $\Omega^3(X)$ (cf. Equation (iii) in the definition of a CDG-ring in Sect. 3.2).

(2) Replacing ∇_L with another connection ∇_L' in L replaces the differential 2-form h_{∇_L} with $h_{\nabla_L'} = h_{\nabla_L} + d(\alpha)$, where $\alpha = \nabla_L' - \nabla_L \in \Omega^1(X)$ is the change-of-connection 1-form (see the discussion in Sect. 10.4; cf. Sect. 3.5). So the cohomology class of the 2-form h_{∇_L}, viewed as an element of the cohomology ring of the DG-ring $(\Omega(X), d)$, does not depend on a connection ∇_L, but only on the line bundle L itself. It is called the *first Chern class* of L and denoted by $c_1(L) \in H^2(\Omega(X), d)$.

(3) For any integer $n \in \mathbb{Z}$, one can consider the line bundle $L^{\otimes n}$ over X, defined in the obvious way for $n \geq 0$ and by the rule $L^{\otimes n} = L^{* \otimes -n}$ for $n \leq 0$. The curvature form of the induced connection $\nabla_{L^{\otimes n}}$ in $L^{\otimes n}$ is given by the rule $h_{\nabla_{L^{\otimes n}}} = n h_{\nabla_L} \in \Omega^2(X)$.

(4) Now assume for a moment that k is a field of characteristic 0, and let $z \in k$ be any element. Then there is *no* such thing as "a line bundle $L^{\otimes z}$ over X." However, one *can* define a filtered ring $\mathrm{Diff}(X, L^{\otimes z})$ of "differential operators acting in the sections of $L^{\otimes z}$." For this purpose, one simply chooses a connection ∇ in L, considers the right finitely projective Koszul CDG-ring $(\Omega(X), d, zh_\nabla)$, and constructs $\mathrm{Diff}(X, L^{\otimes z})$ as the left finitely projective nonhomogeneous Koszul ring corresponding to $(\Omega(X), d, zh_\nabla)$ under the anti-equivalence of categories from Corollary 4.26.

(5) Replacing the connection ∇ with another connection ∇' in L corresponds to a natural change-of-connection isomorphism between the CDG-rings $(\Omega(X), d, h_\nabla)$ and $(\Omega(X), d, h_{\nabla'})$, and consequently also a change-of-connection isomorphism between the CDG-rings $(\Omega(X), d, zh_\nabla)$ and $(\Omega(X), d, zh_{\nabla'})$. This simply means that one has $h_{\nabla_L'} = h_{\nabla_L} + d(\alpha)$, and consequently $zh_{\nabla_L'} = zh_{\nabla_L} + d(z\alpha)$, where $\alpha = \nabla_L' - \nabla_L$ (as $\alpha^2 = 0$ in $\Omega^2(X)$). An isomorphism of right finitely projective Koszul CDG-rings induces an isomorphism of the corresponding left finitely projective nonhomogeneous Koszul rings; so the filtered ring $(\mathrm{Diff}(X, L^{\otimes z}), F)$ is defined uniquely up to a natural isomorphism.

(6) Similarly, given two line bundles L_1 and L_2 over X and two scalars z_1 and $z_2 \in k$, one can define a filtered ring $\mathrm{Diff}(X, L_1^{\otimes z_1} \otimes L_2^{\otimes z_2})$, etc.

Construction 10.22 Quite generally, let X be a smooth affine variety over a field k of arbitrary characteristic, and let $h \in \Omega^2(X)$ be a closed 2-form (i. e., $d(h) = 0$). Then the triple $(\Omega(X), d, h)$, with the standard de Rham differential d and the curvature element h, is a curved DG-ring (because the graded ring $\Omega(X)$ is graded commutative, so in particular h is a central element in $\Omega(X)$ and the equation (ii) of Sect. 3.2 is satisfied). The curved DG-ring $(\Omega(X), d, h)$ is right finitely projective Koszul. The corresponding left finitely projective nonhomogeneous Koszul ring under the anti-equivalence of categories from Corollary 4.26 is denoted by $\mathrm{Diff}^{\mathrm{cr}}(X, h)$ and called the *ring of twisted crystalline differential operators on X* (twisted by h). (See [5, Section 2] or [29, Chapter II] for a much more abstract discussion of twisted differential operators over nonaffine varieties.)

Explicitly, $\mathrm{Diff}^{\mathrm{cr}}(X, h)$ is the filtered ring generated by the functions on X (placed in the filtration component F_0) and the vector fields on X (placed in the filtration component F_1), subject to the same relations as in Sect. 10.3, except that the relation (10.5) is replaced with

$$ v * w - w * v = [v, w] + \langle v \wedge w, h \rangle, \tag{10.16} $$

where $\langle \, , \, \rangle$ denotes the natural pairing $\Lambda^2_{O(X)}(T(X)) \otimes_{O(X)} \Lambda^2_{O(X)}(T^*(X)) \longrightarrow O(X)$ given by the rule $\langle v \wedge w, \alpha \wedge \beta \rangle = \langle v, \alpha \rangle \langle w, \beta \rangle - \langle v, \beta \rangle \langle w, \alpha \rangle$ for $v, w \in T(X)$ and $\alpha, \beta \in T^*(X)$ (cf. formula (3.2) in Sect. 3.3). So $[v, w] \in T(X)$ is a vector field and $\langle v \wedge w, h \rangle \in O(X)$ is a function on X. It is claimed, based on Theorem 4.25, that the associated graded ring $\mathrm{gr}^F \, \mathrm{Diff}^{\mathrm{cr}}(X, h) = \bigoplus_{n=0}^{\infty} F_n \, \mathrm{Diff}^{\mathrm{cr}}(X, h) / F_{n-1} \, \mathrm{Diff}^{\mathrm{cr}}(X, h)$ is naturally isomorphic to the symmetric algebra $\mathrm{Sym}_{O(X)}(T(X))$. The assumption that $d(h) = 0$ is needed for this to be true.

When two closed 2-forms h' and $h'' \in \Omega^2(X)$ represent the same de Rham cohomology class in $H^2(\Omega(X), d)$, the related filtered rings of twisted (crystalline) differential operators $\mathrm{Diff}^{\mathrm{cr}}(X, h')$ and $\mathrm{Diff}^{\mathrm{cr}}(X, h'')$ are isomorphic, but *not yet naturally isomorphic*. The choice of a 1-form $\alpha \in \Omega^1(X)$ such that $h'' - h' = d(\alpha)$ leads to a concrete isomorphism of filtered rings $\mathrm{Diff}^{\mathrm{cr}}(X, h') \simeq \mathrm{Diff}^{\mathrm{cr}}(X, h'')$.

The results of Chaps. 6 and 7 lead to the following theorem.

Theorem 10.23 *For any smooth affine algebraic variety X over a field k and any closed differential 2-form $h \in \Omega^2(X)$, $d(h) = 0$, the construction of Corollary 6.18 provides a triangulated equivalence between the derived category of right modules over the ring of twisted (crystalline) differential operators $\mathrm{Diff}^{\mathrm{cr}}(X, h)$ and the coderived category of right CDG-modules over the de Rham CDG-algebra $(\Omega(X), d, h)$,*

$$ \mathsf{D}(\mathrm{mod}-\mathrm{Diff}^{\mathrm{cr}}(X, h)) \simeq \mathsf{D}^{\mathrm{co}}(\mathrm{mod}-(\Omega(X), d, h)). \tag{10.17} $$

In the same context, the construction of Corollary 7.14 provides an equivalence between the derived category of left $\mathrm{Diff}^{\mathrm{cr}}(X, h)$-modules and the contraderived category of left CDG-modules over $(\Omega(X), d, h)$,

$$ \mathsf{D}(\mathrm{Diff}^{\mathrm{cr}}(X, h)-\mathrm{mod}) \simeq \mathsf{D}^{\mathrm{ctr}}((\Omega(X), d, h)-\mathrm{mod}). \tag{10.18} $$

Proof The proof is similar to that of Theorems 10.6 and 10.10, and the functors providing the triangulated equivalences are constructed similarly.   □

For any closed 2-form h on X, there is a natural equivalence between the abelian categories of left modules over $\mathrm{Diff}^{\mathrm{cr}}(X, h)$ and right modules over $\mathrm{Diff}^{\mathrm{cr}}(X, -h)$, provided by the mutually inverse functors of tensor product with the $O(X)$-modules of top differential forms and top polyvector fields on X. This Morita equivalence can be obtained as a particular case of Theorem 9.9 with Remark 9.10. The results of Chapter 9 provide a comparison between (10.17) and (10.18).

Theorem 10.24 *For any smooth affine algebraic variety X over a field k and any closed differential 2-form $h \in \Omega^2(X)$, the constructions of Theorem 9.13 (with Remark 9.10) provide a commutative square diagram of triangulated equivalences*

$$\mathsf{D}(\mathrm{mod}\text{–}\mathrm{Diff}^{\mathrm{cr}}(X, -h)) \longleftarrow \mathsf{D}(\mathrm{Diff}^{\mathrm{cr}}(X, h)\text{–}\mathrm{mod})$$

$$\mathsf{D}^{\mathrm{co}}((\Omega(X), d, h)\text{–}\mathrm{mod}) = \mathsf{D}^{\mathrm{ctr}}((\Omega(X), d, h)\text{–}\mathrm{mod}) \qquad (10.19)$$

where the equivalence of derived categories in the upper line is induced by the conversion equivalence of abelian categories $\mathrm{mod}\text{–}\mathrm{Diff}^{\mathrm{cr}}(X, -h) \simeq \mathrm{Diff}^{\mathrm{cr}}(X, h)\text{–}\mathrm{mod}$ *(up to a shift by $[-\dim_k X]$), while the co-contra correspondence in the lower line is the result of Theorem 9.6. The vertical equivalences are* (10.17) *and* (10.18). $\qquad\square$

10.6 Smooth Differential Operators

Let X be a smooth compact real manifold. Denote by $O(X)$ the ring of smooth global functions $X \longrightarrow \mathbb{R}$.

We will consider smooth locally trivial vector bundles E on X. Then the $\mathbb{R}$-vector space of smooth global sections $E(X)$ has a natural $O(X)$-module structure. Moreover, the $O(X)$-module $E(X)$ is finitely generated and projective. The correspondence $E \longmapsto E(X)$ is an equivalence between the category of (smooth locally trivial) vector bundles on X and the category of finitely generated projective $O(X)$-modules. In particular, any short exact sequence of vector bundles over X splits (as one can show using a partition of unity on X). In this sense, smooth compact real manifolds are analogues of *affine* algebraic varieties.

Given a vector bundle E on X, the module of global sections of the dual vector bundle E^* can be obtained as the dual finitely generated projective module, $E^*(X) = \mathrm{Hom}_{O(X)}(E(X), O(X))$. Given two vector bundles E' and E'', the global sections of the tensor product bundle $E' \otimes E''$ are the tensor product of the global sections, $(E' \otimes E'')(X) = E'(X) \otimes_{O(X)} E''(X)$. Similarly, the global sections of the symmetric and exterior powers of a vector bundle E are computable as the symmetric and exterior powers of the global sections, $(\mathrm{Sym}^n E)(X) = \mathrm{Sym}^n_{O(X)}(E(X))$ and $(\Lambda^n E)(X) = \Lambda^n_{O(X)}(E(X))$.

Construction 10.25 The *tangent bundle* T and its dual *cotangent bundle* T^* are smooth locally trivial vector bundles on X which can be described as follows. In a local coordinate system $x_1, \ldots, x_m$ defined on an open subset $U \subset X$, the sections of T (called the *vector fields*) are represented by expressions like $\sum_{i=1}^m f_i \, \partial/\partial x_i$ and the sections of T^* (called the *differential 1-forms*) are represented by expressions like $\sum_{i=1}^m f_i \, dx_i$, where $f_i : U \longrightarrow \mathbb{R}$ are local functions.

The natural map $d : O(X) \longrightarrow T^*(X)$ (the differential) is defined locally by the rule $d(f) = \sum_{i=1}^m \partial f/\partial x_i \, dx_i$. The action of vector fields in the functions (the derivative $v(f)$ of a function f along a vector field v) is given locally by the formula $(\sum_{i=1}^m f_i \, \partial/\partial x_i)(f) = \sum_{i=1}^m f_i \, \partial f/\partial x_i$.

Construction 10.26 Let E' and E'' be two vector bundles on X of ranks (the dimensions of the fibers) r' and $r'' \geq 0$. Then a *differential operator* $D : E'(X) \longrightarrow E''(X)$ of *order* $\leq n$ is an $\mathbb{R}$-linear map such that, for any open subset $U \subset X$ with a coordinate system $x_1, \ldots, x_m$ and any chosen trivializations of E' and E'' over U, the operator D can be expressed locally over U as a linear combination of compositions of at most n partial derivatives $\partial/\partial x_i$ (acting in vector functions $U \longrightarrow \mathbb{R}^{r'}$ component-wise) with $r'' \times r'$-matrices of smooth functions $U \longrightarrow \mathbb{R}$ as the coefficients. Notice that this definition presumes, first of all, that any differential operator is *local*, i. e., for any section $e \in E'(X)$, the restriction of $D(e) \in E''(X)$ onto an open subset $U \subset X$ is determined by the restriction of e onto U.

In particular, a differential operator $D : O(X) \longrightarrow O(X)$ of order $\leq n$ is represented locally over U as a linear combination of compositions of at most n partial derivatives $\partial/\partial x_i$, $1 \leq i \leq m$, with smooth local functions $U \longrightarrow \mathbb{R}$ as the coefficients. A differential operator $E'(X) \longrightarrow E''(X)$ of order 0 is the same thing as a global section of the vector bundle $\mathrm{Hom}(E', E'')$, or an $O(X)$-linear map $E'(X) \longrightarrow E''(X)$.

We denote the $\mathbb{R}$-vector space of smooth differential operators $E'(X) \longrightarrow E''(X)$ by $\mathrm{Diff}(X, E', E'')$, and the subspace of differential operators of order $\leq n$ by $F_n \mathrm{Diff}(X, E', E'') \subset \mathrm{Diff}(X, E', E'')$; so $\mathrm{Diff}(X, E', E'') = \bigcup_{n=0}^\infty F_n \mathrm{Diff}(X, E', E'')$. When the two vector bundles $E' = E = E''$ are the same, we write simply $\mathrm{Diff}(X, E)$. Then $\mathrm{Diff}(X, E)$ is a subalgebra in the $\mathbb{R}$-algebra $\mathrm{End}_{\mathbb{R}}(E(X))$ of all $\mathbb{R}$-linear endomorphisms of the vector space $E(X)$ (with respect to the composition). Furthermore, $\mathrm{Diff}(X, E)$ is a filtered ring with an increasing filtration F. When E is the trivial line bundle on X (so $E(X) = O(X)$), we denote simply by $\mathrm{Diff}(X)$ the filtered ring of differential operators $O(X) \longrightarrow O(X)$.

Construction 10.27 Global sections of the exterior power $\Lambda^n(T^*)$ of the cotangent bundle T^* are called the *differential n-forms* on X. The vector space (in fact, $O(X)$-module) of differential n-forms is denoted by $\Omega^n(X) = \Lambda^n_{O(X)}(T^*(X))$. There is a natural differential operator $d\colon \Omega^n(X) \longrightarrow \Omega^{n+1}(X)$ of order 1, defined for every $n \geq 0$ and called the *de Rham differential*. The operator d is uniquely defined by the property of being and odd derivation of the graded algebra $\Omega(X)$ with $d^2 = 0$ whose restriction to the ring of functions, $d\colon O(X) \longrightarrow \Omega^1(X) = T^*(X)$, is the map described in Construction 10.25.

The left action of $\mathrm{Diff}(X)$ by differential operators in $O(X)$ makes $\mathrm{Diff}(X)$ a left augmented ring over its subring $O(X) = F_0\,\mathrm{Diff}(X)$, in the sense of Sect. 3.8. The associated graded ring $\mathrm{gr}^F\,\mathrm{Diff}(X) = \bigoplus_{n=0}^{\infty} F_n\,\mathrm{Diff}(X)/F_{n-1}\,\mathrm{Diff}(X)$ is naturally isomorphic to the graded ring $\mathrm{Sym}_{O(X)}(T(X))$, which is left and right finitely projective Koszul (according to the discussion in Sect. 10.1). Similarly to Sects. 10.2–10.3, the anti-equivalence of categories from Corollary 4.28 assigns the right finitely projective Koszul DG-ring $(\Omega(X), d)$ to the left augmented left finitely projective nonhomogeneous Koszul ring $\mathrm{Diff}(X)$ (with the filtration F).

Hence the results of Chaps. 6 and 7 lead us to the following theorem. It is a rather rough algebraic version of derived relative nonhomogeneous Koszul duality for the ring of smooth differential operators $\mathrm{Diff}(X)$, in that the base ring of smooth functions $O(X)$ is considered as an abstract ring *without* any additional structures (such as a Banach space norm or a topology). Notice that, similarly to the algebraic examples above, the graded ring $\Omega(X)$ has only finitely many grading components (as $\Omega^n(X) = 0$ for $n > \dim_{\mathbb{R}} X$).

Theorem 10.28 *For any smooth compact real manifold X, the construction of Theorem 6.14 provides a triangulated equivalence between the $\mathrm{Diff}(X)/O(X)$-semicoderived category of right modules over the ring of smooth differential operators $\mathrm{Diff}(X)$ and the coderived category of right DG-modules over the de Rham DG-algebra $(\Omega(X), d)$,*

$$\mathsf{D}^{\mathsf{sico}}_{O(X)}(\mathsf{mod}\text{–}\mathrm{Diff}(X)) \simeq \mathsf{D}^{\mathsf{co}}(\mathsf{mod}\text{–}(\Omega(X), d)), \tag{10.20}$$

while Theorem 6.20 establishes a triangulated equivalence between the derived category of right $\mathrm{Diff}(X)$-modules and the reduced coderived category of right DG-modules over $(\Omega(X), d)$ relative to $O(X)$,

$$\mathsf{D}(\mathsf{mod}\text{–}\mathrm{Diff}(X)) \simeq \mathsf{D}^{\mathsf{co}}_{O(X)\text{-red}}(\mathsf{mod}\text{–}(\Omega(X), d)). \tag{10.21}$$

In the same context, the construction of Theorem 7.11 provides a triangulated equivalence between the $\mathrm{Diff}(X)/O(X)$-semicontraderived category of left modules over the ring $\mathrm{Diff}(X)$ and the contraderived category of left DG-modules over the DG-algebra $(\Omega(X), d)$,

$$\mathsf{D}^{\mathsf{sictr}}_{O(X)}(\mathrm{Diff}(X)\text{–}\mathsf{mod}) \simeq \mathsf{D}^{\mathsf{ctr}}((\Omega(X), d)\text{–}\mathsf{mod}), \tag{10.22}$$

while Theorem 7.16 establishes a triangulated equivalence between the derived category of left Diff(X)-*modules and the reduced contraderived category of left DG-modules over* $(\Omega(X), d)$ *relative to* $O(X)$,

$$D(\text{Diff}(X)\text{–mod}) \simeq D^{\text{ctr}}_{O(X)\text{-red}}((\Omega(X), d)\text{–mod}). \tag{10.23}$$

Proof The proof is similar to that of Theorem 10.6, and the functors providing the triangulated equivalences are constructed similarly. The main difference is that there is *no* claim about the homological dimension of the ring of smooth functions on X.

Specifically, the functors in (10.20–10.21) assign to a complex of right Diff(X)-modules $M^\bullet$ the right DG-module $M^\bullet \otimes_{O(X)} \Lambda_{O(X)}(T(X))$ over the de Rham DG-ring $\Omega(X) = \Lambda_{O(X)}(T^*(X))$, and conversely, to a right DG-module $N^\bullet$ over $\Omega(X)$ the complex of right Diff(X)-modules $N^\bullet \otimes_{O(X)} \text{Diff}(X)$ is assigned. The functors in (10.22–10.23) assign to a complex of left Diff(X)-modules $P^\bullet$ the left DG-module $\Omega(X) \otimes_{O(X)} P^\bullet$ over $\Omega(X)$, and conversely, to a left DG-module $Q^\bullet$ over $\Omega(X)$ the complex of left Diff(X)-modules $\text{Hom}_{O(X)}(\text{Diff}(X), Q^\bullet)$ is assigned. $\square$

There is also a classical natural equivalence between the abelian categories of left and right Diff(X)-modules, Diff(X)–mod $\simeq$ mod–Diff(X), which was discussed in Sect. 0.11. It can be obtained as a particular case of Theorem 9.9 with Remark 9.10 (since in fact $\Omega(X)$ is a relatively Frobenius Koszul graded ring).

More generally, the associated graded ring $\text{gr}^F \text{Diff}(X, E) = \bigoplus_{n=0}^{\infty} F_n \text{Diff}(X, E)/ F_{n-1} \text{Diff}(X, E)$ is naturally isomorphic to the tensor product $\text{End}(E)(X) \otimes_{O(X)} \text{Sym}_{O(X)}(T(X))$ of the $O(X)$-algebra $\text{End}(E)(X) = \text{Hom}_{O(X)}(E(X), E(X))$ of sections of the vector bundle $\text{End}(E)$ and the graded $O(X)$-algebra $\text{Sym}_{O(X)}(T(X))$. This graded ring is left and right finitely projective Koszul over its degree-zero component $\text{End}(E)(X)$. So the filtered ring $(\text{Diff}(X, E), F)$ is left finitely projective nonhomogeneous Koszul. In order to describe the corresponding CDG-ring, the notion of a connection in a smooth vector bundle E over X is needed.

Definition 10.29 A (*smooth*) *connection* ∇ in a smooth vector bundle E is a map assigning to every smooth vector field $v \in T(X)$ and any smooth section $e \in E(X)$ a smooth section $\nabla_v(e) \in E(X)$ such that the equations (i) and (ii) from Definition 10.14.C in Sect. 10.4 are satisfied. Similarly to the algebraic case discussed in Sect. 10.4, a connection ∇ in E can be interpreted as an $\mathbb{R}$-linear map $E(X) \longrightarrow \Omega^1(X) \otimes_{O(X)} E(X)$ defined by the rule $\langle v, \nabla(e) \rangle = \nabla_v(e)$. In fact, it is a differential operator $E(X) \longrightarrow (T^* \otimes E)(X)$ of order 1.

Construction 10.30 Using a partition of unity, one can show that a smooth connection exists in any smooth vector bundle over a smooth manifold. The trivial line bundle on X has a canonical *trivial connection* defined by the rule $\nabla_v(f) = v(f)$ for all $v \in T(X)$ and $f \in O(X)$. If vector bundles E' and E'' over X are endowed with connections ∇'

and ∇'', then the vector bundles $E' \oplus E''$, $E' \otimes E''$, and $\mathrm{Hom}(E', E'')$ acquire the induced connections (similarly to Construction 10.15).

Construction 10.31 Similarly to Construction 10.16 in Sect. 10.4, a connection ∇ in a smooth vector bundle E induces an odd derivation d_∇ of degree 1 on the graded left $\Omega(X)$-module $\Omega(X) \otimes_{O(X)} E(X)$ compatible with the de Rham differential d on the graded ring $\Omega(X)$, in the sense of Sect. 6.1. The map $d_\nabla \colon \Omega^n(X) \otimes_{O(X)} E(X) \longrightarrow \Omega^{n+1}(X) \otimes_{O(X)} E(X)$, $n \geq 1$ (in fact, a differential operator of order 1 acting between the global sections of the vector bundles $\Lambda^n(T^*) \otimes E$ and $\Lambda^{n+1}(T^*) \otimes E$ on X) is given by the formula (10.11).

The *curvature 2-form* h_∇ of a smooth connection ∇ in E is an element of the vector space (or $O(X)$-module) $\Omega^2(X) \otimes_{O(X)} \mathrm{End}(E)(X)$ of *differential 2-forms on X with the coefficients in* $\mathrm{End}(E)$, that is, global sections of the vector bundle $\Lambda^2(T^*) \otimes \mathrm{End}(E)$ on X. The element $h_\nabla \in \Omega^2(X) \otimes_{O(X)} \mathrm{End}(E)$ is defined by the identity

$$\nabla_v(\nabla_w(e)) - \nabla_w(\nabla_v(e)) = \nabla_{[v,w]}(e) + \langle v \wedge w, h \rangle(e)$$

holding for all $v, w \in T(X)$ and $e \in E(X)$. Here $[v, w] \in T(X)$ is the commutator of the vector fields v and w on X (defined as in Construction 10.8 in Sect. 10.3), while $\langle \, , \, \rangle$ denotes the $O(X)$-linear map $\Lambda^2_{O(X)}(T(X)) \otimes_{O(X)} \Omega^2(X) \otimes_{O(X)} \mathrm{End}(E)(X) \longrightarrow \mathrm{End}(E)(X)$ induced by the pairing $\Lambda^2_{O(X)}(T(X)) \otimes_{O(X)} \Lambda^2_{O(X)}(T^*(X)) \longrightarrow O(X)$ from Construction 10.22 in Sect. 10.5.

The choice of a smooth connection $\nabla = \nabla_E$ in E defines a left $\mathrm{End}(E)(X)$-submodule of strict generators in $F_1 \mathrm{Diff}(X, E)$ (in the sense of Sect. 3.3) consisting of all the $\mathbb{R}$-linear combinations of differential operators of the form $g \nabla_v \colon E(X) \longrightarrow E(X)$, where $g \in \mathrm{End}(E)(X)$ and $v \in T(X)$. Let $\nabla_{\mathrm{End}(E)}$ denote the connection in the vector bundle $\mathrm{End}(E)$ induced by the connection ∇ in E.

Similarly to the algebraic case considered in Sect. 10.4, the anti-equivalence of categories from Corollary 4.26 assigns the right finitely projective Koszul CDG-ring

$$(\Omega(X) \otimes_{O(X)} \mathrm{End}(E)(X), \, d_{\nabla_{\mathrm{End}(E)}}, \, h_{\nabla_E}) \tag{10.24}$$

as in the formula (10.12) to the left finitely projective nonhomogeneous Koszul ring $\mathrm{Diff}(X, E)$ with the filtration F. Hence we obtain the following Koszul duality theorem.

Theorem 10.32 *For any smooth compact real manifold X and vector bundle E on X, the construction of Theorem 6.14 provides a triangulated equivalence between the* $\mathrm{Diff}(X, E)/\mathrm{End}(E)(X)$*-semicoderived category of right modules over the ring of smooth differential operators* $\mathrm{Diff}(X, E)$ *acting in the sections of E and the coderived category* *of right CDG-modules over the CDG-algebra (10.24) of differential forms on X with the coefficients in* End(E),

$$\mathsf{D}^{\mathsf{sico}}_{\mathrm{End}(E)(X)}(\mathrm{mod-Diff}(X, E))$$

$$\simeq \mathsf{D}^{\mathsf{co}}(\mathrm{mod-}(\Omega(X) \otimes_{O(X)} \mathrm{End}(E)(X), d_{\nabla_{\mathrm{End}(E)}}, h_{\nabla_E})), \qquad (10.25)$$

while Theorem 6.20 establishes a triangulated equivalence between the derived category of right Diff(X, E)*-modules and the reduced coderived category of right CDG-modules over the CDG-ring (10.24) relative to the ring* End(E)(X),

$$\mathsf{D}(\mathrm{mod-Diff}(X, E))$$

$$\simeq \mathsf{D}^{\mathsf{co}}_{\mathrm{End}(E)(X)\text{-red}}(\mathrm{mod-}(\Omega(X) \otimes_{O(X)} \mathrm{End}(E)(X), d_{\nabla_{\mathrm{End}(E)}}, h_{\nabla_E})). \qquad (10.26)$$

In the same context, the construction of Theorem 7.11 provides a triangulated equivalence between the Diff(X, E)/ End(E)(X)*-semicontraderived category of left modules over the ring* Diff(X, E) *and the contraderived category of left CDG-modules over the CDG-algebra (10.24),*

$$\mathsf{D}^{\mathsf{sictr}}_{\mathrm{End}(E)(X)}(\mathrm{Diff}(X, E)\text{-mod})$$

$$\simeq \mathsf{D}^{\mathsf{ctr}}((\Omega(X) \otimes_{O(X)} \mathrm{End}(E)(X), d_{\nabla_{\mathrm{End}(E)}}, h_{\nabla_E})\text{-mod}), \qquad (10.27)$$

while Theorem 7.16 establishes a triangulated equivalence between the derived category of left Diff(X, E)*-modules and the reduced contraderived category of left CDG-modules over the CDG-ring (10.24) relative to the ring* End(E)(X),

$$\mathsf{D}(\mathrm{Diff}(X, E)\text{-mod})$$

$$\simeq \mathsf{D}^{\mathsf{ctr}}_{\mathrm{End}(E)(X)\text{-red}}((\Omega(X) \otimes_{O(X)} \mathrm{End}(E)(X), d_{\nabla_{\mathrm{End}(E)}}, h_{\nabla_E})\text{-mod}). \qquad (10.28)$$

Proof Similar to the proof of Theorem 10.20, with the difference that *no* claim about the homological dimension of the base ring End(E)(X) is presumed. $\square$

For example, the equivalences of categories (10.27–10.28) assign the left CDG-module $(\Omega(X) \otimes_{O(X)} E(X), d_{\nabla_E})$ (as in (10.13)) over the CDG-ring (10.24) to the left Diff(X, E)-module $E(X)$.

Construction 10.33 Similarly to Sect. 10.5, for any closed 2-form $h \in \Omega^2(X)$ (that is a smooth differential 2-form such that $d(h) = 0$), the triple $(\Omega(X), d, h)$ is a right finitely projective Koszul CDG-ring. The corresponding left finitely projective nonhomogeneous Koszul filtered ring Diff(X, h), defined by the relations (10.4) and (10.16), is called the *ring of twisted smooth differential operators on* X (twisted by h). It is claimed, on the basis of Theorem 4.25, that the associated graded ring $\mathrm{gr}^F \, \mathrm{Diff}^{\mathrm{cr}}(X, h) = \bigoplus_{n=0}^{\infty} F_n \, \mathrm{Diff}^{\mathrm{cr}}(X, h) / F_{n-1} \, \mathrm{Diff}^{\mathrm{cr}}(X, h)$ is naturally isomorphic to the symmetric algebra $\mathrm{Sym}_{O(X)}(T(X))$.

Theorem 10.34 *For any smooth compact real manifold* X *and any closed differential 2-form* $h \in \Omega^2(X)$, $d(h) = 0$, *the construction of Theorem 6.14 provides a triangulated equivalence between the* $\mathrm{Diff}(X, h)/O(X)$-*semicoderived category of right modules over the ring of twisted smooth differential operators* $\mathrm{Diff}(X, h)$ *and the coderived category of right CDG-modules over the de Rham CDG-algebra* $(\Omega(X), d, h)$,

$$\mathsf{D}_{O(X)}^{\mathrm{sico}}(\mathrm{mod}\text{–}\mathrm{Diff}(X, h)) \simeq \mathsf{D}^{\mathrm{co}}(\mathrm{mod}\text{–}(\Omega(X), d, h)), \tag{10.29}$$

while Theorem 6.20 establishes a triangulated equivalence between the derived category of right $\mathrm{Diff}(X, h)$-*modules and the reduced coderived category of right CDG-modules over* $(\Omega(X), d, h)$ *relative to* $O(X)$,

$$\mathsf{D}(\mathrm{mod}\text{–}\mathrm{Diff}(X, h)) \simeq \mathsf{D}_{O(X)\text{-red}}^{\mathrm{co}}(\mathrm{mod}\text{–}(\Omega(X), d, h)). \tag{10.30}$$

In the same context, the construction of Theorem 7.11 provides a triangulated equivalence between the $\mathrm{Diff}(X, h)/O(X)$-*semicontraderived category of left modules over the ring* $\mathrm{Diff}(X, h)$ *and the contraderived category of left CDG-modules over the CDG-algebra* $(\Omega(X), d, h)$,

$$\mathsf{D}_{O(X)}^{\mathrm{sictr}}(\mathrm{Diff}(X, h)\text{–}\mathrm{mod}) \simeq \mathsf{D}^{\mathrm{ctr}}((\Omega(X), d, h)\text{–}\mathrm{mod}), \tag{10.31}$$

while Theorem 7.16 establishes a triangulated equivalence between the derived category of left $\mathrm{Diff}(X, h)$-*modules and the reduced contraderived category of left CDG-modules over* $(\Omega(X), d, h)$ *relative to* $O(X)$,

$$\mathsf{D}(\mathrm{Diff}(X, h)\text{–}\mathrm{mod}) \simeq \mathsf{D}_{O(X)\text{-red}}^{\mathrm{ctr}}((\Omega(X), d, h)\text{–}\mathrm{mod}). \tag{10.32}$$

$\square$

Similarly to Sect. 10.5, there is a natural equivalence between the abelian categories of left modules over $\mathrm{Diff}(X, h)$ and right modules over $\mathrm{Diff}(X, -h)$, provided by the mutually inverse functors of tensor product with the $O(X)$-modules $\Omega^m(X)$ and $\Lambda_{O(X)}^m(T(X))$, where $m = \dim_{\mathbb{R}} X$. This equivalence can be obtained as a particular case of the construction of Theorem 9.9 with Remark 9.10.

10.7   Dolbeault Differential Operators

For a reference on the basics of complex analytic geometry, see, e.g., [79, Section I.2].

In this section, we denote by i a chosen imaginary unit (a square root of -1) in the field of complex numbers $\mathbb{C}$. The complex conjugation map is denoted by $z = x + iy \longmapsto \bar{z} = x - iy$ (where $x, y \in \mathbb{R}$).

Construction 10.35 For any vector space V over $\mathbb{C}$, we denote by $V_{\mathbb{R}}$ the underlying real vector space of V. So $\dim_{\mathbb{R}} V_{\mathbb{R}} = 2 \dim_{\mathbb{C}} V$. Similarly, for any vector space U over $\mathbb{R}$, we denote by $U_{\mathbb{C}}$ the $\mathbb{C}$-vector space $U_{\mathbb{C}} = \mathbb{C} \otimes_{\mathbb{R}} U$. So $\dim_{\mathbb{C}} U_{\mathbb{C}} = \dim_{\mathbb{R}} U$. We denote the elements of $U_{\mathbb{C}}$ by $u + iv$, where $u, v \in U$ and i is the imaginary unit.

We also denote by $\overline{V}$ the vector space V with the conjugate complex structure. Denoting by $\bar{v} \in \overline{V}$ the element corresponding to an element $v \in V$, the $\mathbb{C}$-vector space structure on $\overline{V}$ is defined by the rule $z\bar{v} = \overline{\bar{z}v}$ for all $z \in \mathbb{C}$ and $v \in V$, where $z \longmapsto \bar{z} \colon \mathbb{C} \longrightarrow \mathbb{C}$ is the complex conjugation. The vector space $U_{\mathbb{C}}$ is endowed with a natural isomorphism of $\mathbb{C}$-vector spaces $\overline{U_{\mathbb{C}}} \simeq U_{\mathbb{C}}$; the isomorphism is provided by the complex conjugation map $w \longmapsto \overline{w} \colon U_{\mathbb{C}} \longrightarrow U_{\mathbb{C}}$ defined by the obvious rule $\overline{u + iv} = u - iw$ for all $u, v \in U$.

Construction 10.36 For any $\mathbb{C}$-vector space V, the $\mathbb{C}$-vector space $(V_{\mathbb{R}})_{\mathbb{C}}$ decomposes canonically into a direct sum of two $\mathbb{C}$-vector spaces, one of them isomorphic naturally to V and the other one to $\overline{V}$. Specifically, denote by $J \colon V \longrightarrow V$ the action of the imaginary unit $i \in \mathbb{C}$ in V. Let $V^{(1,0)} \subset (V_{\mathbb{R}})_{\mathbb{C}}$ denote the set of all elements of the form $v - iJv$, where $v \in V$, and let $V^{(0,1)} \subset (V_{\mathbb{R}})_{\mathbb{C}}$ denote the set of all elements of the form $v + iJv$, where $v \in V$. Then both $V^{(1,0)}$ and $V^{(0,1)}$ are $\mathbb{C}$-vector subspaces in $(V_{\mathbb{R}})_{\mathbb{C}}$, and $(V_{\mathbb{R}})_{\mathbb{C}} = V^{(1,0)} \oplus V^{(0,1)}$. Furthermore, there are natural isomorphisms of $\mathbb{C}$-vector spaces $v \longmapsto v - iJv \colon V \longrightarrow V^{(1,0)}$ and $\bar{v} \longmapsto v + iJv \colon \overline{V} \longrightarrow V^{(0,1)}$. The complex conjugation map $w \longmapsto \overline{w} \colon (V_{\mathbb{R}})_{\mathbb{C}} \longrightarrow (V_{\mathbb{R}})_{\mathbb{C}}$ switches the two sides of this direct sum decomposition, taking $V^{(1,0)}$ into $V^{(0,1)}$ and $V^{(0,1)}$ into $V^{(1,0)}$.

Construction 10.37 For any $\mathbb{C}$-vector space V, one can consider the dual $\mathbb{C}$-vector space $\mathrm{Hom}_{\mathbb{C}}(V, \mathbb{C})$. One can also consider the dual $\mathbb{R}$-vector space $\mathrm{Hom}_{\mathbb{R}}(V, \mathbb{R})$. Let us endow the $\mathbb{R}$-vector space $\mathrm{Hom}_{\mathbb{R}}(V, \mathbb{R})$ with a $\mathbb{C}$-vector space structure by the rule $\langle v, Jf \rangle = \langle Jv, f \rangle$ for all $v \in V$ and $f \in \mathrm{Hom}_{\mathbb{R}}(V, \mathbb{R})$ (where $\langle \, , \, \rangle \colon V \otimes_{\mathbb{R}} \mathrm{Hom}_{\mathbb{R}}(V, \mathbb{R}) \longrightarrow \mathbb{R}$ is the natural pairing). Then there is a natural isomorphism of $\mathbb{C}$-vector spaces $\mathrm{Hom}_{\mathbb{R}}(V, \mathbb{R}) \simeq \mathrm{Hom}_{\mathbb{C}}(V, \mathbb{C})$ given by the rule $\mathrm{Hom}_{\mathbb{R}}(V, \mathbb{R}) \ni f \longmapsto \hat{f} \in \mathrm{Hom}_{\mathbb{C}}(V, \mathbb{C})$ with $\langle\langle v, \hat{f} \rangle\rangle = \langle v, f \rangle - i \langle Jv, f \rangle \in \mathbb{C}$ for all $v \in V$ (where $\langle\langle \, , \, \rangle\rangle \colon V \otimes_{\mathbb{C}} \mathrm{Hom}_{\mathbb{C}}(V, \mathbb{C}) \longrightarrow \mathbb{C}$ is the notation for the $\mathbb{C}$-valued pairing).

Let X be a complex manifold (which we will eventually assume to be compact). We denote the underlying smooth real manifold of X by $X_{\mathbb{R}}$, and refer to the previous Sect. 10.6 for the discussion of differential operators and differential forms on $X_{\mathbb{R}}$. For any vector bundle V on $X_{\mathbb{R}}$ and a point $s \in X$, we denote by V_s the fiber of V at s. So V_s is a finite-dimensional (real or complex) vector space.

Construction 10.38 Let T and T^* denote the tangent and cotangent bundles to $X_{\mathbb{R}}$; so T_s is the tangent space to $X_{\mathbb{R}}$ at s and $T_s^* = \operatorname{Hom}_{\mathbb{R}}(T_s, \mathbb{R})$ is the cotangent space for every $s \in X$. For every point $s \in X$, the $\mathbb{R}$-vector space T_s has a natural $\mathbb{C}$-vector space structure, which is constructed as follows.

Let $z_1, \ldots, z_m$ be a holomorphic local coordinate system in an open subset $U \subset X$, $s \in U$. Denote by x_k and y_k the real and imaginary parts of the complex variable z_k, $1 \le k \le m$; so $z_k = x_k + i y_k$ and x_k, y_k are $\mathbb{R}$-valued functions on U. Let $w_1, \ldots, w_m$ be another holomorphic coordinate system in U; put $w_l = u_l + i v_l$, where u_l, v_l are $\mathbb{R}$-valued local functions for all $1 \le l \le m$.

Then, for every fixed point $s \in U \subset X$, the $2m \times 2m$-matrix with real entries

$$\begin{pmatrix} \frac{\partial u_l}{\partial x_k}(s) & \frac{\partial u_l}{\partial y_k}(s) \\ \frac{\partial v_l}{\partial x_k}(s) & \frac{\partial v_l}{\partial y_k}(s) \end{pmatrix}$$

can be obtained by applying the functor $V \longmapsto V_{\mathbb{R}}$ to the $m \times m$-matrix with complex entries $\left(\frac{\partial w_l}{\partial z_k}(s) \right)$. Consequently, the $\mathbb{C}$-vector space structure on the vector space T_s defined by the rules $J\left(\frac{\partial}{\partial x_k} \right) = \frac{\partial}{\partial y_k}$ and $J\left(\frac{\partial}{\partial y_k} \right) = -\frac{\partial}{\partial x_k}$ does not depend on the choice of a holomorphic coordinate system $(z_k)_{k=1}^m$ in a neighborhood of s in X.

Construction 10.39 Following the rule of Construction 10.37 for the passage to the dual vector space, there is the induced $\mathbb{C}$-vector space structure on the $\mathbb{R}$-vector space T_s^* for every $s \in X$, defined in holomorphic local coordinates z_k by the formulas $J(dx_k) = -dy_k$ and $J(dy_k) = dx_k$. Notice that T_s^* is the vector space of all differentials $d_s f$ at the point $s \in X$ of *real-valued* smooth functions $f: X \longrightarrow \mathbb{R}$; still it has a natural $\mathbb{C}$-vector space structure. If $g: X \longrightarrow \mathbb{C}$ is a complex-valued smooth function, then the differential $d_s g$ belongs to the *complexified* cotangent space $T_{s,\mathbb{C}}^* = \mathbb{C} \otimes_{\mathbb{R}} T_s^*$.

According to Construction 10.36, there is a natural direct sum decomposition $T_{s,\mathbb{C}}^* = T_s^{*(1,0)} \oplus T_s^{*(0,1)}$. Here the *holomorphic cotangent space* $T_s^{*(1,0)}$ to X at s is a $\mathbb{C}$-vector space with a basis $dz_k = dx_k + i\, dy_k$, $1 \le k \le m$. It consists of the differentials $d_s h$ of holomorphic local functions $h: U \longrightarrow \mathbb{C}$ at the point $s \in U$. The *anti-holomorphic cotangent space* $T_s^{*(0,1)}$ is a $\mathbb{C}$-vector space with a basis $d\bar{z}_k = dx_k - i\, dy_k$. It consists of the differentials $d_s \bar{h}$ of anti-holomorphic local functions $\bar{h}: U \longrightarrow \mathbb{C}$ (where h is holomorphic).

Construction 10.40 The complexified cotangent space $T_{s,\mathbb{C}}^*$ is the dual $\mathbb{C}$-vector space to the complexified tangent space $T_{s,\mathbb{C}} = \mathbb{C} \otimes_{\mathbb{R}} T_s$. The natural pairing $\langle\,,\,\rangle: T_{s,\mathbb{C}} \otimes_{\mathbb{C}} T_{s,\mathbb{C}}^* \longrightarrow \mathbb{C}$ is constructed as the unique $\mathbb{C}$-linear extension of the natural $\mathbb{R}$-linear pairing $\langle\,,\,\rangle: T_s \otimes_{\mathbb{R}} T_s^* \longrightarrow \mathbb{R}$. The dual basis to $(dz_k, d\bar{z}_k \in T_{s,\mathbb{C}}^*)_{k=1}^m$ with respect to this pairing is denoted by $(\partial/\partial z_k, \partial/\partial \bar{z}_k \in T_{s,\mathbb{C}})_{k=1}^m$; so $\langle \partial/\partial z_k, dz_l \rangle = \delta_{k,l} = \langle \partial/\partial \bar{z}_k, d\bar{z}_l \rangle$ and $\langle \partial/\partial z_k, d\bar{z}_l \rangle = 0 = \langle \partial/\partial \bar{z}_k, dz_l \rangle$ for $1 \leq k, l \leq m$. Then one has $dg = \sum_{k=1}^m \frac{\partial g}{\partial z_k} dz_k + \sum_{k=1}^m \frac{\partial g}{\partial \bar{z}_k} d\bar{z}_k$ for any smooth function $g \colon U \longrightarrow \mathbb{C}$. Explicitly, $\partial/\partial z_k = \frac{1}{2}(\partial/\partial x_k - i\,\partial/\partial y_k)$ and $\partial/\partial \bar{z}_k = \frac{1}{2}(\partial/\partial x_k + i\,\partial/\partial y_k)$.

According to the same discussion in Construction 10.36, there is also a natural direct sum decomposition $T_{s,\mathbb{C}} = T_s^{(1,0)} \oplus T_s^{(0,1)}$. Here the *holomorphic tangent space* $T_s^{(1,0)}$ to X at s is a $\mathbb{C}$-vector space with a basis $(\partial/\partial z_k)_{k=1}^m$ and the *anti-holomorphic tangent space* $T_s^{(0,1)}$ is a $\mathbb{C}$-vector space with a basis $(\partial/\partial \bar{z}_k)_{k=1}^m$. Smooth sections of the vector bundle $T^{(1,0)}$ on X are called $(1,0)$-*vector fields*, and smooth sections of the vector bundle $T^{(0,1)}$ on X are called $(0,1)$-*vector fields*. Holomorphic local functions $h \colon U \longrightarrow \mathbb{C}$ are distinguished by the condition that $v(h) = 0$ in U for any $(0,1)$-vector field $v \in T^{(0,1)}(X)$. Similarly, anti-holomorphic local functions $\bar{h} \colon U \longrightarrow \mathbb{C}$ are characterized by the condition that $u(\bar{h}) = 0$ in U for any $(1,0)$-vector field $u \in T^{(1,0)}(X)$.

Lemma 10.41 *For any two vector spaces V and W over a field k, and for every integer $n \geq 0$, there is a natural isomorphism of k-vector spaces*

$$\Lambda_k^n(V \oplus W) \simeq \bigoplus_{\substack{p+q=n \\ }}^{p,q \geq 0} \Lambda_k^p(V) \otimes_k \Lambda_k^q(W).$$

Here the natural map $\Lambda_k^p(V) \otimes_k \Lambda_k^q(W) \longrightarrow \Lambda_k^n(V \oplus W)$ is constructed as the composition $\Lambda_k^p(V) \otimes_k \Lambda_k^q(W) \longrightarrow \Lambda_k^p(V \oplus W) \otimes_k \Lambda_k^q(V \oplus W) \longrightarrow \Lambda_k^{p+q}(V \oplus W)$ of the map induced by the direct summand inclusion maps $V \longrightarrow V \oplus W \longleftarrow W$ and the multiplication map in the exterior algebra $\Lambda_k(V \oplus W)$. □

Construction 10.42 For every $n \geq 0$, the complexified space of exterior forms $\Lambda_{\mathbb{R}}^n(T_s^*)_{\mathbb{C}}$ at a point s in $X_{\mathbb{R}}$ decomposes naturally as

$$\mathbb{C} \otimes_{\mathbb{R}} \Lambda_{\mathbb{R}}^n(T_s^*) \simeq \Lambda_{\mathbb{C}}^n(T_{s,\mathbb{C}}^*) \simeq \Lambda_{\mathbb{C}}^n(T_s^{*(1,0)} \oplus T_s^{*(0,1)})$$

$$\simeq \bigoplus_{\substack{p+q=n \\ }}^{p,q \geq 0} \Lambda_{\mathbb{C}}^p(T_s^{*(1,0)}) \otimes_{\mathbb{C}} \Lambda_{\mathbb{C}}^q(T_s^{*(0,1)})$$

by Lemma 10.41. This is a direct sum decomposition of the smooth complex vector bundle $\Lambda^n(T^*)_{\mathbb{C}}$ on $X_{\mathbb{R}}$. The space of global sections decomposes accordingly,

$$\Omega^n(X_{\mathbb{R}})_{\mathbb{C}} \simeq \Lambda^n(T^*)_{\mathbb{C}}(X_{\mathbb{R}}) \simeq \bigoplus_{\substack{p+q=n \\ }}^{p,q \geq 0} \Omega^{(p,q)}(X). \tag{10.33}$$

This is a decomposition of the $\mathbb{C}$-vector space $\Omega^n(X_{\mathbb{R}})_{\mathbb{C}}$ of smooth $\mathbb{C}$-valued differential n-forms on $X_{\mathbb{R}}$ into the direct sum of the spaces of *differential (p,q)-forms*. In holomorphic local coordinates z_k, a differential (p,q)-form is an expression like

$$\sum_{1 \leq k_1 < \cdots < k_p \leq m} \sum_{1 \leq l_1 < \cdots < l_q \leq m} f_{k_1,\ldots,k_p;l_1,\ldots,l_q} dz_{k_1} \wedge \cdots \wedge dz_{k_p} \wedge d\bar{z}_{l_1} \wedge \cdots \wedge d\bar{z}_{l_q},$$

with p holomorphic differentials $dz_{k_1}, \ldots, dz_{k_p}$ and q anti-holomorphic differentials $d\bar{z}_{l_1}, \ldots, d\bar{z}_{l_q}$ in the exterior product. Here the coefficients $f_{k_1,\ldots,k_p;l_1,\ldots,l_q}$ are smooth complex-valued local functions $U \longrightarrow \mathbb{C}$. In fact, this is even a direct sum decomposition of $\Omega^n(X_{\mathbb{R}})_{\mathbb{C}}$ as a module over the ring $O(X_{\mathbb{R}})_{\mathbb{C}}$ of smooth complex-valued global functions on $X_{\mathbb{R}}$. This direct sum decomposition makes $\Omega(X_{\mathbb{R}})_{\mathbb{C}}$ a *bigraded algebra* over the ring $O(X_{\mathbb{R}})_{\mathbb{C}}$.

Construction 10.43 The de Rham differential d on the graded algebra of differential forms $\Omega(X_{\mathbb{R}})$ was discussed in Sect. 10.6 (see Construction 10.27). Taking the tensor product with $\mathbb{C}$ over $\mathbb{R}$, we obtain an odd derivation of degree 1 on the graded algebra of $\mathbb{C}$-valued differential forms $\Omega(X_{\mathbb{R}})_{\mathbb{C}}$; we denote this differential also by d. With respect to the direct sum decomposition (10.33), the differential $d \colon \Omega^n(X_{\mathbb{R}})_{\mathbb{C}} \longrightarrow \Omega^{n+1}(X_{\mathbb{R}})_{\mathbb{C}}$ has two components:

$$\partial \colon \Omega^{p,q}(X) \longrightarrow \Omega^{p+1,q}(X) \quad \text{and} \quad \bar{\partial} \colon \Omega^{p,q}(X) \longrightarrow \Omega^{p,q+1}(X),$$

making $\Omega^{\bullet,\bullet}(X)$ a *bicomplex*. In particular, the differential $d \colon O(X_{\mathbb{R}})_{\mathbb{C}} \longrightarrow \Omega^1(X_{\mathbb{R}})_{\mathbb{C}} \simeq \Omega^{(1,0)}(X) \oplus \Omega^{(0,1)}(X)$ decomposes as $d = \partial + \bar{\partial}$, where in holomorphic local coordinates z_k one has $\partial(f) = \sum_{k=1}^{m} \frac{\partial f}{\partial z_k} dz_k$ and $\bar{\partial}(f) = \sum_{k=1}^{m} \frac{\partial f}{\partial \bar{z}_k} d\bar{z}_k$ for any smooth complex-valued function f.

The $p = 0$ part of the bicomplex $\Omega^{p,q}(X)$

$$0 \longrightarrow \Omega^{0,0}(X) \xrightarrow{\bar{\partial}} \Omega^{0,1}(X) \xrightarrow{\bar{\partial}} \cdots \xrightarrow{\bar{\partial}} \Omega^{0,m}(X) \longrightarrow 0 \tag{10.34}$$

is called the *Dolbeault complex* of a complex manifold X. Here $m = \dim_{\mathbb{C}} X$ is the dimension of X as a complex manifold. The Dolbeault complex $\Omega^{0,\bullet}(X)$ is a DG-algebra over the field $\mathbb{C}$.

Remarks 10.44 The degree-zero component $\Omega^{0,0}(X)$ of the Dolbeault complex is the $\mathbb{C}$-algebra $O(X_{\mathbb{R}})_{\mathbb{C}}$ of smooth $\mathbb{C}$-valued functions on $X_{\mathbb{R}}$. The kernel $H^0_{\bar{\partial}}(\Omega^{0,\bullet}(X)) = O(X)$ of the differential $\bar{\partial} \colon \Omega^{0,0}(X) \longrightarrow \Omega^{0,1}(X)$ in the Dolbeault complex is the subalgebra $O(X) \subset O(X_{\mathbb{R}})_{\mathbb{C}}$ of *holomorphic* global functions $X \longrightarrow \mathbb{C}$. For comparison, the kernel $H^0_d(\Omega^{\bullet}(X_{\mathbb{R}}))$ of the differential $d \colon \Omega^0(X_{\mathbb{R}}) \longrightarrow \Omega^1(X_{\mathbb{R}})$ in the de Rham complex is the subalgebra of *locally constant* global functions in the $\mathbb{R}$-algebra $O(X_{\mathbb{R}})$ of smooth functions $X_{\mathbb{R}} \longrightarrow \mathbb{R}$.

Construction 10.45 Consider the ring (or $\mathbb{R}$-algebra) of differential operators $\mathrm{Diff}(X_{\mathbb{R}})$ on the smooth real manifold $X_{\mathbb{R}}$, as discussed in Sect. 10.6. The tensor product $\mathbb{C} \otimes_{\mathbb{R}} \mathrm{Diff}(X_{\mathbb{R}}) = \mathrm{Diff}(X_{\mathbb{R}})_{\mathbb{C}}$ is a $\mathbb{C}$-algebra acting naturally in the $\mathbb{C}$-vector space of smooth complex-valued global functions $O(X_{\mathbb{R}})_{\mathbb{C}}$. We are interested in the following subring (or $\mathbb{C}$-subalgebra) $\mathrm{Diff}^{\bar{\partial}}(X) \subset \mathrm{Diff}(X_{\mathbb{R}})_{\mathbb{C}}$, which we call the ring of $\bar{\partial}$-*differential operators* (or *Dolbeault differential operators*) on a complex manifold X.

By the definition, $\text{Diff}^{\bar{\partial}}(X)$ is the subring generated by the subring of smooth $\mathbb{C}$-valued functions $O(X_{\mathbb{R}})_{\mathbb{C}} \subset \text{Diff}(X_{\mathbb{R}})_{\mathbb{C}}$ and the left $O(X_{\mathbb{R}})_{\mathbb{C}}$-submodule of smooth $(0, 1)$-vector fields $T^{(0,1)}(X) \subset T(X)_{\mathbb{C}} \subset \text{Diff}(X_{\mathbb{R}})_{\mathbb{C}}$ in the ring of smooth $\mathbb{C}$-valued differential operators $\text{Diff}(X_{\mathbb{R}})_{\mathbb{C}}$ on $X_{\mathbb{R}}$. The increasing filtration F by the order of differential operators on $\text{Diff}^{\bar{\partial}}(X)$ is induced by the filtration F on $\text{Diff}(X_{\mathbb{R}})_{\mathbb{C}}$ (which, in turn, is induced by the filtration F on $\text{Diff}(X_{\mathbb{R}})$). Alternatively, one can say that the filtration F on $\text{Diff}^{\bar{\partial}}(X)$ is generated by $F_1 \text{Diff}^{\bar{\partial}}(X) = O(X_{\mathbb{R}})_{\mathbb{C}} \oplus T^{(0,1)}(X)$ over $F_0 \text{Diff}^{\bar{\partial}}(X) = O(X_{\mathbb{R}})_{\mathbb{C}}$.

The main property of the ring $\text{Diff}^{\bar{\partial}}(X)$ is that its action in the ring of smooth complex-valued functions $O(X_{\mathbb{R}})_{\mathbb{C}}$ commutes with the operators of multiplication by holomorphic functions. This holds true because holomorphic functions have zero derivatives along $(0, 1)$-vector fields.

More generally, a smooth complex vector bundle E over $X_{\mathbb{R}}$ is said to be *holomorphic* (or "have a holomorphic structure") if the notion of a holomorphic local section of E is defined. For example, the $(1, 0)$-cotangent bundle $T^{*(1,0)}$ on X has a natural holomorphic structure in which $\mathbb{C}$-linear combinations of the expressions $f d(g)$ with holomorphic local functions $f, g : U \longrightarrow \mathbb{C}$ are the holomorphic local sections (these are called the *holomorphic differential 1-forms* on X). Similarly, the $(1, 0)$-tangent bundle $T^{(1,0)}$ on X has a natural holomorphic structure in which the expressions $\sum_{k=1}^{m} f_k \frac{\partial}{\partial z_k}$ with holomorphic local functions $f_k : U \longrightarrow \mathbb{C}$ are the holomorphic local sections (one has to check that the class of *holomorphic vector fields*, defined in this way, does not depend on the choice of a local coordinate system z_k). We denote the $\mathbb{C}$-vector space of smooth sections of a holomorphic vector bundle E on X by $E(X_{\mathbb{R}})$ and the subspace of holomorphic sections by $E(X) \subset E(X_{\mathbb{R}})$.

Construction 10.46 For any holomorphic vector bundle E over X, the ring of $\bar{\partial}$-differential operators $\text{Diff}^{\bar{\partial}}(X)$ acts naturally in the $\mathbb{C}$-vector space of *smooth* sections $E(X_{\mathbb{R}})$; so $E(X_{\mathbb{R}})$ is a left $\text{Diff}^{\bar{\partial}}(X)$-module. This action is constructed as follows. The ring of smooth $\mathbb{C}$-valued functions $O(X_{\mathbb{R}})_{\mathbb{C}}$ acts in $E(X_{\mathbb{R}})$ by the usual multiplications of smooth sections of a vector bundle with smooth functions. To define the action of $(0, 1)$-vector fields in $E(X_{\mathbb{R}})$, choose a local basis of holomorphic sections $e_l \in E(U)$, $1 \leq l \leq n$. This means that e_l are holomorphic sections of E over U and for every point $s \in U$ the vectors $e_l(s)$, $1 \leq l \leq n$, form a basis of the $\mathbb{C}$-vector space E_s.

Let $e = \sum_{l=1}^{n} f_l e_l \in E(U_{\mathbb{R}})$ be a smooth section of E over U; so $f_l \in O(U_{\mathbb{R}})_{\mathbb{C}}$ are smooth complex-valued local functions. Furthermore, let $v = \sum_{k=1}^{m} g_k \, \partial/\partial \bar{z}_k$ be a smooth $(0, 1)$-vector field in U; so $g_k : U \longrightarrow \mathbb{C}$ are smooth complex-valued local functions as well. Then we put

$$v(e) = \sum_{l=1}^{n} v(f_l) e_l = \sum_{l=1}^{n} \sum_{k=1}^{m} g_k \frac{\partial f_l}{\partial \bar{z}_k} e_l \in E(U_{\mathbb{R}}).$$

Using the fact that $v(h) = 0$ for any holomorphic local function $h\colon U \longrightarrow \mathbb{C}$, one can check that the above definition of $v(e)$ does not depend on the choice of a local basis of holomorphic sections e_l of E over U; so the local constructions glue together to a well-defined differential operator of order 1 providing the action of v in $E(X_\mathbb{R})$.

Similarly one can show that, for any morphism $E' \longrightarrow E''$ of holomorphic vector bundles over X (that is, a morphism of smooth complex vector bundles taking holomorphic local sections to holomorphic local sections), the induced map of the spaces of smooth sections $E'(X_\mathbb{R}) \longrightarrow E''(X_\mathbb{R})$ is a morphism of left $\mathrm{Diff}^{\bar\partial}(X)$-modules. This generalizes the above assertion about the commutativity of the action of $\mathrm{Diff}^{\bar\partial}(X)$ in $E(X_\mathbb{R})$ with the multiplications by holomorphic functions.

Now let us assume that X is a *compact* complex manifold. Then, according to the discussion in Sect. 10.6, the category of smooth real vector bundles over $X_\mathbb{R}$ is equivalent to the category of finitely generated projective modules over $O(X_\mathbb{R})$. Similarly, the category of smooth complex vector bundles over $X_\mathbb{R}$ is equivalent to the category of finitely generated projective modules over the ring $O(X_\mathbb{R})_\mathbb{C}$.

Proposition 10.47 *For any compact complex manifold X, the Dolbeault DG-ring $(\Omega^{0,\bullet}(X), \bar\partial)$ (10.34) is the right finitely projective Koszul DG-ring corresponding to the left augmented left finitely projective Koszul filtered ring $(\mathrm{Diff}^{\bar\partial}(X), F)$ under the anti-equivalence of categories from Corollary 4.28.*

Proof The associated graded ring $\mathrm{gr}^F \mathrm{Diff}^{\bar\partial}(X) = \bigoplus_{n=0}^\infty F_n \mathrm{Diff}^{\bar\partial}(X)/F_{n-1}\mathrm{Diff}^{\bar\partial}(X)$ is isomorphic to the symmetric algebra $A = \mathrm{Sym}_{O(X_\mathbb{R})_\mathbb{C}}(T^{(0,1)}(X))$ of the finitely generated projective module $T^{(0,1)}(X)$ of smooth $(0, 1)$-vector fields on X over the ring $O(X_\mathbb{R})_\mathbb{C}$ of smooth complex-valued functions. According to Sect. 10.1, this graded ring is left and right finitely projective Koszul. Hence the filtered ring $\mathrm{Diff}^{\bar\partial}(X)$ is a left finitely projective nonhomogeneous Koszul ring. Furthermore, the natural left action of $\mathrm{Diff}^{\bar\partial}(X)$ in $O(X_\mathbb{R})_\mathbb{C}$ makes $\mathrm{Diff}^{\bar\partial}(X)$ a left augmented ring over its subring $F_0 \mathrm{Diff}^{\bar\partial}(X) = O(X_\mathbb{R})_\mathbb{C}$. So $(\mathrm{Diff}^{\bar\partial}(X), F)$ is a left augmented left finitely projective nonhomogeneous Koszul ring.

The graded ring $B = \bigoplus_{q=0}^m \Omega^{0,q}(X) = \Lambda_{O(X_\mathbb{R})_\mathbb{C}}(T^{*(0,1)}(X))$ is left and right finitely projective Koszul as well. According to Sect. 10.1, it is the right finitely projective Koszul ring quadratic dual to the left finitely projective Koszul ring A, in the sense of Corollary 2.38. This explains the context in which the assertion of the proposition arises; the rest of the proof is a straightforward computation similar to the proofs of the analogous assertions in Sects. 10.2, 10.3, and 10.6. $\square$

Similarly to the smooth real manifold case of Theorem 10.28, we obtain the following rough algebraic version of Koszul duality for the ring of $\bar\partial$-differential operators $\mathrm{Diff}^{\bar\partial}(X)$.

Theorem 10.48 *For any compact complex manifold X, the construction of Theorem 6.14 provides a triangulated equivalence between the $\mathrm{Diff}^{\bar\partial}(X)/O(X_\mathbb{R})_\mathbb{C}$-semicoderived cate-*

gory of right modules over the ring of $\bar{\partial}$-differential operators $\mathrm{Diff}^{\bar{\partial}}(X)$ and the coderived category of right DG-modules over the Dolbeault DG-algebra $(\Omega^{0,\bullet}(X), \bar{\partial})$,

$$\mathsf{D}^{\mathsf{sico}}_{O(X_{\mathbb{R}})_{\mathbb{C}}}(\mathsf{mod}\text{--}\mathrm{Diff}^{\bar{\partial}}(X)) \simeq \mathsf{D}^{\mathsf{co}}(\mathsf{mod}\text{--}(\Omega^{0,\bullet}(X), \bar{\partial})), \tag{10.35}$$

while Theorem 6.20 establishes a triangulated equivalence between the derived category of right $\mathrm{Diff}^{\bar{\partial}}(X)$-modules and the reduced coderived category of right DG-modules over $(\Omega^{0,\bullet}(X), \bar{\partial})$ relative to $O(X_{\mathbb{R}})_{\mathbb{C}}$,

$$\mathsf{D}(\mathsf{mod}\text{--}\mathrm{Diff}^{\bar{\partial}}(X)) \simeq \mathsf{D}^{\mathsf{co}}_{O(X_{\mathbb{R}})_{\mathbb{C}}\text{-red}}(\mathsf{mod}\text{--}(\Omega^{0,\bullet}(X), \bar{\partial})). \tag{10.36}$$

In the same context, the construction of Theorem 7.11 provides a triangulated equivalence between the $\mathrm{Diff}^{\bar{\partial}}(X)/O(X_{\mathbb{R}})_{\mathbb{C}}$-semicontraderived category of left modules over the ring $\mathrm{Diff}^{\bar{\partial}}(X)$ and the contraderived category of left DG-modules over the DG-algebra $(\Omega^{0,\bullet}(X), \bar{\partial})$,

$$\mathsf{D}^{\mathsf{sictr}}_{O(X_{\mathbb{R}})_{\mathbb{C}}}(\mathrm{Diff}^{\bar{\partial}}(X)\text{--}\mathsf{mod}) \simeq \mathsf{D}^{\mathsf{ctr}}((\Omega^{0,\bullet}(X), \bar{\partial})\text{--}\mathsf{mod}), \tag{10.37}$$

while Theorem 7.16 establishes a triangulated equivalence between the derived category of left $\mathrm{Diff}^{\bar{\partial}}(X)$-modules and the reduced contraderived category of left DG-modules over $(\Omega^{0,\bullet}(X), \bar{\partial})$ relative to $O(X_{\mathbb{R}})_{\mathbb{C}}$,

$$\mathsf{D}(\mathrm{Diff}^{\bar{\partial}}(X)\text{--}\mathsf{mod}) \simeq \mathsf{D}^{\mathsf{ctr}}_{O(X_{\mathbb{R}})_{\mathbb{C}}\text{-red}}((\Omega^{0,\bullet}(X), \bar{\partial})\text{--}\mathsf{mod}). \tag{10.38}$$

Proof The proof is similar to that of Theorem 10.28. The functors in (10.35–10.36) assign to a complex of right $\mathrm{Diff}^{\bar{\partial}}(X)$-modules $M^\bullet$ the right DG-module $M^\bullet \otimes_{O(X_{\mathbb{R}})_{\mathbb{C}}} \Lambda_{O(X_{\mathbb{R}})_{\mathbb{C}}}$ $(T^{(0,1)}(X))$ over the Dolbeault DG-ring $\Omega^{0,\bullet}(X) = \Lambda_{O(X_{\mathbb{R}})_{\mathbb{C}}}(T^{*(0,1)}(X))$, and conversely, to a right DG-module $N^\bullet$ over $\Omega^{0,\bullet}(X)$ the complex of right $\mathrm{Diff}^{\bar{\partial}}(X)$-modules $N^\bullet \otimes_{O(X_{\mathbb{R}})_{\mathbb{C}}} \mathrm{Diff}^{\bar{\partial}}(X)$ is assigned. The functors in (10.37–10.38) assign to a complex of left $\mathrm{Diff}^{\bar{\partial}}(X)$-modules $P^\bullet$ the left DG-module $\Omega^{0,\bullet}(X) \otimes_{O(X_{\mathbb{R}})_{\mathbb{C}}} P^\bullet$ over $\Omega^{0,\bullet}(X)$, and conversely, to a left DG-module $Q^\bullet$ over $\Omega^{0,\bullet}(X)$ the complex of left $\mathrm{Diff}^{\bar{\partial}}(X)$-modules $\mathrm{Hom}_{O(X_{\mathbb{R}})_{\mathbb{C}}}(\mathrm{Diff}^{\bar{\partial}}(X), Q^\bullet)$ is assigned. □

In particular, for every holomorphic vector bundle E over X, we have the left $\mathrm{Diff}^{\bar{\partial}}(X)$-module $E(X_{\mathbb{R}})$, as explained above. The corresponding left DG-module over the Dolbeault DG-ring (10.34), under the equivalences of categories (10.37–10.38), is the *Dolbeault complex $(\Omega^{0,\bullet}(X) \otimes_{O(X_{\mathbb{R}})_{\mathbb{C}}} E(X_{\mathbb{R}}), \bar{\partial})$ with the coefficients in a holomorphic vector bundle E on X*. The $\mathbb{C}$-vector spaces $H^*_{\bar{\partial}}(\Omega^{0,\bullet}(X) \otimes_{O(X_{\mathbb{R}})_{\mathbb{C}}} E(X_{\mathbb{R}}))$ of cohomology of the latter complex are the sheaf cohomology spaces $H^*(X, \mathcal{E})$ of the sheaf of holomorphic local sections $\mathcal{E}$ of the vector bundle E.

Furthermore, there is a natural equivalence between the abelian categories of left and right modules over the ring $\mathrm{Diff}^{\partial}(X)$, provided by the mutually inverse functors of tensor product with the invertible modules $\Omega^{0,m}(X)$ and $\Lambda^m_{O(X_{\mathbb{R}})_{\mathbb{C}}}(T^{(0,1)}(X))$ over the ring $O(X_{\mathbb{R}})_{\mathbb{C}}$ (where $m = \dim_{\mathbb{C}} X$). This Morita equivalence can be obtained as a particular case of the construction of Theorem 9.9 with Remark 9.10 (since $\Omega^{0,\bullet}(X)$ is a relatively Frobenius Koszul graded ring by Lemma 10.1).

10.8 Relative Differential Operators

The aim of this section is to replace a ground field k in the context of Sect. 10.3 with a commutative ring S. The role of the ring of functions $O(X)$ on a smooth affine algebraic variety X over k will be played by a commutative ring R. So we let $\iota \colon S \longrightarrow R$ be a homomorphism of commutative rings.

Construction 10.49 Let us consider DG-rings (B, d) endowed with a ring homomorphism $R \longrightarrow B^0$ such that the following conditions are satisfied:

(i) the graded ring $B = \bigoplus_{n \in \mathbb{Z}} B^n$ is *strictly* graded commutative (in the sense of Sect. 10.1);
(ii) B is a DG-algebra over S, that is, in other words, the image of the composition $S \longrightarrow R \longrightarrow B^0$ is annihilated by the differential $d_0 \colon B^0 \longrightarrow B^1$.

The initial object in the category of all such DG-rings (B, d) with ring homomorphisms $R \longrightarrow B^0$ is called the *strictly graded commutative DG-algebra over S freely generated by R* and denoted by $(\Omega_{R/S}, d)$. The other name for $(\Omega_{R/S}, d)$ is the *de Rham DG-algebra of R over S* [17, Tag 0FKF].

The following proposition describes the DG-ring that we have constructed.

Proposition 10.50

(a) *One has $\Omega^n_{R/S} = 0$ for $n < 0$ and $\Omega^0_{R/S} = R$. Furthermore, the graded ring $\Omega_{R/S}$ is generated by its first-degree component $\Omega^1_{R/S}$ over $\Omega^0_{R/S} = R$.*
(b) *The R-module $\Omega^1_{R/S}$ can be constructed as the module of Kähler differentials of R over S [17, Tag 00RM] (cf. the discussion of the particular case when S is a field in Construction 10.4 in Sect. 10.2 above). Specifically, $\Omega^1_{R/S}$ is the R-module generated by the symbols $d(r)$ with $r \in R$ subject to the relations $d(\iota(s)) = 0$ for all $s \in S$ and $d(fg) = f d(g) - g d(f)$ for all $f, g \in R$.*
(c) *The graded ring $\Omega_{R/S}$ is the exterior algebra of the R-module $\Omega^1_{R/S}$, that is $\Omega_{R/S} = \Lambda_R(\Omega^1_{R/S})$.*

Proof Part (a) holds because the graded ring $\Omega_{R/S}$ is generated by the images of elements from R and their differentials. Part (b) is straightforward.

Part (c) claims, in other words, that there exists a well-defined odd derivation d of degree 1 on the graded ring $\Lambda_R(\Omega^1_{R/S})$ whose restriction to $R = \Lambda^0_R(\Omega^1_{R/S})$ is the natural map $d \colon R \longrightarrow \Omega^1_{R/S}$ and whose square vanishes. Such a derivation is clearly unique, because the exterior algebra $\Lambda^0_R(\Omega^1_{R/S})$ is generated by $\Omega^1_{R/S}$ over R and the action of d on $\Omega^1_{R/S}$ is computable as $d(f d(g)) = d(f) \wedge d(g) + f d^2(g) = d(f) \wedge d(g)$.

Concerning the existence, one needs to check that odd derivations defined on the generators and satisfying obvious compatibilities extend well to freely generated strictly graded commutative rings (cf. the discussion in the last paragraph of the proof of Proposition 3.16 and Lemma 3.18). In particular, one can compute that $d(c^2) = d(c)c - cd(c) = 0$ since c commutes with $d(c)$ for any element c of odd degree in $\Lambda_R(\Omega^1_{R/S})$ (as it should be). Furthermore, one has $d^2(ab) = d(d(a)b + (-1)^{|a|}ad(b)) = d^2(a)b + (-1)^{|a|+1}d(a)d(b) + (-1)^{|a|}d(a)d(b) + ad^2(b) = 0$ provided that $d^2(a) = 0 = d^2(b)$ for a given pair of elements $a, b \in \Lambda_R(\Omega^1_{R/S})$; so the square of an odd derivation vanishes whenever it vanishes on the generators.

A simpler alternative approach might be to construct $\Omega_{R/S}$ as the graded commutative graded ring generated by the symbols f of degree 0 and $d(f)$ of degree 1 for all $f \in R$, with the relations that the addition and multiplication of the symbols f in $\Omega_{R/S}$ agrees with their addition and multiplication in R, and also $d(f+g) = d(f)+d(g)$, $d(fg) = d(f)g + fd(g)$ for all $f, g \in R$, and $d(\iota(s)) = 0$ for all $s \in S$. Then the isomorphism $\Omega_{R/S} = \Lambda_R(\Omega^1_{R/S})$ follows simply from the fact that all the relations imposed have degrees 0 or 1. Having observed that, one needs to convince oneself that there exists an (obviously unique) odd derivation d on $\Omega_{R/S}$ taking f to $d(f)$ and $d(f)$ to 0 for all $f \in R$. Indeed, $d(d(fg) - d(f)g - fd(g)) = 0 + d(f)d(g) - d(f)d(g) = 0$, as it should be; so the relation is preserved by the desired odd derivation. $\square$

Lemma 10.51 *Assume that 2 is invertible in R (so there is no difference between graded commutativity and strict graded commutativity for B). Then the condition (i) can be replaced by its weaker form*

(i') the image of the map $R \longrightarrow B^0$ is contained in the (graded) center of B.

The universal graded commutative DG-ring (E, d) with a map $R \longrightarrow E^0$ satisfying (i') and (ii) is the same as the universal graded commutative DG-ring with a similar map satisfying (i) and (ii).

Proof Indeed, the graded ring E, being universal, is clearly generated by $E^0 = R$ and $d(R) \subset E^1$. So it remains to check that $d(f)d(g) + d(g)d(f) = 0$ in B under (i') for all $f, g \in R$. For this purpose, it suffices to compute that $d(f)d(g) + d(g)d(f) = d(fd(g) - d(g)f) = d(0) = 0$. More generally, one shows that the graded center Z

of any graded ring B is preserved by any odd derivation d on B, as one has $d(z)b = d(zb) - (-1)^{|z|}zd(b) = (-1)^{|z||b|}d(bz) - (-1)^{|z||b|}d(b)z = (-1)^{(|z|+1)|b|}bd(z)$ for all $b \in B$ and $z \in Z$.   □

Construction 10.52 Elements of the R-module $T_{R/S} = \mathrm{Hom}_R(\Omega^1_{R/S}, R)$ are interpreted as derivations of the S-algebra R (i. e., S-linear maps $v: R \longrightarrow R$ such that $v(fg) = v(f)g + fv(g)$ for all $f, g \in R$). Specifically, given an R-linear map $v: \Omega^1_{R/S} \longrightarrow R$, the action of v on R is defined by the rule $v(f) = v(df) \in R$. In fact, this rule defines a natural R-module isomorphism between $T_{R/S}$ and the R-module of all S-linear derivations of R. Consequently, the underlying S-module of $T_{R/S}$ acquires the structure of a Lie algebra over S: the bracket $[v, w]$ of two elements $v, w \in T_{R/S}$ is defined by the usual rule $[v, w](f) = v(w(f)) - w(v(f))$.

The construction of the de Rham DG-algebra $\Omega_{R/S}$ is well-behaved for *smooth* morphisms of rings $S \longrightarrow R$. In order not to delve into the intricacies of various definitions of a smooth morphism, let us define and impose the minimal condition that we will actually need. Let us say that a morphism of commutative rings $S \longrightarrow R$ is *weakly smooth of relative dimension m* if $\Omega^1_{R/S}$ is a finitely generated projective R-module everywhere of rank m (in the sense of Sect. 10.1). In this case, $T_{R/S}$ is also a finitely generated projective R-module everywhere of rank m.

Remark 10.53 Notice that the relative dimension of a weakly smooth morphism can exceed the (relative) Krull dimension. For example, given a field k, denote by $k(x)$ the field of rational functions in one variable x with the coefficients in k. Then the natural inclusion $k \longrightarrow k(x)$ is a weakly smooth morphism of relative dimension 1. Moreover, for a field k of prime characteristic p, the inclusion $k(x^p) \longrightarrow k(x)$ is a weakly smooth morphism of relative dimension 1 in the sense of our definition (even though it is a finite, algebraic field extension).

Construction 10.54 The ring of *relative crystalline differential operators* $\mathrm{Diff}^{\mathrm{cr}}_{R/S}$ is now defined similarly to Sect. 10.3. The ring $\mathrm{Diff}^{\mathrm{cr}}_{R/S}$ is generated by elements of the ring R and the R-module $T_{R/S}$, subject to the relations (10.4–10.5) imposed for all $f, g \in R$ and $v, w \in T_{R/S}$. There is a natural structure of left $\mathrm{Diff}^{\mathrm{cr}}_{R/S}$-module on the ring R, with the elements $g \in R \subset \mathrm{Diff}^{\mathrm{cr}}_{R/S}$ acting in R by the multiplication maps $f \longmapsto gf$ and the elements $v \in T_{R/S} \subset \mathrm{Diff}^{\mathrm{cr}}_{R/S}$ acting in R by the derivations $f \longmapsto v(f)$. The operators with which the ring $\mathrm{Diff}^{\mathrm{cr}}_{R/S}$ acts in R are S-linear, but not R-linear. There is also a natural increasing filtration F on the ring $\mathrm{Diff}^{\mathrm{cr}}_{R/S}$ generated by $F_1 \mathrm{Diff}^{\mathrm{cr}}_{R/S} = R \oplus T_{R/S} \subset \mathrm{Diff}^{\mathrm{cr}}_{R/S}$ over $F_0 \mathrm{Diff}^{\mathrm{cr}}_{R/S} = R$.

Assume that the morphism of commutative rings $S \longrightarrow R$ is weakly smooth. Then the filtered ring $\mathrm{Diff}^{\mathrm{cr}}_{R/S}$ together with its action in R can be also defined as the left augmented left finitely projective nonhomogeneous Koszul ring corresponding to the right finitely projective Koszul DG-ring $(\Omega_{R/S}, d)$ under the anti-equivalence of categories from Corollary 4.28. By the Poincaré–Birkhoff–Witt Theorem 4.25, the associated graded ring $\mathrm{gr}^F \mathrm{Diff}^{\mathrm{cr}}_{R/S} = \bigoplus_{n=0}^{\infty} F_n \mathrm{Diff}^{\mathrm{cr}}_{R/S}/F_{n-1} \mathrm{Diff}^{\mathrm{cr}}_{R/S}$ is naturally isomorphic to the symmetric algebra $\mathrm{Sym}_R(T_{R/S})$ of the R-module $T_{R/S}$.

Hence the results of Chaps. 6 and 7 specialize to the following theorem.

Theorem 10.55 *For any weakly smooth morphism of commutative rings $S \longrightarrow R$, the construction of Theorem 6.14 provides a triangulated equivalence between the $\mathrm{Diff}^{\mathrm{cr}}_{R/S}/R$-semicoderived category of right modules over the ring of relative crystalline differential operators $\mathrm{Diff}^{\mathrm{cr}}_{R/S}$ and the coderived category of right DG-modules over the relative de Rham DG-algebra $(\Omega_{R/S}, d)$,*

$$\mathsf{D}^{\mathrm{sico}}_R(\mathrm{mod}{-}\mathrm{Diff}^{\mathrm{cr}}_{R/S}) \simeq \mathsf{D}^{\mathrm{co}}(\mathrm{mod}{-}(\Omega_{R/S}, d)), \tag{10.39}$$

while Theorem 6.20 establishes a triangulated equivalence between the derived category of right $\mathrm{Diff}^{\mathrm{cr}}_{R/S}$-modules and the reduced coderived category of right DG-modules over $(\Omega_{R/S}, d)$,

$$\mathsf{D}(\mathrm{mod}{-}\mathrm{Diff}^{\mathrm{cr}}_{R/S}) \simeq \mathsf{D}^{\mathrm{co}}_{R\text{-red}}(\mathrm{mod}{-}(\Omega_{R/S}, d)). \tag{10.40}$$

In the same context, the construction of Theorem 7.11 provides a triangulated equivalence between the $\mathrm{Diff}^{\mathrm{cr}}_{R/S}/R$-semicontraderived category of left modules over the ring $\mathrm{Diff}^{\mathrm{cr}}_{R/S}$ and the contraderived category of left DG-modules over the DG-algebra $(\Omega_{R/S}, d)$,

$$\mathsf{D}^{\mathrm{sictr}}_R(\mathrm{Diff}^{\mathrm{cr}}_{R/S}{-}\mathrm{mod}) \simeq \mathsf{D}^{\mathrm{ctr}}((\Omega_{R/S}, d){-}\mathrm{mod}), \tag{10.41}$$

while Theorem 7.16 establishes a triangulated equivalence between the derived category of left $\mathrm{Diff}^{\mathrm{cr}}_{R/S}$-modules and the reduced contraderived category of left DG-modules over $(\Omega_{R/S}, d)$,

$$\mathsf{D}(\mathrm{Diff}^{\mathrm{cr}}_{R/S}{-}\mathrm{mod}) \simeq \mathsf{D}^{\mathrm{ctr}}_{R\text{-red}}((\Omega_{R/S}, d){-}\mathrm{mod}). \tag{10.42}$$

Proof The proof is similar to those of Theorems 10.6, 10.28, and 10.48. The functors in (10.39–10.40) assign to a complex of right $\mathrm{Diff}^{\mathrm{cr}}_{R/S}$-modules $M^\bullet$ the right DG-module $M^\bullet \otimes_R \Lambda_R(T_{R/S})$ over the relative de Rham DG-ring $\Omega_{R/S} = \Lambda_R(\Omega^1_{R/S})$, and conversely, to a right DG-module $N^\bullet$ over $\Omega_{R/S}$ the complex of right $\mathrm{Diff}^{\mathrm{cr}}_{R/S}$-modules $N^\bullet \otimes_R \mathrm{Diff}^{\mathrm{cr}}_{R/S}$ is assigned. The functors in (10.41–10.42) assign to a complex of left $\mathrm{Diff}^{\mathrm{cr}}_{R/S}$-modules $P^\bullet$

the left DG-module $\Omega_{R/S} \otimes_R P^\bullet$ over $\Omega_{R/S}$, and conversely, to a left DG-module $Q^\bullet$ over $\Omega_{R/S}$ the complex of left $\mathrm{Diff}^{\mathrm{cr}}_{R/S}$-modules $\mathrm{Hom}_R(\mathrm{Diff}^{\mathrm{cr}}_{R/S}, Q^\bullet)$ is assigned. $\quad\square$

Furthermore, whenever the ring R has finite homological dimension, Corollaries 6.18 and 7.14 are applicable; so there is no difference between (10.39) and (10.40), and similarly there is no difference between (10.41) and (10.42). Moreover, the results of Chap. 9 lead to the following theorem in this case. Here the natural equivalence between the abelian categories of left and right $\mathrm{Diff}^{\mathrm{cr}}_{R/S}$-modules is provided by the mutually inverse functors of tensor product with the invertible R-modules $\Omega^m_{R/S}$ and $\Lambda^m_R(T_{R/S})$, as described in Theorem 9.9.

Theorem 10.56 *For any weakly smooth morphism of commutative rings $S \longrightarrow R$ of relative dimension m such that the ring R has finite homological dimension, the constructions of Theorem 9.13 (with Remark 9.10) provide a commutative square diagram of triangulated equivalences*

$$
\begin{array}{ccc}
\mathsf{D}(\mathrm{mod}\text{--}\mathrm{Diff}^{\mathrm{cr}}_{R/S}) & \xleftrightarrow{\hspace{1.5cm}} & \mathsf{D}(\mathrm{Diff}^{\mathrm{cr}}_{R/S}\text{--}\mathrm{mod}) \\
\Big\updownarrow & & \Big\updownarrow \\
\mathsf{D}^{\mathrm{co}}((\Omega_{R/S}, d)\text{--}\mathrm{mod}) & =\!\!=\!\!= & \mathsf{D}^{\mathrm{ctr}}((\Omega_{R/S}, d)\text{--}\mathrm{mod})
\end{array}
\tag{10.43}
$$

where the equivalence of derived categories in the upper line is induced by the conversion equivalence of abelian categories $\mathrm{mod}\text{--}\mathrm{Diff}^{\mathrm{cr}}_{R/S} \simeq \mathrm{Diff}^{\mathrm{cr}}_{R/S}\text{--}\mathrm{mod}$ *(up to a cohomological shift by $[-m]$), while the co-contra correspondence in the lower line is the result of Theorem 9.6. The vertical equivalences are* (10.39=10.40) *and* (10.41=10.42).

Proof In the assumptions of the theorem, $\Omega_{R/S}$ is a relatively Frobenius Koszul graded ring by Lemma 10.1; so Theorem 9.13 is indeed applicable. $\quad\square$

Given a closed relative 2-form h, i. e., an element $h \in \Omega^2_{R/S}$ such that $d(h) = 0$, one can also consider the right finitely projective Koszul CDG-ring $(\Omega_{R/S}, d, h)$ and construct its nonhomogeneous quadratic dual left finitely projective nonhomogeneous Koszul filtered ring $\mathrm{Diff}^{\mathrm{cr}}_{R/S,h}$ of *twisted relative crystalline differential operators*, using Theorem 4.25 or Corollary 4.26. The results of Theorems 10.55 and 10.56 can be then extended to the twisted case (similarly to Sect. 10.5).

10.9 Lie Algebroids

The setting in this section is a common generalizations of Sects. 10.2–10.3 and 10.6–10.8.

Definition 10.57 A *Lie algebroid* (known also as a *Lie–Rinehart algebra*) [46, 73] is a pair of abelian groups $(R, \mathfrak{g})$ endowed with the following structures:

- R is a commutative ring (with unit);
- $\mathfrak{g}$ is a Lie algebra over $\mathbb{Z}$, i. e., it is endowed with an additive map of *Lie bracket* $[-, -] \colon \Lambda_{\mathbb{Z}}^2 \mathfrak{g} \longrightarrow \mathfrak{g}$ satisfying the Jacobi identity;
- $\mathfrak{g}$ is an R-module, so a commutative ring action map $R \otimes_{\mathbb{Z}} \mathfrak{g} \longrightarrow \mathfrak{g}$ is given, denoted by $a \otimes x \longmapsto ax$ for all $a \in R$ and $x \in \mathfrak{g}$;
- R is a $\mathfrak{g}$-module, so a Lie action map $\mathfrak{g} \otimes_{\mathbb{Z}} R \longrightarrow R$ is given, denoted by $x \otimes a \longmapsto x(a)$ for all $a \in R$ and $x \in \mathfrak{g}$.

In addition to the usual Jacobi identity on the bracket in $\mathfrak{g}$, the identity involved in the notion of a $\mathfrak{g}$-module, the associativity and commutativity equations on the multiplication in R, and the associativity equation involved in the notion of an R-module, the listed structures must also satisfy the following equations:

(i) $\mathfrak{g}$ acts in R by derivations of the commutative multiplication, that is

$$x(ab) = x(a)b + ax(b) \qquad \text{for all } x \in \mathfrak{g} \text{ and } a, b \in R;$$

(ii) the identity

$$(ax)(b) = ax(b) \qquad \text{for all } x \in \mathfrak{g} \text{ and } a, b \in R$$

holds in R, where $ax \in \mathfrak{g}$, $(ax)(b) \in R$, $x(b) \in R$, and $ax(b) \in R$ are the elements obtained using the action of R in $\mathfrak{g}$, the action of $\mathfrak{g}$ in R, the action of $\mathfrak{g}$ in R, and the multiplication in R, respectively;

(iii) the identity

$$[x, ay] = x(a)y + a[x, y] \qquad \text{for all } x, y \in \mathfrak{g} \text{ and } a \in R$$

holds in $\mathfrak{g}$, where $x(a) \in R$ is the element obtained using the action of $\mathfrak{g}$ in R, ay and $x(a)y \in \mathfrak{g}$ are the elements obtained using the action of R in $\mathfrak{g}$, $[x, y]$ and $[x, ay] \in \mathfrak{g}$ are the Lie brackets in $\mathfrak{g}$, and $a[x, y] \in \mathfrak{g}$ is the element obtained using the action of R in $\mathfrak{g}$.

Notice that $\mathfrak{g}$ is *not* a Lie algebra over R, as the Lie bracket in $\mathfrak{g}$ is not R-linear. The identity (iii) describes the obstacle term to R-linearity of the bracket in $\mathfrak{g}$.

Examples 10.58

(1) For any smooth affine algebraic variety X over a field k, the commutative ring of functions $R = O(X)$ and the Lie algebra of vector fields $\mathfrak{g} = T(X)$ form a Lie algebroid with $O(X)$ acting in $T(X)$ as in the module of sections of a vector bundle (namely, the tangent bundle) on X and $T(X)$ acting in $O(X)$ by the derivations of functions along vector fields (cf. Construction 10.8 in Sec. 10.3).

(2) More generally, for any homomorphism of commutative rings $S \longrightarrow R$, the ring R and the Lie algebra $\mathfrak{g} = T_{R/S}$ of S-linear derivations of R (see Construction 10.52 in Sect. 10.8) form a Lie algebroid.

(3) Furthermore, for any smooth real manifold X, the ring $R = O(X)$ of smooth $\mathbb{R}$-valued functions on X together with the Lie algebra $\mathfrak{g} = T(X)$ of smooth vector fields on X (as in Construction 10.25 in Sect. 10.6) form a Lie algebroid.

(4) Similarly, for any real manifold X, the ring $R = O(X)_{\mathbb{C}}$ of smooth $\mathbb{C}$-valued functions on X together with the Lie algebra $\mathfrak{g}' = T(X)_{\mathbb{C}}$ of smooth $\mathbb{C}$-valued vector fields on X form a Lie algebroid $(R, \mathfrak{g}')$.

(5) For a complex manifold X, the same ring of smooth $\mathbb{C}$-valued functions $R = O(X_{\mathbb{R}})_{\mathbb{C}}$ together with the Lie subalgebra $\mathfrak{g} = T^{(0,1)}(X) \subset \mathfrak{g}' = T(X_{\mathbb{R}})_{\mathbb{C}}$ of smooth $(0, 1)$-vector fields on X form a Lie algebroid $(R, \mathfrak{g})$ as well (see Construction 10.40 in Sect. 10.7).

Construction 10.59 The *universal enveloping ring* $U(R, \mathfrak{g})$ of a Lie algebroid $(R, \mathfrak{g})$ is an associative ring defined by generators and relations as follows (see [46, 73] for a differently worded, but equivalent version of this construction). The set of generators $R \sqcup \mathfrak{g}$ is the disjoint union of R and $\mathfrak{g}$. The sum of any two elements of R in $U(R, \mathfrak{g})$ equals their sum in R, and the sum of any two elements of $\mathfrak{g}$ in $U(R, \mathfrak{g})$ equals their sum in $\mathfrak{g}$. Denoting the product in $U(R, \mathfrak{g})$ by $*$, one imposes the multiplicative relations similar to the ones in Sect. 10.3:

$$a * b = ab \qquad \text{for all } a, b \in R,$$

$$a * x = ax \qquad \text{for all } a \in R \text{ and } x \in \mathfrak{g}, \tag{10.44}$$

$$x * a = ax + x(a) \qquad \text{for all } a \in R \text{ and } x \in \mathfrak{g}$$

and

$$x * y - y * x = [x, y] \qquad \text{for all } x, y \in \mathfrak{g}. \tag{10.45}$$

Construction 10.60 The commutative ring R has a natural structure of left module over the associative ring $U(R, \mathfrak{g})$. To define this action, one lets the generators $b \in R$ of the ring $U(R, \mathfrak{g})$ act in R by the multiplication maps $a \longmapsto ba$ and the generators $x \in \mathfrak{g}$ of the ring $U(R, \mathfrak{g})$ act in R as the Lie algebra $\mathfrak{g}$ acts in R, that is $a \longmapsto x(a)$. Then one needs to check that the assignment of such endomorphisms of the abelian group R to the generators of $U(R, \mathfrak{g})$ respects the relations (10.44–10.45), so the resulting action of $U(R, \mathfrak{g})$ in R is well-defined.

Let $(R, \mathfrak{g})$ be a Lie algebroid such that the R-module $\mathfrak{g}$ is projective and finitely generated. Denote by $\mathfrak{g}^\vee = \mathrm{Hom}_R(\mathfrak{g}, R)$ the dual finitely generated projective R-module. Consider the symmetric algebra $A = \mathrm{Sym}_R(\mathfrak{g})$ of the R-module $\mathfrak{g}$ and the exterior algebra $B = \Lambda_R(\mathfrak{g}^\vee)$. According to Sect. 10.1, A and B are left and right finitely projective Koszul graded rings over R. Furthermore, the right finitely projective Koszul graded ring B is quadratic dual to the left finitely projective Koszul graded ring A, in the sense of Chap. 1 and Proposition 2.37.

For any Lie algebroid $(R, \mathfrak{g})$, one can put $\widetilde{A} = U(R, \mathfrak{g})$, and endow the ring $\widetilde{A}$ with the increasing filtration F generated by $F_1\widetilde{A} = \mathrm{im}(R \oplus \mathfrak{g} \to \widetilde{A})$ over $F_0\widetilde{A} = \mathrm{im}(R \to \widetilde{A})$. Consider the associated graded ring $\mathrm{gr}^F \widetilde{A} = \bigoplus_{n=0}^\infty F_n\widetilde{A}/F_{n-1}\widetilde{A}$. Then there is a unique homomorphism of graded rings $A \longrightarrow \mathrm{gr}^F \widetilde{A}$ forming commutative triangle diagrams with the natural isomorphisms $R \simeq A_0$ and $\mathfrak{g} \simeq A_1$ and the natural surjective maps $R \longrightarrow F_0\widetilde{A}$ and $\mathfrak{g} \longrightarrow F_1\widetilde{A}/F_0\widetilde{A}$. This graded ring homomorphism is obviously surjective.

Proposition 10.61 *Let $(R, \mathfrak{g})$ be a Lie algebroid such that the R-module $\mathfrak{g}$ is projective and finitely generated. Then the natural graded ring homomorphism $A \longrightarrow \mathrm{gr}^F \widetilde{A}$ is an isomorphism.*

A more general version of this result (for the case of a projective, but not necessarily finitely generated R-module $\mathfrak{g}$) can be found in the classical paper [73, Theorem 3.1].

Proof This is a particular case of the Poincaré–Birkhoff–Witt Theorem 4.25. In particular, it follows from the proposition that the natural surjective maps $R \longrightarrow F_0\widetilde{A}$ and $R \oplus \mathfrak{g} \longrightarrow F_1\widetilde{A}$ are, in fact, bijective. Let us emphasize that these assertions in the conclusion of the Poincaré–Birkhoff–Witt theorem only have a chance to hold due to (i–iii) and other equations imposed on the structure maps of a Lie algebroid (such as the Jacobi identity for the bracket in $\mathfrak{g}$).

Using the notation of Sect. 3.3, put $V = \mathfrak{g} = A_1$ and let $I \subset V \otimes_R V$ denote the kernel of the multiplication map $A_1 \otimes_R A_1 \longrightarrow A_2$. So I is the R-submodule of skew-symmetric tensors in $V \otimes_R V$. Put $q(x, a) = x(a)$ for every $x \in V = \mathfrak{g}$ and $a \in R$. Then the relations (10.44) take the form (3.1).

Following further the notation of Sect. 3.3, denote by $\widehat{I} \subset V \otimes_\mathbb{Z} V$ the full preimage of the submodule $I \subset V \otimes_R V$ under the natural surjective map $V \otimes_\mathbb{Z} V \longrightarrow V \otimes_R V$. So we get a surjective map $\widehat{I} \longrightarrow I$.

The abelian group $\widehat{I}$ is spanned by the tensors of the form $x \otimes y - y \otimes x$ and $ax \otimes y - x \otimes ay$, where $x, y \in \mathfrak{g}$ and $a \in R$. The rules $p(x \otimes y - y \otimes x) = [x, y] \in \mathfrak{g}$ and $p(ax \otimes y - x \otimes ay) = -x(a)y \in \mathfrak{g}$ define an abelian group homomorphism $p: \widehat{I} \longrightarrow \mathfrak{g}$. Let $h: \widehat{I} \longrightarrow R$ be the zero map, $h = 0$. Then the relation (10.45) takes the form (3.2) for all $\hat{\imath} =$

$x \otimes y - y \otimes x \in \widehat{I}$, and it follows from the relations (10.44) that (3.2) also holds for all $\hat{\imath} = ax \otimes y - x \otimes ay \in \widehat{I}$, i. e., $ax * y - x * ay = -x(a)y$ in $U(R, \mathfrak{g})$.

Using the equations imposed on the structure maps of a Lie algebroid, one can check that the maps q, p, and h satisfy the equations (a–k) in Proposition 3.14. Following the proof of Proposition 3.16, one can then conclude that the formulas (3.4–3.5) define an odd derivation $d \colon B \longrightarrow B$ of degree 1 with zero square, $d^2 = 0$. So one obtains a DG-ring (B, d), called the *cohomological Chevalley–Eilenberg complex* of a Lie algebroid $(R, \mathfrak{g})$ (with trivial coefficients). Alternatively, it may be easier to check directly from the definition of a Lie algebroid that the formulas (3.4–3.5) define a DG-ring structure on the graded ring $B = \Lambda_R(\mathfrak{g}^\vee)$.

Now Theorem 4.25 is applicable to the right finitely projective Koszul DG-ring (B, d); and the filtered ring $(\widetilde{A}, F)$ together with the above action of $\widetilde{A} = U(R, \mathfrak{g})$ in R is the left augmented left finitely projective nonhomogeneous Koszul ring corresponding to (B, d) under the anti-equivalence of categories from Corollary 4.28. Hence the desired isomorphism $A \simeq \mathrm{gr}^F \widetilde{A}$. □

According to Example 7.12(2) and Remark 7.13, the Chevalley–Eilenberg complex $(\Lambda_R(\mathfrak{g}^\vee), d)$ computes the Ext groups/ring $\mathrm{Ext}^*_{U(R,\mathfrak{g})}(R, R)$ (cf. [73, Section 4]).

The results of Chapters 6 and 7 specialize to the following derived nonhomogeneous Koszul duality theorem.

Theorem 10.62 *For any Lie algebroid $(R, \mathfrak{g})$ such that the R-module $\mathfrak{g}$ is projective and finitely generated, the construction of Theorem 6.14 provides a triangulated equivalence between the $U(R, \mathfrak{g})/R$-semicoderived category of right modules over the universal enveloping ring $U(R, \mathfrak{g})$ and the coderived category of right DG-modules over the Chevalley–Eilenberg DG-ring $(\Lambda_R(\mathfrak{g}^\vee), d)$,*

$$\mathsf{D}^{\mathsf{sico}}_R(\mathrm{mod}{-}U(R, \mathfrak{g})) \simeq \mathsf{D}^{\mathsf{co}}(\mathrm{mod}{-}(\Lambda_R(\mathfrak{g}^\vee), d)), \tag{10.46}$$

while Theorem 6.20 establishes a triangulated equivalence between the derived category of right $U(R, \mathfrak{g})$-modules and the reduced coderived category of right DG-modules over $(\Lambda_R(\mathfrak{g}^\vee), d)$,

$$\mathsf{D}(\mathrm{mod}{-}U(R, \mathfrak{g})) \simeq \mathsf{D}^{\mathsf{co}}_{R\text{-red}}(\mathrm{mod}{-}(\Lambda_R(\mathfrak{g}^\vee), d)). \tag{10.47}$$

In the same context, the construction of Theorem 7.11 provides a triangulated equivalence between the $U(R, \mathfrak{g})/R$-semicontraderived category of left modules over the ring $U(R, \mathfrak{g})$ and the contraderived category of left DG-modules over the DG-algebra $(\Lambda_R(\mathfrak{g}^\vee), d)$,

$$\mathsf{D}^{\mathsf{sictr}}_R(U(R, \mathfrak{g}){-}\mathrm{mod}) \simeq \mathsf{D}^{\mathsf{ctr}}((\Lambda_R(\mathfrak{g}^\vee), d){-}\mathrm{mod}), \tag{10.48}$$

while Theorem 7.16 establishes a triangulated equivalence between the derived category of left $U(R, \mathfrak{g})$-modules and the reduced contraderived category of left DG-modules over $(\Lambda_R(\mathfrak{g}^\vee), d)$,

$$\mathsf{D}(U(R, \mathfrak{g})\text{–mod}) \simeq \mathsf{D}^{\mathsf{ctr}}_{R\text{-red}}((\Lambda_R(\mathfrak{g}^\vee), d)\text{–mod}). \tag{10.49}$$

Proof Similar to the proofs of Theorem 10.55 and the previous theorems. The functors in (10.46–10.47) assign to a complex of right $U(R, \mathfrak{g})$-modules $M^\bullet$ the right DG-module $M^\bullet \otimes_R \Lambda_R(\mathfrak{g})$ over the Chevalley–Eilenberg DG-ring $\Lambda_R(\mathfrak{g}^\vee)$, and conversely, to a right DG-module $N^\bullet$ over $\Lambda_R(\mathfrak{g}^\vee)$ the complex of right $U(R, \mathfrak{g})$-modules $N^\bullet \otimes_R U(R, \mathfrak{g})$ is assigned. The functors in (10.48–10.49) assign to a complex of left $U(R, \mathfrak{g})$-modules $P^\bullet$ the left DG-module $\Lambda_R(\mathfrak{g}^\vee) \otimes_R P^\bullet$ over $\Lambda_R(\mathfrak{g}^\vee)$, and conversely, to a left DG-module $Q^\bullet$ over $\Lambda_R(\mathfrak{g}^\vee)$ the complex of left $U(R, \mathfrak{g})$-modules $\mathrm{Hom}_R(U(R, \mathfrak{g}), Q^\bullet)$ is assigned. □

The complex $M \otimes_R \Lambda_R(\mathfrak{g})$ (for a right $U(R, \mathfrak{g})$-module M) can be called the *homological Chevalley–Eilenberg complex* of a Lie algebroid $(R, \mathfrak{g})$ with the coefficients in M, while the complex $\Lambda_R(\mathfrak{g}^\vee) \otimes_R P$ is the *cohomological Chevalley–Eilenberg complex* of $(R, \mathfrak{g})$ with the coefficients in the module P. A discussion of the latter complex and its cohomology can be found in [73, Sections 4–6].

Whenever the ring R has finite homological dimension, Corollaries 6.18 and 7.14 are applicable. So there is no difference between (10.46) and (10.47), and similarly there is no difference between (10.48) and (10.49) in this case.

Assume that the finitely generated projective R-module $\mathfrak{g}$ is everywhere of the same rank $m \geq 0$ (in the sense of the definition at the end of Sect. 10.1). Then there is a natural equivalence between the abelian categories of left and right $U(R, \mathfrak{g})$-modules, provided by the mutually inverse functors of tensor product with the invertible R-modules $T = \Lambda_R^m(\mathfrak{g}^\vee)$ and $\mathrm{Hom}_R(T, R) = \Lambda_R^m(\mathfrak{g})$, as described in Theorem 9.9. Moreover, the results of Chapter 9 lead to the following relative nonhomogeneous Koszul quadrality theorem.

Theorem 10.63 *For any Lie algebroid $(R, \mathfrak{g})$ such that $\mathfrak{g}$ is a finitely generated projective R-module everywhere of rank m and the homological dimension of R is finite, the constructions of Theorem 9.13 (with Remark 9.10) provide a commutative square diagram of triangulated equivalences*

$$\begin{array}{ccc} \mathsf{D}(\mathsf{mod}\text{–}U(R, \mathfrak{g})) & \xleftarrow{\hspace{1.5cm}} & \mathsf{D}(U(R, \mathfrak{g})\text{–mod}) \\ \Big\updownarrow & & \Big\updownarrow \\ \mathsf{D}^{\mathsf{co}}((\Lambda_R(\mathfrak{g}^\vee), d)\text{–mod}) & =\!=\!=\!= & \mathsf{D}^{\mathsf{ctr}}((\Lambda_R(\mathfrak{g}^\vee), d)\text{–mod}) \end{array} \tag{10.50}$$

where the equivalence of derived categories in the upper line is induced by the conversion equivalence of abelian categories $\mathsf{mod}\text{–}U(R, \mathfrak{g}) \simeq U(R, \mathfrak{g})\text{–}\mathsf{mod}$ *(up to a cohomological shift by* $[-m]$*), while the co-contra correspondence in the lower line is the result of Theorem 9.6. The vertical equivalences are* (10.46=10.47) *and* (10.48=10.49). $\qquad\square$

10.10 Noncommutative Differential Forms

Let $\iota\colon S \longrightarrow R$ be a homomorphism of associative rings.

Construction 10.64 Let us consider DG-rings (B, d) endowed with a ring homomorphism $R \longrightarrow B^0$ such that the image of the composition $S \longrightarrow R \longrightarrow B^0$ is annihilated by the differential $d_0\colon B^0 \longrightarrow B^1$.

The initial object in the category of all such DG-rings (B, d) with ring homomorphisms $R \longrightarrow B^0$ is called the *DG-ring over S freely generated by R* and denoted by $(NC_{R/S}, d)$. The other name for $(NC_{R/S}, d)$ is the *DG-ring of noncommutative differential forms for R over S* (cf. [15, proof of Proposition II.1]).

This construction is the noncommutative version of Construction 10.49 in Sect. 10.8. The difference is that, even when the rings S and R happen to be commutative, the DG-rings (B, d) considered in this section do not need to be graded commutative. In other words, we are dropping the condition (i) or (i$'$) of Sect. 10.8 and keeping only the condition (ii).

As in Proposition 10.50, it is clear that the graded ring $NC_{R/S}$ is generated by the images of elements from R and their differentials. It follows that one has $NC_{R/S}^n = 0$ for $n < 0$ and $NC_{R/S}^0 = R$. Furthermore, the graded ring $NC_{R/S}$ is generated by its first-degree component $NC_{R/S}^1$ over $NC_{R/S}^0 = R$. The next lemma provides a precise, explicit description.

Lemma 10.65 *The maps*

$$R \otimes_S R/\iota(S) \otimes_S R/\iota(S) \otimes_S \cdots \otimes_S R/\iota(S) \longrightarrow NC_{R/S}^n \tag{10.51}$$

and

$$R/\iota(S) \otimes_S R/\iota(S) \otimes_S \cdots \otimes_S R/\iota(S) \otimes_S R \longrightarrow NC_{R/S}^n \tag{10.52}$$

given by the formulas $f \otimes \bar{g}_1 \otimes \cdots \otimes \bar{g}_n \longmapsto f\,d(g_1)d(g_2)\cdots d(g_n)$ *and* $\bar{g}_1 \otimes \cdots \otimes \bar{g}_n \otimes f \longmapsto d(g_1)d(g_2)\cdots d(g_n)f$ *are isomorphisms of S-S-bimodules for all* $n \geq 0$. *More precisely, the map* (10.51) *is an isomorphism of R-S-bimodules, while the map* (10.52) *is an isomorphism of S-R-bimodules.*

Here the left-hand side of (10.51) *is the tensor product of one factor R and n factors* $R/\iota(S)$. *The left-hand side of* (10.52) *is the tensor product of n factors* $R/\iota(S)$ *and one factor R. All the tensor products signs in* (10.51–10.52) *mean tensor products of S-S-bimodules. The notation is* f, $g_i \in R$ *for all* $1 \le i \le n$ *and* $\bar{g}_i \in R/\iota(S)$ *is the image of* g_i *under the natural surjection* $R \longrightarrow R/\iota(S)$.

Proof The maps (10.51–10.52) are well-defined due to the condition that $d(\iota(s)) = 0$ in $NC^1_{R/S}$ for all $s \in S$. Having observed that, one can split the assertions of the lemma in two parts. Firstly, it is claimed that the maps $R \otimes_S R/\iota(S) \longrightarrow NC^1_{R/S} \longleftarrow R/\iota(S) \otimes_S R$ given by the formulas $f \otimes \bar{g} \longmapsto f d(g)$ and $\bar{g} \otimes f \longmapsto d(g) f$ are isomorphisms (where f, $g \in R$ and $\bar{g} \in R/\iota(S)$ is the image of g).

Secondly, notice that

$$(R \otimes_S R/\iota(S)) \otimes_R (R \otimes_S R/\iota(S)) \otimes_R \cdots \otimes_R (R \otimes_S R/\iota(S))$$

$$\simeq R \otimes_S R/\iota(S) \otimes_S R/\iota(S) \otimes_S \cdots \otimes_S R/\iota(S).$$

So, in order to deduce the assertions of the lemma for an arbitrary $n \ge 1$ from such assertions for $n = 1$, one needs to show that the multiplication map

$$NC^1_{R/S} \otimes_R \cdots \otimes_R NC^1_{R/S} \longrightarrow NC^n_{R/S} \tag{10.53}$$

(n factors in the left-hand side) is an isomorphism for $n \ge 1$. In other words, this means that the graded ring $NC_{R/S}$ is the tensor ring of the R-R-bimodule $NC^1_{R/S}$, that is $NC_{R/S} \simeq T_R(NC^1_{R/S})$.

Concerning the first part, one observes that the R-R-bimodule $NC^1_{R/S}$ is spanned by the symbols $d(f)$, $f \in R$, with the relations $d(fg) = d(f)g + f d(g)$ for f, $g \in R$ and $d(\iota(s)) = 0$ for $s \in S$. It follows immediately that the maps $R \otimes_S R/\iota(S) \longrightarrow NC^1_{R/S} \longleftarrow R/\iota(S) \otimes_S R$ are surjective. In order to show that these maps are isomorphisms, it suffices to define R-R-bimodule structures on $R \otimes_S R/\iota(S)$ and $R/\iota(S) \otimes_S R$ in such a way that the above relations are satisfied.

Concerning the second part, in order to prove that the maps (10.53) are isomorphisms, it suffices to show that the map $d: R = NC^0_{R/S} \longrightarrow NC^1_{R/S}$ extends to a well-defined odd derivation with zero square on the tensor ring $T_R(NC^1_{R/S})$. Lemma 3.18 is a suitable tool here.

Similarly to the proof of Proposition 10.50 in Sect. 10.8, a simpler alternative approach might be to define $NC_{R/S}$ as the graded associative ring generated by the symbols f and $d(f)$ with $f \in R$, subject to the relations that the addition and multiplication of elements f in R agrees with their addition and multiplication in $NC_{R/S}$, and also $d(f + g) = d(f) + d(g)$, $d(fg) = d(f)g + f d(g)$ for f, $g \in R$, and $d(\iota(s)) = 0$ for $s \in S$. Then the isomorphism $NC_{R/S} \simeq T_R(NC^1_{R/S})$ follows simply from the fact that all the relations imposed have degrees 0 or 1. Subsequently one needs to see that there exists an odd derivation d on $NC_{R/S}$ taking f to $d(f)$ and $d(f)$ to 0 for all $f \in R$. □

Construction 10.66 Consider the following filtered ring $(\widetilde{A}, F)$. Let $\widetilde{A} = \mathrm{Hom}_{S^{op}}(R, R)$ be the ring of endomorphisms of the right S-module R; so R is a left $\widetilde{A}$-module. This is our "ring of noncommutative differential operators acting S-linearly in R." The left action of R in itself defines a natural ring inclusion $R \longrightarrow \widetilde{A}$. So $\widetilde{A}$ is a left augmented ring over its subring R, in the sense of Sect. 3.8. Let $\widetilde{A}^+ \subset \widetilde{A}$ be the augmentation ideal, that is, the subgroup of all elements in $\widetilde{A}$ whose action annihilates the unit element $1 \in R$. So $\widetilde{A}^+$ is a left ideal in $\widetilde{A}$, and the left R-module $\widetilde{A}$ decomposes naturally as the direct sum $\widetilde{A} = R \oplus \widetilde{A}^+$. Put $F_0\widetilde{A} = R$ and $F_n\widetilde{A} = \widetilde{A}$ for all $n \geq 1$; so F is a multiplicative increasing filtration on $\widetilde{A}$ (see Example 3.3(1) for a more general discussion).

Proposition 10.67 *Assume that the map $\iota \colon S \longrightarrow R$ is injective and the right S-module $R/\iota(S)$ is finitely generated and projective. Then the DG-ring of noncommutative differential forms $(NC_{R/S}, d)$ is the right finitely projective Koszul DG-ring corresponding to the left augmented left finitely projective nonhomogeneous Koszul filtered ring $(\widetilde{A}, F)$ under the equivalence of categories from Corollary 4.28.*

Proof It is clear from the isomorphism (10.52) that the right R-module $NC_{R/S}^n$ is finitely generated and projective for every $n \geq 1$. Moreover, $NC_{R/S}^n \simeq T_R(NC_{R/S}^1)$ is the tensor algebra of the R-R-bimodule $NC_{R/S}^1$, as we have seen in (10.53).

The underlying left R-module of the R-R-bimodule $F_1\widetilde{A}/F_0\widetilde{A}$ is isomorphic to $\widetilde{A}^+$. One easily computes this left R-module as $\widetilde{A}^+ \simeq \mathrm{Hom}_{S^{op}}(R/\iota(S), R) \simeq \mathrm{Hom}_{R^{op}}(R/\iota(S) \otimes_S R, R)$. In our assumptions, this is a finitely generated projective left R-module. It follows that the associated graded ring $A = \mathrm{gr}^F \widetilde{A} = F_0\widetilde{A} \oplus F_1\widetilde{A}/F_0\widetilde{A}$ is left finitely projective Koszul and quadratic dual to the right finitely projective Koszul graded ring $B = NC_{R/S}$ (in the sense of Chapter 1 and Proposition 2.37).

The rest of the proof is a straightforward computation; see Examples 3.12(1) and 3.17 for a further discussion. □

The results of Chaps. 6 and 7 specialize to the following theorem. Notice that the graded ring of noncommutative differential forms $NC_{R/S}$ usually has *infinitely many* nonzero grading components in our assumptions. So one has to distinguish $NC_{R/S}$-comodules and $NC_{R/S}$-contramodules (in the sense of Chap. 5) from objects of their ambient categories of arbitrary $NC_{R/S}$-modules.

We use the natural terminology "DG-comodules" and "DG-contramodules" for CDG-comodules and CDG-contramodules, in the sense of Sects. 6.4 and 7.3, over a DG-ring (i. e., a CDG-ring with a vanishing curvature element h).

Theorem 10.68 *For any injective morphism of associative rings $\iota \colon S \longrightarrow R$ such that $R/\iota(S)$ is a finitely generated projective right S-module, the construction of Theorem 6.14 provides a triangulated equivalence between the $\mathrm{Hom}_{S^{\mathrm{op}}}(R, R)/R$-semicoderived category of right modules over the endomorphism ring $\mathrm{Hom}_{S^{\mathrm{op}}}(R, R)$ and the coderived category of right DG-comodules over the nonnegatively graded DG-ring of noncommutative differential forms $(NC_{R/S}, d)$,*

$$\mathsf{D}_R^{\mathsf{sico}}(\mathrm{mod\text{-}Hom}_{S^{\mathrm{op}}}(R, R)) \simeq \mathsf{D}^{\mathsf{co}}(\mathrm{comod\text{-}}(NC_{R/S}, d)), \tag{10.54}$$

while Theorem 6.20 establishes a triangulated equivalence between the derived category of right $\mathrm{Hom}_{S^{\mathrm{op}}}(R, R)$-modules and the reduced coderived category of right DG-comodules over $(NC_{R/S}, d)$,

$$\mathsf{D}(\mathrm{mod\text{-}Hom}_{S^{\mathrm{op}}}(R, R)) \simeq \mathsf{D}^{\mathsf{co}}_{R\text{-red}}(\mathrm{comod\text{-}}(NC_{R/S}, d)). \tag{10.55}$$

In the same context, the construction of Theorem 7.11 provides a triangulated equivalence between the $\mathrm{Hom}_{S^{\mathrm{op}}}(R, R)/R$-semicontraderived category of left modules over the ring $\mathrm{Hom}_{S^{\mathrm{op}}}(R, R)$ and the contraderived category of left DG-contramodules over the nonnegatively graded DG-ring $(NC_{R/S}, d)$,

$$\mathsf{D}_R^{\mathsf{sictr}}(\mathrm{Hom}_{S^{\mathrm{op}}}(R, R)\text{-mod}) \simeq \mathsf{D}^{\mathsf{ctr}}((NC_{R/S}, d)\text{-contra}), \tag{10.56}$$

while Theorem 7.16 establishes a triangulated equivalence between the derived category of left $\mathrm{Hom}_{S^{\mathrm{op}}}(R, R)$-modules and the reduced contraderived category of left DG-contramodules over $(NC_{R/S}, d)$,

$$\mathsf{D}(\mathrm{Hom}_{S^{\mathrm{op}}}(R, R)\text{-mod}) \simeq \mathsf{D}^{\mathsf{ctr}}_{R\text{-red}}((NC_{R/S}, d)\text{-contra}). \tag{10.57}$$

Proof The proof is analogous to that of Theorem 10.55, with suitable adaptations. All the assertions are particular cases of the respective theorems. The functors in (10.54–10.55) assign to a complex of right $\mathrm{Hom}_{S^{\mathrm{op}}}(R, R)$-modules $M^\bullet$ the right DG-comodule $\mathrm{Hom}^\Sigma_{R^{\mathrm{op}}}(NC_{R/S}, M^\bullet)$ over the DG-ring $NC_{R/S}$, and conversely, to a right DG-comodule $N^\bullet$ over $NC_{R/S}$ the complex of right $\mathrm{Hom}_{S^{\mathrm{op}}}(R, R)$-modules $N^\bullet \otimes_R \mathrm{Hom}_{S^{\mathrm{op}}}(R, R)$ is assigned. The functors in (10.56–10.57) assign to a complex of left $\mathrm{Hom}_{S^{\mathrm{op}}}(R, R)$-modules $P^\bullet$ the left DG-contramodule $NC_{R/S} \otimes_R^\Pi P^\bullet$ over $NC_{R/S}$, and conversely, to a left DG-contramodule $Q^\bullet$ over $NC_{R/S}$ the complex of left $\mathrm{Hom}_{S^{\mathrm{op}}}(R, R)$-modules $\mathrm{Hom}_R(\mathrm{Hom}_{S^{\mathrm{op}}}(R, R), Q^\bullet)$ is assigned. (See Sect. 5.7 for the definition of the Σ and Π notation in Hom and tensor product.) $\qquad\square$

Furthermore, whenever the ring R has finite homological dimension, Corollaries 6.18 and 7.14 are applicable; so there is no difference between (10.54) and (10.55), and similarly there is no difference between (10.56) and (10.57).

References

1. J.F. Adams, On the structure and applications of the Steenrod algebra. Commentarii Math. Helvetici **32**, 180–214 (1958)
2. S.M. Arkhipov, Semi-infinite cohomology of associative algebras and bar-duality. Int. Math. Res. Not. **1997**(#17), 833–863. arXiv:q-alg/9602013
3. M. Ballard, D. Deliu, D. Favero, M.U. Isik, L. Katzarkov, Resolutions in factorization categories. Adv. Math. **295**, 195–249 (2016). arXiv:1212.3264 [math.CT]
4. H. Becker, Models for singularity categories. Adv. Math. **254**, 187–232 (2014). arXiv:1205.4473 [math.CT]
5. A. Beilinson, J. Bernstein, A proof of Jantzen conjectures, in *Advances in Soviet Mathematics*, vol. 16, part 1, ed. by I. M. Gelfand Seminar (American Mathematical Society, Providence, 1993), pp. 1–50
6. A. Beilinson, V. Drinfeld, Quantization of Hitchin's integrable system and Hecke eigensheaves. February 2000. Available from http://www.math.utexas.edu/~benzvi/Langlands.html
7. A. Beilinson, V. Ginzburg, W. Soergel, Koszul duality patterns in representation theory. J. Am. Math. Soc. **9**(#2), 473–527 (1996)
8. J. Bernstein, Algebraic theory of D-modules. Available from http://www.math.uchicago.edu/~drinfeld/langlands.html
9. I.N. Bernsein, I.M. Gel'fand, S.I. Gel'fand, Algebraic bundles over $\mathbf{P}^n$ and problems of linear algebra. Funct. Anal. Appl. **12**(#3), 212–214 (1978)
10. R. Bezrukavnikov, I. Mirković, D. Rumynin, Localization of modules for a semisimple Lie algebra in prime characteristic. Ann. Math. **167**(#3), 945–991 (2008). arXiv:math.RT/0205144
11. A. Borel et al., *Algebraic D-Modules*. Perspectives in Mathematics, vol. 2 (Academic, Boston, 1987)
12. R.-O. Buchweitz, Maximal Cohen–Macaulay modules and Tate-cohomology over Gorenstein rings. Manuscript, 1986, 155 pp. Available from http://hdl.handle.net/1807/16682
13. C.-H. Cho, On the obstructed Lagrangian Floer theory. Adv. Math. **229**(#2), 804–853 (2012). arXiv:0909.1251 [math.SG]
14. J. Chuang, A. Lazarev, W. Mannan, Cocommutative coalgebras: homotopy theory and Koszul duality. Homol. Homotopy Appl. **18**(#2), 303–336 (2016). arXiv:1403.0774 [math.AT]
15. A. Connes, Noncommutative differential geometry. Publ. Math. de l'IHES **62**(#2), 257–360 (1985)
16. O. De Deken, W. Lowen, On deformations of triangulated models. Adv. Math. **243**, 330–374 (2013). arXiv:1202.1647 [math.KT]
17. A.J. de Jong et al., The Stacks Project. Available from https://stacks.math.columbia.edu/

© The Author(s), under exclusive license to Springer Nature Switzerland AG 2021
L. Positselski, *Relative Nonhomogeneous Koszul Duality*, Frontiers in Mathematics,
https://doi.org/10.1007/978-3-030-89540-2

18. V.G. Drinfeld, On quadratic quasi-commutational relations in quasi-classical limit. (Russian) *Mat. Fizika, Funkc. Analiz*, 25–34 ("Naukova Dumka", Kiev, 1986). English translation in Selecta Math. Sovietica **11**(#4), 317–326 (1992)

19. W.G. Dwyer, J.P.C. Greenlees, Complete modules and torsion modules. Am. J. Math. **124**(#1), 199–220 (2002)

20. A.I. Efimov, L. Positselski, Coherent analogues of matrix factorizations and relative singularity categories. Algebra Numb. Theory **9**(#5), 1159–1292 (2015). arXiv:1102.0261 [math.CT]

21. S. Eilenberg, J.C. Moore, *Foundations of Relative Homological Algebra*, vol. 55 (Memoirs Am. Math. Soc., New York, 1965)

22. D. Eisenbud, Homological algebra on a complete intersection, with an application to group representations. Trans. Am. Math. Soc. **260**(#1), 35–64 (1980)

23. D. Eisenbud, *Commutative Algebra with a View Towards Algebraic Geometry*. Graduate Texts in Mathematics, vol. 150 (Springer, New York, 1995)

24. P.C. Eklof, J. Trlifaj, How to make Ext vanish. Bull. Lond. Math. Soc. **33**(#1), 41–51 (2001)

25. B.L. Feigin, D.B. Fuchs, Verma modules over the Virasoro algebra, in *Topology (Leningrad, 1982)* Lecture Notes in Mathematics, vol. 1060 (Springer, Berlin, 1984), pp. 230–245

26. R. Fröberg, T. Gulliksen, C. Löfwall, Flat families of local, artinian algebras with an infinite number of Poincaré series, in *Algebra, Algebraic Topology, and their Interactions (Stockholm, 1983)*. Lecture Notes in Mathematics, vol. 1183 (Springer, Berlin, 1986), pp. 170–191

27. K. Fukaya, Deformation theory, homological algebra and mirror symmetry, in *Geometry and Physics of Branes (Como, 2001)*, High Energy Phys. Cosmol. Gravit. (IOP, Bristol, 2003), pp. 121–209

28. K. Fukaya, Y.-G. Oh, H. Ohta, K. Ono, *Lagrangian Intersection Floer Theory: Anomaly and Obstruction. Parts I and II*. Studies in Advanced Mathematics, vol. 46 (American Math. Society/Intern. Press, New York, 2009), pp. 1–2

29. V. Ginzburg, Lectures on $\mathcal{D}$-modules. Available from http://people.math.harvard.edu/~gaitsgde/grad_2009/Ginzburg.pdf

30. J.P.C. Greenlees, J.P. May, Derived functors of I-adic completion and local homology. J. Algebra **149**(#2), 438–453 (1992)

31. V. Hinich, DG coalgebras as formal stacks. J. Pure Appl. Algebra **162**(#2–3), 209–250 (2001). arXiv:math.AG/9812034

32. J. Hirsh, J. Millès, Curved Koszul duality theory. Math. Annalen **354**(#4), 1465–1520 (2012). arXiv:1008.5368 [math.KT]

33. S. Iyengar, H. Krause, Acyclicity versus total acyclicity for complexes over noetherian rings. Documenta Math. **11**, 207–240 (2006)

34. B. Jónsson, Distributive sublattices of a modular lattice. Proc. Am. Math. Soc. **6**(#5), 682–688 (1955)

35. P. Jørgensen, The homotopy category of complexes of projective modules. Adv. Math. **193**(#1), 223–232 (2005). arXiv:math.RA/0312088

36. M. Kapranov, On DG-modules over the de Rham complex and the vanishing cycles functor, in *Algebraic Geometry (Chicago, 1989)*, **1479**. Lecture Notes in Math. (Springer, Heidelberg, 1991), pp. 57–86

37. B. Keller, Deriving DG-categories. Ann. Sci. de l'École Norm. Sup. (4) **27**(#1), 63–102 (1994)

38. B. Keller, Koszul duality and coderived categories (after K. Lefèvre). October 2003. Available from http://webusers.imj-prg.fr/~bernhard.keller/publ/index.html

39. B. Keller, W. Lowen, P. Nicolás, On the (non)vanishing of some "derived" categories of curved dg algebras. J. Pure Appl. Algebra **214**(#7), 1271–1284 (2010). arXiv:0905.3845 [math.KT]

40. H. Krause, The stable derived category of a noetherian scheme. Compositio Math. **141**(#5), 1128–1162 (2005). arXiv:math.AG/0403526

41. K. Lefèvre-Hasegawa, Sur les A$_\infty$-catégories. Thèse de doctorat, Université Denis Diderot – Paris 7, November 2003. arXiv:math.CT/0310337. Corrections, by B. Keller. Available from http://webusers.imj-prg.fr/~bernhard.keller/lefevre/publ.html

42. M. Lieberman, L. Positselski, J. Rosický, S. Vasey, Cofibrant generation of pure monomorphisms. J. Algebra **560**, 1297–1310 (2020). arXiv:2001.02062 [math.CT]

43. S. Mac Lane, *Categories for the Working Mathematician*, vol. 5, 2nd edn. Graduate Texts in Mathematics (Springer, New York, 1998)

44. H. Matsumura, *Commutative Ring Theory*. Translated by M. Reid (Cambridge University Press, Cambridge, 1986–2006)

45. J. Maunder, Koszul duality and homotopy theory of curved Lie algebras. Homol. Homotopy Appl. **19**(#1), 319–340 (2017). arXiv:1512.01975 [math.AT]

46. I. Moerdijk, J. Mrčun, On the universal enveloping algebra of a Lie algebroid. Proc. Am. Math. Soc. **138**(#9), 3135–3145 (2010). arXiv:0801.3929 [math.QA]

47. D. Murfet, The mock homotopy category of projectives and Grothendieck duality. Ph. D. Thesis, Australian National University, September 2007. Available from http://www.therisingsea.org/thesis.pdf

48. R. Musti, E. Buttafuoco, Sui subreticoli distributivi dei reticoli modulari. Bolletino dell'Unione Matem. Italiana (3) **11**(#4), 584–587 (1956)

49. A. Neeman, The homotopy category of flat modules, and Grothendieck duality. Inventiones Math. **174**, 225–308 (2008)

50. L. Ng, Rational symplectic field theory for Legendrian knots. Inventiones Math. **182**(#3), 451–512 (2010)

51. C. Okonek, M. Schneider, H. Spindler, Vector bundles on complex projective spaces. Corrected reprint of the 1988 edition, with an appendix by S.I. Gelfand. (Modern Birkhäuser Classics, Birkhäuser/Springer, Basel, 2011)

52. D. Orlov, Matrix factorizations for nonaffine LG-models. Mathematische Annalen 353(#1), 95–108 (2012). arXiv:1101.4051 [math.AG]

53. A. Polishchuk, L. Positselski, *Quadratic Algebras*. University Lecture Series, vol. 37 (American Mathematical Society, Providence, RI, 2005)

54. A. Polishchuk, L. Positselski, Hochschild (co)homology of the second kind I. Trans. Am. Math. Soc. **364**(#10), 5311–5368 (2012). arXiv:1010.0982 [math.CT]

55. M. Porta, L. Shaul, A. Yekutieli, On the homology of completion and torsion. Algebras Represent. Theory **17**(#1), 31–67 (2014). arXiv:1010.4386 [math.AC]. Erratum in Algebras Represent. Theory **18**(#5), 1401–1405 (2015). arXiv:1506.07765 [math.AC]

56. L. Positselski, Nonhomogeneous quadratic duality and curvature. Funct. Anal. Appl. **27**(#3), 197–204 (1993). arXiv:1411.1982 [math.RA]

57. L. Positselski, Koszul property and Bogomolov's conjecture. Int. Math. Res. Not. **2005**(#31), 1901–1936. arXiv:1405.0965 [math.KT]

58. L. Positselski, Homological algebra of semimodules and semicontramodules: semi-infinite homological algebra of associative algebraic structures. Appendix C in collaboration with D. Rumynin; Appendix D in collaboration with S. Arkhipov. Monografie Matematyczne, vol. 70 (Birkhäuser/Springer, Basel, 2010), xxiv+349 pp. arXiv:0708.3398 [math.CT]

59. L. Positselski, Two kinds of derived categories, Koszul duality, and comodule-contramodule correspondence. Memoirs Am. Math. Soc. **212**(#996), vi+133 pp., (2011). arXiv:0905.2621 [math.CT]

60. L. Positselski, Weakly curved A$_\infty$-algebras over a topological local ring. Mémoires de la Société Mathématique de France **159**, vi+206 pp. (2018). arXiv:1202.2697 [math.CT]

61. L. Positselski, Contramodules. Electronic preprint. arXiv:1503.00991 [math.CT], to appear in Confluentes Math

62. L. Positselski, Dedualizing complexes and MGM duality. J. Pure Appl. Algebra **220**(#12), 3866–3909 (2016). arXiv:1503.05523 [math.CT]

63. L. Positselski, Coherent rings, fp-injective modules, dualizing complexes, and covariant Serre–Grothendieck duality. Selecta Math. (New Ser.) **23**(#2), 1279–1307 (2017). arXiv:1504.00700 [math.CT]

64. L. Positselski, Koszulity of cohomology $= K(\pi, 1)$-ness + quasi-formality. J. Algebra **483**, 188–229 (2017). arXiv:1507.04691 [math.KT]

65. L. Positselski, Contraadjusted modules, contramodules, and reduced cotorsion modules. Moscow Math. J. **17**(#3), 385–455 (2017). arXiv:1605.03934 [math.CT]

66. L. Positselski, Pseudo-dualizing complexes and pseudo-derived categories. Rendiconti Seminario Matematico Univ. Padova **143**, 153–225 (2020). arXiv:1703.04266 [math.CT]

67. L. Positselski, Abelian right perpendicular subcategories in module categories. Electronic preprint. arXiv:1705.04960 [math.CT]

68. L. Positselski, Flat ring epimorphisms of countable type. Glasgow Math. J. **62**(#2), 383–439 (2020). arXiv:1808.00937 [math.RA]

69. L. Positselski, J. Rosický, Covers, envelopes, and cotorsion theories in locally presentable abelian categories and contramodule categories. J. Algebra **483**, 83–128 (2017). arXiv:1512.08119 [math.CT]

70. L. Positselski, O.M. Schnürer, Unbounded derived categories of small and big modules: is the natural functor fully faithful? J. Pure Appl. Algebra **225**(#11), 23 pp. (2021). Article ID 106722. arXiv:2003.11261 [math.CT]

71. L. Positselski, J. Šťovíček, The tilting-cotilting correspondence. Int. Math. Res. Not. **2021**(#1), 189–274 (2021). arXiv:1710.02230 [math.CT]

72. D. Quillen, Rational homotopy theory. Ann. Math. **90**(#2), 205–295 (1969)

73. G.S. Rinehart, Differential forms on general commutative algebras. Trans. Am. Math. Soc. **108**, 195–222 (1963)

74. A. Rocha-Caridi, N. Wallach. Characters of irreducible representations of the Virasoro algebra. Math. Zeitschrift **185**(#1), 1–21 (1984)

75. S. Rybakov, DG-modules over de Rham DG-algebra. Eur. J. Math. **1**(#1), 25–53 (2015). arXiv:1311.7503 [math.AG]

76. A.-M. Simon, Approximations of complete modules by complete big Cohen–Macaulay modules over a Cohen–Macaulay local ring. Algebras Represent. Theory **12**(#2–5), 385–400 (2009)

77. S.P. Smith, Differential operators on commutative algebras, in *Ring Theory (Antwerp, 1985)*. Lecture Notes in Mathematics, vol. 1197 (Springer, Berlin, 1986), pp. 165–177

78. B. Stenström, *Rings of Quotients. An Introduction to Methods of Ring Theory* (Springer, Berlin, Heidelberg, New York, 1975)

79. C. Voisin, Hodge theory and complex algebraic geometry, vol. I. Translated by L. Schneps. Cambridge Studies in Advanced Mathematics, vol. 76 (Cambridge University Press, Cambridge, 2002)

80. R. Vyas, A. Yekutieli, Weak proregularity, weak stability, and the noncommutative MGM equivalence. J. Algebra **513**, 265–325 (2018). arXiv:1608.03543 [math.RA]

81. A. Yekutieli, On flatness and completion for infinitely generated modules over noetherian rings. Communicat. in Algebra **39**(#11), 4221–4245 (2011). arXiv:0902.4378 [math.AC]

Printed in the United States
by Baker & Taylor Publisher Services